Marine Biodiversity of Costa Rica, Central America

MONOGRAPHIAE BIOLOGICAE

VOLUME 86

Series Editor
H.J. Dumont
University of Ghent, Institute of Animal Ecology, Belgium
henri.dumont@rug.ac.be

Aims and Scope

The Monographiae Biologicae provide a forum for top-level, rounded-off monographs dealing with the biogeography of continents or major parts of continents, and the ecology of well individualized ecosystems such as islands, island groups, mountains or mountain chains. Aquatic ecosystems may include marine environments such as coastal ecosystems (mangroves, coral reefs) but also pelagic, abyssal and benthic ecosystems, and freshwater environments such as major river basins, lakes, and groups of lakes. In-depth, state-of-the-art taxonomic treatments of major groups of animals (including protists), plants and fungi are also eligible for publication, as well as studies on the comparative ecology of major biomes. Volumes in the series may include singleauthor monographs, but also multi-author, edited volumes.

For other titles published in this series, go to
www.springer.com/series/6304

Marine Biodiversity of Costa Rica, Central America

Edited by

Ingo S. Wehrtmann and Jorge Cortés
CIMAR and Escuela de Biología, Universidad de Costa Rica, Costa Rica

Springer

Editors

Ingo S. Wehrtmann
Unidad Investigación Pesquera y Acuicultura (UNIP) of the Centro de Investigación en Ciencias del Mar y Limnología (CIMAR), and Museo de Zoología, Escuela de Biología
Universidad de Costa Rica
11501-2060 San José, Costa Rica
ingowehrtmann@gmx.de

Jorge Cortés
Centro de Investigación en Ciencias del Mar y Limnología (CIMAR), and Escuela de Biología
Universidad de Costa Rica
11501-2060 San José, Costa Rica
jorge.cortes@ucr.ac.cr

ISBN: 978-94-007-8944-9 ISBN: 978-1-4020-8278-8 (eBook)

Softcover re-print of the Hardcover 1st edition 2009

Cover page: Seagrass habitat at Parque Nacional Cahuita, Caribbean coast of Costa Rica (Photo: Ingo S. Wehrtmann)
Back cover, photo 1: A green algae, *Acetabularia calyculus*, from the Caribbean coast of Costa Rica (Photo: Ingo S. Wehrtmann)
Back cover, photo 2: *Carpilius corallinus*, a brachyuran crab occurring in the Parque Nacional Cahuita, Caribbean coast of Costa Rica (Photo: Ingo S. Wehrtmann)
Back cover, photo 3: A black turtle, *Chelonia agassizi*, observed in waters around the offshore island, Isla del Coco, Pacific Costa Rica (Photo: Jaime Nivia)

Printed on acid-free paper

9 8 7 6 5 4 3 2 1

springer.com

***In Memoriam* to our parents**

Ulla Irps (1920–1999) and Siegfried Wehrtmann (1915–1997)

Elia Núñez (1926–2005) and Luis Cortés (1926–2000)

Óscar Arias Sánchez

Presidente de la República

Marine Biodiversity of Costa Rica

Óscar Arias Sánches
President of Costa Rica

Anyone who loves Costa Rica, whether a proud citizen or a one-time visitor, is aware that this tiny country holds biological riches few can fathom. From the upper reaches of our cloud forests to the dense wonders of our coastal jungles, thousands of animals and plants have drawn the world's attention, and filled us with pride.

However, these terrestrial species are only the tip of the iceberg – for, like an iceberg, Costa Rica and its diverse habitats lie mostly below the water's surface. With a marine territory 10 times larger than our terrestrial area, our country is home to a vast world of underwater flora and fauna most of us can scarcely imagine. Sadly, much of that vast world is at risk. While Costa Rica has set an example for the world by protecting its wildlife on land, our wildlife at sea is not always as safe.

It is precisely for this reason that this volume is so important. The work of Ingo Wehrtmann, Jorge Cortés and their colleagues, who have created the first-ever compilation of knowledge of our marine organisms, allows the rest of us to imagine underwater life a bit more clearly. Such imagining is crucial if our marine habitats are to be saved. As the old saying goes: out of sight, out of mind. Perhaps most of us will never see the wonders in these pages, but if scientists can help us understand, we are better able to protect.

This work also underscores the urgency of our task as conservationists. It is motivation enough that we must protect these species for our own nation, for our children and grandchildren; but as this book shows, Costa Rica is also responsible for protecting many species for the world entire. A minimum of eighty-five species live in our waters and in our waters only. Not only our own children, but also all children around the world, will be deprived of a chance to coexist with these creatures if we do not act.

Most important of all is not the information contained in these pages, but rather the information not contained there. The English poet William Cowper wrote, "*Knowledge is proud that he has learned so much; wisdom is humble that he knows no more.*" In this book, we have both knowledge and wisdom: its authors have shown us what we do not know, and what remains to be done. They have charted a course for those of us above the water who seek to protect the creatures beneath. For that, I am truly grateful.

EL PRESIDENTE DE LA REPUBLICA
SAN JOSE, COSTA RICA

Foreword

Life began in the sea, and even today most of the deep diversity of the planet is marine. This is often forgotten, especially in tropical countries like Costa Rica, renowned for their rain forests and the multitude of life forms found therein. Thus this book focusing on marine diversity of Costa Rica is particularly welcome.

How many marine species are there in Costa Rica? The authors report a total of 6,777 species, or 3.5% of the world's total. Yet the vast majority of marine species have yet to be formally described. Recent estimates of the numbers of species on coral reefs range from 1–9 million, so that the true number of marine species in Costa Rica is certainly far higher. In some groups the numbers are likely to be vastly higher because to date they have been so little studied. Only one species of nematode is reported, despite the fact that it has been said that nematodes are the most diverse of all marine groups. In better studied groups such as mollusks and crustaceans, reported numbers are in the thousands, but even in these groups many species remain to be described. Indeed the task of describing marine species is daunting – if there really are about 9 million marine species and Costa Rica has 3.5% of them, then the total number would be over 300,000. Clearly, so much remains to be done that new approaches are needed. Genetic methods have enormous promise in this regard.

Like several other countries in Central America, Costa Rica has the additional interest of having not one but two oceans on its borders – the Caribbean and the eastern Pacific. These oceans were only finally separated about three million years ago by the rise of the Central American Isthmus, but are now very different indeed. Though both are tropical and both have coral reefs, the eastern Pacific is typically more nutrient rich and variable in temperature that the Caribbean. Nevertheless, in many groups, close relatives are found on either side – so called geminate species. These geminates (which sometimes have different names and sometimes not) have served as model systems for understanding the formation of new species, a process that starts once genetic connections are broken. To date comparatively few species have been studied in this regard, in part because comprehensive descriptions of the marine diversity are so hard to come by. This book documents 288 species found on both coasts, species that are prime candidates for studying the formation of species, which has endured as a central theme in biology since Darwin.

Sadly, much marine diversity may be disappearing just as we are beginning to study it. We now know that the oceans we see today are but a pale shadow of what they were before people came on the scene. Reports of early explorers indicate that densities of large vertebrates – sharks, turtle, groupers – were once so high as to be almost unimaginable today, and we have also lost many coral reefs, 80% by some estimates. One hopes that the publication of this book will mobilize the additional conservation efforts needed in Costa Rica and its neighbors to protect and restore these rich marine habitats.

Nancy Knowlton
Sant Chair for Marine Science
Department of Invertebrate Zoology
National Museum of Natural History
Smithsonian Institution
Washington, DC, USA

Foreword

I live in the clouds, or more specifically, in the cool temperate Pacific Northwest, where rainclouds cloak the land during the winter months. Being optimistic by nature, I often slip out of my home on chilly nights to look upward. Usually there is a thick, opaque cloudbank. But fortune smiles sometimes on every one of us, and there are intervals when breaks in the clouds afford me a glimpse of the heavens. And, very rarely during our long, wet, gray winters, I get to see a dark sky resplendent with stars. Such a sight fills me with awe. I've enjoyed reading about astronomy for a half century, even briefly flirting with becoming an astronomer. I understand that some stars are brighter because they are far closer than others, or much larger, or burn their hydrogen much faster, sending their brilliance far into the cosmos. Even knowing this, I still wonder why some stars are so bright. And, being a biologist, I often wonder whether these distant places harbor the greatest miracle on our small, blue planet: life.

Anyone who travels knows that seeing new places changes the way we think. I first visited countries outside the United States and Canada in 1970, as a graduate student studying the ecology of blue crabs (genus *Callinectes*) in the Caribbean and tropical East Pacific. Among my first stops were the Smithsonian Institution's laboratories on the coasts of Panama. The more places I visited, the more I realized how little is known about their rich tropical biotas because most biologists reside and do their research in biologically impoverished industrialized countries. Only very scattered information was available about the marine species and ecosystems of Latin America when I did field work there. For example, in 1972 I was told that there had been almost no research on the estuarine biota of three sizable rivers in western Colombia where I trawled for portunid crabs. I wonder how much that has changed in the decades since then.

This knowledge is crucial, especially for scientists trying to understand how ecosystems function on large spatial scales, for people depending heavily on marine life for food, and for officials in places where human activities are affecting marine populations and ecosystems. It is also crucial for many visitors who love nature. Having read about and experienced ecosystems in other nations was essential to the idea of conserving biological diversity that I developed with Roger McManus when we worked at the President's Council on Environmental Quality in 1980. It was even more so when I assembled Global Marine Biological

Diversity: A Strategy for Building Conservation into Decision Making (1993) and my book with Larry Crowder, Marine Conservation Biology: The Science of Maintaining the Sea's Biodiversity (2005). Seeing ecosystems and meeting people from other places made me think about the world beyond my own habitat, my own circumstances. It parted my clouds, allowing me to comprehend that, in the world of nations, as in the universe of stars, some shine far more brightly than others. Costa Rica exemplifies this.

Nature has blessed Costa Rica with extraordinary biological diversity, from its forests to its seas. It is biologically rich because it's in the tropics where more species have evolved or where fewer have disappeared (or, quite possibly, both). It's biologically rich because it's topographically complex and offers an exceptionally broad range of physical conditions for life. It's biologically rich because it's a bridge between two large continents; species originating in both North and South America find suitable habitat there. It's biologically rich because it's bounded by the Earth's two biggest oceans, each of them home to rather different but very diverse assemblages of species.

Costa Rica is rightly famous worldwide for its terrestrial ecosystems, from the dry forests of Guanacaste Province to the rain forests of Cartago Province. But it should be equally acclaimed for the mass-nesting arribadas of its olive ridley sea turtles on the beaches of Ostional, the dazzling fishes of Cocos Island, the abundance of waterbirds in Caribbean wetlands of Limón Province, and the splendid diversity of marine invertebrates in both its Pacific and Atlantic waters.

There is another, more important reason why Costa Rica shines so brightly, and it lies in the hearts of its people (Ticos). Most countries that were once biologically rich are now impoverished; riches are easily squandered and lost forever. Costa Ricans are special for having made a unique commitment to catalogue and conserve their biota.

More than any other country, Costa Rica has encouraged both natives and foreigners who want to study and teach about its species and ecosystems. It is renowned for having welcomed institutions such as the Organization for Tropical Studies, which has trained so many of the world's leading tropical biologists, for having commissioned Dr. Rodrigo Gámez Lobo to build INBio, Costa Rica's National Biodiversity Institute, which has done so much to build science into conservation, and for encouraging scientists such as Dr. Archie Carr, the father of sea turtle biology, who taught so many people to apply marine science to conservation. Educating its people about its biodiversity is a strategic investment, ensuring that Costa Rica will remain a rich coast.

For a small nation in a world dominated by giants, maintaining biological wealth is a struggle when people hear the seductive calls of free trade, bigger houses, cars, appliances, electronics and fancy clothes. But Costa Rica is unique in insisting that its biodiversity is essential to its national economy and the future of its people. It towers above larger nations in its devotion to peaceful democracy and conservation. Rather than amassing military forces to threaten their neighbors, Ticos preserve a substantial portion of their biological legacy in national parks and other protected areas. In providing safe havens for both parrots and parrotfishes, Costa Rica has shown greatness of vision and wisdom. I wish that other nations, including my own, were as wise.

Costa Rica discovered far earlier than other nations that people would come to experience a diversity of life as exceptional in our world as a starry winter night in the Pacific Northwest. As other nations destroyed their biological heritage in exchange for economic development, Ticos realized that growing numbers would pay substantial sums to see miraculous wildlife and ecosystems that were once more widespread. Maintaining biodiversity, as former President José María Figueres once told me, is the essential foundation of Costa Rica's economic development. I have been privileged to meet several world leaders who did very good things for our environment, including US Presidents Jimmy Carter and Bill Clinton. But never before or since my discussion with President Figueres have I met a nation's leader who so thoroughly understood the importance of biodiversity to his people's well-being.

No country has a perfect conservation record, including my country, especially of late. Of course I'd like to see Costa Rica give much stronger protection to the sharks and other large fishes in its waters. Of course I'm troubled that little government support goes toward preventing poaching of endangered leatherback, green and olive ridley sea turtle nests. And of course I'm concerned that too many tourists could harm the places they come to see, essentially loving vulnerable ecosystems to death. But Ticos have reason to be proud as they work to strengthen protection of their beautiful marine waters, as they have their beautiful landscape.

Now we have something more to celebrate: the first book to document the marine biodiversity of a Central American country. It tells us what is known to date, and provides the basis for assessing what needs to be known in the future. It covers a broad range of taxa, including some that have seldom been examined carefully in this region. I suspect it will be a crucial reference for many decades to come.

Life on Earth is a gift that many people overlook as we conduct our daily affairs. But it is a far more enduring gift than so many of the things that we seek to acquire. In staking its future on protecting its diversity of life, on land and in the sea, Costa Rica ranks among the very brightest stars in conservation. Now we have the kind of book long needed by marine scientists working there and in nearby countries. As you savor the writing, facts, ideas and images in this book, I invite you to celebrate both a country consciously protecting its living things as the path to its future, and the scientists who are working to detail its biological wealth.

Elliott A. Norse
President of Marine Conservation Biology Institute
Bellevue, Washington, USA

Preface by the Editors

Central America, and especially Costa Rica is well known for its impressive biodiversity, and in fact many tourists visit the region for that reason. While the terrestrial diversity is fairly well documented, our knowledge of the marine biodiversity is fragmented and dispersed in a huge number of publications, some of them extremely hard to obtain. In 2001, we both agreed that it was time to summarize the current knowledge on the marine biodiversity of Costa Rica. What had to be done to document the marine biodiversity of the region seemed to be straightforward: to compile the existing information. And that was what we initiated. However, what started out as a manageable project which we thought could be accomplished within 2 years or so, was finally completed after many more years.

With the idea of compiling the existing information, we contacted many specialists working directly or indirectly on the taxonomy of marine organisms from Costa Rica. We were overwhelmed by the enthusiastic response of most people, and we are deeply grateful that 51 experts from 11 countries accepted our invitation. Interestingly, the majority of the authors (51%) comes from Costa Rica, which highlights the taxonomic expertise present in this country. An important portion of the contributions is written by specialists from the United States (21.6% of all authors) followed by colleagues from Mexico (9.8%). The remaining authors are associated with European institutions (13.7%), and two contributors come from Panama and Cuba. The contributions of all these experts made the present book possible. The different chapters provide an overview about our current knowledge of the marine biodiversity in Costa Rica, including lists of species and associated references as well as information on experts and collections for each taxonomic group.

Completing this book, it became obvious that the marine flora and fauna of Costa Rica is extremely diverse, representing almost 7,000 species. This number is surprisingly high for such a small country, and makes Costa Rica a hotspot of marine diversity on a worldwide scale. However, such a high number of species also represents an undeniable responsibility of the country to manage, protect and conserve its marine diversity. In this sense, we hope that the book may serve as a guide and scientific basis for decision making (governmental agencies as well as non-governmental organizations) concerning the marine realm of Costa Rica and the Central American region.

On the other hand, the book as a whole as well as each chapter shows what we do not know about this topic, which we consider also as valuable information to guide future research. The number of species reported for the different taxonomic units is in most cases a sub estimation of the actual diversity in the country. The inventory of the marine biodiversity in Costa Rica is an ongoing process, evidenced by the fact that during the editing process we constantly had to adjust the number of species in many chapters.

This book is written for anyone interested in the marine life, especially in Central America. It is also intended to motivate students, colleagues, and the interested public to further study the marine biodiversity in the region, which is the basis for an adequate conservation, management and sustainable use of its marine resources.

We have tried to be exhaustive in the search of information, but it is certainly possible that we might have overlooked existing literature or other relevant documents. Thus, we would like to emphasize that any suggestion, criticism, and corrections are more than welcome.

Ingo S. Wehrtmann
Jorge Cortés

Acknowledgments

First of all, we thank each and all of the authors of the chapters that have made this compilation of the marine biodiversity of Costa Rica possible, illuminating us on our state of knowledge. To all of them, "MUCHAS GRACIAS".

As editors we were obviously concerned about the scientific quality of the submitted contributions. For that reason, we contacted an immense number of international experts, and we are very grateful for their collaboration, critical reviews and valuable suggestions. Their expertise was essential to ensure that each and every of the contribution had the highest scientific standard possible.

The mother tongue of both editors is not English, thus we needed a native speaker to revise the manuscripts. During one of his visits to Costa Rica, we talked with Richard Petersen (Portland State University, USA) about this project, and he kindly accepted to revise a couple of manuscripts. Finally, he ended up going through practically all the manuscripts, revising hundreds of pages and returning them promptly to us, always with constructive comments. We sincerely appreciate his immense support, which greatly facilitated the editing process. His observations were not only important for the editing of the manuscripts, but his comments how he had enjoyed reading the chapters and learning about the marine biodiversity of Costa Rica were very encouraging for us, especially in times, when the project did not progress as expected or desired. Thank you, Richard!

It has been a tremendous amount of work to revise, format, and correct the almost 50 contributions, many of them with extremely extensive tables and reference lists. This immense task needed competent assistants with experience, dedication and a lot of patience with both the manuscripts and the two editors. We were happy to find Silvia Echeverría, Jimena Samper and Carlos Rojas, who dedicated many hours in going through the manuscripts. We especially would like to express our gratitude to Silvia Echeverría, because she accompanied us as an always responsible, reliable and critical assistant during all the ups and downs of the editing process. We might not have finished this book without her help, and we were certainly lucky to have her with us.

Photographs and illustrations form an important part of the book. We would like to express our gratitude to all colleagues and friends, who kindly searched their archives for adequate and high quality images, and allowed us to include some of them in the present book. Jeffry Ortiz scanned dozens of slides, and spent hours in front of the computer formatting the images for the book, which is deeply appreciated.

The Vicerrectoría de Investigación of the Universidad de Costa Rica (UCR) recognized the importance of the preparation of the book and financially supported the editing of the chapters, which is greatly appreciated. In fact, they unconditionally backed the project right from the beginning, which we did not expect considering the financial constraints of the institution.

We feel deeply honored that three highly distinguished personalities accepted our invitation to enrich the book with their foreword. The fact that the President of Costa Rica, Dr. Oscar Arias Sánchez, contributed with a foreword is a clear sign that Costa Rica has realized the ecological and socioeconomic importance of the marine realm, which represents the largest area of the country. Dr. Nancy Knowlton can be considered as one of the most prominent scientists in the area of marine tropical biology. Due to her extensive research in Panama, she is more than familiar with the marine biodiversity in the region. Last but not least, Dr. Elliott Norse has published extensively on the importance of marine biodiversity and has played a key role in promoting the topic of global marine biodiversity through his important publications. We are honored and grateful to all of them for endorsing this book.

We deeply appreciate the confidence and constant support provided by Springer (the initial contact was with the former Kluwer Academic Publishers), especially by Judith Terpos and Tamara Welschot, who accompanied us during the editing process. The communication with them was always smooth and enjoyable, even though we constantly had to postpone our estimate of when the book might be ready for publication.

The contact with the editor of the series Monographiae Biologicae, Henry Dumont, was important for us, especially during the late phase of the project. His advice and suggestions were very valuable; we did not agree on everything, but his comments and observations certainly helped to improve the content and design of the chapters. Moreover, we were happy to communicate with a series editor well familiar with the topic of marine biodiversity (see Dumont, H.J. 2005. Biodiversity: a resource with a monetary value? Hydrobiologia 542: 11–14).

Ingo S. Wehrtmann
Jorge Cortés

Referees List

The editors (and authors) are most grateful to all the referees for the revision of the manuscripts and their valuable suggestions and comments. Although we tried to prepare a complete list, we sincerely apologize if we have forgotten to mention the name of someone else who participated somehow in the revision of the manuscripts.

Name	Country	Affiliation
Acero, Arturo	Colombia	INVEMAR, Santa Marta
Alvarez, Fernando	Spain	Departamento de Geología, Universidad de Oviedo
Alvarez Cadena, José N.	Mexico	Instituto de Ciencias del Mar y Limnología, Universidad Nacional Autónoma de México, México DF
Arauz, Randall	Costa Rica	PRETOMA, San José
Ahyong, Shane T.	Australia	Department of Marine Invertebrates, Australian Museum, Sydney
Aznar, Francisco Javier	Spain	Unidad de Zoología Marina, Instituto Cavanilles de Biodiversidad y Biología Evolutiva, Universidad de Valencia
Baba, Shigeyuki	Japan	University of Ryukyus, Okinawa
Bamber, Roger	United Kingdom	Environment: Coastal & Marine, The Natural History Museum, London
Bayer, Frederick M. (deceased)	USA	National Museum of Natural History, Smithsonian Institution, Washington, DC
Bertsch, Hans	USA	Imperial Beach, California
Blake, James A.	USA	ENSR Marine & Coastal Center , Woods Hole, Massachusetts
Boucot, Arthur J.	USA	Department of Zoology, Oregon State University, Corvallis, Oregon
Bouillon, Jean	Belgium	Jean Laboratoire de Biologie Marine, Université Libre de Bruxelles
Brattegard, Torleiv	Norway	Department of Fisheries and Marine Biology, University of Bergen
Brinton, Edward	USA	Scripps Institution of Oceanography, La Jolla, California

Cairns, Stephen D.	USA	National Museum of Natural History, Smithsonian Institution, Washington, DC
Cheng, Lanna	USA	Scripps Institution of Oceanography, University of California, San Diego, La Jolla, California
Cigliano, John	USA	Cedar Crest College, Allentown, Pennsylvania
Coates, Anthony G.	USA	National Museum of Natural History, Smithsonian Institution, Washington, DC
Díaz, Maria Cristina	Venezuela	Museo Marino de Margarita, Nueva Esparta
Escobar-Briones, Elva	Mexico	Instituto de Ciencias del Mar y Limnología, Universidad Nacional Autonoma de Mexico
Esna, Graciela B.	Argentina	Facultad de Ciencias Exactas y Naturales, Universidad de Buenos Aires
Eernisse, Douglas J.	USA	University of California, Fullerton, California
Fautin, Daphne G.	USA	University of Kansas, Lawrence, Kansas
Freshwater, D. Wilson	USA	Center for Marine Science - UNCW, Wilmington, North Carolina
García Ríos, Cedar I.	Puerto Rico	Universidad de Puerto Rico, Humacao
Gerrodette, Tim	USA	National Marine Fisheries Service, Southwest Fisheries Science Center, La Jolla, California
Giribet, Gonzalo	USA	Department of Organismic & Evolutionary Biology, Museum of Comparative Zoology, Harvard University, Cambridge, Masschusetts
Gochfeld, Michael	USA	Department of Ornithology, American Museum of Natural History, New York, New York
Gómez, Luis Diego	Costa Rica	Organización para Estudios Tropicales, San José
Gosliner, Terrence	USA	California Academy of Sciences, San Francisco, California
Grasshoff, Manfred	Germany	Senckenberg Research Institute and Natural History Museum, Frankfurt a.M.
Guzmán, Hector M.	Panama	Smithsonian Tropical Research Institute, Balboa
Hansknecht, Tom	USA	Barry A. Vittor & Associates, Mobile, Alabama
Hendrickx, Michel E.	Mexico	Universidad Nacional Autónoma de México, Mazatlán
Hernández Trujillo, Sergio	Mexico	Departamento de Plancton, CICIMAR, Instituto Politécnico Nacional, La Paz
Jiménez, Jorge	Costa Rica	MarViva, San José
Krapp, Franz	Germany	Zoologisches Forschungsinstitut und Museum A. Koenig, Bonn
La Barbera-Sánchez, Amelia	Venezuela	Centro de Investigaciones Agrícolas de los Estados Sucre y Nueva Esparta, Cumaná
Lamar, William W.	USA	College of Sciences, University of Texas, Tyler, Texas
Mianzan, Hermes Walter	Argentina	CONICET, Instituto Nacional de Investigación y Desarrollo Pesquero (INIDEP), Mar del Plata
Modlin, Richard	USA	Department of Biological Sciences, The University of Alabama in Huntsville, Alabama
Monks, Scott	Mexico	Centro de Investigaciones Biológicas, Universidad Autónoma del Estado de Hidalgo, Pachuca, Hidalgo

Montu, Mónica (deceased)	Brazil	Federal University of Rio Grande, Rio Grande
Morales Acuña, Juan Alberto	Costa Rica	Escuela de Ciencias Veterinarias Tropicales, Universidad Nacional, Heredia
Nettleship, David N.	Canada	Lundy Environmetal Consulting, Nova Scotia
Pagès, Francesc (deceased)	Spain	Institut de Ciències del Mar (CSIC), Centre Mediterrani d'Investigacions Marines i Ambientals, Barcelona
Painter, David S.	USA	Department of History, Georgetown University, Washington, DC
Petersen, Richard	USA	Department of Biology, Portland State University, Portland, Oregon
Pitombo, Fábio B.	Brazil	Instituto de Biologia, Universidade Federal Rural do Rio de Janeiro
Prieto, Luis	Mexico	Laboratorio de Helmintología, Instituto de Biología, Universidad Nacional Autómoma de México
Pugh, Phil	United Kingdom	Southampton Oceanography Centre, Southampton
Reaka-Kudla, Marjorie L.	USA	Department of Biology, The University of Maryland, College Park, Maryland
Reyes Bonilla, Hector	Mexico	Departamento de Biología Marina, Universidad Autónoma de Baja California Sur, La Paz
Robertson, Ross	Panama	Smithsonian Tropical Research Institute, Balboa
Roccatagliata, Daniel	Argentina	Facultad de Ciencias Exactas y Naturales, Universidad de Buenos Aires
Roldán, Carmen	Spain	Depto. de Biología Animal I, Facultad de Biología, Universidad Complutense, Madrid
Ross, Arnold	USA	Marine Biology Research Division, Scripps Institution of Oceanography, La Jolla, CA
Ruetzler, Klaus	USA	National Museum of Natural History, Smithsonian Institution, Washungton D.C.
Ryland, John S.	Wales	Swansea University, Swansea
Saiz-Salinas, José	Spain	Depto. de Zoología y Dinámica Celular Animal, Universidad del País Vasco, Bilbao
Sanner, JoAnn	USA	National Museum of Natural History, Smithsonian Institution, Washington, DC
Savage, Jay M.	USA	Rana Dorada Enterprises, S.A., San Diego, California
Salazar-Vallejo, Sergio I.	Mexico	Depto. Ecología Acuática, ECOSUR, Chetumal, Q. Roo
Schotte, Marilyn	USA	National Museum of Natural History, Smithsonian Institution, Washington, DC
Schmidt-Effing, Reinhard	Germany	Institut für Geologie und Palaeontologie, Fachbereich Geowissenschaften, Philipps-Universität Marburg
Schreiber, Elizabeth A.	USA	National Museum of Natural History, Smithsonian Institution, Washington, DC
Schrödl, Michael	Germany	Zoologische Staatssammlung München
Skoglund, Carol	USA	Santa Barbara Museum of Natural History, Santa Barbara, California

Sprechmann, Peter	Uruguay	Facultad de Ciencias, Montevideo
Stiles, F. Gary	Colombia	Universidad de los Andes, Bogotá
Stöhr, Sabine	Sweden	Dept. of Invertebrate Zoology, Swedish Museum of Natural History
van Ofwegen, Leen	The Netherlands	Nationaal Natuurhistorisch Museum, Leiden
Vargas-Montero, Maribelle	Costa Rica	Centro de Investigación en Estructuras Microscópicas, Universidad de Costa Rica, San José
Vargas Zamora, José Antonio	Costa Rica	Centro de Investigación en Ciencias del Mar y Limnología, Universidad de Costa Rica, San José
Viales, Ronny	Costa Rica	Centro de Investigaciones Históricas, Universidad de Costa Rica, San José
Viéitez, José M.	Spain	Depto. de Biología Animal, Facultad de Ciencias, Universidad de Alcalá, Alcalá de Henares
Voight, Janet	USA	The Field Museum, Chicago, Illinois
Vonk, Ronald	The Netherlands	Institute for Biodiversity and Ecosystem Dynamics (IBED), Faculty of Science, University of Amsterdam
Voss, Nancy	USA	Invertebrate Museum, Rosenstiel School of Marine and Atmospheric Sciences, University of Miami, Florida
Wetzer, Regina	USA	Natural History Museum of Los Angeles County, Los Angeles, California
Wynne, Michael	USA	University of Michigan, Ann Arbor, Michigan
Zapata, Fernando	Colombia	Departamento de Biología, Universidad del Valle, Cali
Zea, Sven	Colombia	INVEMAR, Santa Marta
Zeidler, Wolfgang	Australia	South Australian Museum (Marine Invertebrates), University of Adelaide, South Australia
Zoppi de Roa, Evelyn	Venezuela	Facultad de Ciencias, Universidad Central de Venezuela, Caracas

Contents

Contributors

Teresita Aguilar
Escuela Centroamericana de Geología and Centro de Investigación en Ciencias del Mar y Limnología (CIMAR); Universidad de Costa Rica. Apdo. 214-2060 San José, Costa Rica
aaguilar@geologia.ucr.ac.cr

Juan José Alvarado
Centro de Investigación en Ciencias del Mar y Limnología (CIMAR), Universidad de Costa Rica, 11501-2060 San José, Costa Rica
juanalva76@yahoo.com

Roger N. Bamber
Department of Zoology, The Natural History Museum, Cromwell Road, London SW7 5BD, United Kingdom
r.bamber@nhm.ac.uk

Gilbert Barrantes
Escuela de Biología, Universidad de Costa Rica. San Pedro, 11501-2060 San José, Costa Rica
gbarrantes@biologia.ucr.ac.cr

Andrea Bernecker
Escuela de Biología, Universidad de Costa Rica, San Pedro, 11501-2060 San José, Costa Rica
aberneck@biologia.ucr.ac.cr

Odalisca Breedy
Museo de Zoología, Escuela de Biología, Universidad de Costa Rica, 11501-2060 San José, Costa Rica
odalisca@racsa.co.cr

Richard C. Brusca
Arizona-Sonora Desert Museum, 2021 N. Kinney Rd., Tucson, Arizona 85743, USA
rbrusca@desertmuseum.org

William A. Bussing
Museo de Zoología, Universidad de Costa Rica & Centro de Investigación en Ciencias del Mar y Limnología (CIMAR), 11501-2060 San José, Costa Rica
wbussing@biologia.ucr.ac.cr

Yolanda E. Camacho-García
Museo de Zoología, Escuela de Biología, Universidad de Costa Rica, 11501-2060 San José, Costa Rica
ycamacho_99@yahoo.com

Iván Castellanos
El Colegio de la Frontera Sur (ECOSUR) Unidad Chetumal, Carr. Chetumal-Bacalar Km 2, A.P. 424, 77000, Chetumal, Quintana Roo, México
ivancast@ecosur-qroo.mx

Gerardo A. Chaves
Escuela de Biología, Universidad de Costa Rica, 11501-2060 San José, Costa Rica
cachi@biologia.ucr.ac.cr

Johel Chaves-Campos
Department of Biological Sciences, Purdue University, West Lafayette, Indiana 47907, USA
jchavesc@costarricense.cr

Jorge Cortés
Centro de Investigación en Ciencias del Mar y Limnología (CIMAR), and Escuela de Biología, Universidad de Costa Rica, 11501-2060 San José, Costa Rica
jorge.cortes@ucr.ac.cr

Harlan K. Dean
Museum of Comparative Zoology (MCZ), Harvard University, 26 Oxford St. Cambridge, Massachusetts 02138, USA
harlandean@comcast.net

Silvia Echeverría-Sáenz
Unidad Investigación Pesquera y Acuicultura (UNIP) of the Centro de Investigación en Ciencias del Mar y Limnología (CIMAR), Universidad de Costa Rica, 11501-2060 San José, Costa Rica
sipama14@yahoo.com

Christian C. Emig
Centre d'Océanologie, CNRS UMR 6540, Station Marine d'Endoume, Rue de la Batterie-des-Lions, F-13007 Marseille, France
christian.emig@com.univ-mrs.fr

José Espinosa
Instituto de Oceanología de la Habana, Cuba. Ave 1[a], N 18406, E. 184 y 186, Playa, La Habana, Cuba
espinosa@oceano.inf.cu

John M. Foster
University of Southern Mississippi, Dept. of Coastal Sciences, 703 East Beach Drive, Ocean Springs, Mississippi 39564, USA
beachbugs@aol.com

Rebeca Gasca
El Colegio de la Frontera Sur (ECOSUR)-Chetumal
A.P. 424. Chetumal, Quintana Roo 77000, México
rgasca@ecosur-qroo.mx

Paul E. Hargraves
Graduate School of Oceanography University of Rhode Island, Narragansett, Rhode Island 02882, USA
pharg@gso.uri.edu

Richard W. Heard
University of Southern Mississippi, Institute of Marine Sciences, Dept. of Coastal Sciences, 703 East Beach Blvd., P.O. Box 7000, Ocean Springs, Mississippi 39566-7000, USA
richard.heard@usm.edu

Rosa Ma. Hernández-Flores
El Colegio de la Frontera Sur-Chetumal. A.P. 424. Chetumal, Quintana Roo 77000. Mexico
rosamahe@ecosur-qroo.mx

Amalia Herrera-Cubilla
Center for Tropical Paleoecology and Archeology, Smithsonian Tropical Research Institute, P.O. Box 2072. Balboa/Ancón, Panamá, República de Panamá
herreraa@ancon.si.edu

Federick G. Hochberg
Department of Invertebrate Zoology, Santa Barbara Museum of Natural History, 2559 Puesta del Sol Road, Santa Barbara, California 93105, USA
fghochberg@sbnature2.org

Nancy Knowlton
Department of Invertebrate Zoology, National Museum of Natural History, Smithsonian Institution MRC 163, PO Box 37012, Washington, DC 20013-7012, USA
nknowlton@popmail.ucsd.edu

Sara E. LeCroy
University of Southern Mississippi, Dept. of Coastal Sciences,
703 East Beach Drive, Ocean Springs, Mississippi 39564, USA
sara.lecroy@usm.edu

Myrna López
Museo de Zoología, Universidad de Costa Rica & Centro de Investigación en Ciencias del Mar y Limnología (CIMAR), 11501-2060 San José, Costa Rica
myrnal@biologia.ucr.ac.cr

Julio A. Magaña-Cubillo
Universidad Estatal a Distancia. Programa de Manejo de Recursos Naturales, Sabanilla, San José Costa Rica
eubranchus@yahoo.com

Simonetta Mattiucci
Department of Public Health Sciences, Section of Parasitology,
University of Rome "La Sapienza", P.le Aldo Moro, 5 00185 Rome, Italy
simonetta.mattiucci@uniroma1.it

Laura May-Collado
Department of Biology, Florida International University. Current address: (FIU) 11200 SW 8th St. Miami, Florida 33199, USA
lmaycollado@gmail.com

Claudia Mora-Baumgartner
Institut de Géologie et Paléontologie, Université de Lausanne,
BFSH-2, CH-1015 Lausanne, Switzerland
Claudia.Baumgartner@igp.unil.ch

Álvaro Morales-Ramírez
Centro de Investigación en Ciencias del Mar y Limnología (CIMAR), and Escuela de Biología, Universidad de Costa Rica, 11501-2060 San José. Costa Rica
alvaro.morales@ucr.ac.cr

Vanessa Nielsen
Unidad Investigación Pesquera y Acuicultura (UNIP) del Centro de Investigación en Ciencias del Mar y Limnología (CIMAR),
Universidad de Costa Rica, 11501-2060 San José, Costa Rica
vnielsen@cariari.ucr.ac.cr

Elliott A. Norse
Marine Conservation Biology Institute, 2122 112th Ave NE,
Suite B-300, Bellevue, Washington 98004, USA
elliott@mcbi.org

Lisa D. Patrick
Department of Biology, Texas Tech University, Lubbock, Texas, USA
lisa_d_patrick@yahoo.com

Iorgu Petrescu
"Grigore Antipa" National Museum of Natural History,
Dept. of Aquatic Fauna, Kiseleff 1, Bucharest, 011341, Romania;
Fax: 4021/312888
iorgup@antipa.ro

W. Wayne Price
University of Tampa, Dept. of Biology, Tampa, Florida 33606, USA
wprice@ut.edu

Karina Rodríguez-Sáenz
Habitat Crítico: Environmental Consulting,
Apdo. 389-1300, San José, Costa Rica
karinarodriguez@racsa.co.cr

Eva Salas
Centro de Investigación en Ciencias del Mar y Limnología (CIMAR),
and Escuela de Biología, Universidad de Costa Rica,
11501-2060 San José, Costa Rica
evamaria_cr@yahoo.com

Mario Santoro
Departamento de Patología, Escuela de Medicina Veterinaria,
Universidad Nacional, PO Box 86, Heredia 3000, Costa Rica
marisant@libero.it

Mahmood Sasa
Instituto Clodomiro Picado, Universidad de Costa Rica,
11501-2060 San José, Costa Rica
msasamarin@gmail.com

Enrico Schwabe
Zoologische Staatssammlung München, Münchhausenstrasse 21,
D-81247 München, Germany
enrico.schwabe@zsm.mwn.de

Lourdes Segura-Puertas
Universidad Nacional Autónoma de México, Instituto de Ciencias
del Mar y Limnología, Unidad Académica Puerto Morelos.
P. O. Box 1152, 77501 Cancún, Quitana Roo, México
lsegura@mar.icmyl.unam.mx

Ana Margarita Silva-Benavides
Escuela de Biología. Universidad de Costa Rica, 11501-2060 San José,
Costa Rica. Centro de Investigación en Ciencias del Mar y Limnología (CIMAR).
Universidad de Costa Rica, 11501-2060 San José, Costa Rica
msilva@biologia.ucr.ac.cr

Leonora Rodríguez-Sevilla
Museo de Zoología, Escuela de Biología, Universidad de Costa Rica,
11501-2060 San José, Costa Rica. Current address: Apdo. 3144-1000,
San José, Costa Rica
leonorars@racsa.co.cr

Monika Springer
Escuela de Biología, Universidad de Costa Rica, 11501-2060 San José, Costa Rica
springer@biologia.ucr.ac.cr

Eduardo Suárez-Morales
El Colegio de la Frontera Sur (ECOSUR) Unidad Chetumal,
Carr. Chetumal-Bacalar Km 2, A.P. 424, 77000, Chetumal, Quintana Roo, México
esuarez@ecosur-qroo.mx

Noam van der Hal
Zoological Museum, University of Amsterdam, P.O. Box 94766,
1090 GT, Amsterdam, Netherlands
nvanderhal@yahoo.com,

R.W.M. Van Soest
Zoological Museum, University of Amsterdam, P.O. Box 94766,
1090 GT, Amsterdam, Netherlands
R.W.M.vanSoest@uva.nl

Robert Van Syoc
Department of Invertebrate Zoology and Geology, California Academy of
Sciences, Golden Gate Park, San Francisco, California 94118-4599, U.S.A
bvansyoc@calacademy.org

José A. Vargas
Centro de Investigación en Ciencias del Mar y Limnología (CIMAR),
Universidad de Costa Rica, 11501-2060 San José, Costa Rica
javargas@cariari.ucr.ac.cr

Rita Vargas
Museo de Zoología, Escuela de Biologia, Universidad de Costa Rica,
11501-2060 San José, Costa Rica
ritav@biologia.ucr.ac.cr

Roxana Víquez
Escuela de Ciencias Biológicas, Universidad Nacional Apdo 86-3000,
Heredia, Costa Rica
rviquezm@una.ac.cr

Ingo S. Wehrtmann
Unidad Investigación Pesquera y Acuicultura (UNIP) of the Centro de
Investigación en Ciencias del Mar y Limnología (CIMAR),
and Museo de Zoología, Escuela de Biología, Universidad de Costa Rica,
11501-2060 San José, Costa Rica
ingowehrtmann@gmx.de

Chapter I
Diversity of Marine Habitats of the Caribbean and Pacific of Costa Rica

Jorge Cortés and Ingo S. Wehrtmann

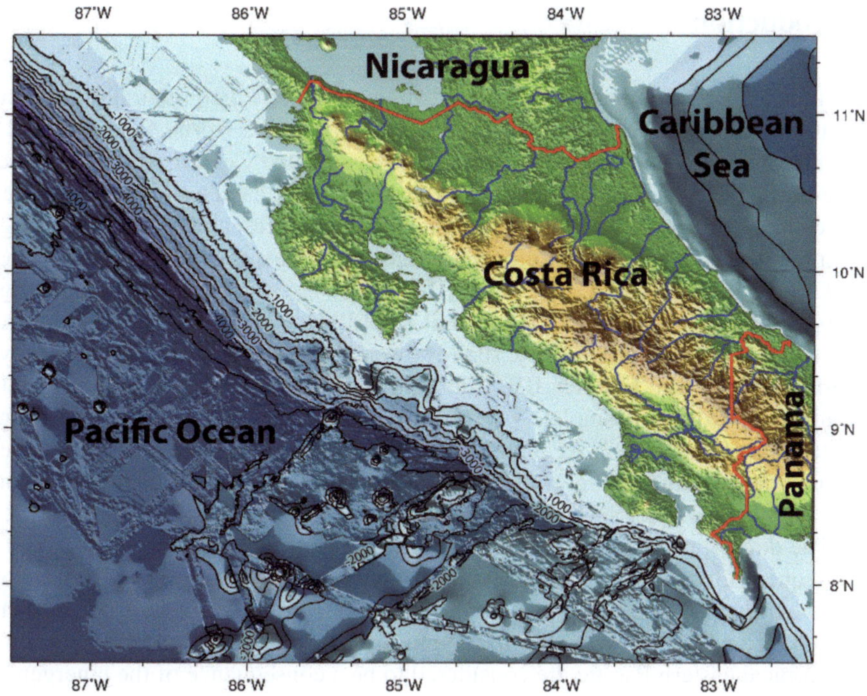

Bottom morphology of Caribbean and Pacific of Costa Rica. (Illustration: Wilhelm Weinrebe)

J. Cortés and I.S. Wehrtmann
Centro de Investigación en Ciencias del Mar y Limnología (CIMAR) and Escuela de Biología, Universidad de Costa Rica, San Pedro, San José 11501-2060, Costa Rica.
e-mails: jorge.cortes@ucr.ac.cr; ingowehrtmann@gmx.de

I.S. Wehrtmann, J. Cortés (eds.) *Marine Biodiversity of Costa Rica, Central America*,

Abstract Costa Rica is privileged to have two coasts: the Caribbean coast is almost a straight line of 212 km, while the coastline along the Pacific is much more complex and substantially longer (1,254 km). This chapter begins with a brief description of the geological processes that led to the emergence of the Central American Isthmus. Subsequently, we present a list of the marine-protected areas in Costa Rica, and continue to describe the main marine habitats which can be found on the Caribbean as well as on the Pacific side of Costa Rica, including Golfo de Nicoya, Golfo Dulce, the deep sea, and Isla del Coco. We summarize the findings of the principal studies carried out in these habitats, and provide relevant information concerning the impact of human activities affecting these systems in Costa Rica.

Introduction

Costa Rica is part of the Central American landbridge, which separates the eastern Pacific Ocean from the Caribbean Sea, and connects the North American and the South American continents. The origin of the so-called Central American Isthmus dates back to 200 million years. This initial period was characterized by the presence of deepwater volcanic rocks that moved into roughly its present-day position, emerging as small islands, around 40 million years ago (Ma) (Coates & Obando 1996; Coates 1997; Denyer *et al.* 2003; see Chapter III). The oldest records are of pelagic organisms deposited in deepwaters. The fossil record between 30 Ma and the closure of the isthmus is characterized by shallow water deposits with a high diversity of marine species (see Chapter III). Subsequently, these islands fused slowly into a continuous landmass, completely separating the Atlantic and Pacific Oceans roughly 3 Ma (Coates *et al.* 1992; Aguilar 2000). The emergence of the Isthmus split marine populations, resulting in speciation and extinctions and also facilitated the evolution of transisthmian sister species (Knowlton 1993; Wehrtmann & Albornoz 2002).

The development of the landbridge between North and South America and the subsequent closure between the Atlantic and Pacific Oceans strongly influenced oceanographic patterns, especially along the Pacific coast. Seasonal coastal upwelling and the occurrence of El Niño-Southern Oscillation (ENSO) events in the tropical eastern Pacific are considered to be a consequence of the emergence of the Isthmus of Central America (Cortés 1997; Jackson & D'Croz 1997). The main consequences of the new oceanographic conditions were extreme thermal events together with changing sea levels during the glaciations, which resulted in extinction of many species (Dana 1975; Cortés 1986; D'Croz & Robertson 1997; Glynn 1997; Glynn & Ault 2000). Associated with the changed current patterns, organisms were transported from the Central Pacific to the coast of the eastern Pacific, thus enriching the region; examples for such events are corals (Dana 1975; Cortés 1986; Glynn & Ault 2000), mollusks (Scheltema 1988; Shasky 1988), echinoderms (Lessios *et al.* 1998), and fishes (Robertson & Allen 1996).

Costa Rica is located between Nicaragua and Panama, covering a total landmass of 51,100 km^2. However, the marine area, including Territorial Seas and Exclusive

Economic Zones, is more than ten times larger, 589,683 km^2 (INCOPESCA 2006), mainly due to the 200 mile-zone on the Pacific coast and the 200 miles around the oceanic Isla del Coco in the Pacific Ocean, which enlarges significantly the Exclusive Economic Zone of Costa Rica. Because of this marine territory the country has borders also with Ecuador and Colombia (Fig. I.1).

The length of the Pacific coast is 1,254 km, while the Caribbean coastline of Costa Rica is considerably shorter (212 km). Marine environments comprise beaches, rocky intertidal areas, mangroves, soft bottoms, estuaries, seagrass beds, coral reefs, a tropical fjord (Golfo Dulce), coastal islands, and an oceanic island (Fig. I.1). The continental shelf is relatively narrow dropping to deepwater zones down to over 4,000 m in the Mesoamerican Trench and abyssal plains with sea mounts (Huene *et al.* 1995; Denyer *et al.* 2003). Ecologically important is the presence of seasonal upwelling in the northern region of the Pacific coast (McCreary *et al.* 1989; Cortés 1996–1997) as well as the Costa Rican thermal dome offshore (Fiedler 2002; Kessler 2006). According to Boschi (2000), the Pacific coast of

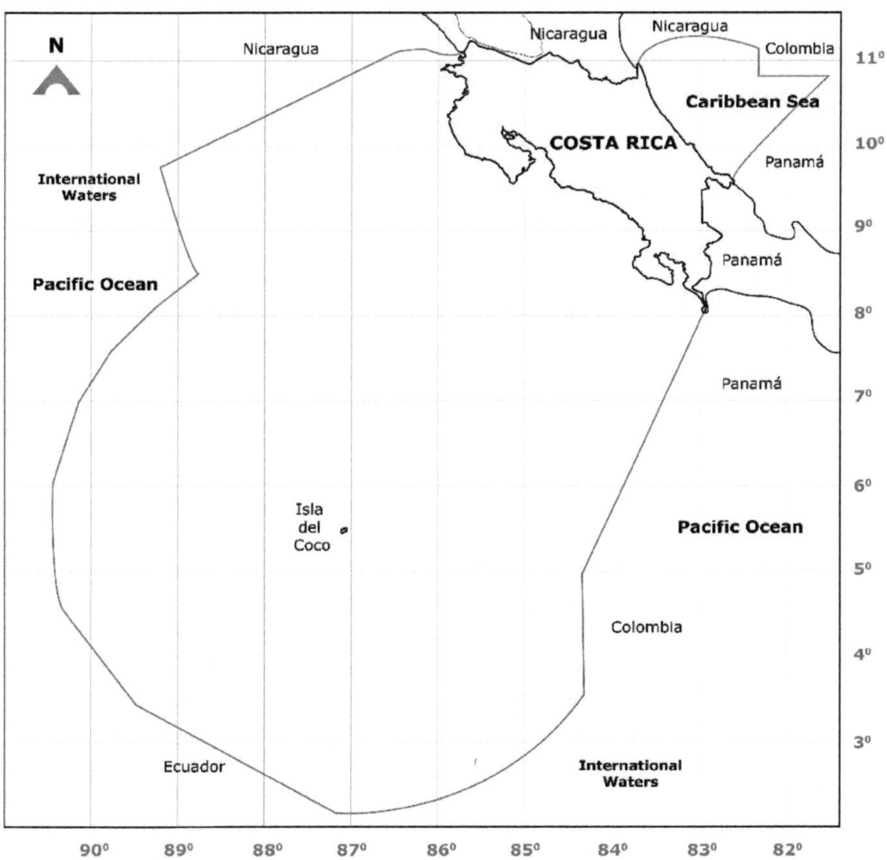

Fig. I.1 Costa Rica and its territorial seas

Costa Rica forms part of the Panamic zoogeographical province, while the Caribbean coast belongs to the Caribbean zoogeographical province.

The marine habitats of Costa Rica are subject to natural and anthropogenic impacts. The main natural event affecting these systems has been the El Niño-Southern Oscillation (ENSO). Several studies documented the massive death of marine invertebrates during the occurrence of strong ENSO events in the last two decades (Cortés *et al.* 1984; Guzmán *et al.* 1987; Guzmán & Cortés 1992, 2001, 2007; Jiménez 2001a; Jiménez & Cortés 2001, 2003a; Jiménez *et al.* 2001). Earthquakes can also have a severe impact on marine systems, as shown by the 1991 Limon earthquake, which uplifted parts of the Caribbean coast in some areas up to 1.9 m (Denyer *et al.* 1994a, b; Denyer 1998). This uplift exposed 2.5 km^2 of subtidal habitats, resulting in the death of the inhabiting organisms (Cortés *et al.* 1992, 1994a).

Human activities affect the marine habitats in Costa Rica. The Golfo de Nicoya, the most productive coastal area and the most important fishery ground in the country, is also one of the most polluted regions along the Pacific coast (Vargas 1995). Principal contaminants are agrochemicals and solid waste (García *et al.* 2006). Fecal pollution (Acuña *et al.* 1998), petroleum (Acuña-González *et al.* 2004), PCB (Spongberg 2004a, b, c), and heavy metals (Rojas *et al.* 1998; García-Céspedes *et al.* 2004) usually play a minor role regarding the pollution. Even coastal areas with low population densities show signs of anthropogenic contamination, as evidenced by studies carried out in Golfo Dulce (pesticides: Spongberg & Davis 1998; hydrocarbons: Acuña-González *et al.* 2004; PCB: Spongberg 2004c). Oil pollution is high near ports, especially around Moín, located close to Limon (Mata *et al.* 1987; Acuña-González *et al.* 2004), while pollutants were under the detection limits in the more isolated Laguna Gandoca (Coll *et al.* 2004).

Several studies along both the Caribbean and the Pacific coasts of Costa Rica have revealed that sedimentation is the principal source of coral reef degradation by reducing both growth rates and recovery from impacts, and causing the death of the corals (Cortés & Risk 1985; Cortés 1990a; Cortés & Jiménez 2003a, b; Alvarado *et al.* 2005). Other sources of anthropogenic impacts on coastal ecosystems are related to the increasing importance of tourism in Costa Rica. Direct impacts of these and related activities result in an increase of: (1) construction and changes in the morphology of coastal areas; (2) waste, especially due to the improper treatment of solid waste and sewage; (3) extraction of marine resources; (4) water and air pollution by tourist boats; (5) visitations of beaches, islands and coral reefs; and (6) disturbance of marine mammals, e.g., by whale watchers.

Costa Rica has been recognized internationally as a leading country in natural habitat protection. The first national parks were established in the early 1970s, and now, roughly 25% of the land surface is in some sort of protected area. Some of these parks include marine habitats, with a majority occurring on the Pacific side of Costa Rica (Fig. I.2 and Table I.1). These marine-protected areas cover 30,308 km^2 or 16.5% of the Internal Waters and Territorial Seas (SINAC-MINAE 2006), but only 0.87% of the 589,683 km^2 of the Exclusive Economic Zone of Costa Rica (INCOPESCA 2006). Recently, the government of Costa Rica signed the executive decree N° 31832-MINAE, of 7 July 2004, to evaluate the possibility of protecting up to 25% of the marine territories of the nation.

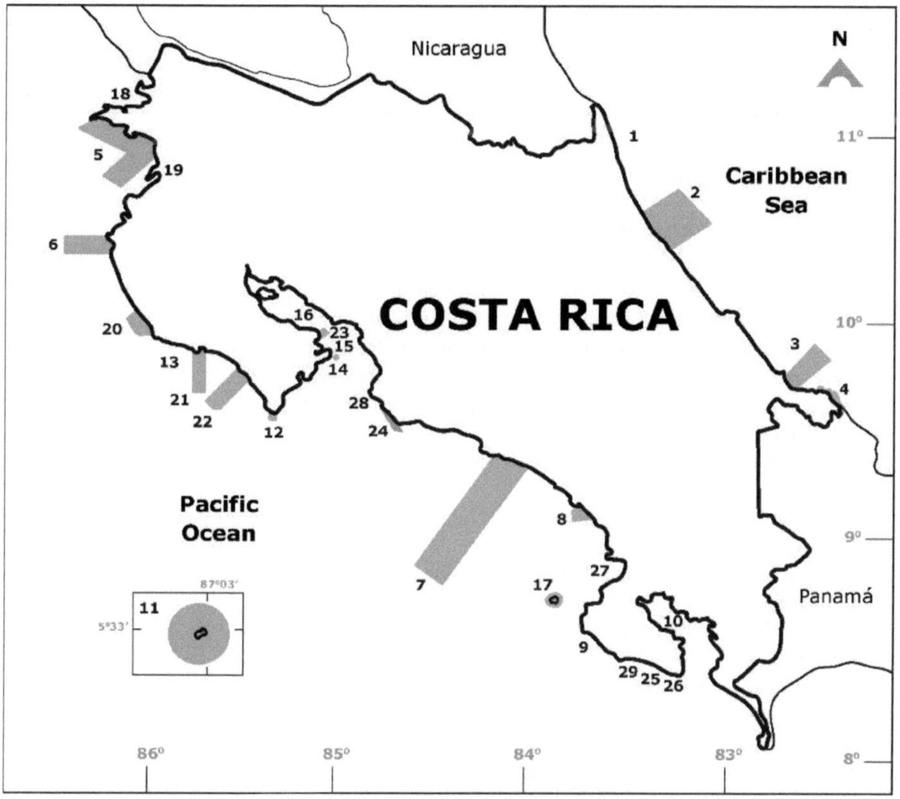

Fig. I.2 Location and extension of marine-protected areas in Costa Rica (see Table I.1 for names of the numbered sites)

Table I.1 Status of marine-protected areas along both the Caribbean and Pacific coasts of Costa Rica. (Mora *et al.* 2006)

Name	Status	Area (ha) Terrestrial	Area (ha) Marine
Caribbean			
(1) Refugio Nacional de Vida Silvestre Barra del Colorado	National Wildlife Refuge	81,211	n.a.
(2) Parque Nacional Tortuguero	National Park	31,187	52,681
(3) Parque Nacional Cahuita	National Park	1,106	23,290
(4) Refugio Nacional de Vida Silvestre Gandoca-Manzanillo	National Wildlife Refuge	5,013	4,983
Pacific			
(5) Parque Nacional Santa Rosa	National Park	38,674	46,391
(6) Parque Nacional Baulas de Guanacaste	National Park	379	25,335
(7) Parque Nacional Manuel Antonio	National Park	1,625	42,016
(8) Parque Nacional Marino Ballena	National Park	162	5,229
(9) Parque Nacional Corcovado	National Park	42,469	2,044

(continued)

Table I.1 (continued)

Name	Status	Area (ha)	
		Terrestrial	Marine
(10) Parque Nacional Piedras Blancas	National Park	14,025	1,356
(11) Área de Conservación Marina Isla del Coco	Marine Conservation Area	2,310	19,483
(12) Reserva Absoluta Cabo Blanco	Absolute Reserve	1,270	1,629
(13) Reserva Biológica Isla Chora	Biological Reserve	5	*
(14) Reserva Biológica Isla Negritos	Biological Reserve	142	*
(15) Reserva Biológica Isla Guayabo	Biological Reserve	6	*
(16) Reserva Biológica Isla Pájaros	Biological Reserve	4	*
(17) Reserva Biológica Isla del Caño	Biological Reserve	226	5,207
(18) Refugio Nacional de Vida Silvestre Junquillal	National Wildlife Refuge	443	**
(19) Refugio Nacional de Vida Silvestre Iguanita	National Wildlife Refuge	114	*
(20) Refugio Nacional de Vida Silvestre Ostional	National Wildlife Refuge	352	8,055
(21) Refugio Nacional de Vida Silvestre Camaronal	National Wildlife Refuge	234	n.a.
(22) Refugio de Vida Silvestre Caletas-Arío	National Wildlife Refuge	313	19,846
(23) Refugio Nacional de Vida Silvestre Isla San Lucas	National Wildlife Refuge	468	725
(24) Refugio Nacional de Vida Silvestre Playa Hermosa	National Wildlife Refuge	44	3,654
(25) Refugio Nacional de Vida Silvestre Pejeperro	National Wildlife Refuge	441	151
(26) Refugio Nacional de Vida Silvestre Río Oro	National Wildlife Refuge	39	1,719
(27) Humedal Nacional Térraba-Sierpe	National Wetland	27,062	5,531
(28) Humedal Nacional Marino de Playa Blanca	National Wetland	3	5
(29) Humedal Nacional Lacustrino Pejeperrito	National Wetland	31	32

Number before the name indicates the location of the protected area in Fig. 2; n.a.: Information not available.

*Officially, no marine-protected area included.

**The Presidential Decree declaring this National Wildlife Refuge is unclear about the marine area.

Marine Habitats

The following section provides a brief overview of marine habitat diversity in Costa Rica. We describe the main ecosystems, and mention relevant publications of studies carried out in the country.

Caribbean Coast

The Caribbean coast of Costa Rica is relatively short, 212 km, and rectilinear (Fig. I.3). The northern section consists of high-energy sandy beaches (Fig. I.4), and linear coastal lagoons, product of sedimentary accretion (Parkinson *et al.* 1998). The southern section of the coast has sandy beaches interrupted by rocky promontories made up mainly by fossil coral reefs (Pleistocene, Holocene) (Cortés & Guzmán 1985a). The present-day reefs are growing on top of these fossil outcrops

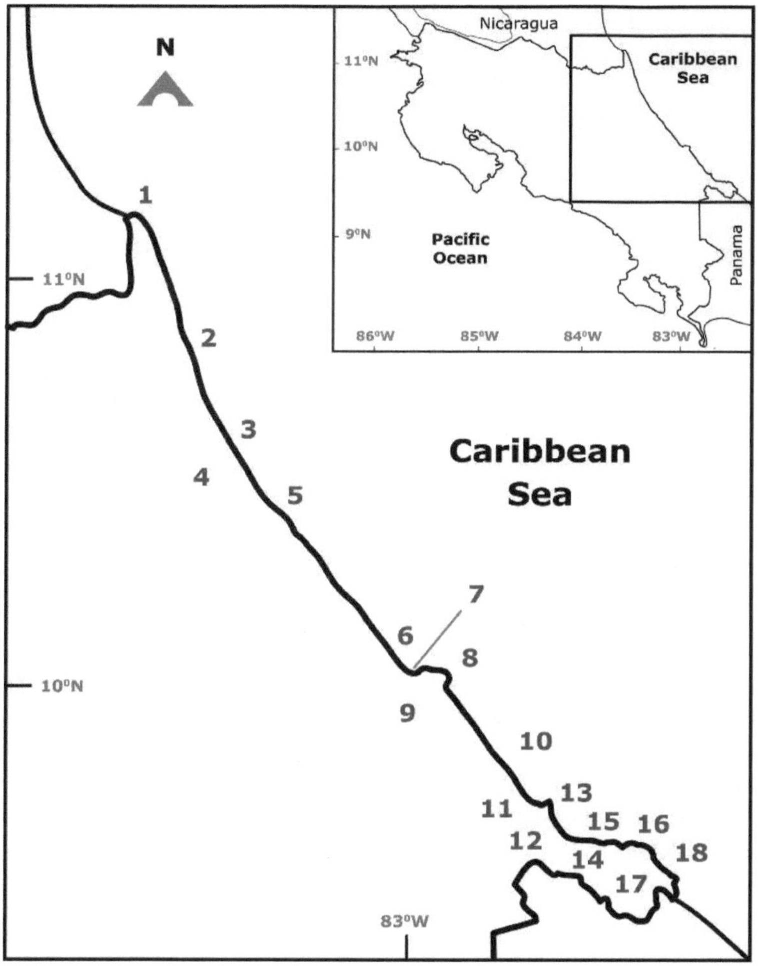

Fig. I.3 Caribbean coast of Costa Rica, with indication of localities mentioned in the text. (1) Río San Juan, (2) Barra del Colorado, (3) Parque Nacional Tortuguero, (4) Canales de Tortuguero, (5) Jalova, (6) Moín, (7) Portete, (8) Isla Uvita, (9) Limón, (10) Cahuita, (11) Parque Nacional Cahuita, (12) Puerto Viejo, (13) Punta Uva, (14) Manzanillo, (15) Refugio Nacional de Vida Silvestre Gandoca-Manzanillo, (16) Punta Mona, (17) Laguna Gandoca, and (18) Río Sixaola

Fig. I.4 High energy sandy beach at Jalova, Caribbean coast. (Photo: Jorge Cortés)

(Cortés & Jiménez 2003a). There are a few islands close to shore consisting of fossil coral reefs and surrounded in most cases by living reefs. Further away from the coast and the islands, the substrate is mainly carbonate hardgrounds and sand. Offshore the bottom consists of soft terrigenous sediments (Cortés 1998; Cortés *et al.* 1998, personal observation).

The climate of the Caribbean coast of Costa Rica is humid and hot, with year-round rains. Precipitation decreases along the coast from the northwest to the southeast (Herrera 1986). Tides are diurnal, with a range of less than 50 cm, and are greatly affected by wind direction and force. The main current runs from the northwest to the southeast, with strong eddies in the opposite direction along the coast. The main NW-SE current and the SE-NW countercurrents transport sediments from the rivers to the reef areas (Cortés & Risk 1985; Cortés *et al.* 1998). Waves are normally from the northeast, although they may come from other directions during hurricanes in the Caribbean, and can be strong, breaking corals and lifting sediments from the reef (Cortés 1981).

Beaches

Beaches are made up by terrigenous sediments and marine carbonates. The black beaches of Cahuita and Puerto Viejo consist of magnetite. Four species of marine turtles use these beaches for nesting: green (*Chelonia mydas*), leatherback (*Dermochelys coriacea*), hawksbill (*Eretmochelys imbricata*), and on rare occasions loggerhead (*Caretta caretta*) turtles (Troëng 2005). Outstanding among the nesting beaches are Parque Nacional Tortuguero as one of the main Caribbean-wide *C. mydas* nesting sites (Troëng & Rankin 2005), and Tortuguero and Gandoca for *D. coriacea* (Chacón 1999; Troëng *et al.* 2004).

Only one paper has been published that studied the organisms of sandy beachs of the Caribbean of Costa Rica (Dexter 1974). She examined the macroscopic infaunas of five beaches and found that the most abundant species were isopods and polychaetes. Other groups found were: nematods, nemerteans, oligochaetes, cumaceans, amphipods, decapods, bivalves, and gastropods.

Rocky Intertidal

Rocky intertidal zones are among the least studied habitats on the Caribbean coast of Costa Rica (Fig. I.5). The rocky platforms tend to be narrow because the tidal range on this coast is only 50 cm. One paper published on rocky intertidal assemblages is about the endolithic fauna of rocky substrates in Cahuita (Pepe 1985). He found that the sipunculid *Phascolosoma antillarum* and the polychaete *Eunice aphroditois* were the most abundant species in stable intertidal rocky platforms. He reported the presence of five other species of sipunculids and three more of polychaetes. Other papers reported on mollusks collected from the rocky intertidal (Houbrick 1968; Robinson & Montoya 1987; Espinosa & Ortea 2001).

During the 1991 Limon Earthquake (7.5 Richter scale) the entire intertidal zone was uplifted, in some areas up to 1.9 m (Fig. I.6). This resulted in the death of intertidal and shallow subtidal organisms (Cortés *et al.* 1992, 1994a; Denyer *et al.*

Fig. I.5 Rocky intertidal zone of Isla Uvita, Caribbean coast of Costa Rica. (Photo: Jorge Cortés)

Fig. I.6 Uplifted coastal areas after the 1991 Limón earthquake. Punta Piuta, 25 May 1991. (Photo: Jorge Cortés)

1994a, b; Denyer 1998). A significant impact was experienced by the intertidal and shallow subtidal burrowing sea urchin, *Echinometra lucunter*. Skeletons of hundreds of this urchin were collected and measured to determine population structure (unpublished data, Soto *et al.* 1992). Evidence of previous uplift, similar to the 1991 event, is recorded in notches in the rocks of the Limon area (Denyer *et al.* 1994a; Denyer 1998, personal observation).

Mangroves

Only two mangrove forests exist on the Caribbean coast. One highly impacted mangrove area is that near to the main port of the coast, Moín (Cortés 1991a) (Fig. I.3), and the other is Laguna Gandoca, located within the Refugio Nacional de Vida Silvestre Gandoca-Manzanillo (National Wildlife Refuge) (Cortés 1991a; Coll *et al.* 2001) (Fig. I.3). Isolated mangrove trees have been found at Cahuita National Park and in Puerto Viejo (Sánchez-Vindas 2001, personal observation). One paper published on the mangroves at Moín (Pool *et al.* 1977) found that the riverine mangroves there were more structurally developed than either the riverine or fringe mangroves of the drier Pacific coast sites. The trees at Moín consisted mainly of *Pterocarpus officinalis* Jacq. (not a mangrove forest nuclear species) (Pool *et al.* 1977).

The mangrove forest at Laguna Gandoca is the largest on the Caribbean coast of Costa Rica (12.5 ha), and its area has increased threefold since 1976 (Coll *et al.* 2001). A tropical rain forest surrounds the mangroves. The dominant species were

Fig. I.7 Mangrove forest at Laguna Gandoca, the predominant species is *Rhizophora mangle*. (Photo: Ingo S. Wehrtmann)

Rhizophora mangle (red mangrove) (Fig. I.7), followed by *Avicennia germinans* (black mangrove), *Laguncularia racemosa* (white mangrove), and *Conocarpus erectus* (buttonwood) (Coll *et al.* 2001; Fonseca *et al.* 2007a). The levels of pollution at Laguna Gandoca were found to be very low (Coll *et al.* 2004).

Seagrasses

Seagrasses are shallow, productive environments located mainly in the lagoons of the reefs (Paynter *et al.* 2001; Krupp 2006; Fonseca *et al.* 2007b). Many species of algae and animals are associated with the seagrasses, either living above them, on the leaves, on the substrate between the plants, or buried in the sediments. They serve as food, shelter, or both to a large number of species. The most extensive seagrass beds of Costa Rica are found on the Caribbean coast, and are made up mainly by two species: turtle grass, *Thalassia testudinum* (Fig. I.8), and manatee grass, *Syringodium filiformis*. Other species of seagrass present are: *Halophila decipiens* and *Halodule wrightii*.

Coral Reefs

The Caribbean coast of Costa Rica can be divided into three separate sections (Cortés & Guzmán 1985a; Cortés & Jiménez 2003a): (1) fringing and patch reefs between

Fig. I.8 Seagrass bed at Parque Nacional Cahuita, exposed during extreme low tide. The predominant species is *Thalassia testudinum*, with a few *Syringodium filiformis* plants in between. (Photo: Ingo S. Wehrtmann)

Moín and Limón; (2) fringing reef (the largest of the Caribbean coast, Fig. I.9), patch reefs, and carbonate banks at Cahuita National Park; and (3) fringing reefs, patch reefs, carbonate banks, and algal ridges between Puerto Viejo and Punta Mona (Fig. I.2). Twenty-six species of octocorals (Guzmán & Cortés 1985; Guzmán & Jiménez 1989), and 41 species of scleractinian corals are known from the Caribbean coast of Costa Rica (Cortés & Guzmán 1985b; Cortés 1992c; Cortés & Jiménez 2003a).

Coral reefs of the Caribbean have been damaged by natural impacts such as the warming events of 1983 and 1995 (Cortés *et al.* 1984; Cortés 1991a; Jiménez 2001a) and by the 1991 Limon earthquake (Cortés *et al.* 1992, 1994a; Denyer *et al.* 1994a, b). However, the main impacts to the reefs are of human origin, mainly high terrigenous sedimentation resulting from the deforestation of the watersheds and coastal alteration (Cortés & Risk 1985; Cortés 1994; Fonseca & Cortés 2002; Cortés & Jiménez 2003a).

Offshore

The offshore areas of the Caribbean coast of Costa Rica are practically unknown. As far as we know, there are only two reports on organisms from offshore. The first is the report of cruise P-7101 of the RV *John Elliott Pillsbury* (Voss 1971). The research vessel operated between 20 January and 5 February 1971 along the

Fig. I.9 Aerial view of the coral reef at Parque Nacional Cahuita, Caribbean coast of Costa Rica. (Photo: Federico Chavarría Kopper)

Caribbean coast off Central America, but only 2 days (26–27 January 1971) in Costa Rican territory, where they visited 11 stations (Sta. P-1315–1325) along the entire coast. Depths varied greatly: six trawling stations covered a depth range between 22 and 73 m, two stations were in 245 and 290 m depth; another three stations (P-1318–1320) were carried out in depths between 715 and 770 m; however, only plankton tows were performed, and the report mentioned principally fish larvae. The cruise report (Voss 1971) indicates only preliminary identifications of some of the collected material, and to our knowledge, results of this project with information on the flora and fauna collected in Costa Rica Rican waters have not been published yet. The other report was by Michel & Foyo (1976), as part of a research on plankton from the Caribbean Sea, they reported their finding from a single tow off Costa Rica in 1968.

Pacific Coast

The Pacific coast of Costa Rica is 1,254 km long and irregular (Figs. I.1 and I.10). It has a high diversity of habitats: rocky shores of a wide variety of rock types, sandy beaches of several compositions and grain sizes, mangrove forests, estuaries, a tropical fjord, islands of various sizes, and several gulfs and bays (Jiménez & Soto 1985; Denyer & Cárdenas 2000; Cortés & Jiménez 2003b). The northern section of the coast is characterized by a dry tropical forest, with a dry season that extends from December to April, and a rainy season from May to November (Cortés 1996–1997; Jiménez 2001b). The southern end of the coast is covered with tropical rain forest, and it rains year-round, with a low between December and April. The central section of the coast is a transitional area from dry to humid climate

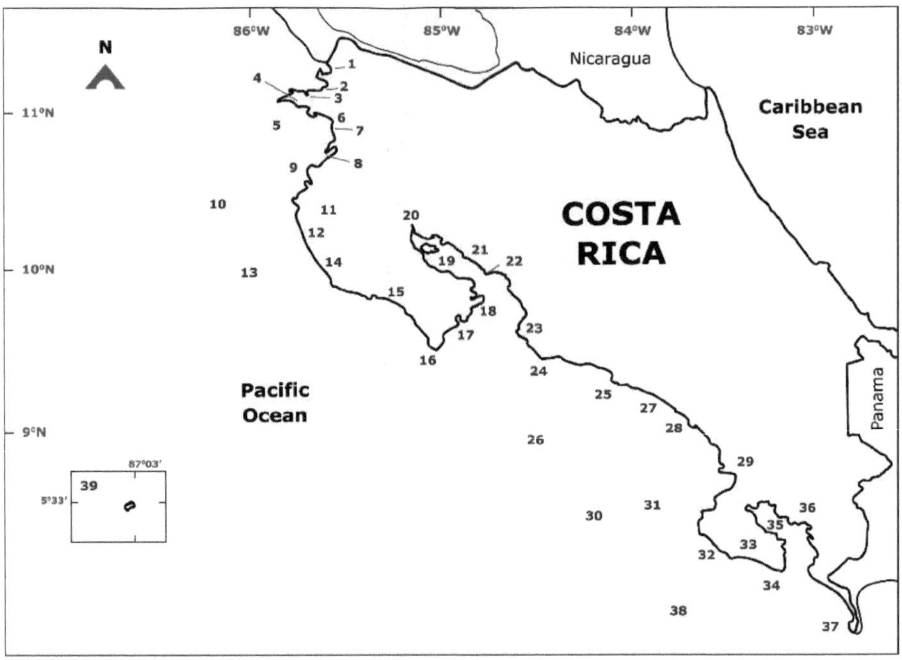

Fig. I.10 Pacific coast of Costa Rica, with indication of the locations mentioned in the text. (1) Bahía Salinas, (2) Cuajiniquil, (3) Bahía Santa Elena, (4) Península de Santa Elena, (5) Islas Murciélago, (6) Potrero Grande, (7) Playa Nancite, (8) Bahía Culebra, (9) Islas Catalinas, (10) Mound Culebra, (11) Parque Nacional Marino Las Baulas, (12) Tamarindo, (13) Mound 10, (14) Refugio Nacional de Vida Silvestre Ostional, (15) Sámara, (16) Reserva Absoluta Cabo Blanco, (17) Bahía Ballena, (18) Islas Tortuga, (19) Golfo de Nicoya, (20) Río Tempisque, (21) Punta Morales, (22) Puntarenas, (23) Herradura, (24) Punta Mala, (25) Parque Nacional Manuel Antonio, (26) Quepos Landslide (8°51′N 84°13′W), (27) Dominical, (28) Parque Nacional Marino Ballena, (29) Humedal Nacional Térraba-Sierpe, (30) Mound, 11 and 12, (31) Reserva Biológica Isla del Caño, (32) Parque Nacional Corcovado, (33) Península de Osa, (34) Cabo Matapalo, (35) Golfo Dulce, (36) Golfito, (37) Punta Burica, (38) Cocos Ridge, and (39) Isla del Coco

(Herrera 1986). Tides are semidiurnal, with a range of about 3 m. The Costa Rican Current runs from the southeast to the northwest, paralleling the coastline. Closer to shore, eddies move in the opposite direction. The northern section of the coast experiences upwelling of cool, nutrient-rich waters during the dry season, when the NE Trade Winds cross the lowlands from the Caribbean to the Pacific (Legeckis 1988; McCreary *et al.* 1989). On the other hand, the high mountains of the central and especially the southern parts of the country block the Trade Winds, preventing coastal upwelling in the area. The Pacific of Costa Rica also include open ocean where 19 species of cetaceans have been observed (May-Collado *et al.* 2005).

Beaches

There are many beaches along the Pacific coast of Costa Rica and these are made up of many types of sediments, from carbonates to basalts, and with grain sizes from fine, almost muddy, to large cobbles (Denyer & Cárdenas 2000). Only one paper has been published on the fauna of sandy beaches (Dexter 1974). She examined the macroscopic infaunas of eight beaches on the Pacific coast during the spring of 1971. Pacific beaches averaged seven times the density of individuals, had significantly larger numbers of species, and had significantly finer sand than the Atlantic beaches. The most abundant species, the isopods *Cirolana salvadorensis* and *Exosphaeroma diminutum*, occurred on both coasts. The polychaetes *Scolepis agilis* and *Hemipodus armatus* were found on the majority of beaches on both coasts. Other groups found were platyhelminthes, nematods, nemerteans, oligochaetes, cumaceans, amphipods, mysidaceans, decapods, bivalves, gastropods, echinoids, ophiuroids, and cephalochordates. In general, the faunas of the Pacific and Atlantic sandy beaches were composed of closely related species (Dexter 1974).

Rocky Intertidal

This type of habitat can be found along the entire Pacific coast of Costa Rica (Fig. I.11), with the exception of coastal areas dominated by mangroves. Despite the relative abundance of rocky intertidal habitats, our knowledge concerning the ecology of its flora and fauna is patchy and fairly limited. Growth patterns, spatial distribution and behavior of the intertidal pulmonate gastropod *Siphonaria gigas* have been studied in some detail (Lahmann & González 1982; Crisp *et al.* 1990), and Ortega (1987a) documented the effect of human extraction activities on the size distribution of this intertidal gastropod. Other investigations concern competitive interactions among intertidal limpets (Ortega 1985), habitat segregation and seasonal changes in density of the most common intertidal organisms (Ortega 1987b), census and diversity of tropical prosobranch gastropods assemblages (Playas del Coco), and comparisons to temperate assemblages (Washington State) (Spight 1976, 1977), temporal changes in a rocky shore snail community in Guanacaste (Spight 1978), intensity of fish predation by comparing loss of gastropod shells in four

Fig. I.11 Collecting material in the rocky intertidal zone of Parque Nacional Marino Ballena, southern Pacific coast of Costa Rica. (Photo: Ingo S. Wehrtmann)

intertidal shores on the Pacific coast (Ortega 1986), and Sutherland & Ortega (1986) measured competitive interactions between barnacles and limpets, and the effect of predation. Diversity of rocky intertidal organisms was studied in the Golfo de Nicoya (Paine 1966), while Pepe (1985) described the endolithic fauna of rocky substrates in Playas del Coco. Willis & Cortés (2001) made an inventory of the mollusks in Parque Nacional Manuel Antonio and found a total of 74 species, the rocky intertidal being the most diverse habitat. Different aspects of the ecology of the intertidal barnacles *Tetraclita stalactifera* and *T. panamensi* were studied by Villalobos (1980a, b) and Sutherland (1987), respectively. More recently, Jörger *et al.* (2008) analyzed species composition and vertical distribution of chitons in tide pools and on exposed rock areas of the intertidal zone of Sámara, Guanacaste; they recorded a total of nine species, with *Ischnochiton dispar* as the predominant species. The uplift rate of the Pacific coast, measured on rocky outcrops is between 0.4 and 2.1 mm/year (Denyer & Cortés 2001).

Mud Flats

There are only a few mud flats which have been studied in Pacific Costa Rica (Fig. I.12). In fact, all research has focused on the Golfo de Nicoya area. Vargas (1987) described the benthic community of an intertidal mud flat, and also published on the results of a survey of the meiofauna of this system (Vargas 1988a). The same author demonstrated the effects of macropredator exclusion experiments

Fig. I.12 Mudflat at Isla Chira in the Golfo de Nicoya. (Photo: Ulrich Saint-Paul)

on the macrobenthos community structure of an intertidal mud flat in the inner Golfo de Nicoya (Vargas 1988b). A summary of studies on the ecological dynamics of tropical intertidal mudflat community from Pacific Costa Rica was published by Vargas (1996).

Mangroves

Mangrove forests are present all along the Pacific coast of Costa Rica, from dwarf forest (Fig. I.13) on the north to extensive forest with large tress on the south (Jiménez & Soto 1985). The mangrove forests cover only 0.8% of the surface of Costa Rica, but their biological and economic importance is significant. The Pacific coast has 99% of the mangrove forest of Costa Rica (Pizarro & Angulo 1993; Polanía 1993). The main species are *Rhizophora mangle* and *Rhizophora harrisoni* along the canal edges, backed by *Avicennia germinans*, and farther inland *Avicennia bicolor, Conocarpus erectus*, and *Laguncularia racemosa* (Chapter IV, Part 4).

Seagrasses

Four species of seagrasses have been reported from the eastern Pacific (Davidse *et al.* 1994), but of those only two have been found in Costa Rica (Cortés 2001): *Ruppia maritima* Linnaeus, 1753, Family Potamogetonaceae and *Halophila baillonii* Ascherson, 1874, Family Hydrocharitaceae. In 1994 a 5,000 m^2 patch formed by *R. maritima* was found at Bahía Culebra, north Pacific coast of Costa Rica

Fig. I.13 Jorge Cortés inspecting a drawf mangrove tree at Puerto Soley, Bahía Salinas, northern Costa Rica. (Photo: Andrea Bernecker)

Fig. I.14 *Ruppia maritima* at Bahía Culebra, Guanacaste. (Photo: Jorge Cortés)

(Fig. I.14). Ecological parameters were recorded as well as 44 invertebrate species associated with the seagrass bed. The studied patch as well as other smaller patches found in Bahía Culebra and their associated organisms disappeared after a severe storm in June 1996 (Cortés 2001).

Coral Reefs

Coral communities, reefs, and solitary coral colonies can be found along the Pacific coast of Costa Rica (Cortés & Jiménez 2003b). Most coral reefs are relatively small, isolated, and made up by a few species of reef-building corals, which is characteristic of eastern tropical Pacific reefs (Cortés 1997). The main reef builder is *Porites lobata* (Fig. I.15), followed by several species of the genus *Pocillopora (elegans, damicornis*, and *eydouxi*) (Fig. I.16) and by two species of *Pavona (clavus* and *gigantea*) (Fig. I.17).

The extreme northern section of the coast, around Bahía Salinas (Fig. I.10), which experiences a seasonal upwelling of cold, nutrient-rich waters, has limited coral development, and *P. gigantea* is the main reef builder (Cortés 1996–1997, Cortés, in preparation). South of this area, Santa Elena, Murciélago and Bahía Culebra (Fig. I.10), the corals are marginally influenced by the upwelling, but apparently it is not strong which permits the development of other species of corals and the formation of larger reefs (Cortés & Murillo 1985; Jiménez 1997, 1998). Additionally, growth rates of corals in Bahía Culebra are the highest in the eastern tropical Pacific for most reef-building species (Jiménez & Cortés 2003b). The coral reefs and communities in this area were severely impacted between 150 to 400 years ago, an effect attributed by Glynn and coworkers (1983) to the intensification of upwelling during the Little Ice Age. The main anthropogenic impacts have been construction of roads and tourist complexes, extraction of corals for the local curio trade, and inappropriate tourist and fishing practices (Cortés & Murillo 1985; Jiménez 1997, 1998). In recent years the increase in populations of the green algae, *Caulerpa sertulariodes*, in Bahía Culebra has been smothering the corals (Fernández & Cortés 2005; Bezy *et al.* 2006).

Fig. I.15 *Porites lobata*, Pacific Costa Rica. (Photo: Jorge Cortés)

Fig. I.16 *Pocillopora* spp., Islas Murciélago, north Pacific Costa Rica. (Photo: Carlos E. Jiménez)

Along the central Pacific coast of Costa Rica the best development coral reefs and communities are in Parque Nacional Marino Ballena (Alvarado 2004, 2006; Alvarado *et al.* 2005). However, recent studies showed that they have been impacted by sediments (Alvarado *et al.* 2005), and by the El Niño-Southern Oscillation (Jiménez & Cortés 2001, 2003a).

Some of the largest coral reefs on the Pacific coast of Costa Rica are in the south (Fig. I.10): Isla del Caño (Guzmán 1986; Guzmán & Cortés 1989, 2001; Fonseca *et al.* 2006), Peninsula de Osa (Cortés & Jiménez 1996), and Golfo Dulce (Cortés 1990a, b, 1992b; Cortés *et al.* 1994b). The main reef builder in this region is *P. lobata*, followed by several species of *Pocillopora* and *Pavona*. These reefs have been impacted by salinity changes due to changes in river flow regimes (Cortés 1990b; Cortés *et al.* 1994b), El Niño-Southern Oscillation warming events (Cortés *et al.* 1984; Guzmán *et al.* 1987; Glynn *et al.* 1988; Guzmán & Cortés 2001), phytoplankton blooms (Guzmán *et al.* 1990), low tide exposures (Guzmán 1986, personal observation), storms (Guzmán 1986), and excessive terrigenous sedimentation (Cortés 1990a, b, 1991b, 1992b; Cortés *et al.* 1994b).

Isla del Coco, located offshore (see below), has well developed coral reefs built mainly by massive corals, *P. lobata* and *P. clavus*, with some pocilloporids and other pavonids (Bakus 1975; Cortés & Murillo 1985; Cortés & Guzmán 1998).

Fig. I.17 *Pavona clavus*, Islas Murciélago, north Pacific Costa Rica. (Photo: Jorge Cortés)

The reefs were greatly impacted by El Niño-Southern Oscillation warming events (Guzmán & Cortés 1992) but are now recovering (Guzmán & Cortés 2007).

Golfo de Nicoya

Introduction

The Golfo de Nicoya is one of the most productive estuaries in the world (Cordoba-Muñoz 1998; Gocke *et al.* 2001a), and constitutes the most important fishing ground in the country. The gulf is located in the north-central part of the

Pacific coast of Costa Rica (Figs. I.10 and I.18), and Puntarenas is the main city and also the most important fishery port in the country. Isla Chira, located in the northern part of the inner gulf, is the largest island of Costa Rica. One characteristic feature is the influence of the Río Tempisque with its estuary in the northern part of the gulf. This river as well as other riverine sources, including extensive mangrove areas, provides a considerable amount of organic material to the estuarine system (Gocke *et al.* 2001a; Jiménez 2004).

The Golfo de Nicoya is a tidally influenced shallow water estuary with an average depth of less than 20 m in its inner part (Lizano 1998; Vargas & Mata 2004). The outer section of the Golfo de Nicoya deepens gradually to 200 m and is influenced by equatorial subsurface waters (Voorhis *et al.* 1983; Chaves & Birkicht 1996). Water characteristics and faunal assemblages are strongly influenced by a dry (December–April) and a rainy season (May–November) (Vargas 1995). Water temperatures generally vary between 28 °C and 30 °C, and salinities range from 22 and 32 PSU, although some near-shore areas show seasonally considerably lower values due to freshwater input (Lizano 1998; Brenes *et al.* 2001). More information regarding oceanographic features of the gulf has been published by Peterson (1958), Epifanio *et al.* (1983), Voorhis *et al.* (1983), Brenes & León (1996), Chaves & Birkicht (1996), Lizano (1998), and Brenes *et al.* (2001).

Fig. I.18 Golfo de Nicoya, northern Pacific coast of Costa Rica (shuttle Russian infrared photo) (Kohlmann *et al.* 2002).

According to Vargas & Mata (2004), the Golfo de Nicoya is the best known tropical estuary in the world; more than 100 scientific papers have been published, mainly focusing on the ecology of invertebrates (e.g., benthic soft bottom communities; see Vargas 1995 and references therein) and commercially exploited fish species (e.g., Bartels *et al.* 1983; Araya 1984; Campos 1992). Overall, approximately 400 species of benthic invertebrates and 200 species of fish have been reported for the Golfo de Nicoya (Vargas 1995). The field station "Estación Nacional de Ciencias Marino-Costeras" (ECMAR), located in Punta Morales and currently managed by the Universidad Nacional, facilitated the development of a series of research projects, which greatly improved our knowledge concerning the Golfo de Nicoya and its adjacent areas.

Mangroves

The Golfo de Nicoya comprises roughly 38% of the 41,289 ha of mangrove areas along the Pacific coast of Costa Rica (Polanía 1993). Main concentrations of mangroves are located in the innermost part of the gulf around the mouth of the river Tempisque, and around the eastern section between Puntarenas and Punta Morales (Fig. I.19). The principal species are *Rhizphora mangle, R. racemosa, Avicennia germinans, A. bicolor*, and *Laguncularia racemosa* (Soto & Jiménez 1982). The field station ECMAR served as a platform for numerous studies related to different organisms associated with the surrounding mangroves and adjacent estuarine habitats of the Golfo de Nicoya: marine fungi (Ulken *et al.* 1990), soft bottom fauna (Vargas 1988a, b), polychaetes (Dean 1996a, 1988), decapods (Wehrtmann & Dittel 1990; Dittel 1991; Dittel *et al.* 1991), and fishes (Szelistowski 1990). Other investigations carried out in Punta Morales concerned primary production (Córdoba-Muñoz 1998; Gocke *et al.* 2001a, b) and plankton (Brugnoli-Olivera & Morales-Ramírez 2001; Morales-Ramírez & Brugnoli-Oliveira 2001).

Phytoplankton

The phytoplankton community in the Golfo de Nicoya has been studied: Víquez & Hargraves (1995) and Brugnoli-Oliveira & Morales-Ramírez (2001) provided information on the taxonomic composition (Chapter IV, Part 1), whereas Hargraves & Víquez (1985) focused on the spatial and temporal distribution of the phytoplankton. Literature is available concerning harmful algal blooms in the Golfo de Nicoya and its effects on human health (Mata *et al.* 1990). Víquez & Hargraves (1995) described the annual cycle of potentially toxic dinoflagellates, and Vargas-Montero & Freer (2004a, b, c) reported on the presence of toxic dinoflagellates and cyanobacteria.

Zooplankton

There are relatively few studies focusing on zooplankton in the Golfo de Nicoya: Morales & Vargas (1995) listed the common pelagic copepods of the Golfo de Nicoya.

Fig. I.19 The city of Puntarenas and adjacent mangrove areas, Golfo de Nicoya, central Pacific coast of Costa Rica (aerial color photo) (Kohlmann *et al.* 2002).

Morales-Ramírez (1996) presented a checklist of the copepods encountered in the gulf, and provided information on the distribution of these species. Our knowledge of arrow worms (Chaetognatha) is limited to the study carried out by Hossfeld (1996), where she described the species composition, distribution, and biomass of these planktonic animals in the Golfo de Nicoya. The paper published by von Wangelin & Wolff (1996) concerned biomass and species composition of the zooplankton communities in the gulf. Studies on the composition and abundance of ichthyoplankton in the Golfo de Nicoya were published by Ramírez *et al.* (1990) and Molina-Ureña (1996). Several studies concerned the ecology and development of the early life stages of some decapods (Dittel & Epifanio 1984a, b; Epifanio & Dittel 1984; Wehrtmann & Dittel 1990).

Benthos

One of the first and most complete benthic invertebrate surveys was published by Maurer *et al.* (1984). Subsequently, several authors provided additional information on the taxonomic diversity, distribution, and ecology of benthic invertebrates of the gulf, including polychaete worms (Dean 1996b), mollusks (Cruz & Palacios 1983; Cruz 1996; Koch & Wolff 1996), decapods (Wehrtmann 1990; Vargas *et al.* 1996; Díaz-Ferguson & Vargas-Zamora 2001), and zonation of epifauna and algae in the rocky intertidal (Sibaja-Cordero & Vargas-Zamora 2006). Bartels *et al.* (1983), Bussing & López (1996), and Wolff (1996) provided information on the taxonomic composition and demersal fish assemblages in the Golfo de Nicoya.

Fisheries

Most fishing activities in Costa Rica are carried out along the Pacific coast, and the Golfo de Nicoya is considered the most important fishing ground in the country (Vargas 1995), accounting for roughly 30% of annual landings. Commercial bottom trawling is prohibited in the inner part of the gulf, and thus small-scale fishery is the prevailing type of fishing activity in the Golfo de Nicoya. More than 2,000 fishermen operate in the gulf area; however, the real number of fishermen might be considerably higher than officially reported (Fischer & Wolff 2006).

Fishery studies concerning commercially exploited species within the Golfo de Nicoya are sparse and focus principally on decapods. Tabash & Palacios (1996) provided information on the stock assessment of two penaeid shrimps (*Litopenaeus occidentalis* and *L. stylirostris*). Subsequently, Tabash-Blanco & Chávez (2006) analyzed the fishing efforts of the white shrimp fishery (*L. occidentalis* and *L. stylirostris*) in the Golfo de Nicoya and concluded that a further increase of fishing efforts should not be encouraged to maintain the relatively high level of the spawning stock of the two shrimp species for successive years. Moreover, Tabash-Blanco (2007) analyzed the official shrimp fishery statistics covering the period from 1991 to 1999. The results indicated that all the six shrimp species studied by him reached the Maximum Sustainable Yield in the decades of 1970 and 1980, and are now found at overexploitation levels. The author recommended closing down these shrimp trawl fisheries.

During recent years, a small-scale blue crab (*Callinectes arcuatus*) trap-fishery has been established in the gulf area. This fishery is run by roughly 10–15 fishermen, and managed by the "Instituto Costarricense de Pesca y Acuicultura" (INCOPESCA) which allowed the use of a maximum of 1,600 traps distributed over 40 licenses (Fischer & Wolff 2006). These authors suggested that *C. arcuatus* is still underexploited in the Golfo de Nicoya. However, since male size of *C. arcuatus* already has decreased and the local market has shown signs of oversaturation, Fischer & Wolff (2006) advised allowing only a slight increase in the fishing effort (from currently 300 to 360 traps). According to Fischer & Wolff (2006), a maximum effort of 1,600 traps, as recommended by INCOPESCA, is unlikely to be sustained by the population.

Our knowledge about fish stocks in the Golfo de Nicoya is even more limited. Araya (1984) studied fin fish of the Sciaenidae and described the long-term overexploitation of the main fishery resources. Campos (1992) estimated the length at first maturity of *Cynoscion* spp. ("corvinas") in the Golfo de Nicoya, and Baltz & Campos (1996) identified the spawning sites of Sciaenidae in the gulf. On a more general basis, Campos & Corrales (1986) presented preliminary results on the trophic dynamics of the Golfo de Nicoya, and discussed its impacts on the fisheries and coastal management in the area. More recently, Wolff *et al.* (1998) developed a trophic flow model for the Golfo de Nicoya and included also fishery-related aspects.

The mud cockle *Anadara tuberculosa* (locally called "piangua") is another commercially exploited resource in the Golfo de Nicoya as well as in Térraba-Sierpe and Golfo Dulce. Campos *et al.* (1990) studied the population dynamics of this mud cockle in Térraba-Sierpe, and recently Stern-Pirlot & Wolff (2006) provided information on the fisheries potential of *A. tuberculosa* along the Pacific coast of Costa Rica, including Golfo de Nicoya. These authors demonstrated that estimated exploitation rates exceed sustainable levels at all study sites, and recommended the implementation of conservation measures to avoid local extinction of this species.

Pollution

The first paper on water quality in the Golfo de Nicoya was published by Brunker & Fernández (1965), and at that time they found that in 10% of their samples the coliform levels were above the acceptable levels for swimming waters. It's encouraging that sampling in 2000 and 2002 indicated lower levels of coliform contamination (García *et al.* 2006). The same has been observed in the Estero de Puntarenas, where levels of coliforms have dropped in the last two decades coinciding with the operation of the water treatment plant (Mora *et al.* 1989; Acuña *et al.* 1998; García *et al.* 2006).

Dean *et al.* (1986) determined trace metal concentrations in sediments and invertebrates throughout the gulf including four stations at the mouth of the Río Grande de Tárcoles. At these last stations concentrations of trace metals were highest. Fuller *et al.* (1990) found that chromium was bound to fine grain sediments upriver and even with dilution down river, the levels were several times above background levels at the mouth of Río Tárcoles. Also in this area, Loría *et al.* (2002) found the highest levels of potassium and relate it to anthropogenic pollution.

Pesticide levels in sediments from Golfo de Nicoya were low (Spongberg & Davis 1998), as were PCB concentrations (Spongberg 2004a, b). Higher PCB levels were found in fine grain sediments than in sandy sediments (Spongberg 2004b). Sipunculids from the Golfo de Nicoya had relatively high concentrations of PCBs compared to samples from other parts of the country, but those levels are still low compared to other countries (Spongberg 2006). Oil pollution was the highest around the port city of Puntarenas, while low to undetectable in other parts of Golfo de Nicoya (Acuña-González *et al.* 2004). The highest concentrations of solid waste

Fig. I.20 A typical landing from shrimp trawling off Tárcoles at 30 m depth with the presence of a considerable amount of plastics. (Photo: Ingo S. Wehrtmann)

within Golfo de Nicoya were found in beaches with heavy tourist use (García *et al.* 2006). The presence of solid waste (especially plastic) is a common feature in shallow water trawling close to large rivers as shown in Fig. I.20.

Golfo Dulce

The Golfo Dulce, located on the southern Pacific coast of Costa Rica (Figs. I.10 and I.21), is unique in being a tropical fjord-like embayment (Hebbeln *et al.* 1996; Dalsgaard *et al.* 2003), characterized by a deep, periodically anoxic inner basin (down to 200 m) and a shallow sill (max. 60 m). The Golfo is situated in the most humid region of the country (Cortés 1990a) and is surrounded by lush tropical vegetation, which is attracting more and more national and foreign tourists to visit the area (Kappelle *et al.* 2002).

The physico-chemical environment of the Golfo Dulce is fairly well studied. The first report concerning the oceanography of the Golfo was published by Richards *et al.* (1971), and on its benthic fauna by Nichols-Driscoll (1976). Subsequently, the RV *Victor Hensen* expedition (1993–1994) provided an excellent platform for studying the geology (Hebbeln *et al.* 1996; Hebbeln & Cortés 2001), chemistry (Thamdrup *et al.* 1996; Córdoba & Vargas 1996; Acuña-González *et al.* 2006), microbiology (Kuever *et al.* 1996), and biology of different organisms, especially zooplankton (Morales-Ramírez 1996; von Wangelin & Wolff 1996; Morales-Ramírez & Nowaczyk 2006; Quesada-Alpízar & Morales-Ramírez 2006), benthic

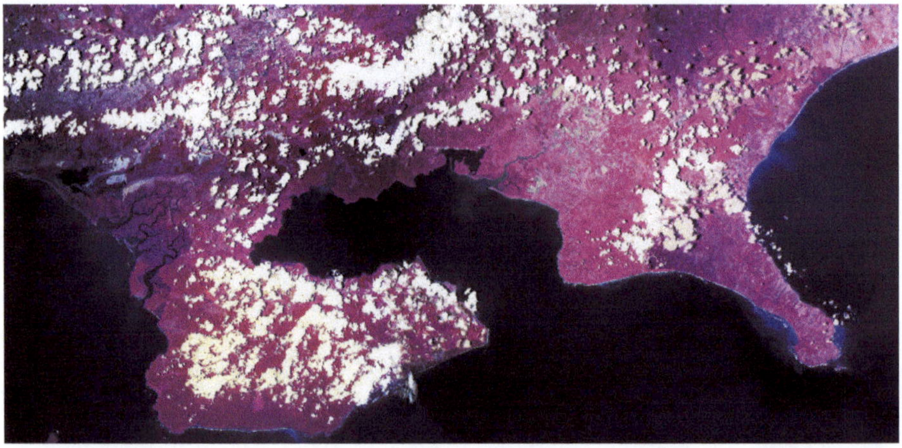

Fig. I.21 Golfo Dulce, southern Pacific coast of Costa Rica (shuttle infrared photo) (Kohlmann *et al.* 2002).

macrofauna (León-Morales & Vargas 1998), polychaete worms (Dean 1996b), copepods (Morales-Ramírez 1996), decapods and stomatopods (Castro & Vargas 1996; Vargas *et al.* 1996; Jesse 1996), mollusks (Cruz 1996; Høisæter 1998), chaetognaths (Hossfeld 1996), fish (Bussing & López 1996), and ichthyoplankton (Molina-Ureña 1996). A summary of the research carried out with the RV *Victor Hensen* in the Golfo Dulce has been published by Vargas & Wolff (1996). Independent of the RV *Victor Hensen* expedition, papers have been published concerning zooplankton (Morales-Ramírez & Nowaczyk 2006; Quesada-Alpízar & Morales-Ramírez 2004, 2006), coral reefs (Cortés 1990a, b; Cortés *et al.* 1994b; Fonseca *et al.* 2006), and marine mammals (Acevedo 1996; Acevedo & Smultea 1995; Acevedo & Burkhart 1998; Cubero-Pardo 1998). Recent studies focused on pollution (Spongberg 2004c, 2006; Acuña-González *et al.* 2004; García *et al.* 2006), physical oceanography (Svendsen *et al.* 2006), and nutrient cycling (Dalsgaard *et al.* 2003; Acuña-González *et al.* 2006). Moreover, the analyses of bottom samples recently taken in the Golfo Dulce revealed the presence of macrobacterial mats (Gallardo & Espinoza 2007) similar to those described from the oxygen-minimum zone off Chile (Gallardo 1977).

During the last few years, the entire region of the Golfo Dulce has received an increasing interest for ecotourism and marine resource exploitation (e.g., fish farming). The ongoing forest clearing and construction activities are affecting the terrestrial as well as the aquatic systems, mainly due to sedimentation processes (Cortés 1990a, b; Umaña 1998; Kappelle *et al.* 2002; Lobo *et al.* 2007). The city of Golfito is located in the inner part of Golfo Dulce, and is the main port of southern Pacific region of Costa Rica. Offshore sport fishing is a lucrative and common tourist activity in the region. Due to an increasing interest in these activities, several plans exist to expand existing marinas or to build new ones, which certainly will affect the water quality and as a possible consequence the entire functioning of the fragile ecosystem of Golfo Dulce. Currently, no marine-protected areas have been declared in Golfo Dulce; however, several proposals are being discussed to introduce some kind of management for the area (Quesada-Alpízar & Cortés 2006; Quesada-Alpízar *et al.* 2006).

Deep-Sea

Our knowledge concerning deep-sea habitats along the Pacific coast of Costa Rica stems from expeditions and research programmes carried out by foreign institutions. The first deep-sea dredging (mainly around Isla del Coco) was done between 1893 and 1900 (Townsend 1901), and information on the diversity of Mollusca, Decapoda, and Echinodermata were published by Dall (1908), Faxon (1893), and Agassiz (1904), respectively. The more recently published information concerns mainly geochemical processes (e.g., McAdoo *et al.* 1996; Meschede *et al.* 1999; von Huene *et al.* 2000; Bohrmann *et al.* 2002; Han *et al.* 2004; Linke *et al.* 2005; Mau *et al.* 2006). Deep-sea environments reach from the continental slope to hadal depths, and its biota, especially that of vent ecosystems (Suess *et al.* 1985), is characterized by highly specialized organisms. Most species constituting these chemosynthesis-based communities rely on chemoautotrophy, and some of them (e.g., vesicomyid bivalves) harbor chemoautotroph bacteria in their gills which are indispensable for their nutrition (Fisher 1990). One representative of these mollusks is *Calyptogena costaricana*, recently described by Krylova & Sahling (2006), and recorded off the Pacific coast of Costa Rica in depths between 2,258 and 2,263 m.

Another remarkable feature of the deep seafloor off Costa Rica is the presence of widespread white, blue/grey, and orange bacterial mats (Fig. I.22), which may completely cover the bottom between 220 and 410 m depth; many smaller-sized bacterial mats can be encountered downslope (Bohrmann *et al.* 2002). Below 550 m, bottom assemblages are composed by numerous vesicomyid clams, to a lesser extent *Acharax* shells, as well as dispersed snails, crabs, and crinoids (Bohrmann *et al.* 2002). Extensive clam fields associated with bacterial mats and Pogonophora can be observed between 1,400 and 1,900 m. Other relatively common organisms can be found on or close to the seafloor at these depths, including sea anemones, crinoids, corals, brittle stars, crabs, and rattail fish. On the other hand, sponges and sea cucumbers appear only in the upper part of this depth zone (Bohrmann *et al.* 2002). Recently, Neuhaus (2004), and Neuhaus & Blasche (2006) reported Kinorhyncha from the Central American East Pacific Deep and from the continental shelf of Costa Rica, respectively.

Mud diapirs or mud volcanoes at the Costa Rican continental margin have been reported at depths between 1,800 and 3,100 m (Shipley *et al.* 1992; Ranero *et al.* 2000; Bohrmann *et al.* 2002). A mud volcano, located at 2,000 m depth off the Pacific coast of Costa Rica, was closely studied by Bohrmann *et al.* (2002). They found that the fauna of the fluid expulsion sites in the center of the mud volcano was composed by bacterial mats, vesicomyid, and solemyid clams, to a lesser extent by *Bathymodio*-like mussels, as well as by huge bunches of pogonophoran

Fig. I.22 Bacterial mats at 400 m water depth at Quepo Slide, a landslide ~ 50 nm west of Osa Peninsula (8°51′N, 84°13′W). Seepage of fluids from the sediment into the oxygen-depleted water of the Equatorial East Pacific results in areas covered by bacterial mats (mostly *Beggiatoa*, but also *Thiotrix*) stretching over several hundereds of meters, while other vent organisms are almost completely absent. (Photo: Gregor Rehder)

tubeworms. The authors concluded that the dense populations as well as the diversity of chemosynthetic assemblages indicate the presence of concentrated and/or episodic fluid and mud release from the center of the volcano.

Schleicher (2006) studied vent-specific fauna associations along the Central American continental slope. Her study focused on two geological structures located off Pacific Costa Rica: (1) Mound Iguana (11°12′N 87°09′W) in 1,200 m depth, and (2) Quepos Landslide (8°51′N 84°13′W), covering a depth range from 270 to 420 m (Fig. I.10). The presence of a vent-specific fauna at Mound Iguana was concentrated in a small area with high carbonate coverage. Typical organisms were mytilid and vesicomyide bivalves, vestimentiferans as well as small bacterial mats. Quepos Landslide was characterized by three different forms of bacterial mats: white, yellow-orange, and gray mats of filamentous bacteria, most probably of the genus *Beggitoa*; however, other genera (e.g., *Thiotrix*) might have been present also. Schleicher (2006) concluded that differences in geochemical conditions at cold seeps result in a different composition of its chemosynthetic fauna communities.

Four other mud volcanoes at depths between 1,000 and 2,400 m, located at the erosive subduction zone off Costa Rica, were studied by Mau *et al.* (2006). Their results revealed that bacterial mats were the predominant community at two mud volcanoes (Mound 11 and Mound 12; Fig. I.10), while vesocomyid clams constituted the main faunal element at Mound Culebra and Mound 10. The typical cold

Fig. I.23 Bottom community (pogonophoran tubeworms and mussels) off Pacific Costa Rica at a carbonated seamount located in roughly 1,600 m water depth. (Photo: Gregor Rehder)

vent fauna communities (mytilid mussels, pogonophoran tubeworms and bacterial mats; Fig. I.23) indicate a strong venting activity which is supported by high bottom water methane concentrations (Linke *et al.* 2005).

Despite of these scattered, but valuable data regarding the deep-sea ecosystem off the Pacific coast of Costa Rica, much work is needed to obtain a more complete picture of the biodiversity and functioning of these biological associations. The collaboration with foreign institutions will be essential to achieve this goal.

Isla del Coco

Isla del Coco (also known as Cocos Island) is located at 5°32′N and 87°04′W in the Pacific Ocean, approximately 530 km south-southwest of Costa Rica and 680 km north-northeast of the Galapagos Islands (Ecuador) (Fig. I.1). The island has a land surface of approximately 24 km^2 and it was declared National Park in 1978 and a UNESCO World Heritage Site in 1997 (Fig. I.24). At present, the marine-protected area extends 22.2 km (12 nautical miles) around the island, covering an area of 1,977 km^2.

The island is covered by thick tropical rainforest and has an average annual rainfall is 5,000–6,000 mm (Garrison 2005). Isla del Coco is surrounded by warm surface waters from the western Pacific that are carried by the North Equatorial Countercurrent (NECC), the position of which varies seasonally in accordance with

Fig. I.24 Isla del Coco. (Photo: Jorge Cortés)

the position of the Intertropical Convergence Zone (ITCZ). Between February and April, when the ITCZ is located more to the south, the NECC is weakened and reaches Isla del Coco to a lesser degree, whereas between August and September, when the ITCZ migrates north, the island is affected by the strong current which may carry coral larvae of marine organisms (Guzmán & Cortés 1993; Cortés 1997; Garrison 2005).

During the 18th and 19th centuries, the island was frequently visited by pirates, who supposedly buried treasures on it. Between 1793 and 1860, the abundant resources of fresh water, fish, and wood as well as the abundance of coconuts made Isla del Coco a common stop for all kinds of ships operating in the area. The island has not only attracted the attention of treasure hunters, but due to its unique flora and fauna numerous scientists and scientific expeditions have visited Isla del Coco (Chapter II), contributing significantly to our present knowledge of its marine and terrestrial biodiversity. The scientific exploration of Isla del Coco started with the US Fish Commission Steamer *Albatross* 1888 expedition led by Alexander Agassiz. Between 1920 and 1940 extensive collections were done at Isla del Coco, and many species were reported for the first time and new species described for the island. In recent years there has been more collecting, and the number of new species continues to increase (see Chapter II). Without any doubt, Isla del Coco is a center of marine diversity of the eastern tropical Pacific (see Chapter V).

Isla del Coco has sandy and cobble beaches, rocky intertidal zones, sandy and rocky bottoms at various depths, coral communities and reefs, and deep communities. The coral reefs were devastated by the 1982–1983 El Niño-Southern Oscillation (90% mortality; Guzmán & Cortés 1992), and to a lesser extend by subsequent El Niño-Southern Oscillation events (Guzmán & Cortés 2007). The reefs have been recovering significantly (Guzmán & Cortés 2007).

More than 1,100 species of marine organisms have been reported from Isla del Coco; of those more than 30 are endemic species representing roughly 3% of the island's flora and fauna. The best studied groups are: fishes (Bussing & López 2005; Garrison 2005), corals (Cairns 1991; Cortés & Guzmán 1998), and mollusks, especially gastropods (Shasky 1988; Rodríquez *et al.*, Part 28, Chapter IV). There are groups that have been observed at Isla del Coco but no papers have been published describing them (e.g., flatworms and cyanobacteria). Of other groups, such as algae and sponges, only one paper has been published, Setchell & Mason (1943), and Wilson (1904) respectively.

Conclusions

Costa Rica has a high diversity of coastal and marine habitats on both coasts. The Caribbean coast, being much shorter and with a lower complexity than the Pacific, has less habitat, but species numbers are high. The Pacific coast, including the open waters and Isla del Coco have many different habitats and diversity is high. Some of these habitats have been studied in detail while others have never been sampled. Many habitats, especially around populated coastal areas, e.g., Golfo de Nicoya and Limón, are being severely impacted by human activity.

Acknowledgments The information presented in this chapter is the product of many years of research, for which we are grateful to theVicerrectoría de Investigación, Universidad de Costa Rica (UCR), Consejo Nacional para Investigaciones Científicas y Tecnológicas (CONICIT), Centro de Investigación en Ciencias del Mar y Limnología (CIMAR), UCR, and the Escuela de Biología, UCR. Silvia Echeverría Sáenz helped all along the preparation of this chapter and the entire book. Jeffry Ortiz collaboration with the photographs is very much appreciated. Xochilt Lezama Cáceres drew the maps. Juan José Alvarado and Cindy Fernández provided useful information regarding the status of marine-protected areas in Costa Rica. And special thanks go to Gregor Rehder, Heiko Sahling, and Wilhelm Weinrebe who kindly cooperated in many ways with the preparation of the deep-sea section. Bert Kohlmann provided the images from space, and we are grateful for his valuable collaboration.

References

Acevedo A (1996) Lista de mamíferos marinos en Golfo Dulce e Isla del Coco. Rev Biol Trop 44:933–934

Acevedo A, Burkhart S (1998) Seasonal distribution of bottlenose (*Tursiops truncatus*) and pan-tropical spotted (*Stenella attenuata*) dolphins (Cetacea: Delphinidae) in Golfo Dulce, Costa Rica. Rev Biol Trop 46(Suppl 6):91–102

Acevedo A, Smultea MA (1995) First records of humpback whales including calves at Golfo Dulce and Isla del Coco, Costa Rica, suggesting geographical overlap of northern and southern hemisphere populations. Mar Mamm Sci 11:554–560

Acuña J, García V, Mondragón J (1998) Comparación de algunos aspectos físico-químicos y calidad sanitaria del Estero de Puntarenas, Costa Rica. Rev Biol Trop 46(Suppl 6):1–10

Acuña-González J, Vargas-Zamora J, Gómez-Ramírez E, García-Céspedes J (2004) Hidrocarburos de petróleo, disueltos y dispersos, en cuatro ambientes costeros de Costa Rica. Rev Biol Trop 52(Suppl 2):43–50

Acuña-González J, Vargas-Zamora J, Córdoba-Muñoz R (2006) A snapshot view of some vertical distributions of water parameters at a deep (200 m) station in the fjord-like Golfo Dulce embayment, Costa Rica. Rev Biol Trop 54(Suppl 1):193–200

Agassiz A (1904) Reports on an exploration off the west coast of Mexico, Central and south America, and off the Galapagos Islands, in charge of Alexander Agassiz, by the U.S. Fish Commission Steamer "Albatross2, during 1891, Lieut. Commander Z.L. Tanner, U.S. Commanding. XXXII. The Panamic deep sea echini. Mem Mus Comp Zool Harvard Coll XXXI:1–246

Aguilar T (2000) Paleontología. *In*: Denyer P, Kussmaul S (eds) Geología de Costa Rica. Ed. Tecnol. Costa Rica, Cartago, pp 87–100

Alvarado JJ (2004) Descripción de las comunidades arrecifales del Parque Nacional Marino Ballena, Pacífico Central-Sur de Costa Rica. Thesis, University of Costa Rica, San Pedro, Costa Rica. p 80

Alvarado JJ (2006) Factores físico-químicos y biológicos que median en el desarrollo de los arrecifes y comunidades coralinas del Parque Nacional marino Ballena, Pacífico sur, Costa Rica. M.Sc. Thesis, University of Costa Rica, San Pedro, Costa Rica. p 85

Alvarado JJ, Cortés J, Fernández C, Nivia J (2005) Coral communities and reefs of Ballena Marine National Park, Pacific coast of Costa Rica. Cienc Mar 31:641–651

Araya HA (1984) Los sciaenidos (corvinas) del Golfo de Nicoya, Costa Rica. Rev Biol Trop 32:179–196

Bakus GJ (1975) Marine zonation and ecology of Cocos Island, off Central America. Atoll Res Bull 179:1–11

Baltz C, Campos J (1996) Hydrophone identification and characterization of *Cynoscion squamipinnis* (Perciformes: Sciaenidae) spawning sites in the Gulf of Nicoya, Costa Rica. Rev Biol Trop 44:743–751

Bartels C, Price K, López M, Bussing WA (1983) Occurrence, distribution, abundance and diversity of fishes in the Gulf of Nicoya, Costa Rica. Rev Biol Trop 31:75–101

Bezy MB, Jiménez C, Cortés J, Segura A, León A, Alvarado JJ, Gillén C, Mejía E (2006) Contrasting *Psammocora*-dominated coral communities in Costa Rica, tropical eastern Pacific. Proceedings of the 10th International Coral Reef Symposium, Okinawa, pp 376–381

Bohrmann G, Heeschen K, Jung C, Weinrebe W, Baranov B, Cailleau B, Heath R, Hühnerbach V, Hort M, Masson D, Trummer I (2002) Widespread fluid expulsion along the seafloor of the Costa Rica convergent margin. Terra Nova 14:69–79

Boschi EE (2000) Species of decapod crustaceans and their distribution in the American marine zoogeographical provinces. Rev Invest Des Pesq 13:7–136

Brenes CL, León S (1996) Influence of coastal waters on some physical and chemical oceanographic characteristics of Gulf of Nicoya, Costa Rica. Tóp Meteor Oceanogr 3:65–72

Brenes C, León S, Chaves J (2001) Variación de las propiedades termohalinas en el Golfo de Nicoya, Costa Rica. Rev Biol Trop 49(Suppl 2):145–152

Brugnoli-Olivera E, Morales-Ramírez Á (2001) La comunidad fitoplánctica de Punta Morales, Golfo de Nicoya, Costa Rica. Rev Biol Trop 49(Suppl 2):11–17

Brunker T, Fernández B (1965) Calidad sanitaria de las aguas de la playa de Puntarenas. Rev Biol Trop 13:259–269

Bussing WA, López M (1996) Fishes collected during the R.V. Victor Hensen Costa Rica Expedition (1993/1994). Rev Biol Trop 44(Suppl 3):183–186

Bussing WA, López MI (2005) Peces de la Isla del Coco y peces arrecifales de la costa Pacífica de América Central meridional/Fishes of Cocos Island and reef fishes of the Pacific coast of lower Central America. Rev Biol Trop 53(Suppl 2):192 p

Cairns SD (1991) A revision of the ahermatypic Scleractinia of the Galápagos and Cocos Islands. Smithson Cont Zool 504:1–33

Campos J (1992) Estimates of length at first sexual maturity in *Cynoscion* spp. (Pisces: Sciaenidae) from the Gulf of Nicoya, Costa Rica. Rev Biol Trop 40:239–241

Campos J, Corrales A (1986) Preliminary results on the trophic dynamics of the Gulf of Nicoya, Costa Rica. An Inst Cienc Mar Limnol, UNAM 13:329–334

Campos JA, Fournier ML, Soto R (1990) Estimación de la población de *Anadara tuberculosa* (Bivalvia: Arcidae) en Sierpe-Térraba, Costa Rica. Rev Biol Trop 38:477–480

Castro M, Vargas R (1996) Annotated list of species of marine crustaceans (Decapoda and Stomatopoda) from Golfo Dulce, Costa Rica. Rev Biol Trop 44(Suppl 3):87–96

Chacón CD (1999) Anidación de la tortuga *Dermochelys coriacea* (Testudines: Dermochelyidae) en Playa Gandoca, Costa Rica (1990 a 1997). Rev Biol Trop 47:225–236

Chaves J, Birkicht M (1996) Equatorial subsurface water and the nutrient seasonality distribution of the Gulf of Nicoya, Costa Rica. Rev Biol Trop 44(Suppl 3):41–48

Coates AG (ed) (1997) Central America: A Natural and Cultural History. Yale Univ. Press, New Haven and London, p 277

Coates AG, Obando JA (1996) The geologic evolution of the Central American Isthmus. *In*: Jackson JBC, Budd AF, Coates A.G. (eds) Evolution and Environment in Tropical America. University of Chicago Press, Chicago, pp. 21–56

Coates AG, Jackson JBC, Collins LS, Cronin TM, Dowsett HJ, Bybell LM, Jung P, Obando JA (1992) Closure of the Isthmus of Panama: The near-shore marine records of Costa Rica and western Panama. Geol Soc Am Bull 104:14–28

Coll M, Fonseca AC, Cortés J (2001) El manglar y otras asociaciones vegetales de la laguna Gandoca, Limón, Costa Rica. Rev Biol Trop 49(Suppl 2):321–329

Coll M, Cortés J, Sauma D (2004) Características físico-químicas y determinación de plaguicidas en el agua de la laguna de Gandoca, Limón, Costa Rica. Rev Biol Trop 52(Suppl 2):33–42

Córdoba R, Vargas JA (1996) Temperature, salinity, oxygen, and nutrient profiles at a 200 m deep station in Golfo Dulce, Pacific coast of Costa Rica. Rev Biol Trop 44(Suppl 3):233–236

Córdoba-Muñoz R (1998) Primary productivity in the water column of Estero Morales, a mangrove system in the Gulf of Nicoya, Costa Rica. Rev Biol Trop 46(Suppl 6):257–262

Cortés J (1981) The coral reef at Cahuita, Costa Rica, a reef under stress. M.Sc. thesis, McMaster University, Hamilton, Ontario, Canada, 176 p

Cortés J (1986) Biogeografía de corales hermatípicos: El istmo centroamericano. Anales Inst Cienc Mar Limnol, UNAM 13:297–304

Cortés J (1990a) Coral reef decline in Golfo Dulce, Costa Rica, Eastern Pacific: anthropogenic and natural disturbances. Ph.D. dissertation, Rosenstiel School of Marine and Atmospheric Science, University of Miami, Miami, FL, 147 p

Cortés J (1990b) The coral reefs of Golfo Dulce, Costa Rica: distribution and community structure. Atoll Res Bull 344:1–37

Cortés J (1991a) Ambientes y organismos marinos del Refugio Nacional de Vida Silvestre Gandoca-Manzanillo, Limón, Costa Rica. Geoistmo 1–2:61–68

Cortés J (1991b) Los arrecifes coralinos de Golfo Dulce, Costa Rica: aspectos geológicos. Rev Geol Am Central 13:15–24

Cortés J (1992a) Los arrecifes coralinos del Refugio Nacional de Vida Silvestre Gandoca-Manzanillo, Limón, Costa Rica. Rev Biol Trop 40:325–333

Cortés J (1992b) Los arrecifes coralinos de Golfo Dulce, Costa Rica: aspectos ecológicos. Rev Biol Trop 40:19–26

Cortés J (1992c) Nuevos registros de corales (Anthozoa: Scleractinia) para el Caribe de Costa Rica: *Rhizosmilia maculata* y *Meandrina meandrites*. Rev Biol Trop 40:243–244

Cortés J (1994) A reef under siltation stress: a decade of degradation. *In*: Ginsburg RN (compiler), Proceedings of the Colloquium of Global Aspects Coral Reefs: Health, Hazards and History, 1993. RSMAS, University of Miami, Miami, Fl, pp 240–246

Cortés J (1996–1997) Comunidades coralinas y arrecifes del Área de Conservación Guanacaste, Costa Rica. Rev Biol Trop 44(3)/45(1):623–625

Cortés J (1997) Biology and geology of coral reefs of the eastern Pacific. Coral Reefs 16(Suppl): S39–S46

Cortés J (1998) Capítulo 1: Selección y evaluación del estado de especies importantes de corales, moluscos, crustáceos decápodos y equinodermos. *In*: Planificación y manejo de bahías y áreas costeras fuertemente contaminadas del Gran Caribe. Estudio de caso: Puerto Limón, Costa Rica. Final Report, UNEP Regional Project GEF/RLA/G41, pp 1–24

Cortés J (2001) Requiem for an eastern Pacific seagrass bed, Bahía Culebra, Costa Rica. Rev Biol Trop 49(Suppl 2):273–278

Cortés J, Guzmán HM (1985a) Arrecifes coralinos de la costa Atlántica de Costa Rica. Brenesia 23:275–292

Cortés J, Guzmán HM (1985b) Organismos de los arrecifes coralinos de Costa Rica. III: descripción y distribución geográfica de corales escleractinios (Cnidaria: Anthozoa: Scleractinia) de la costa Caribe. Brenesia 24:63–124

Cortés J, Guzmán HM (1998) Organismos de los arrecifes coralinos de Costa Rica: descripción, distribución geográfica e historia natural de los corales zooxantelados (Anthozoa: Scleractinia) del Pacífico. Rev Biol Trop 46:55–92

Cortés J, Jiménez CE (1996) Coastal-marine environments of Parque Nacional Corcovado, Puntarenas, Costa Rica. Rev Biol Trop 44(Suppl 3):35–40

Cortés J, Jiménez CE (2003a) Past, present and future of the coral reefs of the Caribbean coast of Costa Rica. *In*: Cortés J (ed) Latin American Coral Reefs, Elsevier Science B.V., Amsterdam, pp. 223–239

Cortés J, Jiménez CE (2003b) Corals and coral reefs of the Pacific of Costa Rica: history, research and status. *In*: Cortés J (ed) Latin American Coral Reefs. Elsevier Science B.V., Amsterdam, pp. 361–385

Cortés J, Murillo MM (1985) Comunidades coralinas y arrecifes del Pacífico de Costa Rica. Rev Biol Trop 33:197–202

Cortés J, Risk MJ (1985) A reef under siltation stress: Cahuita, Costa Rica. Bull Mar Sci 36:339–356

Cortés J, Murillo MM, Guzmán HM, Acuña J (1984) Pérdida de zooxantelas y muerte de corales y otros organismos arrecifales en el Caribe y Pacífico de Costa Rica. Rev Biol Trop 32:227–232

Cortés J, Soto R, Jiménez C, Astorga A (1992) Earthquake associated mortality of intertidal and coral reef organisms. Proc 7th Int Coral Reef Symp, Guam 1:235–240

Cortés J, Soto R, Jiménez C (1994a) Efectos ecológicos del Terremoto de Limón. Revista Geológica de América Central, volumen especial sobre el Terremoto de Limón, pp 187–192

Cortés J, Macintyre IG, Glynn PW (1994b) Holocene growth history of an eastern Pacific fringing reef, Punta Islotes, Costa Rica. Coral Reefs 13:65–73

Cortés J, Fonseca AC, Barrantes M, Denyer P (1998) Type, distribution and origin of sediments of the Gandoca-Manzanillo National Wildlife Refuge, Limón, Costa Rica. Rev Biol Trop 46(Suppl 6):251–256

Crisp DJ, Wieghell JG, Richardson CA (1990) Tidal microgrowth bands in *Siphonaria gigas* (Gastropoda, Pulmonata) from the coast of Costa Rica. Malacology 31:229–236

Cruz RA (1996) Annotated checklist of marine molluscs collected during the R.V. Victor Hensen Costa Rica Expedition 1993/1994. Rev Biol Trop 44(Suppl 3):59–68

Cruz RA, Palacios JA (1983) Biometría del molusco *Anadara tuberculosa* (Pelecypoda: Arcidae) en Punta Morales, Puntarenas, Costa Rica. Rev Biol Trop 31:175–180

Cubero-Pardo P (1998) Patrones de comportamiento diurnos y estacionales de *Tursiops truncatus* y *Stenella attenuata* (Mammalia: Delphinidae) en el Golfo Dulce, Costa Rica. Rev Biol Trop 46(Suppl 6):103–110

Dall WH (1908) Reports on the dredging operations off the west coasts of Central America to the Galapagos, to the west coast of Mexico, and in the Gulf of California, in charge of Alexander Agassiz, carried on by the U.S. Fish Commission steamer "Albatross", during 1891, Lieut. Commander Z.L. Tanner, U.S.N., commanding. XXXVII, and Reports on the scientific results of the expedition to the eastern tropical Pacific, in charge of Alexander Agassiz, by the U.S. Fish Commission steamer "Albatross," from October, 1904, to March, 1905, Lieut. Commander L.M. Garrett, U.S.N., commanding. XIV. The Mollusca and Brachiopoda. Bull Mus Comp Zoöl Harvard Coll XLIII:205–487

Dalsgaard T, Canfield DE, Peterson J, Thamdrup B, Acuña-González J (2003) N_2 production by the anammox reaction in the anoxic water column of Golfo Dulce, Costa Rica. Nature 422:606–608
Dana TF (1975) Development of contemporary eastern Pacific coral reefs. Mar Biol 33:355–374
Davidse G., M. Sousa S. & A.O. Chater (eds) (1994) Flora Mesoamericana: Vol 6. Alismataceae a Cyperaceae. Inst. Biología, UNAM, México, D.F, 543 p
D'Croz L, Robertson DR (1997) Coastal oceanographic conditions affecting coral reefs on both sides of the Isthmus of Panama. Proc 8th Int Coral Reef Symp, Panama 2:2053–2058
Dean HK (1996a) Subtidal benthic polychaetes (Annelida) of the Gulf of Nicoya, Costa Rica. Rev Biol Trop 44(Suppl 3):69–80
Dean HK (1996b) Polychaete worms (Annelida) collected in Golfo Dulce during the Victor Hensen Costa Rica expedition (1993/1994). Rev Biol Trop 44(Suppl 3):81–86
Dean HK (1998) A new species of Hesionidae, *Glyphohesione nicoyensis* (Annelida: Polychaeta), from the Gulf of Nicoya, Costa Rica. Proc Biol Soc Wash 111:257–262
Dean HK, Maurer D, Vargas JA, Tinsman CH (1986) Trace metal concentrations in sediment and invertebrates from the Gulf of Nicoya, Costa Rica. Mar Pollut Bull 17:128–131
Denyer P (1998) Historic-prehistoric earthquakes, seismic hazards, and Tertiary and Qauternary geology of the Gandoca-Manzanillo National Wildlife Refuge, Limón, Costa Rica. Rev Biol Trop 46(Suppl 6):237–250
Denyer P, Cárdenas G (2000) Costas marinas. *In*: Denyer P, Kussmaul S (eds) Geología de Costa Rica. Ed. Tecnol. Costa Rica, Cartago, pp 185–218
Denyer P, Cortés J (2001) Levantamiento tectónico de la costa pacífica central de Costa Rica. Rev Biol Trop 49(Suppl 2):179–183
Denyer P, Arias O, Personius S (1994a) Efectos tectónicos del terremoto de Limón. Rev Geol Am Central, Vol. Esp. Terremoto de Limón:39–52
Denyer P, Personius S, Arias O (1994b) Generalidades sobre los efectos geológicos del terremoto de Limón. Rev. Geol. Am. Central, Vol. Esp. Terremoto de Limón:29–38
Denyer P, Montero W, Alvarado GE (2003) Atlas tectónico de Costa Rica. Edit. University of Costa Rica, San Pedro, Costa Rica, 81 p.
Dexter DM (1974) Sandy-beach fauna of the Pacific and Atlantic coasts of Costa Rica and Colombia. Rev Biol Trop 22:51–66
Díaz-Ferguson E, Vargas-Zamora JA (2001) Abundancia de *Petrolisthes armatus* (Crustacea: Porcellanidae) on a tropical estuarine intertidal rocky beach, Gulf of Nicoya estuary, Costa Rica. Rev Biol Trop 49(Suppl 2):97–102
Dittel AI (1991) Distribution, abundance and sexual composition of stomatopod Crustacea in the Gulf of Nicoya, Costa Rica. J Crust Biol 11:269–276
Dittel AI, Epifanio CE (1984a) Desarrollo larval de *Ozius verreauxii* Saussure (Brachyura: Xanthidae) en el laboratorio. Rev Biol Trop 32:171–172
Dittel AI, Epifanio CE (1984b) Growth and development of the portunid crab *Callinectes arcuatus* Ordway: zoeae, megalopae, and juveniles. Rev Biol Trop 4:491–494
Dittel AI, Epifanio CE, Lizano O (1991) Flux of crab larvae in a mangrove creek in the Gulf of Nicoya, Costa Rica. Estuar Coast Shelf Sci 32:129–140
Epifanio CE, Dittel AI (1984) Seasonal abundance of brachyuran crab larvae in a tropical estuary: Gulf of Nicoya, Costa Rica, Central America. Estuaries 7:501–505
Epifanio CE, Maurer D, Dittel AI (1983) Seasonal changes in nutrients and dissolved oxygen in the Gulf of Nicoya, a tropical estuary on the Pacific coast of Central America. Hydrobiologia 101:231–238
Espinosa J, Ortea J (2001) Moluscos del mar Caribe de Costa Rica: desde Cahuita hasta Gandoca. Avicennia (Suppl 4):1–76
Faxon W (1893) Reports on the dredging operations off the west coasts of Central America to the Galapagos, to the west coast of Mexico, and in the Gulf of California, in charge of Alexander Agassiz, carried on by the U.S. Fish Commission steamer "Albatross", during 1891, Lieut. Commander Z.L. Tanner, U.S.N., commanding. VI. Preliminary descriptions of new species of Crustacea. Bull Mus Comp Zool XXIV:149–220

Fernández C, Cortés J (2005) Reef Site: *Caulerpa sertularioides*, a green alga spreading aggressively over coral reef communities in Culebra Bay, north Pacific of Costa Rica. Coral Reefs 24:10

Fiedler PC (2002) The annual cycle and biological effects of the Costa Rica Dome. Deep-Sea Res I 49:321–338

Fischer S, Wolff M (2006) Fisheries assessment of *Callinectes arcuatus* (Brachyura, Portunidae) in the Gulf of Nicoya, Costa Rica. Fish Res 77:301–311

Fisher CR (1990) Chemoautotrophic and methanotrophic symbioses in marine invertebrates. Rev Aquat Sci 2:399–436

Fonseca E, A.C. J. Cortés (2002) Land use in the La Estrella River basin and soil erosion effects on the Cahuita reef system, Costa Rica, pp. 68–82 *In*: Kjerfve B, Wiebe WJ, Kremer HH, Salomons W, Marshall JI Crossland (Caribbean) (eds) Caribbean Basins: LOICZ Global Change Assessment and Synthesis of River Catchment/ Island-Coastal Sea Interaction and Human Dimension. LOICZ Reports & Studies No. 27. LOICZ, Texel, The Netherlands, pp. 114–118

Fonseca AC, Dean HK, Cortés J (2006) Non-colonial macroborers as indicators of coral ref. stress in the south Pacific of Costa Rica. Rev Biol Trop 54:101–115

Fonseca AC, Cortés J, Zamora P (2007a) Monitoreo del manglar de Gandoca, Costa Rica (Sitio CARICOMP). Rev Biol Trop 55:23–31

Fonseca AC, Nielsen V, Cortés J (2007b) Monitoreo de pastos marinos en Perezoso, sitio CARICOMP en Cahuita, Costa Rica. Rev Biol Trop 55:55–66

Fuller CC, Davis JA, Cain DJ, Lamothe PJ, Fries TL, Fernández G, Vargas JA, Murillo MM (1990) Distribution and transport of sediment-bound metal contaminants in the Río Grande de Tárcoles, Costa Rica (Central America). Water Res 24:805–812

Gallardo VA (1977) Large benthic microbial communities in sulphide biota under Peru-Chile countercurrent. Nature 268:331–332

Gallardo VA, Espinoza C (2007) New communities of large filamentous sulphur bacteria in the eastern South Pacific. Int Microbiol 10:97–102

García V, Acuña-González J, Vargas-Zamora JA, García-Céspedes J (2006) Calidad bacteriológica y desechos sólidos en cinco ambientes costeros de Costa Rica. Rev Biol Trop 54(Suppl 1):35–48

García-Céspedes J, Acuña-González JA, Vargas-Zamora JA (2004) Metales traza en sedimentos de cuatro ambientes costeros de Costa Rica. Rev Biol Trop 52(Suppl 2):51–60

Garrison G (2005) Isla del Coco Fishes, 2nd ed. INBio, Santo Domingo de Heredia, Costa Rica, 429 p

Glynn PW (1997) Eastern Pacific reef coral biogeography and faunal flux: Durham's dilemma revisited. Proc 8th Int Coral Reefs Symp 1:371–378

Glynn PW, Ault J (2000) A biogeographic analysis and review of the far eastern Pacific coral reef region. Coral Reefs 19:1–23

Glynn PW, Druffel EM, Dunbar RB (1983) A dead Central American coral reef tract: possible link with the Little Ice Age. J Mar Res 41:605–637

Glynn PW, Cortés J, Guzmán HM, Richmond RH (1988) El Niño (1982–83) associated coral mortality and relationship to sea surface temperature deviations in the tropical eastern Pacific. Proc 6th Int Coral Reef Symp, Australia 3:237–243

Gocke K, Cortés J, Murillo M (2001a) The annual cycle of primary production in a tropical estuary: the inner regions of the Golfo de Nicoya, Costa Rica. Rev Biol Trop 49(Suppl 2):289–306

Gocke K, Cortés J, Murillo MM (2001b) Planktonic primary production in a tidally influenced mangrove forest on the Pacific coast of Costa Rica. Rev Biol Trop 49(Suppl 2):279–288

Guzmán HM (1986) Estructura de la comunidad arrecifal de la Isla del Caño, Costa Rica, y el efecto de perturbaciones naturales severas. M.Sc. thesis, Universidad de Costa Rica, San Pedro, Costa Rica, 179 p

Guzmán HM, Cortés J (1985) Organismos de los arrecifes coralinos de Costa Rica. IV. Descripción y distribución geográfica de octocoralarios (Anthozoa: Octocorallaria) de la costa Caribe. Brenesia 24:125–173

Guzmán HM, Jiménez CE (1989) *Pterogorgia anceps* (Pallas) (Octocorallia: Gorgoniidae): nuevo informe para la costa caribeña de Costa Rica. Rev Biol Trop 37:231–232

Guzmán HM, Cortés J (1989) Coral reef community structure at Caño Island, Pacific Costa Rica. P.S.Z.N.I: Mar Ecol 10:23–41

Guzmán HM, Cortés J (1992) Cocos Island (Pacific of Costa Rica) coral reefs after the 1982–83 El Niño disturbance. Rev Biol Trop 40:309–324

Guzmán HM, Cortés J (1993) Los arrecifes coralinos del Pacífico oriental tropical: revisión y perspectivas. Rev Biol Trop 41:535–557

Guzmán HM, Cortés J (2001) Changes in reef community structure after fifteen years of natural disturbances in the eastern Pacific (Costa Rica). Bull Mar Sci 69:133–149

Guzmán HM, Cortés J (2007) Reef recovery 20-yr after the 1982–83 El Niño massive mortality. Mar Biol 151:401–411

Guzmán HM, Cortés J, Richmond RH, Glynn PW (1987) Efectos del fenómeno de "El Niño-Oscilación Sureña" 1982/83 en los arrecifes de la Isla del Caño, Costa Rica. Rev Biol Trop 35:325–332

Guzmán HM, Cortés J, Glynn PW, Richmond RH (1990) Coral mortality associated with dynoflagellate blooms in the eastern Pacific (Costa Rica and Panama). Mar Ecol Prog Ser 60:299–303

Han X, Suess E, Sahling H, Wallmann K (2004) Fluid venting activity on the Costa Rica margin: new results from authigenic carbonates. Intl J Earth Sci 93:596–611

Hargraves PE, Víquez R (1985) Spatial and temporal distribution of phytoplankton in Gulf of Nicoya, Costa Rica. Bull Mar Sci 37:577–585

Hebbeln D, Cortés J (2001) Sedimentation in a tropical fjord: Golfo Dulce, Costa Rica. Geo-Mar Letters 20:142–148

Hebbeln D, Beese D, Cortés J (1996) Morphology and sediment structures in Golfo Dulce, Costa Rica. Rev Biol Trop 44(Suppl 3):1–10

Herrera W (1986) Clima de Costa Rica. Edit. UNED, San José, Costa Rica, 118 p

Høisæter T (1998) Preliminary check-list of the marine, shelled gastropods (Mollusca) of Golfo Dulce, on the Pacific coast of Costa Rica. Rev Biol Trop 46(Suppl 6):263–270

Hossfeld B (1996) Distribution and biomass of arrow worms (Chaetognatha) in Golfo de Nicoya and Golfo Dulce, Costa Rica. Rev Biol Trop 44(Suppl 3):157–172

Houbrick JR (1968) A survey of the littoral marine molluscs of the Caribbean coast of Costa Rica. Veliger 11:4–23

Huene R von, Bialas J, Flüeh E, Cropp B, Csernok T, Fabel E, Hoffman XX, Emeis K, Holler P, Jeschke G, Leandro G, Pérez MI, Chavarría J, Flores A, Ecobedo Z, León R, Barrios O (1995) Morphotectonics of the Pacific convergent margin of Costa Rica. *In*: Mann P (ed) Geological and Tectonic Development of the Caribbean Plate Boundaryin Southern Central America. Geol Soc Am Special Paper 295:291–307

INCOPESCA (2006) Memoria Institucional 2002–2006: Instituto Costarricense de Pesca y Acuicultura. Imprenta Nacional, San José, Costa Rica. 92 p

Jackson JBC, D'Croz L (1997) The ocean divided. *In*: Coates AG (ed) Central America: A Natural and Cultural History. Yale University Press, New Haven/London, pp 38–71

Jesse S (1996) Demersal crustacean assemblages along the Pacific coast of Costa Rica: a quantitative and multivariate assessment based on the Victor Hensen Costa Rica Expedition (1993/1994). Rev Biol Trop 44(Suppl 3):103–114

Jiménez CE (1997) Corals and coral reefs of Culebra Bay, Pacific coast of Costa Rica: anarchy in the reef. Proc 8th Int Coral Reef Symp, Panamá 1:329–334

Jiménez CE (1998) Arrecifes y comunidades coralinas de Bahía Culebra, Pacífico norte de Costa Rica (Golfo de Papagayo). M.Sc. thesis, University of Costa Rica, San Pedro. 218 p

Jiménez C (2001a) Bleaching and mortality of reef organism during a warming event in 1995 on the Caribbean coast of Costa Rica. Rev Biol Trop 49(Suppl 2):233–238

Jiménez C (2001b) Arrecifes y ambientes coralinos de bahía Culebra, Pacífico de Costa Rica: aspectos biológicos, económico-recreativos y de manejo. Rev Biol Trop 49(Suppl 2): 215–231

Jiménez CE, Cortés J (2001) Effects of the 1991–92 El Niño on scleractinian corals of the Costa Rican central Pacific coast. Rev Biol Trop 49(Suppl 2):239–250

Jiménez CE, Cortés J (2003a) Coral cover change associated to El Niño, eastern Pacific, Costa Rica, 1992–2001. P.S.Z.N.: Mar Ecol 24:179–192

Jiménez CE, Cortés J (2003b) Growth of seven species of scleractinian corals in an upwelling environment of the eastern Pacific (Golfo de Papagayo, Costa Rica). Bull Mar Sci 72:187–198

Jiménez C, Cortés J, León A, Ruiz E (2001) Coral bleaching and mortality associated with El Niño 1997/98 event in an upwelling environment in the eastern Pacific (Gulf of Papagayo, Costa Rica). Bull Mar Sci 69:151–169

Jiménez J (2004) Mangrove forests under dry seasonal climates in Costa Rica. *In*: Frankie GW, Mata A, Vinson SB (eds) Biodiversity Conservation in Costa Rica: Learning the Lessons in a Seasonal Dry Forest. University of California Press, California, pp 136–145

Jiménez J, Soto R (1985) Patrones regionales en la estructura y composición florística de los manglares de la costa Pacífica de Costa Rica. Rev Biol Trop 33:25–37

Jörger KM, Meyer R, Wehrtmann IS (2008). Species composition and vertical distribution of chitons (Mollusca: Polyplacophora) in a rocky intertidal zone of the Pacific coast of Costa Rica. J Mar Biol Ass UK 88:807–816

Kappelle M, Castro M, Acevedo H, González L, Monge H (2002) Ecosistemas del área de Conservación Osa (ACOSA). Edit. INBio, Sto. Domingo de Heredia, Costa Rica, 500 p.

Kessler WS (2006) The circulation of the eastern tropical Pacific: a review. Prog Oceonogr 69:181–217

Knowlton N (1993) Sibling species in the sea. Annu Rev Ecol Syst 24:189–216

Koch V, Wolff M (1996) The mangrove snail *Thais kioskiformis* Duclos: a case of life history adaptation to an extreme environmental. J Shell Res 15:421–432

Kohlmann B, Wilkinson J, Lulla K (2002) Costa Rica from Space. EARTH, NASA & UNESCO. Neotrópica Foundation Publ., San José Costa Rica, 227 p.

Krupp LS (2006) Distribution, ecology and state of the seagrass beds in the Gandoca-Manzanillo National Wildlife Refuge, Caribbean Costa Rica. M.Sc. thesis, University of Bremen, Germany, 87 p..

Krylova EM, Sahling H (2006) Recent bivalve molluscs of the genus *Calyptogena* (Vesicomyidae). J Mollusc Stud 72:359–395

Kuever J, Wawer C, Lillebæk R (1996) Microbiological observations in the anoxic basin Golfo Dulce, Costa Rica. Rev Biol Trop 44(Suppl 3):49–58

Lahmann EJ, González W (1982) Observaciones sobre la distribución espacial y el comportamiento de *Siphonaria gigas* Sowerby, 1825 en la costa Pacífica de Costa Rica (Gastropoda: Siphonariidae). An Inst Cien Mar Limnol, UNAM 9:101–110

Legeckis R (1988) Upwelling off the Gulfs of Panamá and Papagayo in the tropical Pacific during March 1985. J Geophys Res 93:15489–15489

León-Morales R, Vargas JA (1998) Macroinfauna of a tropical fjord-like embayment: Golfo Dulce, Costa Rica. Rev Biol Trop 46(Suppl 6):81–90

Lessios HA, Kessing BD, Robertson DR (1998) Massive gene flow across the world's most potent marine biogeographic barrier. Proc R Soc London, Ser B 265:583–588

Linke P, Wallmann K, Suess E, Hensen C, Rehder G (2005) In situ benthic fluxes from an intermittently active mud volcano at the Costa Rica convergent margin. Earth Planet Sci Lett 235:79–95

Lizano OG (1998) Dinámica de las aguas en la parte interna del Golfo de Nicoya ante altas descargas del Río Tempisque. Rev Biol Trop 46(Suppl 6):11–20

Lobo J, Barrantes G, Castillo M, Quesada R, Maldonado T, Fuchs EJ, Solís S, Quesada M (2007) Effects of selective logging on the abundance, regeneration and short term survival of *Caryocar costaricense* (Caryocaceae) and *Peltogyne purpurea* (Caesalpinaceae), two endemic timber species in southern Central America. Forest Ecol Manag 245:88–95

Loría LG, Jiménez R, Lizano O (2002) Radionucleídos naturales y antropogénicos en el estuario del Golfo de Nicoya, Costa Rica. Top Meteor Oceanogr 9:74–78

Mata A, Acuña JA, Murillo MM, Cortés J (1987) La contaminación por petróleo en el Caribe de Costa Rica: 1981–1985. Carib J Sci 23:41–49

Mata L, Abarca G, Maranghello L, Víquez R (1990) Intoxicación paralítica por mariscos (IPM) por *Spondylus calcifer* contaminado con *Pyrodinium bahamense*, Costa Rica, 1989–1990. Rev Biol Trop 38:129–136

Mau S, Sahling H, Rehder G, Suess E, Linke P, Soeding E (2006) Estimates of methane output from mud extrusions at the erosive convergent margin off Costa Rica. Mar Geol 225:129–144

Maurer D, Epifanio C, Harlan D, Howe S, Vargas J, Dittel A, Murillo M (1984) Benthic invertebrates of a tropical estuary: Gulf of Nicoya, Costa Rica. J Nat Hist 18:47–61

May-Collado L, Gerrodette T, Calambokidis J, Rasmussen K, Sereg I (2005) Patterns of cetacean sighting distribution in the Pacific Exclusive Economic Zone of Costa Rica based on data collected from 1979–2001. Rev Biol Trop 53:249–263

McAdoo BG, Orange DL, Silver EA, McIntosh K, Abott L, Galewsky J, Kahn L, Protti M (1996) Seafloor structural observations, Costa Rica accretionary prism. Geophys Res Lett 23:883–886

McCreary JP, Lee HS, Enfield DB (1989) The response of the coastal ocean to strong offshore winds: With application to circulation in the gulfs of Tehuantepec and Papagayo. J Mar Res 47:81–109

Meschede M, Zweigel P, Kiefer E (1999) Subsidence and extension at a convergent plate margin: Evidence for subduction erosion of Costa Rica. Terra Nova 11:112–117

Michel HB, Foyo M (1976) Caribbean Zooplankton. Part I. Siphonophora, Heteropoda, Copepoda, Euphausiacea, Chaetognatha and Salpidae. Off Naval Res, Dept. Navy, Washington, DC, pp 1–549

Molina-Ureña H (1996) Ichthyoplancton assemblages in the Gulf of Nicoya and Golfo Dulce embayments, Pacific coast of Costa Rica. Rev Biol Trop 44(Suppl 3):173–182

Mora A, Fernández C, Guzmán AG (2006) Áreas Marinas Protegidas y Áreas Marinas de Uso Múltiple de Costa Rica: notas para una discusión. MarViva, San José, Costa Rica, 104 p.

Mora DA, Rojas JC, Sequeira M, Mata A, Coto M (1989) Criterios bacteriológicos y calidad sanitaria de las aguas de las playas de Costa Rica. Período 1986–1987. Tecnología en Marcha 9:45–59

Morales A, Vargas JA (1995) Especies comunes de copépodos (Crustacea: Copepoda) pelágicos del Golfo de Nicoya, Costa Rica. Rev Biol Trop 43:207–218

Morales-Ramírez A (1996) Checklist of copepods from Gulf of Nicoya, Coronado Bay and Golfo Dulce, Pacific coast of Costa Rica, with comments on their distribution. Rev Biol Trop 44(Suppl 3):103–114

Morales-Ramírez A, Brugnoli-Oliveira E (2001) El Niño 1997–1998 impact on the plankton dynamics in the Gulf of Nicoya, Pacific coast of Costa Rica. Rev Biol Trop 49(Suppl 2):103–114

Morales-Ramírez A, Nowaczyk J (2006) Gelatinous zooplankton in the Golfo Dulce, Pacific coast of Costa Rica during transition from rainy season to dry season 1997–1998. Rev Biol Trop 54(Suppl 1):201–223

Neuhaus B (2004) Description of *Campyloderes* cf. *vanhoeffeni* (Kinorhyncha, Cyclorhagida) from the Central American East Pacific Deep with a review of the genus. Meiofauna Mar 13:3–20

Neuhaus B, Blasche T (2006) *Fissuroderes*, a new genus of Kinorhyncha (Cyclorhagida) from the deep sea and continental shelf of New Zealand and from the continental shelf of Costa Rica. Zool Anz 245:19–52

Nichols-Driscoll J (1976) Benthic invertebrate communities in Golfo Dulce, Costa Rica, na anoxic basin. Rev Biol Trop 24:281–298

Ortega S (1985) Competitive interactions among tropical intertidal limpets. J Exp Mar Biol Ecol 90:11–25

Ortega S (1986) Fish predation on gastropods on the Pacific coast of Costa Rica. J Exp Mar Biol Ecol 97:181–191

Ortega S (1987a) The effects of human predation on the size distribution of *Siphonaria gigas* (Mollusca: Pulmonata) on the Pacific coast of Costa Rica. Veliger 29:251–255

Ortega S (1987b) Habitat segregation and temporal variation in some tropical intertidal populations. J Exp Mar Biol Ecol 113:247–265

Paine RT (1966) Food web complexity and species diversity. Am Nat 100:65–75

Parkinson RW, Cortés J, Denyer P (1998) Passive margin sedimentation on Costa Rica's north Caribbean coastal plain, Río Colorado. Rev Biol Trop 46(Suppl 6):221–236

Paynter CK, Cortés J, Engels M (2001) Biomass, productivity and density of the seagrass *Thalassia testudinum* at three sites in Cahuita National Park, Costa Rica. Rev Biol Trop 49(Suppl 2):265–272

Pepe PJ (1985) Littoral endolithic fauna of the Central American Isthmus. Rev Biol Trop 33:191–194

Pizarro F, Angulo H (1993) Diagnóstico de los manglares de la costa Pacífica de Costa Rica: Informe para la Comisión Nacional de Manglares. UICN, Costa Rica, 82 p

Polanía J (1993) Mangroves of Costa Rica. *In*: Lacerda LD (ed) Conservation and Sustainable Utilization of Mangrove Forests in Latin America and Africa Regions. ITTO/ISME Project PD 114/90 (F), Part I. Okinawa, Japan, pp. 129–137

Pool DJ, Snedaker SC, Lugo AE (1977) Structure of mangrove forests in Florida, Puerto Rico, Mexico, and Costa Rica. Biotropica 9:195–212

Quesada-Alpízar MA, Cortés J (2006) Los ecosistemas marinos del Pacífico sur de Costa Rica: estado del conocimiento y perspectivas del manejo. Rev Biol Trop 54(Suppl 1):101–145

Quesada-Alpízar MA, Morales-Ramírez A (2004) Comportamiento de las masas de agua en el Golfo Dulce, Costa Rica durante "El Niño" (1997–1998). Rev Biol Trop 52(Suppl 2):95–104

Quesada-Alpízar MA, Morales-Ramírez A (2006) Posible efecto de El Niño en el zooplancton no gelatinoso del Golfo Dulce, Pacífico de Costa Rica, 1997–1998. Rev Biol Trop 54(Suppl 1):225–240

Quesada-Alpízar MA, Cortés J, Alvarado JJ, Fonseca AC (2006) Características hidrográficas y biológicas de la zona marino-costera del área de conservación OSA. Serie Técnica: apoyando los esfuerzos en manejo y protección de la biodiversidad tropical. The Nature Conservancy, San José, Costa Rica, 80 p

Ramírez AR, López MI, Szelistowski WA (1990) Composition and abundance of ichthyoplancton in a Gulf of Nicoya mangrove estuary. Rev Biol Trop 38:463–466

Ranero CR, von Huene R, Weinrebe W, McIntosh K, Reichert C (2000) Mass transfer and fluid flow paths related to subduction erosion at the Middle America convergent margin (Fall Meeting Supplement). EOS, Trans Am Geophys Union 81:48

Richards FA, Anderson JJ, Cline JD (1971) Chemical and physical observations in Golfo Dulce, an anoxic basin on the Pacific coast of Costa Rica. Limnol Oceanogr 16:43–50

Robertson DR, Allen GA (1996) Zoogeography of the shorefish fauna of Clipperton Atoll. Coral Reefs 15:121–131

Robinson DG, Montoya M (1987) Los moluscos marinos de la costa Atlántica de Costa Rica. Rev Biol Trop 35:375–400

Rojas M, Acuña JA, Rodríguez OM (1998) Metales traza en pepinos de mar *Holothuria mexicana* del Caribe de Costa Rica. Rev Biol Trop 46(Suppl 6):215–220

Sánchez-Vindas PE (2001) Flórula arborescente del Parque Nacional Cahuita. Ed. University of Estatal Distancia (EUNED), San José, 376 p

Scheltema RS (1988) Initial evidence for the transport of teleplanic larvae of benthic invertebrates across the east Pacific barrier. Biol Bull 174:145–152

Schleicher T (2006) Bestimmung von ventspezifischen Faunenvergesellschaftungen am mittelamerikanischen Kontinentalrand mit Hilfe quantitativer Videoauswertung. Diplom thesis, Leibniz-Institut für Meereswissenschaften, Christian-Albrechts-Univ., Kiel, Germany, 68 p

Setchell WA, Mason LR (1943) New or little known crustaceous corallines of Pacific North America. Proc Natl Acad Sci USA 29:92–97

Shasky DR (1988) Update on mollusks with Indo-Pacific faunal affinities in the tropical eastern Pacific VIII. Festivus 20:104–105.

Shipley TH, McIntosh KD, Silver EA, Stoffa PL (1992) Three dimensional imaging of the Costa Rica accretionary prism: structural diversity in a small volume of the lower slope. J Geophys Res 97:4439–2259

Sibaja-Cordero JA, Vargas-Zamora JA (2006) The vertical zonation of epifauna and algae species in rocky substrates of the Gulf of Nicoya, Costa Rica. Rev Biol Trop 54(Suppl 1):49–67

SINAC-MINAE (2006) El Sistema de Áreas Silvestres Protegidas de Costa Rica. Report prepared for presentation at the II Congreso Mesoamericano de Áreas Protegidas, Panamá City, 24–26 April 2006. Ministerio del Ambiente y Energía, San José, Costa Rica, 96 p

Soto R, y J. Jiménez (1982) Análisis fisonómico estructural del manglar de Puerto Soley, La Cruz, Guanacaste, Costa Rica. Rev Biol Trop 30:161–168
Spight TM (1976) Censuses of rocky shore prosobranchs from Washington and Costa Rica. Veliger 18:309–317
Spight TM (1977) Diversity of shallow-water gastropod communities on temperate and tropical beaches. Am Nat 111:1077–1097
Spight TM (1978) Temporal changes in a tropical rocky shore snail community. Veliger 21:137–143
Spongberg AL (2004a) PCB contamination in surface sediments in the coastal waters of Costa Rica. Rev Biol Trop 52(Suppl 2):1–10
Spongberg AL (2004b) PCB concentrations in sediments from the Gulf of Nicoya estuary, Pacific coast of Costa Rica. Rev Biol Trop 52(Suppl 2):11–22
Spongberg AL (2004c) PCB contamination in marine sediments from Golfo Dulce, Pacific coast of Costa Rica. Rev Biol Trop 52(Suppl 2):23–32
Spongberg AL (2006) PCB concentrations in intertidal sipunculan (Phylum Sipuncula) marine worms from the Pacific coast of Costa Rica. Rev Biol Trop 54(Suppl 1):27–33
Spongberg A, Davis P (1998) Organochlorinated pesticide contaminants in Golfo Dulce, Costa Rica. Rev Biol Trop 46(Suppl 6):111–124
Stern-Pirlot A, Wolff M (2006) Population dynamics and fisheries potencial of *Anadara tuberculosa* (Bivalvia: Arcidae) along the Pacific coast of Costa Rica. Rev Biol Trop 54(Suppl 1):87–99
Suess E, Carson B, Ritger SD, Moore JC, Jones M, Kulm LD, Cochrane G (1985) Biological communities at vent sites along the subduction zone off Oregon. Biol Soc Wash Bull 6:475–484
Sutherland JP (1987) Recruitment limitation in a tropical intertidal barnacle: *Tetraclita panamensis* (Pilsbry) on the Pacific coast of Costa Rica. J Exp Mar Biol Ecol 113:267–282
Sutherland JP, Ortega S (1986) Competition conditional on recruitment and temporary escape from predators on a tropical rocky shore. J Exp Mar Biol Ecol 95:155–166
Svendsen H, Rosland R, Myking S, Vargas JA, Lizano O, Alfaro E (2006) A physical-oceanographic study of Golfo Dulce, Costa Rica. Rev Biol Trop 54(Suppl 1):147–170
Szelistowski WA (1990) A new clingfish (Teleostei: Gobiesocidae) from the mangroves of Costa Rica, with notes on its ecology and early development. Copeia 1990:500–507
Tabash-Blanco FA (2007) Exploración de la pesquería de arrastre de camarón durante el período 1991–1999 en el Golfo de Nicoya, Costa Rica. Rev Biol Trop 55:207–218
Tabash FA & Palacios JA (1996) Stock assessment of two penaeid prawn species, *Penaeus occidentalis* and *Penaeus stylirostris* (Decapoda: Penaeidae), in Golfo de Nicoya, Costa Rica. Rev Biol Trop 44:595–602
Tabash-Blanco FA, Chávez EA (2006) Optimizing harvesting strategies of the white shrimp fisheries in the Gulf of Nicoya, Costa Rica. Crustaceana 79:327–343
Thamdrup B, Canfield DE, Ferdelman TG, Glud RN, Gundersen JK (1996) A biogeochemical survey of the anoxic basin Golfo Dulce, Costa Rica. Rev Biol Trop 44(Suppl 3):19–34
Townsend CH (1901) Dredging and other records of the United States Fish Commission Steamer Albatross, with bibliography relative to the work of the vessel. US Fish Comm Rep for 1900:387–562
Troëng S (2005) Migration of sea turtles from Caribbean Costa Rica: implications for management. Ph.D. dissertration, Lund University, Lund, Sweden, 104 p
Troëng S, Rankin E (2005) Long-term conservation efforts contribute to positive green turtle *Chelonia mydas* nesting trend in Tortuguero, Costa Rica. Biol Conserv 121:111–116
Troëng S, Chacón D, Dick B (2004) Possible decline in leatherback turtle *Dermochelys coriacea* nesting along the coast of Caribbean Central America. Oryx 38:395–403
Ulken A, Víquez R, Valiente C, Campos M (1990) Marine fungi (Chytridiomycetes and Thraustochytriales) from a mangrove area at Punta Morales, Golfo de Nicoya, Costa Rica. Rev Biol Trop 38:243–250
Umaña G (1998) Characterization of some Golfo Dulce drainage basin rivers (Costa Rica). Rev Biol Trop 46(Suppl 6):125–136
Vargas JA (1987) The benthic community of an intertidal mud flat in the Gulf of Nicoya, Costa Rica. Description of the community. Rev Biol Trop 35:299–316

Vargas JA (1988a) A survey of the meiofauna of an eastern Pacific intertidal mudflat. Rev Biol Trop 36:541–544

Vargas JA (1988b) Community structure of macrobenthos and the result of macropredator exclusion on a tropical intertidal mud flat. Rev Biol Trop 36:287–308

Vargas JA (1995) The Gulf of Nicoya estuary: past, present and future cooperative research. Helgol Meeresunters 49:821–828

Vargas JA (1996) Ecological dynamics of a tropical intertidal mudflat community. *In*: Nordstrom KF, Roman CT (eds) Estuarine Shores: Evolution, Environments and Human Alterations. Wiley, London, pp 355–371

Vargas JA, Mata A (2004) Where the dry forest feeds the sea: the Gulf of Nicoya Estuary. *In*: Frankie GW, Mata A, Vinson SB (eds) Biodiversity Conservation in Costa Rica: Learning the Lessons in a Seasonal Dry Forest. University of California Press, Berkeley, LA/London, pp 126–135

Vargas JA, Wolff M (1996) Preface. Pacific coastal ecosystems of Costa Rica with emphasis on the Golfo Dulce and adjacent areas: a synoptic view based on the R.V. Victor Hensen – expedition 1993/94 and previous studies. Rev Biol Trop 44(Suppl 3)

Vargas R, Jesse S, Castro M (1996) Checklist of crustaceans (Decapoda and Stomatopoda), collected during the Victor Hensen Costa Rica expedition (1993/1994). Rev Biol Trop 44(Suppl 3):97–102

Vargas-Montero M, Freer E (2004a) Presencia de los dinoflagelados *Ceratium dens, C. fusus* y *C. furca* (Gonyaulacales: Ceratiaceae) en el Golfo de Nicoya, Costa Rica. Rev Biol Trop 52(Suppl 1):115–120

Vargas-Montero M, Freer E (2004b) Proliferaciones algales nocivas de cianobacterias (Oscillatoriaceae) y dinoflagelados (Gymnodiniaceae) en el Golfo de Nicoya, Costa Rica. Rev Biol Trop 52(Suppl 1):121–125

Vargas-Montero M, Freer E (2004c) Proliferaciones algales de la diatomea toxigénica *Pseudo-Nitzschia* (Bacillariophyceae) en el Golfo de Nicoya, Costa Rica. Rev Biol Trop 52(Suppl 1):127–132

Villalobos CR (1980a) Variations in population structure in the genus *Tetraclita* (Crustacea: Cirripedia) between temperate and tropical populations. III. *T. stalactifera* in Costa Rica. Rev Biol Trop 28:193–201

Villalobos CR (1980b) Variations in population structure in the genus *Tetraclita* (Crustacea: Cirripedia) between temperate and tropical populations. IV. The age structure of *T. stalactifera* and concluding remarks. Rev Biol Trop 28:353–359

Víquez R, Hargraves PE (1995) Annual cycle of potentially toxic dinoflagellates in the Golfo de Nicoya, Costa Rica. Bull Mar Sci 57:467–475

von Huene R, Ranero CR, Weinrebe W (2000) Quaternary convergent margin tectonics of Costa Rica, segmentation of the Cocos Plate, and Central American volcanism. Tectonics 19:314–334

von Wangelin M, Wolff M (1996) Comparative biomass spectra and species composition of the zooplancton communities in Golfo Dulce and Golfo de Nicoya, Pacific coast of Costa Rica. Rev Biol Trop 44(Suppl 3):135–156

Voorhis AD, Epifanio CE, Maurer D, Dittel AI, Vargas J (1983) The estuarine character of the Gulf of Nicoya, an embayment on the Pacific coast of Central America. Hydrobiologia 99:225–237

Voss GL (1971) Narrative of R/V John Elliott Pillsbury cruise P-7101 to Central America, 20 January – February 1971. School of Marine and Atmospheric Science, Univ. Miami, Florida, 31 p.

Wehrtmann IS (1990) Distribution and reproduction of *Ambidexter panamense* and *Palaemonetes schmitti* in Pacific Costa Rica (Crustacea, Decapoda). Rev Biol Trop 38:327–329

Wehrtmann IS, Albornoz L (2002) Evidence of different reproductive traits in the transisthmian sister species, *Alpheus saxidomus* and *A. simus* (Decapoda, Caridea, Alpheidae): Description of the first postembryonic stage. Mar Biol 140:605–612

Wehrtmann IS, Dittel A (1990) Utilization of floating mangrove leaves as a transport mechanism of estuarine organisms, with emphasis on decapod Crustacea. Mar Ecol Progr Ser 60:67–73

Willis S, Cortés J (2001) Mollusks of Manuel Antonio National Park, Pacific Costa Rica. Rev Biol Trop 49(Suppl 2):25–36

Wilson HV (1904) Reports on an exploration off the west coasts of Mexico, Central and South America, and off the Galapagos Islands, in charge of Alexander Agassiz, by the U.S. Fish Commission Steamer Albatross, during 1891, Lieut. Commander Z.L. Tanner U.S.N., commanding. XXX. The sponges. Mem Mus Comp Zoöl Harvard Coll XXX:1–164
Wolff M (1996) Demersal fish assemblages along the Pacific coast of Costa Rica: a quantitative and multivariate assessment based on the Victor Hensen Costa Rica Expedition (1993/1994). Rev Biol Trop 44(Suppl 3):187–214
Wolff M, Chavarría J, Koch V, Vargas JA (1998) A trophic flow model of the Golfo de Nicoya, Costa Rica. Rev Biol Trop 46(Suppl 6):63–79

Chapter II
A History of Marine Biodiversity Scientific Research in Costa Rica

Jorge Cortés

The RV *Skimmer* of the University of Delaware was used in the Golfo de Nicoya between 1979 and 1981 (Photo: Charles E. Epifanio)

Abstract Marine biodiversity scientific research in Costa Rica started in the 19th century. I have divided its history into three phases based on the chronology, methods, and focus of the research. The first phase, from the early 19th to the early 20th century, was a time of exploration of the seas worldwide. The first publications

J. Cortés
Centro de Investigación en Ciencias del Mar y Limnología (CIMAR), and Escuela de Biología, Universidad de Costa Rica, San Pedro, San José 11501-2060, Costa Rica.
e-mail: jorge.cortes@ucr.ac.cr

I.S. Wehrtmann, J. Cortés (eds.) *Marine Biodiversity of Costa Rica, Central America,*

on marine organisms collected in Costa Rica were published in 1832. During this phase, collections for museums were made by ship captains and professional collectors, and by the first expeditions that visited Costa Rica in the late 19th century. The second phase, from 1924 to the 1990s, was a time of intense collection by foreign scientists, mainly from the United States, and especially between 1925 and 1939. Publications based on those collections are still appearing to the present day. The third phase, starting in the early 1970s, sees the contributions of the first Costa Rican marine scientists and resident foreign scientists. The creation of the Marine Sciences and Limnology Research Center (CIMAR) at the Universidad de Costa Rica (UCR) in 1979 is a turning point in marine research in Costa Rica. It is characterized by the publication of papers on many different groups, by ecological studies, and by the collaboration between national and foreign scientists. The importance of a strong local scientific community is emphasized for the generation and the assimilation of new knowledge and collections, and for the dissemination of that knowledge.

Introduction

Marine organisms and environments were mentioned in travelers' descriptions of Costa Rica starting in the 16th century, for example, the use of muricid snails to dye textiles by indigenous populations and the extraction of pearl oysters (Meléndez 1974). Even though these antique records provide interesting information on past environments and organisms, I will concentrate on research about marine animals and plants from Costa Rica that is published in scientific journals.

The first scientific publications on the natural history of Costa Rica were from the early 19th century, and were done by visiting or resident foreign "scientists," most of who were naturalists without formal training (Eakin 1999). This author presents an excellent reconstruction of the first scientific activities in Costa Rica, and covers the creation (1887) and functioning (until 1904) of the Instituto Físico-Geográfico Nacional (National Physical Geographic Institute), the first scientific institution in the country. Peraldo-Huertas (2003) compiled a series of papers on the history of research in Costa Rica during the 19th and early 20th centuries; all these publications were on terrestrial investigations.

Marine research in Costa Rica also started in the 19th century, and in this paper I will concentrate on research related to marine biodiversity, i.e., papers with lists of species and description of new species. I have divided the historical development of marine biodiversity scientific research into three phases based on chronology, logistics for doing the collections, and the focus of the research. The first two phases differ in time and, in some cases, in the logistics to obtain the specimens, but all the research was carried out by foreign scientists with their own agendas, and with no connection to local scientists or interests. During the third phase, research was done by both local scientists and resident foreign scientists, and sometimes through collaborative projects with foreign institutions.

From the early 19th to the early 20th century, there were two sources of marine organisms for study and publication. The first source was from ship captains, professional collectors and Costa Rican scientists who sent specimens to foreign specialists, and the second source was from organisms collected during oceanographic expeditions. The first expeditions organized with the sole purpose of scientifically exploring the oceans began in the mid- to late 19th century (e.g., the HMS *Challenger* Expedition, 1872–1876: Corfield 2003). Towards the end of the 19th century the United States mounted a number of marine scientific expeditions, especially to the Atlantic and eastern Pacific oceans. During these expeditions, large numbers of organisms were collected, and their descriptions published during the following decades.

The phase from 1924 to the 1990s was a time of intense collection, especially in the eastern Pacific, by foreign scientists, mainly from the United States, both in organized expeditions and by individuals, most of them funded by philanthropists. Publications based on those collections are still appearing to the present day.

The third phase started in the early 1970s with Manuel M. Murillo, the first Costa Rican marine scientist with a Ph.D., and the creation of the Marine Sciences and Limnology Research Center (Centro de Investigación en Ciencias del Mar y Limnología; CIMAR) at the Universidad de Costa Rica (UCR) in 1979. This period is characterized by the publication of papers on many different taxonomic groups, ecological studies, and the collaboration between national and foreign scientists. Also, the focus of the research is more locally oriented.

First Phase: Early 19th to Early 20th Century

The first scientific papers on marine organisms from Costa Rica were published by Sowerby (1832) who described species of mollusks collected by Hugh Cuming in Bahía Potrero, Golfo de Nicoya, and Golfo Dulce. Later, Mörch (1859–1861) described more species of mollusks, collected in the Golfo de Nicoya. These collections were done by ship captains or by professional collectors (e.g., Hugh Cuming collected in South and Central America between 1827 and 1830 for British scientists) who sent the specimens to museums in Europe and in the United States. Between 1845 and 1848, marine specimens were collected in the West Indies and Central America, and from those collections the first sipunculid from Costa Rica was described, with specimens from Puntarenas (Grube 1859). Another early contribution was by A.E. Verrill who described six species of octocorals and one species of scleractinian coral, all collected in the Golfo de Nicoya by J.A. McNeil (Verrill 1870).

Resident foreign scientists, like Paul Biolley, and Costa Rican scientists, like José Fidel Tristán, sent specimens between 1902 and 1907 to the United States National Museum (USNM) in Washington, DC, and several new species were described (Stebbing 1903, 1906; Rathbun 1906). Henri Pittier, a Swiss scientist who lived in Costa Rica from 1887 to 1904, also sent marine organisms to the US National Museum (Rathbun 1930).

Towards the end of the 19th century and early in the 20th century, there were six expeditions that visited Costa Rican waters, specifically Isla del Coco (Cocos Island) (Table II.1), all of them but one were funded by government institutions. The first scientific expedition to the region was in 1888 with the US Fish Commission Steamer *Albatross* (Fig. II.1) and led by Alexander Agassiz (Fig. II.2). During that occasion four stations south of Isla del Coco were sampled in 1 and 2 April, and hydrographic data recorded; but in only one of the stations, a few samples were collected using scoop nets and electric lights (Townsend 1901). Apparently they stopped at Isla del Coco, because two species of snails collected from the beaches of the island are reported in the only reference in which specimens from Costa Rica are mentioned (Dall 1900). The next major expedition made significant contributions to the knowledge of marine biodiversity of the region: the US Fish Commission Steamer *Albatross* 1891 Expedition to the west coast of Central America and the Galapagos, again led by Alexander Agassiz (Faxon 1895; Townsend 1901). The Steamer *Albatross* 1891 Expedition was in Costa Rican waters from 26 February to 2 March 1891, and 15 stations were sampled, most were around Isla del Coco. Dredging and trawling was done at 12 stations (stations number 3362–3373), hydrographic soundings at three additional stations, six surface tow-net stations (where dredging was also done), and six serial temperature stations, with records from the surface to the bottom, ranging from 29.9 °C to 2.5 °C (Townsend 1901). Many papers were published and new species described (Cortés in preparation), including the only paper on sponges from the Pacific of Costa Rica (Wilson 1904).

Towards the end of the 19th century the Instituto Físico-Geográfico Nacional de Costa Rica (IFGN, National Physical-Geographical Institute) was created by the resident Swiss scientists Henri Pittier. One of the objectives of this institute was the scientific study of natural history in Costa Rica (Eakin 1999). In early 1898, the first scientific expedition from Costa Rica was to Isla del Coco, aboard the steamer *Poás*, and was organized by the IFGN (Pittier 1899). The expedition was led by Henri Pittier, the Director of the Institute (Table II.1). All the research was on land, but Anastacio Alfaro from the Museo Nacional de Costa Rica mentioned the presence of seabirds on the island (Alfaro 1899).

The fourth major expedition that touched Isla del Coco was the Hopkins Stanford Galapagos Expedition of 1898–1899 (Heller 1903) (Table II.1); the only expedition from this phase partly funded by a philanthropist. This expedition was organized by the Zoology Department at Stanford University under the patronage of Timothy Hopkins, using a sealing schooner. The primary interest was in collecting vertebrates, but many groups of animals and plants were collected, including coastal marine organisms. On the way back to the United States they stopped at Isla del Coco on 29 June 1899 and stayed there 4 days. Many papers were published mainly on terrestrial vertebrates and plants, a few publications were on marine organisms (Rathbun 1902; Snodgrass & Heller 1905), and papers were published until the early 20th century (e.g., Dall 1908, 1920).In 1902, under the leadership of Henri Pittier, the government organized an expedition to Isla del Coco, aboard the steamer *Turrialba*. Paul Biolley participated as a naturalist from the Instituto Físico-Geográfico Nacional. He collected mollusks and listed 34 species, including

Table II.1 List of expeditions that collected marine organisms on the Pacific and Caribbean of Costa Rica. Only the years that Costa Rica was visited are indicated. Codes: BO Buque Oceanográfico (Research Vessel); BP Buque Pesquero (Fishing Vessel); DSRV Deep Submersible Research Vehicle; FS Forschungsschiff (Research Vessel); MY Motor Yacht; NMNH National Museum of Natural History, Washington, DC; RV Research Vessel; Sch Schooner; SSch Sealing Schooner; SSV Sailing School Vessel; SY SteamYacht; USFC United States Fish Commission; USS United States Ship; V Vapor (Steamer); n.a. = information not available

Name of expedition/Organizer	Vessel	Expedition leader	Years	Reference
Pacific				
U.S. Fish Commission	USFC Steamer *Albatross*	Alexander Agassiz	1888	Townsend (1901)
U.S. Fish Commission	USFC Steamer *Albatross*	Alexander Agassiz	1891	Faxon (1895)
Instituto Físico-Geográfico Nacional	V *Poás*	Henri Pittier	1898	Pittier (1899)
Hopkins Stanford Galapagos Expedition	SSch *Julia E. Whalen*	R.E. Snodgrass/Edmund Heller	1898–1899	Heller (1903)
Instituto Físico-Geográfico Nacional	V *Turrialba*	Henri Pittier	1902	Biolley (1907)
U.S. Fisheries Commission	USFC Steamer *Albatross*	Alexander Agassiz	1904–1905	Bigelow (1909)
Expedition to the Galapagos Islands/ California Academy of Sciences	Sch *Academy*	Joseph R. Slevin	1905–1906	Slevin (1931)
Scripps Institution of Oceanography	RV *Alexander Agassiz*	W.C. Crandall	1920	Luke (1995)
Zoology Laboratory/Cambridge University	SY *St. George*	Cyrill Crossland	1924	Kramp (1956)
Arcturus Oceanographic Expedition/ New York Zoological Society	SY *Arcturus*	William Beebe	1925	Beebe (1926)
William K. Vandervilt	SY *Eagle & Ara*	William K. Vandervilt	1926 and 1928	Boone (1933)
Pinchot South Seas Expedition	n.a.	Gifford Pinchot	1929	Fowler (1932)
Austrian Biological Expedition	–	–	1930	Pesta (1931)
Templeton Crocker Expedition/ California Academy of Science	SY *Zaca*	Templeton Crocker	1932	Hertlein (1935)
Expeditions to the Eastern Pacific/ Allan Hancock Foundation	MY *Velero III*	Allan Hancock	1932–1935	Fraser (1943)
Eastern Pacific Expeditions/ New York Zoological Society	SY *Zaca*	William Beebe	1937–1938	Beebe (1938)
Presidential Cruise/President of the USA	USS *Houston*	Waldo L. Schmitt	1938	Schmitt (1939)
U.S. Navy Galapagos Expedition	n.a.	Waldo L. Schmitt	1941	Smithsonian Inst. Archives

(continued)

Table II.1 (continued)

Name of expedition/Organizer	Vessel	Expedition leader	Years	Reference
Wooldrow G. Krieger Expedition	n.a.	Bruce W. Halstead	1952–1953	Halstead & Schall (1955)
Galapagos Expedition	RV *Xarifa*	Hans Hass	1954	Durham (1962)
Lamont-Doherty Earth Observatory/Columbia University	RV *Vema*	Joseph Worzel	1958	Child (1992)
Beaudette Foundation	MY *Stella Polaris*	E. Yale Dawson	1959	Dawson & Beaudette (1959)
Stanford Oceanographic Expedition 18/Stanford University	RV *Te Vega*	Donald P. Abbott	1968	Ball (1972)
Stanford Oceanographic Expedition 20/Stanford University	RV *Te Vega*	Ellsworth Wheeler	1968	Stanford University Archives
Cruise 35/Department of Oceanography, University of Washington	RV *T.G. Thompson*		1969	Richards *et al.* (1971)
Central American Expedition/Janss Foundation	RV *Searcher*	William A. Bussing	1972	Reaka & Manning (1980)
Allan Hancock/Los Angeles County Museum	MY *Velero IV*	Richard Pieper	1973	Jean Crampon personal communication, 2005
MV73-I, Fisheries/Scripps Institution of Oceanography	RV *Alexander Agassiz*	Carl L. Hubbs	1973	Luke (1977)
Exploratory fisheries	BP *Macuro*	n.a.	1977	López (1980)
Caribbean-Pacific Expedition Phase VI/Scripps Institution of Oceanography	RV *Alpha Helix*	William Newman	1978	Luke (1995)
Golfo de Nicoya Expeditions/Universidad de Costa Rica and University of Delaware	RV *Skimmer*	Manuel M. Murillo	1979–1981	Vargas (1995)
Domo I – II/Universidad Nacional Autónoma de México	BO *El Puma*	José Barberán	1979 (x2)	Segura-Puertas (1991)
Domo III/Universidad Nacional Autónoma de México	BO *El Puma*	José Barberán	1981	Gasca & Suárez (1992)
Domo IV/Universidad Nacional Autónoma de México	BO *El Puma*	José Barberán	1982	Suárez-Morales & Gasca (1989)
SeaPharm Project 7 Harbor Branch Oceanographic Institution	RV Seward Johnson/DSRV Johnson-Sea-Link I	K.L. Rinehart/S.A. Pomponi	1986	Cairns (1991a)

n.a.	RV *F. Nansen*	n.a.	1987	Bianchi (1991)
Cocos Island Expedition/Smithsonian Tropical Research Institute	RV *Benjamin*	Harilaos A. Lessios	1987	Lessios *et al.* (1996)
Natural History Museum of Los Angeles County	Sch *Victoria af Carlstat*	n.a.	1988	Camp & Kuck (1990)
Nicoya Peninsula	RV *Nishin Maru*	Manuel M. Murillo	1988–1989	J.A. Vargas personal communication, 2006
Eastern Pacific Expedition	RV *Moana Wave*	Peter W. Glynn	1989	Macintyre *et al.* (1992)
Eastern Pacific Expedition	RV *Gyre*	Gerard M. Wellington	1990	Lessios *et al.* (1996)
Victor Hensen Costa Rica Expedition/ Universidad de Costa Rica and Universität Bremen	FS *Victor Hensen*	José A. Vargas/Matthias Wolff	1993–1994	Vargas & Wolff (1996)
Costa Rica Accretionary Prism/Woods Hole Oceanographic Institution	DSRV *Alvin*	Eli Silver	1994	McAdoo *et al.* (1996), and NMNH Catalog
Cocos Island Expedition/Smithsonian Tropical Research Institute	RV *Urracá*	D. Ross Robertson	1997	Ross Robertson personal communication, 2007
S-177 cruise/Sea Education Association	SSV *Robert C. Seamans*	Lisa Graziano	2001	Cruise report prepared by L. Graziano
Costa Rica Accretionary Prism/Woods Hole Oceanographic Institution	DSRV *Alvin*	n.a.	2005	J.A. Vargas personal communication, 2006
Survey of coastal fishes of northen and central Costa Rica/Smithsonian Tropical Research Institute	RV *Urracá*	D. Ross Robertson	2005	Ross Robertson personal communication, 2007
Benthic survey of northen and central Costa Rica/Smithsonian Tropical Research Institute	RV *Urracá*	Rachel Collins	2005	Rachel Collins personal communication, 2007
First Scientific Expedition to Isla del Coco of the MV *Proteus*/Universidad de Costa Rica-MarViva	MV *Proteus*	Jorge Cortés	2006	Cortés 2008

(continued)

Table II.1 (continued)

Name of expedition/Organizer	Vessel	Expedition leader	Years	Reference
Isla del Coco Marine Biodiversity	MV *Proteus*	Jorge Cortés	2007	Cortés (2008)
Caribbean				
Austrian Biological Expedition	–	–	1930	Pesta (1931)
U.S. Navy	n.a.	n.a.	1968	Michel & Foyo (1976)
Cruise P-7101 to Central America/Institute of Marine and Atmospheric Science, University of Miami	RV *John Elliott Pillsbury*	Gilbert L. Voss	1972	Voss (1971)
Belém to Belize City Expedition/ Scripps Institution of Oceanography	RV *Alpha Helix*	Abraham Fleminger	1977	Ferrari & Bowman (1980)

23 marine gastropods (Biolley 1907). He also collected an amphipod that was sent to Thomas R.R. Stebbing, and was subsequently described as a new species (Stebbing 1903). That was one of the several new species that Stebbing (1906, 1908) described from samples sent by Biolley from several coastal areas of Costa Rica.

Once more, from 1904 to 1905 the US Fish Commission Steamer *Albatross* was used in the eastern Pacific. No information on this cruise has been found and only two papers have been located in which samples from Costa Rica are mentioned; one with species of medusae and the other with siphonophorans (Bigelow 1909, 1911).

The last expedition at the beginning of the 20th century to visit Isla del Coco was the California Academy of Sciences Expedition to the Galapagos Islands. This expedition visited the island from 3 to 13 September 1905, and collected insects, birds, lizards, and freshwater animals. All the marine animals mentioned (not

Fig. II.1 The US Fish Commission Steamer *Albatross* was used in the eastern tropical Pacific, and sampled around Isla del Coco in 1888, 1891, and 1904–1905 (Used with permission of the Museum of Comparative Zoology Archives, Harvard University)

collected) were large vertebrates: fishes, turtles, sharks, devilfish, and whales. Most of the marine animals that they collected were eaten (Slevin 1931).

On 22 March 1920, the RV *Alexander Agassiz*, of the Scripps Institution of Oceanography, visited Costa Rica and dredged near the Islas Murciélago (northern Pacific of Costa Rica). Several mollusks were collected by Captain W.C. Crandall and E.L. Michael (Luke 1995) (Table II.1). I have not found any other reference to this cruise or seen any publication in which the specimens collected are mentioned. In summary, during this first phase of scientific exploration, from 1832 to 1920, at least 45 papers were published, which include information on specimens collected in Costa Rica. Of these papers about one third was based on collections by individuals and the rest associated with expeditions (Cortés in preparation).

Fig. II.2 Alexander Agassiz, in 1896, was the Chief Scientist of the US Fish Commission Steamer *Albatross* when it explored the eastern tropical Pacific in the late 1890s (Used with permission of the Museum of Comparative Zoology Archives, Harvard University)

Second Phase: 1924 to the 1990s

Another intense phase of exploration occurred between 1924 and 1939 when many expeditions visited Costa Rica (details of all the expeditions are given in Table II.1); the larger ones were funded mainly by philanthropists. In 1924, there was a British expedition, one of only three European Expeditions to Costa Rica during this phase lead by Cyril Crossland aboard the *St. George*. Mollusks from Isla del Coco were reported by Tomlin (1927) and medusae, collected at two tow-netting stations south of the island by Kramp (1956). An important expedition during this phase was the Arcturus Oceanographic Expedition (Figs. II.3 and II.4), in 1925, organized by the New York Zoological Society and led by William Beebe. It undertook plankton tows south and north of Isla del Coco, and made collections on and around the island for 10 days in May 1925 (Beebe 1926). Tee-Van (1926) presents a detail account of the equipment used and the operations. At least eight papers were published that include information from Costa Rica (Cortés in preparation).

The next expeditions were the cruises with the yachts *Eagle* and *Ara* in 1926 and 1928, commanded by William K. Vanderbilt. So many specimens were collected during these expeditions that W.K. Vanderbilt founded the Vanderbilt Marine Museum in Centerport, New York to house them. They collected at Isla del Coco in 1926 and around Puntarenas and Isla del Caño in 1928 (Boone 1930a, b, 1933).

Fig. II.3 Wafer Bay, Isla del Coco in May 1925, during William Beebe's visit on board the SY *Arcturus* (Used with permission of the Bibliographisches Institut & F.A. Brockhaus AG)

Fig. II.4 SY *Arcturus* at Isla del Coco in May1925 (Used with permission of the Bibliographisches Institut & F.A. Brockhaus AG)

Lee Boone reported and described species of many crustacean groups, coelenterates, echinoderms, and mollusks. In 1929, Governor Gifford Pinchot traveled to the South Seas, across the Caribbean, the Panama Canal, then to Isla del Coco and the Galápagos, before heading to the Marquesas, Tahiti, and Society islands. During this so-called Pinchot South Seas Expedition, A.K. Fisher of the United States Biological Survey, and Henry A. Pilsbry of the Academy of Natural Sciences of Philadelphia, collected fishes, including five species of marine fishes and one freshwater species from Isla del Coco (Fowler 1932).

The second European expedition to visit Costa Rica during this phase was the Austrian Biological Expedition to Costa Rica in 1930. This brief expedition collected nothing on the Caribbean coast, but obtained eight marine crabs from Puntarenas and Golfo Dulce, on the Pacific coast (Pesta 1931).

In 1932, the Templeton Crocker Expedition of the California Academy of Sciences set sail for the eastern Pacific (Table II.1). During this expedition the following sites in Costa Rica were visited: Puntarenas (spelled Punta Arenas in the Appendix) from 22 to 26 June and on 30 June, in transit from Puntarenas to Isla del Coco and at the island from 26 to 30 June; Bahía Brasilito (indicated as Braxilito Bay) on 1 and 2 July; Bahía Murciélago on 2 and 3 July; and Bahía de Santa Elena (designated as Port Parker) on 3 and 4 July (Crocker 1933). Many papers were published, from terrestrial fungi to birds; and at least nine contain information on marine organisms collected in Costa Rica (Cortés in preparation). A sample of three of the publications are: Hertlein (1935) on pectinidid bivalves, Setchell (1937) on algae of the genus Sargassum, and Bigelow (1940) on medusae.During 1937 and 1938, there was an expedition of the New York Zoological Society with the yacht *Zaca* to the eastern Pacific, under the scientific leadership of William Beebe (Table II.1). It collected at Islas Murciélago, Santa Elena, Potrero Grande, Culebra, Brasilito, Sámara, Ballena, Uvita, and Pavones bays, off Cabo Blanco, around several islands in the Golfo de Nicoya, southeast of Punta Judas, south of the Osa Peninsula, at the entrance to Golfo Dulce, and within the gulf and in Golfito (Beebe 1938). At least 22 papers have been published in which specimens collected in Costa Rica during these expeditions were mentioned (Cortés in preparation). Two groups, mollusks, and crustaceans, were studied in great detail, and there is one paper on echinoderms.

The Allan Hancock Foundation *Velero III* expeditions to the eastern Pacific that visited Costa Rica took place between 1932 and 1939 (Table II.1). In February 1932, Bahía Chatham at Isla del Coco was sampled; in February–March 1933: Chatham and Wafer bays at Isla del Coco, and Culebra and Cocos bays on the mainland; in February 1934: points around Bahía Culebra; in February 1935: Playa Blanca, Port Parker (now Bahía de Santa Elena) and Bahía Salinas; in January 1938: Nuez Island (now Isla Manuelita), Wafer and Chatham bays all at Isla del Coco; and finally in March 1939: Bahía de Santa Elena and Golfo Dulce (Fraser 1943). At least 58 papers have been published on the material collected (Cortés in preparation), covering algae (Taylor 1945) to echinoderms (Ziesenhenne 1942), including corals, mollusks, crustaceans, and fish parasites (Manter 1940). Specimens from these collections are still being analyzed and papers published (e.g., Barnard 1980; Castro 1996).

One of the last expeditions of the 1930s was the Presidential Cruise (Franklin D. Roosevelt, US President, on board the USS *Houston*) from July to August 1938, under the scientific leadership of Waldo L. Schmitt of the Smithsonian Institution (Table II.1). It collected samples on 3 August 1938 from Isla del Coco (Schmitt 1939). At least five papers (Cortés in preparation), covering five different phyla, were published on species from Isla del Coco. Additionally, Schmitt (1940), in his review of stomatopods mentioned a species from Isla San Lucas sent to the US National Museum by Manuel Valerio in 1930.

During the next 15 years, which included the Second World War, there was only the US Navy Galapagos Expedition (1941) that touched Isla del Coco. Waldo L. Schmitt collected some specimens (Smithsonian Institution Archives), but no reports on them have been published yet. The next expeditions (Table II.1) occurred in the 1950's. Between 1952 and 1953 the Wooldrow G. Krieger Expedition to the eastern Pacific, collected fish in Galápagos and Isla del Coco. Later, Halstead & Schall (1955) published on the toxicity of venenous fish collected at Isla del Coco. In 1953 and 1954 the RV *Xarifa*, led by Hans Hass, the third European expedition, visited the eastern Pacific. Corals were collected from Isla del Coco by Georg Scheer in 1954 (Durham 1962). Apparently the RV *Vema* visited Isla del Coco on 30 November 1958 (Child 1992), but I have found no information about this expedition or any other publication in which it is mentioned. In 1959, the *Stella Polaris* visited the eastern Pacific for the Beaudette Foundation (Dawson & Beaudette 1959). Dawson (1960) published a list of the species of algae, including new species from Mexico and Central America, with a significant number of species from Costa Rica.

In the decade between 1968 and 1978 there were five expeditions to the eastern Pacific and collections were done in Costa Rica. The Stanford Oceanographic RV

Te Vega expeditions #18 and #20 in 1968, cruise 76 of the RV *T.G. Thompson* in 1969, and two expeditions of the RV *Agassiz* (Scripps Institution of Oceanography), one in 1973 and another in 1978 (Table II.1). The Stanford Oceanographic Expedition 18, between April and June 1968, sampled intertidal fauna and flora from Paita, Perú to Bahía Magdalena, Baja California, México. In Costa Rica they collected in the area of Negritos and Cedros islands, Golfo de Nicoya (11–12 May) and Bahía Brasilito (13 May; Ball 1972). From this expedition three papers were published in which samples from Costa Rica were mentioned, two on hermit crabs (Ball 1972; Ball & Haig 1974) and one on barnacles (Zullo 1991). One of Zullo's reports is from Isla del Coco, collected during Cruise 18 of the *Te Vega*; however, this is a mistake, since Ball (1972), in his station map of that cruise, did not indicate that Isla del Coco was visited. The Stanford Oceanographic Expedition 20 (September–December 1968) sampled two stations: one just off Isla del Coco (Z-78: 5°34′ N 86°58′ W, 22 October 1968) and one northwest of Cocos (G-36: 7°29′ N 87°58′ W, 27 October 1968), where cephalopods were collected (Fields & Gauley 1972). During that expedition Isla del Coco was visited on 22 October, but weather conditions were not good. Trawls were done with no success due to clumps of corals, which snagged the net. John S. Pearse collected *Diadema* by diving (Stanford University Archives), but no publication has been found in which any sample collected was mentioned. The first scientific exploration of Golfo Dulce was in 1969 Cruise 35 aboard the RV *T.G. Thompson*. Richards *et al.* (1971) described the physicochemical characteristics of this tropical fjord, while Nicholls-Driscoll (1976) worked samples of benthic fauna.

The Scripps Institution of Oceanography conducted expeditions in the eastern Pacific and did some collections in Costa Rica with the RV *Agassiz* in 1973 and with RV *Alpha Helix* in 1978. Wicksten (1979) reported a new record of a benthic bathyal shrimp off Punta Guiones during the 1973 cruise. All other records of specimens collected are in the catalogs of the Benthic Invertebrate Collections of the Scripps Institution of Oceanography (Luke 1977, 1995).

The Instituto de Ciencias del Mar y Limnología of the Universidad Nacional Autónoma de México collected plankton samples at the Costa Rica Dome with their research vessel *PUMA* twice in 1979, and once in 1981 and 1982 (Table II.1). The collected material formed the basis for many theses (Cortés in preparation) and the following papers: Sánchez-Nava & Segura-Puertas (1987) on pelagic gastropods; Suárez-Morales & Gasca (1989) on epipelagic calanoid copepods; Segura-Puertas (1991) on medusae; Gasca & Suárez (1992) on siphonophores, and finally Segura-Puertas *et al.* (1992) on chaetognaths.

Between 1983 and 1995, malacological collections were made at Isla del Coco by a group consisting of Kirstie L. Kayser, Michel Montoya, and Donald R. Shasky, aboard the Schooner *Victoria af Carlstat* (Montoya 2007). Many collection methods were used: hand-picking in the intertidal zone, snorkeling, SCUBA, tangle nets, and dredges. More than 50 papers were published, significantly increasing the number of mollusk species reported for the island, including a large number of new species (Cortés in preparation). A small sample of some of the papers is presented in Table II.2.

Table II.2 Malacological contributions to the knowledge of marine biodiversity of Isla del Coco, generated by collecting trips between 1982 and 1992. This list is by no means exhaustive

Topic	Reference
New records of Indo-Pacific species	Shasky (1983, 1987)
New species of muricid	D'Attilio *et al.* (1987)
List of chitons including two new species	Ferreira (1987)
Biogeography of the genus *Terebra*	Montoya & Kaiser (1988)
Described their collections at Isla del Coco	Shasky (1989)
Indicated the species found during 1992	Chaney (1992)
Reported on two Indo-Pacific gastropods	Kaiser (1998)
New records of mollusks for the island	Hertz & Kaiser (1998), Kaiser & Hertz (2001)

In 1986, the Harbor Branch Oceanographic Institution at Fort Pierce, Florida, used their submarine, the Johnson-Sea-Link I, to collect around the Galápagos Islands and Isla del Coco (Table II.1). Two papers were published, one on ahermatypic corals (Cairns 1991a) and another on stylasters (Cairns 1991b). In 1987, the fisheries RV *F. Nansen* visited the Golfo de Papagayo, and Bianchi (1991) reported collecting 14 invertebrates (two cephalopods, two stomatopods, and ten decapod crustaceans), but gave no details on the species. The Natural History Museum of Los Angeles County organized an expedition to Isla del Coco in 1988 aboard the Schooner *Victoria af Carlstat*. Two papers have information on organisms from the island collected during that expedition: the first on four species of stomatopod crustaceans, three of them are new reports for the island (Camp & Kuck 1990), and the second on brachyuran crabs (Zimmerman & Martin 1999).

In 1987 aboard the RV *Benjamin* and again in 1990 on the RV *Gyre*, scientists from the Smithsonian Tropical Research Institute visited Isla del Coco (Table II.1), and later published several papers on the distribution of echinoids in the eastern Pacific (e.g., Lessios *et al.* 1996, 1998, 1999, 2003). Macintyre *et al.* (1992) as part of the 1990 RV *Gyre* expedition took coral cores to reconstruct the Holocene history of the islands' coral reefs, at Isla del Caño and Isla del Coco.

In 1994, during a geophysical expedition of the submersible *Alvin* operated by the Woods Hole Oceanographic Institution (Table II.1) collected several organisms that are deposited at the National Museum of Natural History, Smithsonian Institution in Washington, DC. No report has been published regarding these specimens. Again in 2005, specimens were collected using the *Alvin* (Table II.1), but this time they were donated to the Universidad de Costa Rica (J.A. Vargas, 2005, personal communication). Scientists from the Smithsonian Tropical Research Institution and other organizations collected fishes at Isla del Coco aboard the RV *Urracá* in 1997 (D.R. Robertson, 2007, personal communication) (Table II.1). In 2001, the SSV *Robert Seamans* visited Isla del Coco and plankton samples were collected (Table II.1), but no paper has been published based on those samples.

All the major expeditions and the main collections have been made on the Pacific side of Costa Rica, with only four short expeditions to the Caribbean

(Table II.1). The first was the Austrian Biological Expedition in 1930, when a few freshwater crabs were collected (Pesta 1931). The second expedition was by the US Navy for a study of the zooplankton of the Caribbean Sea; they collected at one station off Limón in 1968 (Michel & Foyo 1976). The third expedition was the RV *John Elliott Pillsbury* Cruise P-7101 to Central America (20 January–5 February 1971) from the School of Marine and Atmospheric Science, University of Miami, with Gilbert L. Voss as Chief Scientist (Voss 1971). They visited 11 stations from off the Río Sixaola to offshore between Río Tortuguero and Río Colorado (26 and 27 January 1971). Trawls and plankton tows collected sponges, mollusks, crabs, crinoids, sea stars, holothurians, a few polychaetes, pennatulids, and nemerteans, and "typical surface plankton" (Voss 1971). Lastly, the RV *Alpha Helix* (Belém to Belize City Expedition) in 1977 generated one paper noting copepods collected (9 July) off the Port of Limon (Ferrari & Bowman 1980).

Between the late 1950s and the 1990s individual foreign scientists continued to contribute to the knowledge of Costa Rican marine biodiversity both in the Caribbean and the Pacific. These scientists were not connected to any national institution. A few examples are: Elmer Yale Dawson who was invited by a Mr. Maurice A. Marchis to visit Golfo de Nicoya and Golfo Dulce in June 1957, where he collected 108 species of algae (Dawson 1957). Dawson again visited Costa Rica in 1962, while attending the second annual Advanced Seminar in Tropical Biology at the University of Costa Rica. He published on the marine algae of the Caribbean coast, and expanded the list of algae for Costa Rica to 196 species (Dawson 1962). Banta & Carson (1977) described 24 species of bryozoans, from Portete and Playas del Coco, collected in 1964. Houbrick (1968) collected 250 species of mollusks in 1966 at Portete and Barra del Colorado, Caribbean. Charles Birkeland and Tom M. Spight (Smithsonian Tropical Research Institute) collected ascidians (= tunicates) from Playas del Coco in February–March 1970. With that collection Tokioka (1971, 1972) described a new species and listed 14 ascidians (= tunicates) as present in Costa Rica. Dexter (1974) published on the macroinfauna of beaches from the Caribbean and Pacific of Costa Rica collected in 1971. Child (1979) reported on sea spiders collected in Golfo de Nicoya and Playas del Coco by C.E. Dawson and C.A. Child in 1972. David G. Robinson worked on fossil formations from the Caribbean, but also collected and identified 395 living mollusks from the Caribbean coast (Robinson & Montoya 1987). Finally, Jung (1989) through the Panama Paleontology Project of the Smithsonian Tropical Research Institute published on strombinids collected in Costa Rica.

Third Phase: Early 1970s to the Present

From the early 1970s Costa Ricans, resident foreign scientists, and foreign scientists in association with local institutions have started to contribute significantly to the knowledge of the country's marine biodiversity. The return of Manuel M. Murillo and Carlos Villalobos to Costa Rica was important for the motivation and recruitment of young students into the marine sciences. Manuel Murillo (Fig. II.5) was also a key figure in the design and implementation of the research system at the Universidad de Costa Rica (UCR), through the creation of research centers and institutes and the Graduate School. In 1979, the Marine Sciences and Limnology Research Center (Centro de Investigación en Ciencias del Mar y Limnología; CIMAR, Fig. II.6) at UCR was founded. A significant contribution to the knowledge of Costa Rica's marine biodiversity during this phase has been made by William A. Bussing, a resident foreign scientist, who started working on the Central American fish fauna in the 1970s (Bussing 1980). He has continued describing marine and freshwater species to the present (Bussing & Lavenberg 2003), and now has at least 40 papers on fishes, including many new species (Tables II.3 and II.4).

The first expeditions in which Costa Rican or foreign resident scientists participated were conducted in 1972 and 1973. The first was the RV *Searcher* Central American Expedition in 1972, and the second was the *Velero IV* Costa Rica Expedition in 1973.

The Janss Foundation vessel, RV *Searcher*, visited Costa Rica in 1972 and the Chief Scientist was William A. Bussing (UCR) (Table II.1). Between 29 March and 8 April, the expedition visited Isla del Coco, and Bakus (1975) published the first report on the underwater zonation of its coral reefs. Manning & Reaka (1979) and Reaka & Manning (1980) published on stomatopod crustaceans from Bahía Herradura, Quepos, Isla del Caño, Puerto Jiménez, and Golfo Dulce, and Reid & Kaiser (2001) described a species of gastropod dredged from Bahía Herradura.

Fig. II.5 Dr. Manuel M. Murillo (on the left next to the tree, with yellow shirt) with students and teacher assistants on a field trip to Playa Conchal, Guanacaste in March 1984 (Photo: Jorge Cortés)

Fig. II.6 Building of the Centro de Investigación en Ciencias del Mar y Limnología (CIMAR) at the Ciudad de la Investigación (Research Campus) of the Universidad de Costa Rica 2002 (Photo: Jorge Cortés)

From 5 May to 1 July 1973, the RV *Velero IV* cruises 1244–1247 took place with Richard Pieper as cruise leader (J. Crampon and K. Fauchald, 2005, personal communication) (Table II.1), and they worked primarily off the Pacific coast of Costa Rica; some sampling was also done off Panama and Isla del Coco; M.M. Murillo and W.A. Bussing from the Universidad de Costa Rica participated in this cruise. The sampling devices included a box corer, beam trawl, hydrocast, Isaacs Kid Mid Water Trawl (IKMWT), acoustic trawl, Campbell benthic grab, biological dredge, and plankton tow-net. Sampling depths were from 18 to 3,750 m; the majority of the sampled were collected between 1,000 and 2,000 m with only a few from shallow waters. In Costa Rica, the following sites were sampled: north of Cabo Velas, Punta Guiones, Punta Coyote, Cabo Blanco, Negritos Adentro and Isla Aves/Golfo de Nicoya, Isla Herradura, Isla del Caño, Cabo Matapalo, Punta Arenitas/Golfo Dulce, Isla Burica, Isla del Coco, and off Isla Manuelita (J. Crampon, 2005, personal communication).

The creation of the CIMAR at the UCR in 1979 is a turning point in marine research in Costa Rica. The first large-scale project was with the University of Delaware, and a research vessel, the RV *Skimmer*, was operated in the Golfo de Nicoya from 1979 to 1981 (Vargas 1995) (Figs. II.7 and II.8). Based on the research carried out using that vessel, more than 100 papers have been published by CIMAR or collaborating scientists on biodiversity, processes, ecology, physical, and chemical characteristics, pollution, and ecosystems of the Golfo de Nicoya (*www.cimar.ucr.ac.cr*).

Table II.3 Contributions by CIMAR (Universidad de Costa Rica) scientists to the knowledge of marine biodiversity.

Taxonomic group	References
Fungi	Ulken *et al.* (1990)
Algae	Soto (1982, 1983), Soto & Ballantine (1986)
Ciliates	Camacho & Chinchilla (1989a, b)
Sponges	Risk *et al.* (1980, Cortés (1996)
Hydroids	Kelmo & Vargas (2002)
Calcified hydroids	Cortés (1992a)
Octocorals	Guzmán & Cortés (1985), Breedy & Guzman (2002, 2007)
Stony corals	Cortés & Guzmán (1985, 1998), Cortés (1992b)
Cnidarians	Cortés (1996/1997)
Mollusks	Høisæter (1998), Willis & Cortés (2001), Rodríguez-Sevilla *et al.* (2003), Ajtai *et al.* (2003)
Sipunculans	Cutler *et al.* (1992), Dean (2001c)
Polychaetes	Maurer *et al.* (1988), Dean (1996a, b, 1998a, 2001a, b, 2004)
Echiurans	Dean & Cutler (1998), Dean (2001c)
Copepods	Mielke (1995, 1997), Morales-Ramírez & Vargas (1995), Morales-Ramírez (1996, 2001), Suárez-Morales & Morales-Ramírez (2001)
Isopods	Brusca & Iverson (1985), Breedy & Murillo (1995)
Stomatopods	Dittel (1991), Vargas *et al.* (1996), Vargas & Cortés (1997)
Decapods	Dittel & Epifanio (1984), Moran & Dittel (1993), Vargas *et al.* (1996, Vargas & Cortés (1999a, b, 2006), Wehrtmann & Vargas (2003)
Brachiopods	Emig & Vargas (1990)
Chaetognaths	Hossfeld (1966)
Echinoderms	Alvarado (2004), Alvarado & Cortés (2004), Alvarado & Fernández (2005), Bolaños *et al.* (2005), Hendler (2005), Alvarado (in press)
Fishes	López & Bussing (1982, 2005), Bartels *et al.* (1983), Bussing (1985), Bussing & López (1993, 1996), Dominici-Arosemena *et al.* (2005)
Marine mammals	Cubero-Pardo (1988), May-Collado & Morales-Ramírez (2005), May-Collado *et al.* (2005)
Phytoplankton	Brugnoli-Olivera & Morales-Ramírez (2001)
Zooplancton	Morales & Murillo (1996), Brugnoli-Olivera *et al.* (2004), Morales-Ramírez & Nowaczyk (2006)
Ictioplankton	López (1983), Ramírez *et al.* (1990), Molina-Ureña (1996), Dominici-Arosemena *et al.* (2000)
Benthic invertebrates	Maurer *et al.* (1984), Maurer & Vargas (1984), Vargas *et al.* (1985), Vargas (1987), León-Morales & Vargas (1998)
Meiofauna	De la Cruz & Vargas (1987), Guzmán *et al.* (1987), Vargas (1988)

Between 1993 and 1994, an expedition was organized between CIMAR and the Center for Tropical Marine Ecology (ZMT), University of Bremen, Germany (Vargas & Wolff 1996). The RV *Victor Hensen* (Fig. II.9) collected samples along the Pacific coast, mainly from Golfo Dulce, Bahía de Coronado, and Golfo de Nicoya. Many papers were published from that expedition and some were directly related to marine biodiversity (e.g., Bussing & López 1996; Morales-Ramírez 1996; Vargas *et al.* 1996; León-Morales & Vargas 1998). Scientist from the Smithsonian Tropical Research

Table II.4 New species described by CIMAR or associated scientists. In parenthesis the number of new species described.

Taxonomic group	References
Octocorals (11)	Breedy (2001), Breedy & Guzman (2003, 2005)
Cestoda (10)	Berman & Brooks (1994), Marques *et al.* (1995, 1996), McCorquodale & Brooks (1995)
Nematodes (1)	Hoberg *et al.* (1998)
Polychaetes (11)	Dean (1998a, b, 2001a), Dean & Blake (2007)
Copepods (6)	Mielke (1992, 1994, 1995, 1997), Suárez-Morales & Morales-Ramírez (2003)
Isopods (2)	Brusca & Iverson (1985)
Cumaceans (1)	Watling & Breedy (1988)
Decapods (3)	Vargas (2000), Williams & Vargas (2000), Wicksten & Vargas (2001)
Fishes (22)	Bussing (1980, 1981, 1983, 1990, 1991a, b, 1997, 1998, 2001), López (1980), Szelistowski (1990), López & Bussing (1998), Bussing & Lavenberg (2003)
Total: 67 new species	

Fig. II.7 Collecting benthic samples with a Smith-McIntyre grab (from left to right: Tere Quesada, Gustavo Fernández, and Sonia Hernández) aboard the RV *Skimmer* in the Golfo de Nicoya in 1980 (Photo: José Antonio Vargas)

Fig. II.8 Separating fishes (William A. Bussing, left and Myrna López, right) collected aboard the RV *Skimmer* in the Golfo de Nicoya in 1980 (Photo: José Antonio Vargas)

Fig. II.9 The FS *Victor Hensen* sampling in Costa Rica (1993–1994) (Photo: José Antonio Vargas)

Institution together with scientists from CIMAR aboard the RV *Urracá* collected fishes between 30 June and 13 July 2005 between the northern and central Pacific coast (D.R. Robertson, 2007, personal communication), and 14 and 18 July, benthic samples were collected in the same area (R. Vargas, 2007, personal communication).

In late 2006 and again in early 2007, Isla del Coco was visited by scientist from CIMAR aboard the MV *Proteus* (Fig. II.10, Table II.1); the objective of the first expedition was the evaluation and establishment of monitoring sites, but organisms were also collected. The second expedition's objective was the collection of organisms from the intertidal down to 40 m depth. New species are being discovered of even conspicuous groups such as octocorals (probably one new species collected in January 2007; Breedy & Cortés 2008); and several new records of species were reported (e.g., echinoderms) (Alvarado in press).

Most of the papers published on marine biodiversity since the 1980s have been done through three national institutions: Universidad de Costa Rica (UCR), Universidad Nacional (UNA), and Instituto Nacional de Biodiversidad (INBio). The taxonomic groups studied are mentioned below, but only publications related to biodiversity are mentioned. There is an extensive bibliography on research on fishes, marine invertebrates, and ecology (Cortés in preparation).

The UCR has contributed to the knowledge of many groups that has amplified our knowledge of marine biodiversity; over 100 papers have been published on the marine biodiversity of Costa Rica. The largest contribution has been on fish by W.A. Bussing and M.I. López (Tables II.3 and II.4). A list of papers published by CIMAR scientists or their associates is presented in Tables 3 and 4, in which a total of 67 new species were described. Other contributions by UCR scientists from the Museo de Zoología have contributed to the knowledge of marine biodiversity of Costa Rica (e.g., Camacho-García & Gosliner 2004, 2006; Camacho-García *et al.* 2005).

Fig. II.10 The MV *Proteus* of the non-governmental organization "MarViva" has been used to collect material around Isla del Coco starting in 2006 (Photo: Jorge Cortés)

Researchers of the UNA have contributed to the study of marine biodiversity mainly on mollusks through the contributions of Rafael A. Cruz (e.g., Cruz 1985, 1996; Cruz & Jiménez 1994). One paper has been published on the taxonomy of sponges (Loaiza 1991).

The INBio has published on mollusks (The Malacology Department was closed in September 2004 and the collections transferred to the Museo de Zoología, Escuela de Biología, UCR). In collaboration with Cuban and Spanish malacologists about 20 papers were published, but just a few are mentioned here. Papers on the mollusks of the Caribbean coast of Costa Rica include the description of new species published by Espinosa & Ortea (1999, 2000, 2001) and Ortea *et al.* (2001). Another series of papers is on nudibranchs from the Caribbean (Ortea *et al.* 1999; Ortea 2001) and from the Pacific (Camacho-García & Ortea 2000; Camacho-García & Valdés 2003).

Conclusions

Three phases are recognized as productive in the generation of marine biodiversity information for Costa Rica: the late 19th century, between 1930 and 1940, and after 1980. The knowledge of marine biodiversity increased significantly during those phases, but there is a difference between research conducted by foreign scientists with no connection to local institutions, and research done by local and resident scientists or by foreign scientists in true collaboration with local institutions. In the first case, a large amount of new knowledge was generated, but little of that came back to the country, and no collections were left in the country or returned afterward. This was normal practice at the time, but also, it could have been due to the lack of national scientific institutions interested in marine organisms or to the absence of local scientists working on marine biodiversity. Until the early 1970s no Costa Rican was trained in the study of marine organisms.

In the second case, Costa Rica scientists had been trained and collections were deposited in Costa Rica. In the early 1970s, the Universidad de Costa Rica organized its research activities. Professors went abroad for advanced degrees, the Museo de Zoología at the Escuela de Biología was set up (1966), and research centers were created. During that phase the first research on marine organisms started. In 1976, the new building of the Museo de Zoología was completed, creating an area and environment conducive for research. In 1979, the CIMAR was founded, significantly advancing marine sciences in Costa Rica, including biodiversity studies. At present, most foreign scientists work in collaboration with Costa Rican scientists, resulting in a direct transfer of information to Costa Rica, training of students and researchers, and enlargement of research collections in Costa Rica. It is important to have a strong local scientific community and institutions for the reception of information and collections produced by foreign scientists, for the generation of new knowledge, and for the use of that experience and information in teaching, advancing research, and for the management and rational use of the marine resources of the country.

Acknowledgments This paper is the product of many years of accumulating information on the marine biodiversity of Costa Rica. Financial support through several projects was received from the Vicerrectoría de Investigación, Universidad de Costa Rica. The libraries of the Smithsonian Institution and the Smithsonian Tropical Research Institute were visited with Short-Term Fellowships from that Institution and in 2005 with a Fulbright Visiting Scholar Fellowship. Most of the old literature was obtained through Kay Hale and Helen Albertson previously at the Rosenstiel School of Marine and Atmospheric Science (RSMAS) Library, University of Miami; Harlan Dean, Museum of Comparative Zoology, Harvard University, and Frederick M. Bayer, Smithsonian Institution also helped to obtain some old publications. Deborah Day, Archivist of the Scripps Institution of Oceanography and Patricia E. White, Department of Special Collections, Stanford University Libraries, helped with information on expeditions. Michel Montoya helped with literature on Isla del Coco. The librarians at the Natural History Library, Smithsonian Institution were very helpful in locating old and obscure publications. I thank L.D. Gómez, R. Petersen, R. Viales, D.S. Painter, A.G. Coates, F.M. Bayer, S.D. Cairns, and J.A. Vargas for reviewing the manuscript.

References

Ajtai P, Camacho-García Y, Wehrtmann IS (2003) Range extension of *Crosslandia daedali* (Nudibranchia: Scyllaeidae). Rev Biol Trop 51:274

Alfaro A (1899). Flora y fauna de la Isla del Coco. Memoria Secretaría de Fomento, San José, Costa Rica, pp 31–36

Alvarado JJ (2004) Abundancia del erizo de mar *Centrostephanus coronatus* (Echinoidea: Diadematidae) en el Pacífico de Costa Rica. Rev Biol Trop 52:911–913

Alvarado JJ (2008) Isla del Coco (Costa Rica) Echinoderms: state of knowledge. Proceedings of the 12th International Echinoderm Conference, Durham, New Hampshire (in press)

Alvarado JJ, Cortés J (2004). The state of knowledge on echinoderms of Costa Rica and Central America. In: T Heinzeller, JH Nebelsick (eds) Echinoderms: München. Balkema, Leiden, pp 149–155

Alvarado JJ, Fernández C (2005) Equinodermos del Parque Nacional Marino Ballena, Pacífico, Costa Rica. Rev Biol Trop 53(Suppl 3):275–284

Bakus GJ (1975) Marine zonation and ecology of Cocos Island, off Central America. Atoll Res Bull 179:1–11

Ball EE (1972) Observations on the biology of the hermit crab, *Coenobita compresus* H. Milne Edwards (Decapoda, Anomura) on the west coast of the Americas. Rev Biol Trop 20:265–273

Ball EE, Haig J (1974) Hermit crabs from the tropical eastern Pacific. I. Distribution, color and natural history of some common shallow-water species. Bull South Calif Acad Sci 73:95–104

Banta WC, Carson RJM (1977) Bryozoa from Costa Rica. Pacific Sci 31:381–424

Barnard JL (1980) Revision of *Metharpinia* and *Microphoxus* (marine Phoxocephalid Amphipoda from the Americas). Proc Biol Soc Wash 93:104–135

Bartels C, Price K, López M, Bussing WA (1983) Occurrence, distribution, abundance and diversity of fishes in the Gulf of Nicoya, Costa Rica. Rev Biol Trop 31:75–101

Beebe W (1926) The Arcturus Oceanographic Expedition. Zoologica 8:1–45

Beebe W (1938) Eastern Pacific expeditions of the New York Zoological Society, XIV. Introduction, itinerary, list of stations, nets and dredges of the eastern Pacific *Zaca* expedition, 1937–1938. Zoologica 23:287–298

Berman R, Brooks DR (1994) *Escherbothrium molinae* n.gen. et n.sp.(Eucestoda: Tetraphyllidea: Triloculariidae) in *Urotrygon chilensis* (Chondrichthyes: Myliobatiformes: Urolophidae) from the Gulf of Nicoya, Costa Rica. J Parasitol 80:775–780

Bianchi G (1991) Demersal assemblages of the continental shelf and slope edge between the Gulf of Tehuantepec (Mexico) and the Gulf of Papagayo (Costa Rica). Mar Ecol Prog Ser 73:121–140

Bigelow HB (1909) Reports of the scientific results of the Expedition to the Eastern Tropical Pacific, in charge of Alexander Agassiz, by the U.S. Fish Commission Steamer "Albatross," from October, 1904, to March, 1905, Lieut. Commander L.M. Garrett, U.S.N., commanding. XVI. The Medusae. Mem Mus Comp Zoöl 37:1–243

Bigelow HB (1911) Reports of the scientific results of the Expedition to the Eastern Tropical Pacific, in charge of Alexander Agassiz, by the U.S. Fish Commission Steamer "Albatross," from October, 1904, to March, 1905, Lieut. Commander L.M. Garrett, U.S.N., commanding. XXIII. The Siphonophorae. Mem Mus Comp Zoöl 38:369–408

Bigelow HB (1940) Eastern Pacific Expeditions of the New York Zoological Society. XX. Medusae of the Templeton Crocker and Eastern Pacific *Zaca* Expeditions, 1936–1938. Zoologica XXV:281–321

Biolley P (1907) Mollusques de l'Isla del Coco. Museo Nacional de Costa Rica, Tipografía Nacional, San José, Costa Rica, 30 p (published in Spanish: Biolley P (1935). Moluscos de la Isla del Coco. Revista del Colegio de Señoritas (San José), 2(6):2–18)

Bolaños N, Bourg A, Gómez J, Alvarado JJ (2005) Diversidad y abundancia de equinodermos en la laguna arrecifal del Parque Nacional Cahuita, Caribe de Costa Rica. Rev Biol Trop 53(Suppl 3):285–290

Boone L (1930a) Scientific results of the cruises of the yachts "Eagle" and "Ara," 1921–1928, William K. Vanderbilt, commanding. Crustacea: Stomatopoda and Brachyura. Bull Vanderbilt Mar Mus II:1–228

Boone L (1930b) Scientific results of the cruises of the yachts "Eagle" and "Ara," 1921–1928, William K. Vanderbilt, Commanding. Crustacea: Anomura, Macrura, Schizopoda, Isopoda, Amphipoda, Mysidacea, Cirripedia, and Copepoda. Bull. Vanderbilt Mar Mus III:1–221

Boone L (1933) Scientific results of cruises of the yachts "Eagle" and "Ara," 1921–1928, William K. Vanderbilt, commanding. Coelenterata, Echinodermata and Mollusca. Bull Vanderbilt Mar Mus IV:1–217

Breedy O (2001) A new species of *Pacifigorgia* from the eastern Pacific (Coelenterata: Octocorallia: Gorgoniidae). Bull Biol Soc Wash 10:181–187

Breedy O, Cortés J (2008) Octocorals (Coelenterata: Anthozoa: Octocorallia) of Isla del Coco, Costa Rica. Rev Biol Trop 56(Suppl 2):71–77

Breedy O, Guzman HM (2002) A revision of the genus *Pacifigorgia* (Coelenterata: Octocorallia: Gorgoniidae). Proc Biol Soc Wash 115:782–839

Breedy O, Guzman HM (2003) Octocorals from Costa Rica: The genus *Pacifigorgia* (Coelenterata: Octocorallia: Gorgoniidae). Zootaxa 281:1–60

Breedy O, Guzman HM (2005) A new species of *Leptogorgia* (Coelenterata: Octocorallia: Gorgoniidae) from the shallow waters of the eastern Pacific. Zootaxa 899:1–11

Breedy O, Guzman HM (2007) A revision of the genus *Leptogorgia* Milne Edwards & Haime, 1857 (Coelenterata: Octocorallia: Gorgoniidae) in the eastern Pacific. Zootaxa 1407:1–90

Breedy O, Murillo MM (1995) Isópodos (Crustacea: Peracarida) de un arrecife del Caribe de Costa Rica. Rev Biol Trop 43:219–229

Brugnoli-Olivera E, Morales-Ramírez A (2001) La comunidad fitoplánctica de Punta Morales, Golfo de Nicoya, Costa Rica. Rev Biol Trop 49(Suppl 2):11–17

Brugnoli-Olivera E, Díaz-Ferguson E, Delfino-Machin M, Morales-Ramírez A, Dominici-Arosemena A (2004) Composition of the zooplankton community, with emphasis in copepods, in Punta Morales, Golfo de Nicoya, Costa Rica. Rev Biol Trop 52:897–902

Brusca RC, Iverson EW (1985) A guide to the marine isopod crustacean of Pacific Costa Rica. Rev Biol Trop 33(Suppl 1):1–77

Bussing WA (1980) *Liopropoma fasciatum*, a new serranid fish and only known member of the genus from the tropical eastern Pacific Ocean. Rev Biol Trop 28:147–151

Bussing WA (1981) *Elacatinus janssi*, a new gobiid fish from Costa Rica. Rev Biol Trop 29:251–256

Bussing WA (1983) A new tropical eastern Pacific labrid fish, *Halichoeres discolor* endemic to Isla del Coco. Rev Biol Trop 31:19–23

Bussing WA (1985) Los peces de la Familia Labridae de la costa Pacífica de Costa Rica. Rev Biol Trop 33:81–98
Bussing WA (1990) New species of gobiid fishes of the genera *Lythrypnus, Elacatinus* and *Chriolepis*. Rev Biol Trop 38:99–118
Bussing WA (1991a) A new genus and two new species of tripterygiid fishes from Costa Rica. Rev Biol Trop 39:77–85
Bussing WA (1991b) A new species of eastern Pacific moray eel (Pisces: Muraenidae). Rev Biol Trop 39:97–102
Bussing WA (1997) *Chriolepis atrimelum* (Gobiidae) a new species of gobiid fish from Isla del Coco, Costa Rica. Rev Biol Trop 45:1547–1552
Bussing WA (1998) *Gymnothorax phalarus*, a new eastern Pacific moray eel (Pisces: Muranidae). Rev Biol Trop 46:439–446
Bussing WA (2001) *Ptereleotris carinata*, a new species of hovering goby (Perciformes: Microdesmidae) from the tropical eastern Pacific. Rev Biol Trop 49(Suppl 1):133–139
Bussing WA, Lavenberg RJ (2003) Four new species of eastern tropical Pacific jawfishes (*Opistongathus*: Opistognathidae). Rev Biol Trop 51:529–550
Bussing WA, López MI (1993) Peces demersales y pelágicos costeros del Pacífico de Centro América Meridional. Guía Ilustrada. Rev Biol Trop (Spec. Publ.): 164 p
Bussing WA, López MI (1996) Fishes collected during the RV Victor Hensen Expedition (1993/1994). Rev Biol Trop 44(Suppl 3):183–186
Bussing WA, López MI (2005) Peces de la Isla del Coco y peces arrecifales de la costa Pacífica de América Central meridional/Fishes of Cocos Island and Reef Fishes of the Pacific Coast of Lower Central America. Rev Biol Trop 53(Suppl 2):192 p
Cairns SD (1991a) A revision of the ahermatypic Scleractinia of the Galápagos and Cocos Islands. Smithson Contr Zool 504:1–33
Cairns SD (1991b) New records of Stylasteridae (Hydrozoa: Hydroida) from the Galápagos and Cocos Islands. Proc Biol Soc Wash 104:209–228
Camacho L, Chinchilla M (1989a) Clave taxonómica de ciliados epibiontes de Decapoda (Natantia). Rev Biol Trop 37:15–22
Camacho L, Chinchilla M (1989b) Ciliados epibiontes en *Macrobrachium rosenbergii* (De Man) cultivados en Limón, Costa Rica Rev. Biol Trop 37:105–106
Camacho-García Y, Ortea J (2000) A new species of *Trapania* (Nudibranchia: Goniodorididae) from the Pacific coast of Central America. Rev Biol Trop 48:317–322
Camacho-García YE, Gosliner TM (2004) A new species of *Okenia* (Gastropoda: Nudibranchia) from the Pacific coast of Costa Rica. Proc Calif Acad Sci 55:431–438
Camacho-García YE, Gosliner TM (2006) A new species of the zephyrinid nudibranch genus *Janolus* (Mollusca: Nudibranchia) from North America and Costa Rica. Rev Biol Trop 54:1295–1305
Camacho-García YE, Valdés Á (2003) Caryophyllidia-bearing dorid nudibranchs (Mollusca, Nudibranchia, Doridacea) from Costa Rica. Proc Calif Acad Sci 54:65–79
Camacho-García YE, Gosliner TM, Valdés Á (2005) Guía de campo de las babosas marinas del Pacífico este tropical/Field Guid to the Sea Slugs of the Tropical Eastern Pacific. Calif Acad Sci, San Francisco, California, 129 p
Camp DK, Kuck HG (1990) Additional records of stomatopods crustaceans from Isla del Coco and Golfo de Papagayo, east Pacific Ocean. Proc Biol Soc Wash 103:847–853
Castro P (1996) Eastern Pacific species of *Trapezia* (Crustacea, Brachyura: Trapeziidae), sibling species symbiotic with reef corals. Bull Mar Sci 58:531–554
Chaney HW (1992) Interesting gastropods from the 1992 Cocos Island Expedition. Festivus 24:86–89
Child CA (1979) Shallow-water Pycnogonida of the Isthmus of Panama and the coast of Middle America. Smithson Contr Zool 293:1–86
Child CA (1992) Pycnogonida of the Southeast Pacific Biological Oceanographic Project (SEPBOP). Smithson Contr Zool 526:43 p
Corfield R (2003) The silent landscape: the scientific voyage of the HMS *challenger*. Joseph Henry, Washington, DC, 285 p

Cortés J (1992a) Organismos de los arrecifes coralinos de Costa Rica. V. Descripción y distribución geográfica de Hydrocorales (Cnidaria: Hydrozoa: Milleporina & Stylasterine) de la costa Caribe. Brenesia 38:45–50
Cortés J (1992b) Nuevos registros de corales (Anthozoa: Scleractinia) para el Caribe de Costa Rica: *Rhizosmilia maculata* y *Meandrina meandrites*. Rev Biol Trop 40:243–244
Cortés J (1996) Biodiversidad marina de Costa Rica: Filo Porifera. Rev Biol Trop 44:911–914
Cortés J (1996/1997) Biodiversidad marina de Costa Rica: Filo Cnidaria. Rev Biol Trop 44(3)/45(1):323–334
Cortés J (2008) Historia de la investigación marina de la Isla del Coco, Costa Rica. Rev Biol Trop 56(Suppl 2):1–18
Cortés J, Guzmán HM (1985) Organismos de los arrecifes coralinos de Costa Rica. III. Descripción y distribución geográfica de los corales escleractinios (Anthozoa: Scleractinia) de la Costa Caribe. Brenesia 24:63–123
Cortés J, Guzmán HM (1998) Organismos de los arrecifes coralinos de Costa Rica: VI. Descripción, distribución geográfica e historia natural de los corales zooxantelados (Anthozoa: Scleractinia) del Pacífico. Rev Biol Trop 46:55–92
Crocker T (1933) The Templeton Crocker Expedition of the California Academy of Sciences, 1932, No. 2: introductory statement. Proc Calif Acad Sci 4th Ser 21:3–9
Cruz RA (1985) Revisión taxonómica sobre la familia Arcidae. Uniciencia 2:71–76
Cruz RA (1996) Costa Rican marine gastropods: range extension. Rev Biol Trop 44(Suppl 3):237–238
Cruz RA, Jiménez JA (1994) Moluscos asociados a las áreas de manglar de la costa Pacífica de América Central: Guía. Editorial Fundación UNA, Heredia, 182 p
Cubero-Pardo P (1998) Patrones de comportamiento diurnos y estacionales de *Tursiops truncatus* y *Stenella attenuata* (Mammalia: Delphinidae) en el Golfo Dulce, Costa Rica. Rev Biol Trop 46(Suppl 6):103–110
Cutler N, Cutler E, Vargas JA (1992) Peanut worms (Phylum Sipuncula) from Costa Rica. Rev Biol Trop 40:153–158
Dall WH (1900) Additions to the insular land-shell faunas of the Pacific Coast, especially of the Galápagos and Cocos islands. Proc Acad Nat Sci Phil 52:88–106
Dall WH (1908) Reports on the dredging operations off the west coast of Central America to the Galápagos, to the west coast of Mexico, and in the Gulf of California, in charge of Alexander Agassiz, by the U.S. Fish Commission steamer Albatross, during 1891, Lieut. Commander Z.L. Tanner U.S.N., commanding. Reports on the Mollusca and Brachiopoda. Bull Mus Comp Zool 43:205–487
Dall WH (1920) Annotated list of the recent Brachiopoda in the collection of the United States National Museum, with descriptions of thirty-three new forms. Proc US Natl Mus 57(2314):261–377
D'Attilio A, Myers BW, Shasky DR (1987) A new species of *Phyllonotus* (Muricidae: Muricinae) from Isla del Coco, Costa Rica. Nautilus 101:62–165
Dawson EY (1957) Marine algae from the Pacific Costa Rican gulfs. Los Angeles County Mus Contr Sci 15:1–28
Dawson EY (1960) New records of marine algae from Pacific Mexico and Central America. Pac Nat 1(20):31–52
Dawson EY (1962) Additions to the marine flora of Costa Rica and Nicaragua. Pac Nat 3:375–395
Dawson EY, Beaudette PT (1959) Field notes from the 1959 eastern Pacific cruise of the Stella Polaris. Pac Nat 1(13):1–24
De la Cruz E, Vargas JA (1987) Abundancia y distribución vertical de la meiofauna en la playa fangosa de Punta Morales, Golfo de Nicoya, Costa Rica. Rev Biol Trop 35:363–367
Dean HK (1996a) Subtidal benthic polychaetes (Annelida) of the Gulf of Nicoya, Costa Rica. Rev Biol Trop 44(Suppl 3):69–80
Dean HK (1996b) Polychaete worms (Annelida) collected in Golfo Dulce, during the Victor Hensen, Costa Rica expedition (1993/1994). Rev Biol Trop 44(Suppl 3):81–86

Dean HK (1998a) The Pilargidae (Annelida: Polychaeta) of the Pacific coast of Costa Rica. Rev Biol Trop 46(Suppl 6):47–62

Dean HK (1998b) A new species of Hesionidae, *Glyphohesione nicoyensis* (Annelida: Polychaeta), from the Gulf of Nicoya, Costa Rica. Proc Biol Soc Wash 111:257–262

Dean HK (2001a) Some Nereididae (Annelida: Polychaeta) from the Pacific Coast of Costa Rica. Rev Biol Trop 49(Suppl 2):37–67

Dean HK (2001b) Capitellidae (Annelida: Polychaeta) from the Pacific coast of Costa Rica. Rev Biol Trop 49(Suppl 2):69–84

Dean HK (2001c) Marine biodiversity of Costa Rica: the phyla Sipuncula and Echiura. Rev Biol Trop 49(Suppl 2):85–90

Dean HK (2004) Marine biodiversity of Costa Rica: Class Polychaeta (Annelida). Rev Biol Trop 52(Suppl. 2):131–181

Dean HK, Cutler EB (1998) Range extension of *Nephasoma pellucidum*, and new records of *Apionsoma (Edmonsius) pectinatum* (Sipuncula) and *Thalassema steinbecki* (Echiura) from the Pacific of Costa Rica. Rev Biol Trop 46(Suppl 6):279–280

Dean HK, Blake JA (2007) *Chaetozone* and *Caulleriella* (Polychaeta: Cirratulidae) from the Pacific coast of Costa Rica, with description of eight new species. Zootaxa 1451:41–68

Dexter DM (1974) Sandy-beach fauna of the Pacific and Atlantic coasts of Costa Rica and Colombia. Rev Biol Trop 22:51–66

Dittel A, Epifanio CE (1984) Desarrollo larval de *Ozius verreauxii* Saussure (Brachyura: Xanthidae) en el laboratorio. Rev Biol Trop 32:171–172

Dittel AI (1991) Distribution, abundance and sexual composition of stomatopod Crustacea in the Gulf of Nicoya, Costa Rica. J Crust Biol 11:269–276

Dominici-Arosemena A, Brugnoli-Olivera E, Solano-Ulate S, Molina-Ureña H, Ramírez-Coghi AR (2000) Ictioplancton en la zona portuaria de Limón, Costa Rica. Rev Biol Trop 48:439–442

Dominici-Arosemena A, Brugnoli-Olivera E, Cortés-Núñez J, Molina-Ureña H, Quesada-Alpízar M (2005) Community structure of eastern Pacific reef fishes (Gulf of Papagayo, Costa Rica). Tecnociencia 7:19–41

Durham JW (1962) Corals from the Galapagos and Cocos Islands. Proc Calif Acad Sci 4th Ser 32:41–56

Eakin MC (1999) The origins of modern science in Costa Rica: The Instituto Físico-Geográfico Nacional, 1887–1904. Latin Am Res Rev 34:123–150

Emig C, Vargas JA (1990) *Glottidia audebarti* (Broderip) (Brachiopoda, Lingulidae) from the Gulf of Nicoya, Costa Rica. Rev Biol Trop 38:251–258

Espinosa J, Ortea J (1999) Descripción de nuevas Marginelas (Mollusca: Neogastropoda: Marginellidae) de Cuba y del Caribe de Costa Rica y Panamá. Avicennia 10/11:165–176

Espinosa J, Ortea J (2000) Descripción de un género y once especies nuevas de Cystiscidae y Marginellidae (Mollusca: Neogastropoda) del Caribe de Costa Rica. Avicennia 12/13:95–114

Espinosa J, Ortea J (2001) Moluscos del mar Caribe de Costa Rica: desde Cahuita hasta Gandoca. Avicennia (Suppl 4):1–76

Faxon W (1895) Reports on an exploration off the west coasts of Mexico, Central and South America, and off the Galapagos Islands, in charge of Alexander Agassiz, by the U.S. Fish Commission steamer Albatross, during 1891, Lieut. Commander Z.L. Tanner U.S.N., commanding. XV. The stalk-eyed Crustacea. Mem Mus Comp Zool Harvard Coll 18:1–292

Ferreira AJ (1987) The chiton fauna of Cocos Island, Costa Rica (Mollusca: Polyplacophora) with the description of two new species. Bull Sci Calif Acad Sci 86:41–53

Ferrari F, Bowman TEB (1980) Pelagic copepods of the Family Oithonidae (Cyclopoida) from the east coast of Central and South America Smithson. Contr Zool 312:27 p

Fields WG, Gauley VA (1972) A report on the cephalopods collected by Stanford Oceanographic Expedition 20 to the eastern tropical Pacific Ocean, September to November, 1968. Veliger 15:113–118

Fowler HW (1932) The fishes obtained by the Pinchot South Seas Expedition of 1929, with description of one new genus and three new species. Proc US Natl Mus 80(6):1–16

Fraser CM (1943) General account of the scientific work of the *Velero III* in the eastern Pacific, 1931–1941, Part III: a ten-year list of the *Velero III* collecting stations (Charts 1–115). With an appendix of collecting stations of the Allan Hancock Foundation for the year 1942. Allan Hancock Pac Exp 1(3):259–431

Gasca R, Suárez E (1992) Sifonóforos (Cnidaria: Siphonophora) del Domo de Costa Rica. Rev Biol Trop 40:125–130

Grube E (1859) Annulata Oerstediana. Enumeratio Annulatorum quae in itinere der Indiam Occidentalem et American Centralis annis 1845–1848. Vidensk Medd Dansk naturh Foren Part 3(1859):105–120

Guzmán HM, Cortés J (1985) Organismos de los arrecifes coralinos de Costa Rica. IV: Descripción y distribución geográfica de octocoralarios (Anthozoa: Octocorallia) de la costa Caribe. Brenesia 24:125–174

Guzmán HM, Obando V, Cortés J (1987) Meiofauna associated with a Pacific coral reef in Costa Rica. Coral Reefs 6:107–112

Halstead, BW, Schall DW (1955) A report on poisonous fishes captured during the Wooldrow G. Krieger Expedition to Cocos Island. Pac Sci 10:103–109.

Heller E (1903) Papers from the Hopkins Stanford Galapagos Expedition, 1898–1899. XIV. Reptiles. Proc Wash Acad Sci 5:39–98

Hendler G (2005) Two new brittle star species of the genus *Ophiothrix* (Echinodermata: Ophiuroidea: Ophiotrichidae) from coral reefs in the southern Caribbean Sea, with notes on their biology. Carib J Sci 41:583–599

Hertlein LG (1935) The Templeton Crocker Expedition of the California Academy of Sciences, 1932. No. 25. The Recent Pectinidae. Proc Calif Acad Sci 4th Ser 21(25):301–328

Hertz J, Kaiser KL (1998) *Colubraria (Colubraria) ochsneri* and *Colubraria (C.) lucasensis* from Isla del Coco, Costa Rica. Festivus 30:35–39

Hoberg EP, Brooks DR, Molina-Ureña H, Erbe E (1998) *Echinocephalus janzeni* n.sp. (Nematoda: Gnathostomatidae) in *Himantura pacifica* (Chondrichthyes: Myliobatiformes) from the Pacific coast of Costa Rica and Mexico, with historical biogeographic analysis of the genus. J Parasitol 84:571–581

Høisæter T (1998) Preliminary check-list of the marine, shelled gastropods (Mollusca) of Golfo Dulce, on the Pacific coast of Costa Rica. Rev Biol Trop 46(Suppl 6):263–270

Hossfeld B (1996) Distribution and biomass of arrow worms (Chaetognatha) in Golfo de Nicoya and Golfo Dulce, Costa Rica. Rev Biol Trop 44(Suppl 3):157–172

Houbrick JR (1968) A survey of the littoral marine molluscs of the Caribbean coast of Costa Rica. Veliger 11:4–23

Jung P (1989) Revision of the Strombina-group (Gastropoda: Columbellidae), fossil and living: distribution, biostratigraphy and systematics. Schweiz Paläont Abhandl 111:1–298

Kaiser KL (1998) *Cypraea talpa* on eggs and a live *Cymatium (Gutturnium) muricinum*, both at Isla del Coco, Costa Rica. Festivus 30:28–29

Kaiser KL, Hertz CM (2001) Notes on *Rhizochilus antipathum* Steenstrup, 1850 (Gastropoda: Coralliophilidae) with new records from Île Clipperton and Isla del Coco (tropical eastern Pacific). Festivus 33:87–94

Kelmo F, Vargas R (2002) Anthoathecatae and Leptothecatae hydroids from Costa Rica (Cnidaria: Hydrozoa). Rev Biol Trop 50:599–627

Kramp PL (1956) Medusae collected in the eastern tropical Pacific by Cyril Crossland in 1924–1925. Vidensk Medd Dansk naturh Foren 118:1–6

León-Morales R, Vargas JA (1998) Macroinfauna of a tropical fjord-like embayment, Golfo Dulce, Costa Rica. Rev Biol Trop 46(Suppl 6):81–90

Lessios HA, Kessing BD, Wellington GM, Graybeal A (1996) Indo-Pacific echinoids in the tropical eastern Pacific. Coral Reefs 15:133–142

Lessios HA, Kessing BD, Robertson DR (1998) Massive gene flow across the world's most potent marine biogeographic barrier. Proc R Soc London, Ser B 265:583–588

Lessios HA, Kessing BD, Robertson DR, Pauly G (1999) Phylogeographic of the pantropical sea urchin *Eucidaris* in relation to land barriers and ocean currents. Evolution 53:806–817

Lessios HA, Kane J, Robertson DR (2003) Phylogeography of the pantropical sea urchin *Tripneustes*: contrasting patterns of population structure between oceans. Evolution 57:2026–2036

Loaiza B (1991) Estudio taxonómico de las esponjas del Parque Nacional Cahuita, sector Puerto Vargas e Isla Uvita, Limón, Costa Rica. Brenesia 36:21–62

López M (1980) *Umbrina bussingi*, a new sciaenid fish from the tropical eastern Pacific Ocean. Rev Biol Trop 28:202–208

López M (1983) *Lycodontis verrilli* (Pisces: Muraenidae) descripción de su larva leptocéfala del Golfo de Nicoya, Costa Rica. Rev Biol Trop 31:343–344

López M, Bussing WA (1982) Lista provisional de los peces marinos de la costa Pacífica de Costa Rica. Rev Biol Trop 30:5–26

López MI, Bussing WA (1998) *Urotrygon cimar*, a new eastern Pacific stingray (Pisces: Urolophidae). Rev Biol Trop 46(Suppl 6):271–277

Luke SR (1977) Catalog of the benthic invertebrate collections. I. Decapod Crustacea and Stomatopoda. Scripps Inst Oceanogr Ref Ser 77–9:72 p

Luke SR (1995) Catalog of the Benthic Invertebrate Collections of the Scripps Institution of Oceanography. Mollusca. Scripps Inst Oceanogr Ref Ser 95–24:477 p

Macintyre IG, Glynn PW, Cortés J (1992) Holocene reef history in the eastern Pacific: mainland Costa Rica, Caño Island, Cocos Island, and Galápagos Islands. Proceedings of the 7th International Coral Reef Symposium, Guam 2:1174–1184

Manning RB, Reaka ML (1979) Three new stomatopod crustaceans from the Pacific coast of Costa Rica. Proc Biol Soc Wash 92:634–639

Manter HW (1940) The geographic distribution of Digenetic trematodes of marine fishes of the tropical American Pacific. Allan Hancock Pac Exped 2(16):531–547

Marques F, Brooks DR, Monks S (1995) Five new species of *Acanthobothrium* van Beneden, 1849 (Eucestoda: Tetraphyllidea: Onchobothriidae) in stingrays from the Gulf of Nicoya, Costa Rica. J Parasitol 81:942–951

Marques F, Brooks DR, Molina H (1996) Two new species of Tetraphyllidean cestodes in *Himantura pacifica* (Chondrichthyes: Myliobatiformes: Dasyatididae) from the northwest coast of Costa Rica. J Parasitol 82:302–306

Maurer D, Vargas JA (1984) Diversity of soft-bottom benthos in a tropical estuary: Gulf of Nicoya, Costa Rica. Mar Biol 81:97–106

Maurer D, Epifanio CE, Dean HK, Howe S, Vargas JA, Dittel AI, Murillo MM (1984) Benthic invertebrates of a tropical estuary: Gulf of Nicoya, Costa Rica. J Natl Hist 18:47–61

Maurer D, Vargas JA, Dean HK (1988) Polychaetous annelids from the Gulf of Nicoya, Costa Rica. Int Rev Hydrobiol 73:43–59

May-Collado L, Morales-Ramírez A (2005) Presencia y patrones de comportamiento del delfín manchado costero, *Stenella attenuata* (Cetacea: Delphinidae) en el Golfo de Papagayo, Costa Rica. Rev Biol Trop 53:265–276

May-Collado L, Gerrodette T, Calambokidis J, Rasmussen K, Sereg I (2005) Patterns of cetacean sighting distribution in the Pacific Exclusive Economic Zone of Costa Rica based on data collected from 1979–2001. Rev Biol Trop 53:249–263

McAdoo BG, Orange DL, Silver EA, McIntosh K, Abott L, Galewsky J, Kahn L, Protti M (1996) Seafloor structural observations, Costa Rica accretionary prism. Geophys Res Lett 23:883–886

McCorquodale S, Brooks DR (1995) *Acanthobothrium nicoyaense* n. sp. (Eucestoda: Tetraphyllidea: Onchobothriidae) in *Aetobatus narinari* (Euphrasen) (Chondrichthyes: Myliobatiformes: Myliobatidae) from the Gulf of Nicoya, Costa Rica. J Parasitol 81:244–246

Meléndez C (1974) Viajeros por Guanacaste. Ministerio de Cultura, Juventud y Deportes, Costa Rica, Ser. NOS VEN 4, 92 p

Michel HB, Foyo M (1976) Caribbean Zooplankton. Part I. Siphonophora, Heteropoda, Copepoda, Euphausiacea, Chaetognatha and Salpidae. Office of Naval Research, Department of the Navy, Washington, DC, pp 1–549

Mielke W (1992) Six representatives of the Tetragonicipitidae (Copepoda) from Costa Rica. Microfauna Mar 7:101–146

Mielke W (1994) *Microcanuella bisetosa* gen. n., sp. n., a new taxon of Canuellidae (Crustacea, Copepoda) from the Costa Rican Pacific coast, with remarks on *Galapacanuella beckeri*. Bijdr Dierk 64:55–64

Mielke W (1995) Species of the taxon *Schizopera* (Copepoda) from the Pacific coast of Costa Rica. Microfauna Mar 10:89–116

Mielke W (1997) New findings of interstitial Copepoda from Punta Morales, Pacific coast of Costa Rica. Microfauna Mar 11:271–280

Molina-Ureña H (1996) Ichthyoplankton assemblages in the Gulf of Nicoya and Golfo Dulce embayments, Pacific coast of Costa Rica. Rev Biol Trop 44(Suppl 3):173–182

Montoya M (2007) Conozca la Isla del Coco. Una guía para su visitación. BioCurso Isla del Coco: la isla del tesoro. Organización para Estudios Tropicales, San Pedro, Costa Rica, 176 p

Montoya M, Kaiser KL (1988) Biogeographical notes on the genus *Terebra* (Gastropoda) at Isla del Coco, Costa Rica. Rev Biol Trop 36:569–574

Morales A, Murillo MM (1996) Distribution, abundance and composition of coral reef zooplankton, Cahuita National Park, Limón, Costa Rica. Rev Biol Trop 44:619–630

Morales-Ramírez A (1996) Checklist of copepods from Gulf of Nicoya, Coronado Bay and Golfo Dulce, Pacific coast of Costa Rica, with comments on their distribution. Rev Biol Trop 44(Suppl 3):103–113

Morales-Ramírez A (2001) Biodiversidad marina de Costa Rica, los microcrustáceos: subclass Copepoda (Crustacea: Maxillopoda). Rev Biol Trop 49(Suppl 2):115–133

Morales-Ramírez A, Nowaczyk J (2006) El zooplancton gelatinoso del Golfo Dulce, Pacífico de Costa Rica, durante la transición de la estación lluviosa a la seca 1997–1998. Rev Biol Trop 54(Suppl 1):201–223

Morales-Ramírez A, Vargas JA (1995) Especies comunes de copépodos (Crustacea: Copepoda) pelágicos del Golfo de Nicoya, Costa Rica. Rev Biol Trop 43:207–218

Moran DA, Dittel AI (1993) Anomuran and brachyuran crabs of Costa Rica: annotated list of species. Rev Biol Trop 41:599–617

Mörch OAL (1859–1861) Beiträge zur Molluskenfauna Central Amerika's. Malakozool Blät 6(4):102–126; 7(2):66–96; 7(3):97–106; 7(4):170–192; 7(5):193–213

Mörch OAL (1860) Etudes sur la famille des Vermets. J Conchyl 8:27–48

Nicholls-Driscoll J (1976) Benthic invertebrate communities in Golfo Dulce, Costa Rica, an anoxic basin. Rev Biol Trop 24:281–298

Ortea J (2001) El género *Doto* Oken, 1815 (Mollusca: Nudibranchia) en el mar Caribe: Historia natural y descripción de nuevas especies. Avicennia (Suppl 3):1–46

Ortea J, Espinosa J, Camacho Y (1999) Especies del género *Polycera* Cuvier, 1816 (Mollusca: Nudibranchia) recolectados en la epifauna de algas rojas del Caribe de Costa Rica y Cuba. Avicennia 10/11:157–164

Ortea J, Espinosa J, Moro L (2001) Descripción de una nueva especie de *Philine* Ascanius, 1772. Avicennia (Suppl 4):38–40

Peraldo-Huertas G (Compiler) (2003) Ciencia y técnica en la Costa Rica del Siglo XIX. Ed. Tecnol. Costa Rica, Cartago, Costa Rica, 408 p

Pesta O (1931) Ergebnisse der Österreichischen Biologischen Costa-Rica-Expedition 1930. I. Teil. Crustacea Decapoda aus Costa-Rica. Ann Naturhist Mus Wien 45:173–181

Pittier HF (1899) Apuntamientos preliminares sobre la Isla del Cocos, posesión costarricense en el Océano Pacífico. Memoria Secretaría de Fomento, San José, Costa Rica 1899, pp 15–28

Ramírez AR, López MI, Szelistowski WA (1990) Composition and abundance of ichthyoplankton in a Gulf of Nicoya mangrove estuary. Rev Biol Trop 38:463–466

Rathbun MJ (1902) Papers from the Hopkins Stanford Galápagos Expedition, 1898–1899. VIII. Brachyura and Macrura. Proc Wash Acad Sci 4:275–292

Rathbun MJ (1906) Descriptions of three new mangrove crabs from Costa Rica. Proc Biol Soc Wash 19:99–100

Rathbun MJ (1930) The cancroid crabs of America of the families Euryalidae, Portunidae, Ateleeyelidae, Cancridae and Xanthidae. US Natl Mus Bull 152:I–XVI, 1–609

Reaka ML, Manning RB (1980) The distributional ecology and zoogeographical relationships of stomatopod crustacea from Pacific Costa Rica. Smithson Contr Mar Sci 7:1–29

Reid DG, Kaiser KL (2001) Three new records of Indo-Pacific Littorinidae (Gastropoda) in the tropical eastern Pacific. Veliger 44:381–388

Richards FA, Anderson JJ, Cline JD (1971) Chemical and physical observations in Golfo Dulce, an anoxic basin on the Pacific coast of Costa Rica. Limnol Oceanogr 16:43–50

Risk MJ, Murillo MM, Cortés J (1980) Observaciones biológicas preliminares sobre el arrecife coralino en el Parque Nacional de Cahuita, Costa Rica. Rev Biol Trop 28:361–382

Robinson DG, Montoya M (1987) Los moluscos marinos de la costa Atlántica de Costa Rica. Rev Biol Trop 35:375–400

Rodríguez-Sevilla L, Vargas R, Cortés J (2003) Biodiversidad marina de Costa Rica: gastrópodos (Mollusca: Gastropoda) de la costa Caribe. Rev Biol Trop 51(Suppl 3):305–399

Sánchez-Nava S, Segura-Puertas L (1987) Los moluscos pelágicos (Gastropoda: Heteropoda y Pteropoda) recolectados en el Domo de Costa Rica y regiones adyacentes. Mem Soc Méx Malacol 3:232–240

Schmitt WL (1939) Decapod and other Crustacea collected on the Presidential Cruise of 1938 (with introduction and station data). Smithson Mis Col 98 (6):1–29

Schmitt WL (1940) The Stomatopods of the West Coast of America, based on collections made by the Allan Hancock Expeditions, 1933–38. Allan Hancock Pac Exped 5 (4):129–225

Segura-Puertas L (1991) Medusas (Cnidaria: Hydrozoa y Scyphozoa) de la región del Domo de Costa Rica. Rev Biol Trop 39:159–163

Segura-Puertas L, Hernández RM, Morones L (1992) Distribución y abundancia de los quetognatos (Chaetognatha) en la región del Domo de Costa Rica. Rev Biol Trop 40:35–42

Setchell WA (1937) The Templeton Crocker Expedition of the California Academy of Sciences, 1932. No. 34: report on the *Sargassums*. Proc Calif Acad Sci 4th Ser 22:127–158

Shasky DR (1983) New records of Indo-Pacific Mollusca from Cocos Island, Costa Rica. Nautilus 97:144–145

Shasky DR (1987) Update on mollusks with Indo-Pacific faunal affinities in the tropical eastern Pacific VI. Festivus 19:100–101

Shasky DR (1989) My last seven years – at Cocos Island. Festivus 21:72–75

Slevin JR (1931) Log of the Schooner "Academy" on a voyage of scientific research to the Galápagos Islands, 1905–1906. Occ Pap Calif Acad Sci 17:1–162

Snodgrass RE, Heller E (1905) Papers from the Hopkins-Stanford Galapagos Expedition, 1898–1899. XVII. Shore fishes of the Revillagigedo, Clipperton, Cocos and Galapagos islands. Proc Wash Acad Sci 6:333–427

Soto R (1982) *Caloglossa ogassuarensis* Skuja (Rhodophyta, Ceramiales, Delesseriaceae) en Costa Rica. Brenesia 19/20:251–253

Soto R (1983) Nuevos informes para la flora bentónica marina de Costa Rica. Brenesia 21:365–370

Soto R, Ballantine DL (1986) La flora bentónica marina del Caribe de Costa Rica. Brenesia 25/26:123–162

Sowerby GB (1832) Characters of new species of Mollusca and Conchifera, collected by Hugh Cuming. Proceedings of the Zoological Society of London, Part II, 1832, pp 25–33 (February 28); 50–61 (June 5); 113–120 (June 12)

Stebbing TRR (1903) Amphipoda from Costa Rica. Proc US Natl Mus 26 (1341):925–931

Stebbing TRR (1906) A new Costa Rica amphipod. Proc US Natl Mus 31 (1490):501–504

Stebbing TRR (1908) A new amphipod crustacean, *Orchestoidea biolleyi*, from Costa Rica. Proc US Natl Mus 34 (1609):241–244

Suárez-Morales E, Gasca R (1989) Copépodos calanoides epipelágicos del Domo de Costa Rica (junio-agosto 1982). Cienc Mar 15:89–102

Suárez-Morales E, Morales-Ramírez A (2001) Nuevo registro de *Acartia (Planktacartia) negligens* (Copepoda, Calanoida) en el Pacífico Tropical Oriental. Rev Biol Trop 49:1286

Suárez-Morales E, Morales-Ramírez A (2003) A new species of *Cymbasoma* (Crustacea: Copepoda: Monstrilloida) from the Pacific coast of Costa Rica, Central America. Proc Biol Soc Wash 116:206–214

Szelistowski WA (1990) A new clingfish (Teleostei: Gobiesocidae) from the mangroves of Costa Rica, with notes on its ecology and early development. Copeia 1990:500–507

Taylor WR (1945) Pacific marine algae of the Allan Hancock Expeditions to the Galapagos Islands. Allan Hancock Pac Exp 12:1–528

Tee-Van J (1926) The Arcturus: equipment and operation. Zoologica 8:47–106

Tokioka T (1971) A new species of *Rhopalaea* from the Pacific coast of Costa Rica (Tunicata, Ascidiacea). Publ Seto Mar Biol Lab XIX (2/3):119–122

Tokioka T (1972) On a small collection of ascidians from the Pacific coast of Costa Rica. Publ Seto Mar Biol Lab XIX (6):383–408

Tomlin JR Le B (1927) The Mollusca of the 'St. George' expedition (I). The Pacific coast of S. Am J Conchol 18:153–170

Townsend CH (1901) Dredging and other records of the United States Fish Commission Steamer Albatross, with bibliography relative to the work of the vessel. US Fish Comm Rep 1900:387–562

Ulken A, Víquez R, Valiente C, Campos M (1990) Marine fungi (Chytridiomycetes and Thraustochytriales) from a mangrove area at Punta Morales, Golfo de Nicoya, Costa Rica, America Central. Rev Biol Trop 38:243–250

Vargas JA (1987) The benthic community of an intertidal mud flat in the Gulf of Nicoya, Costa Rica. Description of the community. Rev Biol Trop 35:229–316

Vargas JA (1988) A survey of the meiofauna of an eastern tropical Pacific intertidal mud flat. Rev Biol Trop 36:541–544

Vargas JA (1995) The Gulf of Nicoya estuary, Costa Rica: past, present, and future cooperative research. Helgoländer Meeresunters 49:821–828

Vargas JA, Wolff M (1996) Pacific coastal ecosystems of Costa Rica with emphasis on the Golfo Dulce and adjacent areas: a synoptic view based on the RV Victor Hensen expedition 1993/1994 and previous studies. Rev Biol Trop 44(Suppl 3):238 p

Vargas JA, Dean HK, Maurer D, Orellana P (1985) Lista preliminar de invertebrados asociados a los sedimentos del Golfo de Nicoya, Costa Rica. Brenesia 24:327–342

Vargas R (2000) *Periclimenes murcielaguensis*, a new species of shrimp (Crustacea: Decapoda: Palaemonidae), living on black coral from the Pacific coast of Costa Rica. Proc Biol Soc Wash 113:17–23

Vargas R, Cortés J (1997) Biodiversidad marina de Costa Rica: Orden Stomatopoda (Crustacea: Hoplocarida). Rev Biol Trop 45:1531–1539

Vargas R, Cortés J. (1999a) Biodiversidad marina de Costa Rica: Crustacea: Decapoda (Penaeoidea, Sergestoidea, Stenopodidae, Caridea, Thalassinidea, Palinura) del Caribe. Rev Biol Trop 47:877–885

Vargas R, Cortés J. (1999b) Biodiversidad marina de Costa Rica: Crustacea. Decapoda (Penaeoidea, Sergestoidea, Caridea, Astacidea, Thalassinidea, Palinura) del Pacífico. Rev Biol Trop 47:887–911

Vargas R, Cortés J (2006) Biodiversidad marina de Costa Rica: Crustacea: Infraorden Anomura. Rev Biol Trop 54:461–488

Vargas R, Jesse S, Castro M (1996) Checklist of crustaceans (Decapoda and Stomatopoda), collected during the Victor Hensen Costa Rica expedition (1993/1994). Rev Biol Trop 44(Suppl 3):97–102

Verrill AE (1870) Notes on Radiata in the Museum of Yale College, Number 6: review of the corals and polyps of the west coast of America. Trans Connect Acad Arts Sci 1:377–558, pls 4–10

Voss GL (1971) Narrative of RV John Elliot Pillsbury Cruise P-7101: Central America, January 20–February 5, 1971. School of Marine and Atmospheric Science, University of Miami, 31 p

Watling L, Breedy O (1988) A new cumacean (Crustacea) genus from beaches of Golfo de Nicoya, Costa Rica. Rev Biol Trop 36:527–533

Wehrtmann IS, Vargas R (2003) New records and range extensions of shrimps (Decapoda: Penaeoidea, Caridea) from the Pacific and Caribbean coasts of Costa Rica, Central America. Rev Biol Trop 51:268–274

Wicksten MK (1979) New records of the species of *Glyphocrangon* in the northeastern Pacific Ocean (Caridea: Glyphocrangonidae). Proc Biol Soc Wash 92:217–224

Wicksten MK, Vargas R (2001) A new species of *Thor* (Crustacea: Decapoda: Caridea) from the Tropical Eastern Pacific. Proc Biol Soc Wash 114:139–14

Williams AB, Vargas R (2000) A new species of mud shrimp, *Upogebia cortesi*, from Pacific Costa Rica (Decapoda: Thalassinidea: Upogebiidae). Proc Biol Soc Wash 113:13–16

Willis S, Cortés J (2001) Mollusks of Manuel Antonio National Park, Pacific Costa Rica. 2001. Rev Biol Trop 49(Suppl 2):25–36

Wilson HV (1904) Reports on an exploration off the west coasts of Mexico, Central and South America, and off the Galapagos Islands, in charge of Alexander Agassiz, by the U.S. Fish Commission Steamer Albatross, during 1891, Lieut. Commander Z.L. Tanner U.S.N., commanding. XXX. The sponges. Mem Mus Comp Zoöl Harvard Coll 30:1–164

Ziesenhenne FC (1942) New eastern Pacific sea stars. Allan Hancock Pac Exped 8(4):197–223

Zimmerman TL, Martin JW (1999) Brachyuran crabs of Cocos Island (Isla del Coco), Costa Rica: Leucosiidae, Calappidae, and Parthenopidae, with descriptions of two new species. J Crust Biol 19:643–668

Zullo VA (1991) Zoogeography of the shallow-water cirriped fauna of the Galápagos Islands and the adjacent regions in the tropical eastern Pacific, pp 173–192. *In*: James MJ (ed.) Galápagos Marine Invertebrates: Taxonomy, Biogeography, and Evolution in Darwin's Islands. Plenum, New York/London

Chapter III
Marine Fossils

Teresita Aguilar

Macrochlamys pittieri. Miocene, Valle Central, Costa Rica. (Photo: Percy Denyer)

Abstract The geological and tectonic evolution of the southern Central American isthmus resulted in the establishment of different marine and continental biotopes along its distinct stages of development during the last 200 Ma (million years ago).

T. Aguilar
Escuela Centroamericana de Geología and CIMAR; Universidad de Costa Rica.
Apdo. 214-2060 San José, Costa Rica.
e-mail: aaguilar@geologia.ucr.ac.cr

I.S. Wehrtmann, J. Cortés (eds.) *Marine Biodiversity of Costa Rica, Central America*,

Geological, geographical, and ecological conditions in these biotopes subsequently varied from deep oceanic up to continental environments. From Early Jurassic to Late Cretaceous, the fossil record was almost exclusively made of pelagic organisms – mainly radiolarians and foraminiferans – deposited in deep oceanic basins. During the Late Cretaceous, the first carbonate platforms with bioconstructions of rudists were installed; they were similar to Campanian-Maastrichtian faunas from Mexico and the Caribbean. The mass extinction at the terminal Cretaceous affected several invertebrate groups, among them ammonites, rudists, inoceramids, nerineids, and trigonids. With the exception of some shallow reef-like carbonate platforms of Paleocene age, the fossil record was poor until the Early Eocene, mainly because of a deepening of the sedimentary basins. Widely spread carbonate platforms were present again in the Mid–Late Eocene, and corals and mollusks were the most abundant fossils. There was a predominance of shallow marine environments during the Miocene–Early Pliocene, therefore the fossil record presents a higher diversity and abundance of marine invertebrates, especially mollusks (120 species of bivalves and 100 species of gastropods recognized). Ichnofossils, traces of animals are in some cases the only record of organisms without mineralized skeletons, and are quite valuable for paleoenvironmental reconstructions. The continental bridge between North and South America was set about 3.5 Ma, starting a process of both extinction and diversification of marine invertebrates because of the separation of the Pacific Ocean and the Caribbean Sea.

Introduction

The geologic-tectonic evolution of the Central American Isthmus caused rapid changes in the morphology of the ocean floor and the continental masses, and had a direct influence on geographic, climatic, and oceanographic conditions (Denyer *et al.* 2000). These events have driven the development of the different continental and marine biotopes. The evolution of flora and fauna of the isthmus and the surrounding seas are thus the result of the interaction between endogenous processes (tectonism, volcanism, seismic activity), and exogenous processes (sedimentation, erosion, climate, and sea-level changes).

The most complete studies on invertebrate paleontology in Central America were carried out in Costa Rica. In the rest of the Central American countries, there are much fewer studies of invertebrate fossil diversity, except within the Panama Canal Zone. There, the detailed studies conducted by Woodring (1925–1982) and the Panama Paleontology Project (Smithsonian Tropical Research Institute, starting in 1988) present an extended and comprehensible account of fossil marine invertebrates, particularly mollusks (Nehm 1994; Fortunato & Jung 1995), bryozoans (Jackson & Cheetham 1994; Cheetham *et al.* 1999), ostracoda (Borne 1995; Borne *et al.* 1996, 1999), corals (Budd *et al.* 1994, 1996, 1999; Johnson *et al.* 1995), Foraminifera (Collins *et al.* 1995; Collins 1996, 1999a, b; Cotton 1999) and nannofossils (Bybell 1999).

The record of fossil invertebrates of Central America is incomplete. There is no information about sponges, non-anthozoan coelenterates, annelids, non-crustacean arthropods, poly- and monoplacophorans, and non-echinozoan echinoderms. A few papers discuss cephalopods, scaphopods, brachiopods, and trace fossils. The diversity of Neogene bryozoans is well documented (Cheetham *et al.* 1999), but a systematic presentation is still lacking. Monographs exist on gastropods and bivalves (Olsson 1922, 1942; Woodring 1957, 1959, 1964, 1970, 1973a, b, 1982), but nearly exclusively on those of the Miocene and Pliocene. In spite of exposed marine sedimentary deposits in Central America, there is little or no information about fossil invertebrates of Permian to Cretaceous age, or of the Paleogene in Nicaragua, El Salvador, Honduras, and to a lesser degree Guatemala (Fischer & Aguilar 2007).

The Fossil Record

Early Jurassic–Late Cretaceous: The geological history of Costa Rica begins with a prolonged period of oceanic history (Fig. III.1), which extends from Early Jurassic to Late Cretaceous (Campanian, 200–80 Ma) (Fischer & Aguilar 1994;

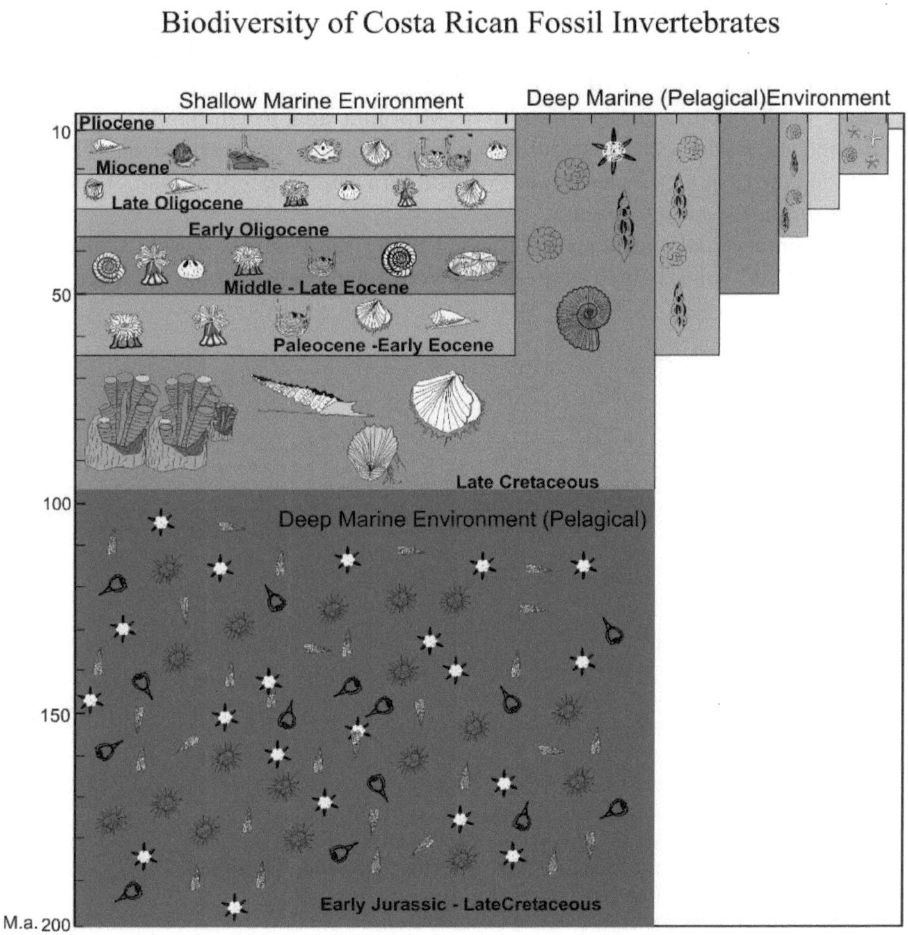

Fig. III.1 Fossil diversity of Costa Rica related with depositional environment and age

Denyer *et al.* 2000). During this period, the present territory of Costa Rica was part of deep marine basins; therefore the majority of the fossils were planktonic organisms: radiolarians and foraminiferans. Radiolarians have siliceous skeletons that accumulate in large quantities, forming pelagic rocks known as radiolarites. These radiolarites were deposited below the carbonate compensation depth (CCD) under conditions of reduced sedimentation, without terrigenous influence and in low energy environments (Gursky 1989). The fossil genera of radiolarians so far recognized in Costa Rica include: *Gigi*, *Gorgansium*, *Jacus*, *Paleosatunalis*, *Protopsium*, *Saitoumi*, and *Triactoma* (see De Wever *et al.* 1985). A well-preserved radiolarian fauna with 23 species of Mid-Cretaceous age from Península de Santa Elena was described by Schmidt-Effing (1980). Radiolarian faunas from Callovian to Santonian (164.4–85.8 Ma) were described from different levels of siliceous rocks within the "Complejo de Nicoya" (Baumgartner 1984).

In depressions with low oxygen circulation, sediments with high organic matter content were deposited. These include the remains of cephalopod mollusks as ammonites of the genera *Neokentroceras* and *Pseudokosmaticeras*) (Schmidt-Effing 1974; Azema *et al.* 1978).

Foraminifera were also abundant, with a predominance of planktonic species in pelagic facies, where the genus *Globotruncana* dominated. In shallower facies, small benthic foraminiferans of Miliolida and Rotaliida (Nodosariacea, Buliminacea, Rotaliacea, Cassidulinacea, and Noniacea) were abundant. The most common genera of larger calcareous Foraminifera were: *Sulcoperculina*, *Asterorbis*, *Pseudorbitoides*, and *Orbitoides*.

Sedimentary rocks are currently found along the Pacific coast of Costa Rica (Fig. III.2) and Panama, and range in age from Late Pliensbachian (Astorga 1997)

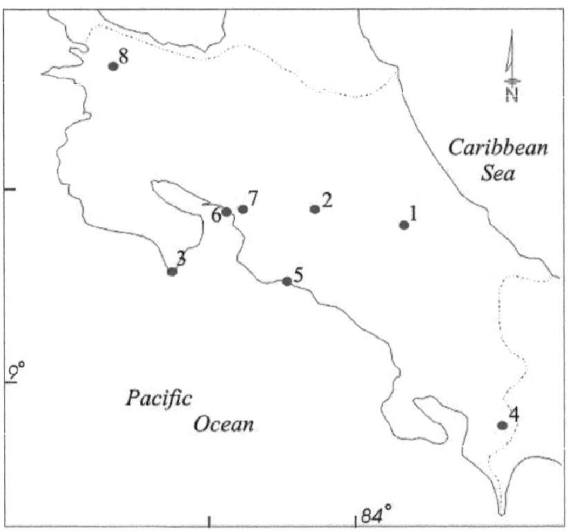

1.Turrialba, 2.Parritilla, 3.Cabo Blanco, 4.Fila de Cal, 5.Punta Judas, 6.Punta Carballo, 7.Esparza, 8.Bagaces

Fig. III.2 Main marine fossil outcrops in Costa Rica

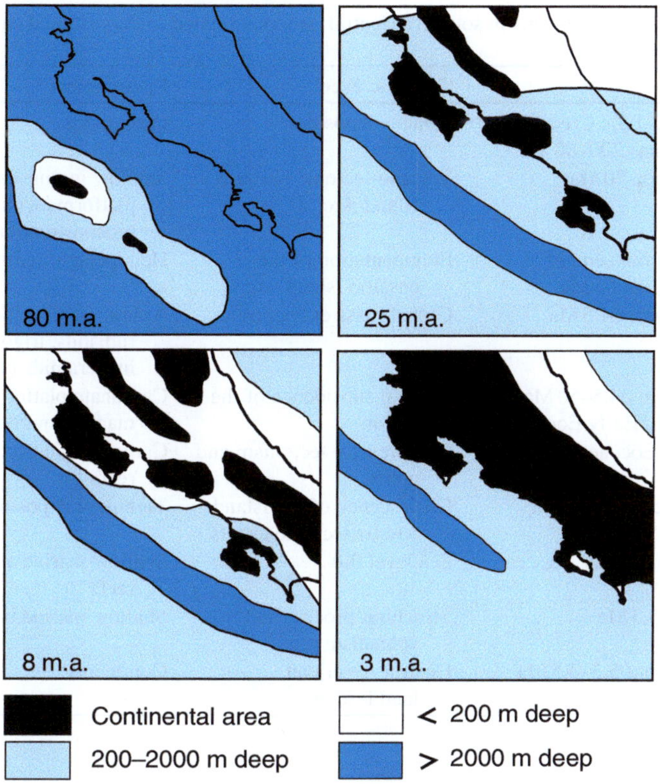

Fig. III.3 Paleogeographic reconstruction of Costa Rica showing the relationship between continental areas and marine depositional areas during different epochs

and Callovian to Early Campanian, between 200 and 80 Ma (Schmidt-Effing 1979; Baumgartner 1984) (Fig. III.3). The species recognized are typical of the Central Atlantic-Tethys Seaway, which was an equatorial east to west stream (Alencaster 1978).

Latest Cretaceous: As a consequence of a series of volcano-tectonic events (Table III.1), during the Late Cretaceous (Campanian and Maastrichtian), an arc of basaltic isles was formed in the southern region of Central America. By the end of the Late Cretaceous (Maastrichtian: 70 Ma), the archipelago of basaltic islands was fragmented (Fig. III.3; Table III.1) and, subsided in large areas, forming different sedimentary basins. These basins were filled by hemipelagic and episodic sedimentation, with sporadic and catastrophic, high sedimentation rate events, alternating with periods of no sedimentation that inhibited the development of benthic associations. Therefore, the fossil records consist of allochthonous remains of radiolarians, foraminiferans, ammonites, inoceramids, amber, and wood.

Diverse ecosystems were generated in response to the uplift of the sea floor and the creation of shallow environments surrounding the emerged areas. These areas

Table III.1 Geologic events in southern Central America related to depositional environments (Jurassic–Recent)

Age	Geologic event	Environment
Early Jurassic–Late Cretaceous (Campanian) 200–80 Ma	Pelagic deposition	Deep marine
Late Cretaceous 70 Ma	Volcano-tectonic basaltic Island Arc	Shallow marine carbonate platforms with rudist bioconstructions
Latest Cretaceous: end of Maastrichtian 65 Ma	Fragmentation of the basaltic Island Arc	Hemipelagic sedimentation
End of Mesozic: 64.5 Ma K/T	Global mass extinction	Ammonites, nerineas, rudistids, trigoniids and inoceramids became extant
Early Cenozoic 64.5–50 Ma: Paleocene– Early Eocene	General subsidence of the region	Carbonate platforms and deep marine environments
Middle–Late Eocene 50–36 Ma	Compressive tectonism and uplift	Carbonate platforms and patch reefs
Early Oligocene 35–30 Ma	Coalescence of the Island Arc by tectonic events	Turbiditic deposition
Late Oligocene-Early Miocene 30–23 Ma	Sea level fall	Shallow marine and patch reefs
Miocene 23–5.3 Ma	Structural process and volcanism	Shallow marine basins
Pliocene–Recent 5.3–3.5 Ma	Central American land bridge	Continental

allowed the development of carbonate platforms where rudists (bivalve mollusks adapted to reef existence) created bioconstructions (Fig. III.3). The main genera known are: *Coralliochama* and *Mitrocaprina* (Caprinidae), *Barretia* and *Parastroma* (Hippuritidae), *Biradiolites*, *Chiapasella*, *Tampsia*, and *Sauvagesia* (Radiolitidae). This fauna is similar to Campanian–Masstrichtian faunas from Mexico and the Caribbean (Alencaster 1978; Pons & Schmidt-Effing 1998).

There are other mollusks associated with the rudists: cephalopods, bivalves (inoceramids, trigoniids), scaphopods (dentaliids), and gastropods (nerineids); small colonies of corals, hydrozoans, echinoderms (*Conulus* sp., *Micraster* sp., *?Holaster* sp.), algae, radiolarians, and foraminiferans (Galli & Schmidt-Effing 1977). The presence of bivalves of the genus *Pterotrigonia* (*Pterotrigonia*) in shallow warm marine environments proves an origin related to the ancient Tethys Sea (Aguilar & Denyer 2002).

In general, the fauna recognized for the Late Cretaceous consists of stenohaline organisms of shallow seas that occupied different habitats. There were genera that colonized hard substrates like *Plicatula*, *Ostrea*, *Spondylus*, and *Clavagella* and the rudists among the bivalves, and *Catophragmus* from the cirripeds. Endo and epibenthic bivalves and sea urchins inhabited soft and hard substrates: inocermids and trigoniids among the bivalves, and representatives of the echinoid genus *Conulus*. Nektonic organisms were represented by ammonites: *Canadoceras* sp. (*?Pachydiscus*

Fig. III.4 Diversity of fossils from Costa Rica. a. *Dichocoenia* sp., Scleractinia; Pliocene, Limón; b. *Asterosmilia hilli*, Scleractinia; Pliocene, Limón. c. *Ficus carbacea carbacea*, Gastropoda; Pliocene, Península de Nicoya (Montezuma Formation); d. *Prionocidaris cojimarensis*, Echinoidea; Miocene, Turrúcares Formation (Photos: Percy Denyer)

sp.), *Bostrychoceras* cf. *polyplocum*, and ?*Neokossmaticeras* sp., while planktonic species were mainly foraminiferans of the genus *Globotruncana*. The outcrops that contain this fauna can be found in northwestern Costa Rica (Fig. III.2).

The global mass extinction at the end of the Mesozoic (some 65 Ma) (Cowen 2000) affected several marine groups in the Costa Rica area, including ammonites, some gastropods such as *Nerinea*, and bivalves such as rudists, trigonids, and inoceramids (Ward 1990; Fischer & Aguilar 1994).

Paleogene: At the beginning of the Cenozoic (Lower Tertiary: Paleocene–Early Eocene (between 65 and 50 Ma) sediments were mainly carbonates. They developed in platform environments, while in deep marine regions, with little continental influence, sediments were of the fine siliciclastic type. Both were affected by a general subsidence of the area.

The predominant biocoenosis in carbonate environments included a monospecific patch reef, built by the coral *Euphyllia donatoi* (Scleractinia) (Aguilar & Denyer 2001) in association with algae, Foraminifera, and mollusk fragments. The presence of red coralline algae and squamariaceas (*Polystrata alba*) demonstrated the existence of a restricted open platform (Jaccard *et al.* 2001). In the micritic facies there were rotaliid planktonic foraminiferans (*Thalmannita madrugaensis, Morozovella velascoensis*) and benthic foraminiferans, among them miliolid and alveolinid (Jaccard et al. 2001). The presence of the open platform crinoid *Pentacrinites* sp., which has not been previously reported for Costa Rica, must be pointed out.

During the Mid–Late Eocene, another compressive tectonic event generated the establishment of shallow areas that served as suitable environments for the development of carbonate platforms and patch reefs (Table III.1). These platforms contained the fossilized remains of macroforaminiferans, mainly of the genus *Nummulites*, corals, mollusks, sea urchins, and ichnofossils. These fossils were found in different parts of Costa Rica, including Turrialba, Parritilla, Cabo Blanco, and Fila de Cal (Fig. III.2) (Malavassi 1960, 1975; Hoffteter *et al.* 1961).

During the Early Oligocene, southern Central America suffered a morphologic transformation due to tectonic events that caused the islands of the arc to coalesce and form great continental areas in northern Costa Rica. In the south, the environments became deeper, promoting a strong turbiditic sedimentation in response to the erosion of the emerged lands in both the outer and fore arcs. No benthic associations from this period have been recorded (Winsemann 1992). This sedimentation process was probably increased by a rise of the global sea level, which occurred between 30 and 36 Ma (Haq *et al.* 1988).

The Upper Oligocene–Lower Miocene strata were characterized by the existence of shallow marine ecosystems, probably generated by the combined effect of falling sea level and the input of material to the basins. The development of patch reefs was initiated during this time as a low diversity bioherm constructed mainly by *Antiguastrea celullosa*, though associated with many other organisms, mainly mollusks, algae, and echinoderms (Aguilar 1999; Aguilar & Cortés 2001).

Neogene: Beginning in the Miocene, various relatively shallow marine basins developed, controlled by structural factors that yielded the existence of a great diversity of invertebrate species, mainly mollusks, echinoderms, and arthropods (Fig. III.1). This high diversity present throughout most of Costa Rica necessarily implies the presence of geographic or environmental barriers, which promoted endemism and inhibited the colonization of all the basins by the same species (Fig. III.3). In spite of their geographical and chronological proximity, for example, the associations of mollusks of Punta Judas and Punta Carballo, along the Central Pacific of Costa Rica (Fig. III.2), present different composition, diversity, and preservation. This cannot be explained as a result of the predominant ecological conditions in the time of their colonization, but by different substrate features like texture (grain size) and composition (Seyfried *et al.* 1985; Aguilar & Alvarado 1996).

The fauna associations within these basins were exposed to high rates of clastic sedimentation in episodic events. In some areas a strong subsidence was combined with a rapid elevation of adjacent continental areas and a reactivation of vulcanism, which provided more material and initiated basin filling since the Late Miocene (Fischer & Aguilar 2007).

The Miocene invertebrate communities were composed by mobile benthic organisms (gastropods, sea urchins), opportunistic colonizers of sandy substrates (bivalves: donacid, tellinid, and lucinid), and epibenthic bivalves such as pectinids and arcids (Fig. III.4). Epi and endolithic organisms were restricted to secondary substrates (shells from other mollusks, wood, rock fragments) or found as allochthonous elements (barnacles, regular sea urchins) (Fischer & Aguilar 1994). Sedimentophobic invertebrates (corals) were almost completely absent while brachiopods and sponges were scarce. Cheilostomata bryozoans were found mainly growing over mollusk shells (Fischer & Aguilar 1994, 2001). In some localities (Esparza and Punta Judas, central Pacific coast; Fig. III.2), communities of estuarine environments developed in calm water associated with the accumulation of fine grain sediments, barely oxygenated. These were low diversity communities, where *Nuculana fundationis* was the most representative element of the association (Fischer & Aguilar 1995). Other estuarine fauna included the first inarticulate brachiopods reported from Central America found interbedded within the ignimbrites of the Bagaces Formation (Aguilar & Alvarado 2004).

Because of the predominance of marine intertidal–subtidal environments, the mollusks showed the highest diversity (120 and 100 species of bivalves and gastropods, respectively) during the Miocene (Aguilar 1990) (Fig. III.1).

At the beginning of the Pliocene, with the closure of the Central American isthmus some 3.5 Ma (Fig. III.3), the environmental conditions changed drastically, starting with the modifications of sublittoral environments that predominated since the Late Miocene, and ended as a continental environment. This land bridge connected North and South Americas and thus interrupted the prevailing marine communication. The development of the Isthmus caused an increase in evolutionary divergence of species, the extinction of some groups (Jackson *et al.* 1996, 1999), and marked the beginning of the differentiation of the faunas of the Caribbean and the Pacific. Moreover, it resulted in the creation of two new marine provinces, the Caribbean to the east and the Panamic to the west (Woodring 1966; Aguilar & Fischer 1986; Aguilar 1990; Aguilar & Alvarado 1996; Jackson *et al.* 1996, 1999) with dramatic changes in the oceanic circulation pattern of the region (Cortés 1986; Jackson & D'Croz 1997).

Ichnofossils: An important group, in spite of its low diversity, is that of the ichnofossils (traces of vital activity of organisms in sediments or rocks), because they may constitute the only evidence of the existence of organisms without mineralized skeletons. Additionally, they have a great value for studies concerning the sedimentological, paleoecological, or taphonomical conditions of geological formations. Among the most abundant ichnogenera in Costa Rica are: *Thalassinoides*, *Ophiomorpha*, *Chondrites*, *Zoophicus*, *Echinocardium*, and *Gastrochaenolites*. These traces are characteristic of unstable environments with great input of sediments. Shallow water traces (skolithos ichnofacies) were abundant during the Miocene and Pliocene (Laurito 1988). The deposits accumulated in the continental slope (turbiditic) presented sometimes nereites ichnofacies, mainly from the Paleogene.

Conclusions

- The geologic and tectonic evolution of the Central American Isthmus had a direct influence in the development of different marine ecosystems and the associated organisms (Table III.1).
- Besides Costa Rica, in the remaining Central American countries there are no studies of invertebrate fossil diversity, except those of the Panama Canal Zone.
- During the Late Cretaceous, the species were typical of the Central Atlantic-Tethys Seaway, which was an equatorial east-to-west current.
- At the end of the Mesozoic, the global mass extinction affected several groups of marine organisms in the Costa Rica area, including ammonites, some gastropods (such as *Nerinea*) and bivalves (such as rudists, trigoniids, and inoceramids).
- During the Mid–Late Eocene, carbonate platforms developed extensively. This was due to a regional tectonic uplift. *Nummulite* limestones document this episode.
- During the Miocene, shallow marine basins developed with extensive littoral to sublittoral environments, allowing the existence of a great diversity of invertebrate species, mainly mollusks, echinoderms, and arthropods. This high diversity necessarily implied the presence of geographic barriers, which limited the dispersal of some species.
- At the beginning of the Pliocene, conditions changed drastically from marine sublittoral to continental environment (some 3.5 Ma). This event was the result of the formation of the land bridge between North and South America that interrupted the prevailing marine communication. As a consequence, there was an increment of speciation and extinction rates. The rise of the Isthmus marked the beginning of the differentiation of the Caribbean and Panamic faunas.
- Ichnofossils may constitute the only evidence of the existence of organisms without mineralized skeletons. They are of great value for the reconstruction of sedimentological, paleoecological, or taphonomical conditions of geological formations.

Specialists

RUDOLF FISCHER: University of Hannover, Germany (retired) – Mollusks.
PETER JUNG: Natural History Museum, Basel, Switzerland (retired) – Mollusks.
REINHART SCHMIDT-EFFING: University of Marburg, Germany – Microfossils and Rudists. schmidte@staff.uni-marburg.de
PETER BAUMGARTNER: University of Laussane, Switzerland – Radiolarians. Peter.Baumgartner@unil.ch
JEREMY JACKSON: Smithsonian Tropical Research Institution, Panama – Bryozoan. jacksonj@si.edu

Collections

The main collections containing fossil invertebrates and some type material of Costa Rican species are:

Museo de Paleontología, Escuela Centroamericana de Geología, Universidad de Costa Rica; http://www.geologia.ucr.ac.cr

Museo Nacional de Costa Rica; http://www.museocostarica.go.cr

National Museum of Natural History, Smithsonian Institution, Washington, DC, USA; http://www.nmnh.si.edu/paleo

Acknowledgments G.E. Alvarado, J. Cortés, K. Flores, G.J. Soto and A. Granados helped in reviewing the manuscript and providing ideas. K. Flores and J. Méndez elaborated the figures. S. Kruse checked the English. Thanks to them all. This manuscript was prepared with the support of the research project "Arrecifes fósiles de Costa Rica": 808-99-237 of the Vicerrectoría de Investigación, Universidad de Costa Rica.

References

Aguilar T (1990) La evolución de las comunidades de moluscos y su relación con el desarrollo geológico del sur de América Central. Rev Geol Am Central 11:51–57

Aguilar T (1999) Organismos de un arrecife fósil (Oligoceno Superior-Mioceno Inferior) del Caribe de Costa Rica. Rev Biol Trop 47:453–474

Aguilar T. & Alvarado G. (1996). A synopsis of the Neogene events in Costa Rica. Proc Intl Symp IGCP-355. La Paz, B.C.S., Mexico, pp 51–68

Aguilar T, Alvarado G (2004) Tafonomía y sedimentología de la paleobiota estuarina en los Cerros Barbudal (Costa Rica) sepultada por el vulcanismo Neógeno. Rev Geol Am Central 30:111–116

Aguilar T, Cortés J (2001) Arrecifes coralinos del Oligoceno Superior-Mioceno Inferior de Turrialba, Costa Rica. Rev Biol Trop 49(Suppl 2):203–213

Aguilar T, Denyer P (2001) Una especie nueva de *Euphyllia* (Scleractinia: Caryophylliidae) en las calizas de Barra Honda (Paleógeno), Costa Rica. Rev Biol Trop 49:(Suppl 2):195–201

Aguilar T, Denyer P (2002) Primer hallazgo de Trigonias (Mollusca: Bivalvia) cretácicas de Costa Rica (Tortugal, Guanacaste): geología, estratigrafía y paleontología. Rev Geol Am Central 26:53–63

Aguilar T, Fischer R (1986) Moluscos de la Formación Montezuma (Plioceno-Pleistoceno; Costa Rica. Geol Paleontol 20:209–241

Alencaster G (1978) Distribución de faunas marinas del sur de México y del norte de América Central durante el Cretácico. *In*: Ferrusquía-Villafranca I (ed) Conexiones terrestres entre Norte y Sudamérica, UNAM. Bol. 101, pp 47–65

Astorga A (1997) El puente-istmo de América Central y la evolución de la Placa Caribe (con énfasis en el Mesozoico). Profil 12:1–201

Azéma J, Glaçon G, Tournon J, Sormay J (1978) Presencia de amonites en las Formaciones del "Complejo de Nicoya". El yacimiento de loma Chumico, provincia de Guanacaste, Costa Rica. Inf Sem Inst Geogr Nac II-78:77–90

Baumgartner P (1984) El Complejo Ofiolítico de Nicoya (Costa Rica): modelos estructurales analizados en función de las edades de radiolarios (Calloviense a Santoniense). Pp. 115–123In: Sprechmann P (ed) Manual de geología de Costa Rica 1. Edit Universidad de Costa Rica, San José

Borne P (1995) Evidence of Late Pliocene-Early Pleistocene environmental change and possible coastal upwelling based on marine ostracodes from the Moin Formation, Costa Rica. Gulf Coast Ass Geol Soc Transact 45:61–64

Borne P, Hazel J, Cronin T, McNeill D (1996) Characteristic ostracode assemblages of the Rio Banano Formation (Pliocene) of Costa Rica. Geol Soc Am Ann Meet, Abstr Progr 28(6): 129

Borne P, Cronin T, Hazel J (1999) Neogene-Quaternary Ostracoda and paleoenvironments of the Limon Basin, Costa Rica, and Bocas del Toro Basin, Panama, pp 213–232. *In*: Collins LS, Coates AG (eds) A paleobiotic survey of Caribbean faunas from the Neogene of the Isthmus of Panama. Bull Am Paleontol 357

Budd A, Stemann T, Johnson K (1994) Stratigraphic distributions of genera and species of Neogene to Recent Caribbean reef corals. J Paleont 68:951–977

Budd A, Foster CT, Jr, Fortunato H, Golden J, Petersen RA (1996) Neogene marine invertebrates of tropical America: a world-wide web taxonomy database. Sixth North Am Paleontol Convent Abstr Paleontol Soc Spec Publ 8:52

Budd A, Johnson K, Stemann T, Tompkins B (1999) Pliocene to Pleistocene reef coral assemblages in the Limon Group of Costa Rica, pp. 119–158. *In*: Collins LS, Coates AG (eds) A paleobiotic survey of Caribbean faunas from the Neogene of the Isthmus of Panama. Bull Am Paleont 357

Bybell L (1999) Neogene calcareous nannofossil biostratigraphy of the Caribbean coast of Panama and Costa Rica, pp. 41–59. *In*: Collins LS, Coates AG (eds) A paleobiotic survey of Caribbean faunas from the Neogene of the Isthmus of Panama. Bull Am Paleont 357

Cheetham A, Jackson J, Sanner J, Ventocilla Y (1999) Neogene cheilostome Bryozoa of tropical America: comparison and contrast between the Central American isthmus (Panama, Costa Rica) and the North-Central Caribbean (Dominican Republic), pp 159–192. *In*: Collins LS, Coates AG (eds) A paleobiotic survey of Caribbean faunas from the Neogene of the Isthmus of Panama. Bull Am Paleont 357

Collins L (1996) When did the Isthmus of Panama emerge? Biogeographic, paleobathymetric and evolutionary evidence from benthic Foraminifera. Sixth North Am Paleontol Conv Abstr Paleont Soc Spec Publ 8:78

Collins L (1999a) The Miocene to recent diversity of Caribbean benthic Foraminifera from the Central American Isthmus, pp 91–107. *In*: Collins LS, Coates AG (eds) A paleobiotic survey of Caribbean faunas from the Neogene of the Isthmus of Panama. Bull Am Paleont 357

Collins L (1999b) Increasing diversity in Caribbean benthic Foraminifera over the last 10 million years. Geol Soc Am Ann Meet, Abstr Progr 31:A-355

Collins, L, Geary D, Lohmann K (1995). A test of the prediction of decreased Caribbean coastal upwelling caused by emergence of the Isthmus of Panama, using stable isotopes of neritic Foraminifera. Geol Soc Am Ann Meet, Abstr Progr 27:A-156

Cortés J (1986) Biogeografía de corales hermatípicos: el istmo centroamericano. Anales Inst Cien Mar Limnolog, U.N.A.M. 13:297–304

Cotton M (1999) Neogene planktic foraminiferal biochronology of the southern Central American Isthmus, pp 61–80. *In*: Collins LS, Coates AG (eds) A paleobiotic survey of Caribbean faunas from the Neogene of the Isthmus of Panama. Bull Am Paleont 357

Cowen R (2000) History of life. Blackwell Science, Malden, MA, 432 p

Denyer P, Alvarado G, Aguilar T. (2000). Historia geológica. *In*: Denyer P, Kussmaul S (eds) Geología de Costa Rica. Editorial Tecnológica, Cartago, Costa Rica, pp 155–167

De Wever P, Azema J, Tournon J, Desmet A. 1985. Découverte de materiel du Lias-Dogger inferieur dans la péninsula de Santa Elena (Costa Rica, Amérique Centrale). Acad Sci Paris II(15):759–765

Fischer R, Aguilar T (1994) Paleontology of an evolving Island Arc. Profil 7:391–400

Fischer R, Aguilar T (1995) La asociación bentónica del bivalvo miocénico *Nuculana fundationis*: índice de condiciones disaeróbicas en sedimentos clásticos finos. Rev Geol Am Central 18:65–76

Fischer R, Aguilar T (2001) Bryozoan borings in shells of Costa Rican Miocene oysters. Geol Beitr Hannover 2:71–77

Fischer R, Aguilar T (2007). Chapter 17, pp. 453–466. Invertebrate Paleontology. *In*: Bundschuh J, Alvarado G (eds) Central America Geology, Resources and Hazards, Vol 1, Taylor & Francis, London

Fortunato H, Jung P (1995) The *Strombina* group (Neogastropoda: Columbellidae): a case study of evolution in the Neotropics. Geol Soc Am Ann Meet, Abstr Progr 27:A-52

Galli C. & Schmidt-Effing R (1977) Estratigrafía de la cubierta sedimentaria supra-ofiolítica cretácica de Costa Rica. Cienc Tecnol 1:87–96

Gursky H (1989) Presencia y origen de rocas sedimentarias en el basamento ofiolítico de Costa Rica. Rev Geol Am Central 10:19–66

Haq B, Hardenbol J, Vail P (1988) Mesozoic and Cenozoic chronostratigraphy and eustatic cycles in sea level changes: an integrated approach. Soc Econ Paleont Mineral, Spec Publ 42:71–108

Hoffstetter R, Dengo G, Dixon CG, Meyer-Abich H, Weyl R, Woodring WP, Zoppis Bracci L (1961) Amérique Latine: Fascicule IIa: Amérique Centrale. International de la Recherche Scientifique, Paris, pp 224–354

Jaccard S, Münster M, Baumgartner P, Baumgartner C, Denyer P. (2001) Barra Honda (Upper Paleocene-Lower Eocene) and El Viejo (Campanian-Maastrichtian) carbonate platforms in the Tempisque area (Guanacaste, Costa Rica). Rev Geol Am Central 24:9–28

Jackson J, Cheetham A (1994) Phylogeny reconstruction and the tempo of speciation in cheilostome Bryozoa. Paleobiology 20:407–423

Jackson JBC, D'Croz L (1997) The ocean divided, pp. 38–71. *In*: Coates AG (ed) Central America: A Natural and Cultural History. Yale University Press, New Haven/London

Jackson J, Jung P, Fortunato H (1996). Paciphilia revisited: transisthmian evolution of the Strombina group (Gastropoda: Columbellidae), pp 234–270. *In*: Jackson JBC Budd AF, Coates AG (eds) Evolution and Environment of Tropical America. University of Chicago Press, Chicago

Jackson J, Todd J, Fortunato H, Jung P (1999) Diversity and assemblages of Neogene Caribbean mollusca of lower Central America, pp 193–230. *In*: Collins LS, Coates AG (eds) A paleobiotic survey of Caribbean faunas from the Neogene of the Isthmus of Panama. Bull Am Paleontol 357

Johnson K, Budd A, Stemann T (1995) Extinction selectivity and ecology of Neogene Caribbean reef corals. Paleobiology 21:52–73

Laurito C (1988) Análisis sedimentológico de las Formaciones El Fraile, en la costa suroeste de Nicaragua, Carmen y Punta Carballo, en la costa Pacífica de Costa Rica. Tesis Lic. Esc. Centroamer. Geología, Universidad de Costa Rica, San Pedro, p. 117.

Malavassi E (1960) Algunas localidades de Costa Rica con foraminíferos grandes. Ministerio de Agricultura e Industrias. Dept Geol, Minas Petróleo 1:(3):1–5

Malavassi E (1975) Nota geopaleontológica preliminar sobre el Valle del Río Reventazón en su curso medio. Rev Geol Am Central 2:35–48

Nehm R (1994) A biogeographic comparison of evolutionary turnover in *Prunum* (Gastropoda: Marginellidae) from the Neogene American tropics. Geol Soc Am Ann Meet, Abstr Progr 26: A-375

Olsson A (1922) The Miocene of northern Costa Rica. With notes on its general stratigraphic relations. Bull Am Paleontol 9(39):173–481

Olsson A (1942) Tertiary and Quarternary fossils from the Burica Peninsula of Panama and Costa Rica. Bull Am Paleontol 27:1–106

Pons J, Schmidt-Effing R (1998) Upper Cretaceous rudists from Guanacaste Province, Costa Rica, Central América. Terra Nostra 16:124–125

Schmidt-Effing R (1974) El primer hallazgo de amonites en América Central Meridional y notas sobre las facies cretácicas en dicha región. Costa Rica. Inf Sem Inst Geogr Nac I-74:53–61

Schmidt-Effing R (1979) Alter und Genese des Nicoya-Komplexes, einer ozeanischen Paläokruste (Oberjura bis Eozän) des südlichen Zentralamerikas. Geol Rundsch 68:457–494

Schmidt-Effing R (1980) Radiolarien der Mittel-Kreide aus dem Santa Elena-Massiv von Costa Rica. N Jb Geol Paläont Abh 160:241–257

Seyfried H, Sprechmann P, Aguilar T (1985) Sedimentología y paleoecología de un estuario del litoral Pacífico del Istmo Centroamericano primordial (Mioceno Medio, Costa Rica). Rev Geol Am Central 3:1–68

Ward PD (1990) The Cretaceous/Tertiary extinctions in the marine realm; a 1990 perspective. Geol Soc Am Spec Pap 247:425–432

Winsemann J (1992) Tiefwasser-Sedimentationsprozesse und -produkte in dem Forearc-Becken des mittelamerikanischen Inselbogensystems: eine sequenzstratigraphische Analyse. Profil 2:1–218.

Woodring W (1957) Geology and paleontology of Canal Zone and adjoining parts of Panama. Geology and description of Tertiary molluscs (Gastropods: Trochidae to Turritellidae). Geol Surv Prof Pap 306-A:1–145

Woodring W (1959) Geology and paleontology of Canal Zone and adjoining parts of Panama. Description of Tertiary molluscs (Gastropods: Vermetidae to Thaididae). Geol Surv Prof Pap 306-B:147–239

Woodring W (1964) Geology and paleontology of Canal Zone and adjoining parts of Panama. Description of tertiary molluscs (Gastropods: Columbellidae to Volutidae). Geol Surv Prof Pap 306-C:241–297

Woodring W (1966) The Panama land bridge as a sea barrier. Proc Am Phil Soc 110:425–433

Woodring W (1970) Geology and paleontology of Canal Zone and adjoining parts of Panama. Description of Tertiary molluscs (Gastropods: Eulimidae, Marginellidae to Helminthoglyptidae). Geol Surv Prof Pap 306-D:299–452

Woodring W (1973a) The Miocene Caribbean faunal province and its subprovinces. Verhandl naturforsch Ges Basel 84:209–213

Woodring W (1973b) Geology and paleontology of Canal Zone and adjoining parts of Panama. Description of tertiary molluscs (Additions to gastropods, pelecypods: Nuculidae to Mallaeidae). Geol Surv Prof Pap 306-E:453–539

Woodring W (1982) Geology and paleontology of Canal Zone and adjoining parts of Panama. Description of tertiary molluscs (Pelecypods: Propeamussidae to Cuspidariidae. Additions to families covered in P-306-E; additions to Gastropods; Cephalopods). Geol Surv Prof Pap 306-F:541–759

Chapter IV
Taxonomic Groups

Part 1
Phytoplankton

Roxana Víquez and Paul E. Hargraves

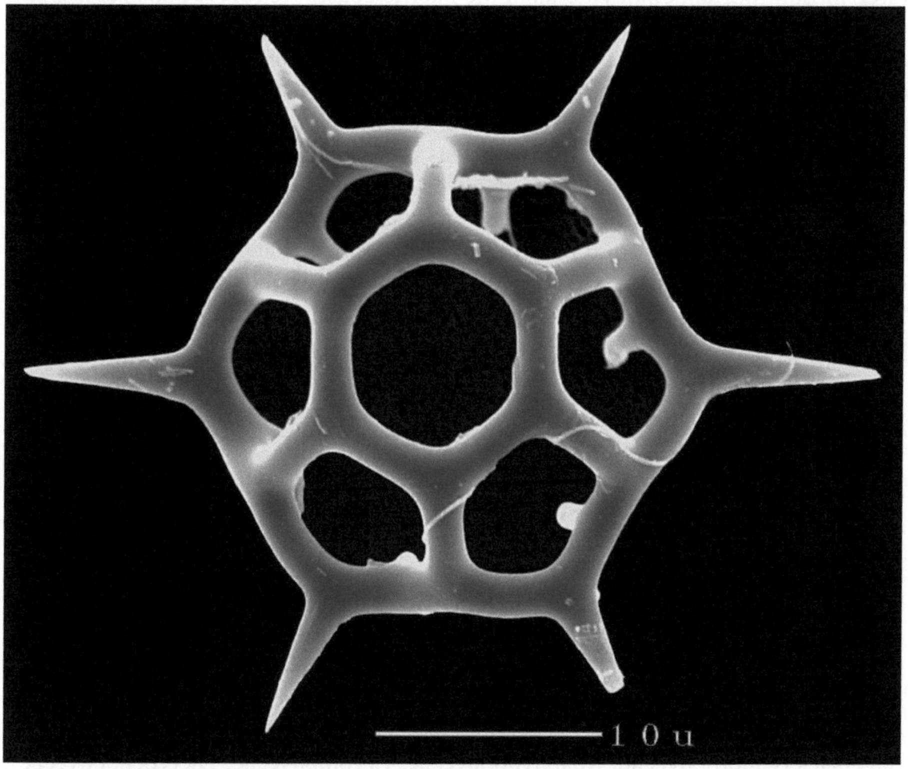

Dictyocha speculum, a silicoflagellate, from the Golfo de Nicoya, Pacific Costa Rica. (Photo: Paul E. Hargraves)

R. Víquez
Escuela de Ciencias Biológicas, Universidad Nacional Apdo 86-3000, Heredia, Costa Rica.
e-mail: rviquezm@una.ac.cr

P.E. Hargraves
Graduate School of Oceanography, University of Rhode Island, Narragansett Rhode Island 02881, USA and Harbor Branch Oceanographic Institution at Florida Atlantic University, Ft. Pierce, Florida 34946, USA
e-mail: pharg@gso.uri.edu

Abstract Phytoplankton are a complex group, composed of several taxonomic classes of unicellular or colonial species with great morphological and color variation, generally drifting with water movement. Some have one or several benthic life stages. They are mostly photosynthetic but some require organic nutrients. Color is important for taxonomic classification and depends on the chloroplast pigment composition. Studies on the biodiversity of marine phytoplankton in Costa Rican coasts are sparse and mostly confined to the Pacific coast, more specifically to the Golfo de Nicoya. New records presented here come from the three most important embayments in the Pacific coast. We report 268 species, 258 for the Pacific coast, 53 for the Caribbean coast, and 43 species common to both coasts. There are no reports of coccolithophorids, and only few species of flagellates other than dinoflagellates have been identified. Some taxonomic classes possibly occurring in local waters have never been examined. Phytoplankton listed for both Caribbean and Pacific coasts is basically neritic and cosmopolitan. When compared to the floras described for other parts of the world it is evident that much more work is needed on both Costa Rican coasts. It is important to monitor the occurrence of potentially toxic species that have proven to be a danger to human health in Costa Rica, to evaluate possible extinction events of local endemic flora, and to intensify the search of potentially valuable marine pharmaceuticals.

Introduction

The great majority of phytoplankton are eukaryotic cells that spend all or most of their life stages drifting in the water column of the oceans (and lakes). Some may have a resting stage that sink to the bottom where they await favorable environmental conditions to restart vegetative growth; some of these, so-called meroplanktonic species, can have complex life cycles with several bottom-dwelling and/or free-floating stages. The main components of phytoplankton are diatoms (Bacillariophyceae), dinoflagellates (Dinophyceae), coccolithophores (Prymnesiophyceae), and some flagellates (Chlorophyceae, Rhaphydophyceae, Cryptophyceae, Dictyochophyceae). Some ciliates (*Myrionecta*) or flagellates (*Hermesinum*) are regarded as heterotrophic but are autotrophic because they have endosymbiotic algae. Most of them are mixotrophic, i.e., they can photosynthesize but also require organic matter (mostly vitamins and amino acids) which they get by absorption from the surrounding water or by consuming entire cells, as some dinoflagellates do. There are some planktonic photosynthetic prokaryotes: purple and green sulfur bacteria and the blue green algae (Cyanophyceae) (Raymond 1980).

Perhaps the most striking feature of phytoplankton is the variability in shape and cell wall constitution (Raymond 1980): cellulose in many dinoflagellates, opal-like glass in diatoms, calcium carbonate in coccolithophores, organic or silicon scales in some flagellates. Some show elaborated cell wall projections, such as spines, horns, or wings; some can live as solitary cells, other can form long chains or colonies. Color depends on the chloroplast pigments. The pigment composition, along with cell wall composition and structure, type of nucleus, chloroplast envelope, thylakoid organization, and type of flagella are taxonomic criteria used for class separation.

Studies of phytoplankton species richness on the Pacific coast of Costa Rica started with the collections of Allen (1925, 1939) and Cupp (1934). In a study of the annual cycle of potentially toxic dinoflagellates, Víquez & Hargraves (1995) documented the occurrence of species that were reported as toxic in other areas such as *Gymnodinium catenatum* and *Pyrodinium bahamense var. compressum.* Brugnoli (1998) and Brugnoli Olivera & Morales Ramírez (2001) studied the dynamics and structure of the phytoplankton community but did not consider species, only higher taxa levels such as genus and family. Vega (1999) records 151 taxa, including 73 species of diatoms, 18 species of dinoflagellates, and 2 other flagellates that were identified to the species level. Records of single species are provided by Morales *et al.* (2001) who reported a red tide by *Lingulodinium polyedrum*, and Mata *et al.* (1990) documented the occurrence of human poisoning due to *Pyrodinium bahamense var. compressum.* Vargas (2001) included descriptions, ultrastructure, and electron microscope photographs of 38 taxa of dinoflagellates, some of them identified to genus. More recently, Vargas-Montero & Freer (2004a) provided information on the presence of dinoflagellates of the genus *Ceratium* spp. in the Golfo de Nicoya, and they reported also on recent toxic algal bloom in the Golfo de Nicoya (Vargas-Montero & Freer 2004b, c).

Studies on the phytoplankton in the Costa Rican Caribbean are even sparser, and limited information exists on the flora. The only published annual cycle analysis of the phytoplankton assemblage of the Caribbean coast of Costa Rica was provided by Vargas (1991), and this information is included in Species List 1.1 (on CD-Rom). Another study (Silva 1986) on the photosynthesis and standing crop of phytoplankton in Cahuita identified phytoplankton to the genus level.

Most of the phytoplankton records for the Pacific coast (Species List 1.2 is included on the CD-Rom) come from Bahía Culebra (Golfo de Papagayo) in the north, Golfo de Nicoya in the central part of the country (Fig. 1.1), and Golfo Dulce in the south. Species reported by us as new records (indicated as NR in the first column in Species Lists 1.1 and 1.2, both on CD-Rom) are from bottle samples and net tow collections obtained from 1980 to 2001 (projects UNA-921002, UNA-021066 and UNA-023433 granted to R. Víquez). These samples were preserved with Lugol's solution. Dinoflagellates were identified from water mounts and diatoms from permanent resin mounts (Hyrax, Aroclor, or Melmount). All samples and resin mounts on duplicate glass slides were catalogued and deposited in the Plankton Collection at Estación de Biología Marina, Universidad Nacional, Puntarenas (R. Víquez collection), or at the Graduate School of Oceanography, URI (PEH collection). Here we report 268 species, 258 species for the Pacific coast, 53 species for the Caribbean coast, and 43 species occurring along both coasts (Table 1.1).

As in most compilations of species lists and surveys, the accuracy of identifications depends on the individual skills of the surveyor, which is unfortunately variable, and the adequacy of available literature. This compilation is no exception, and we have chosen not to judge the accuracy of the identifications of others. In addition, advances in taxonomic opinion may change rapidly, and the names in this list are always subject to modification.

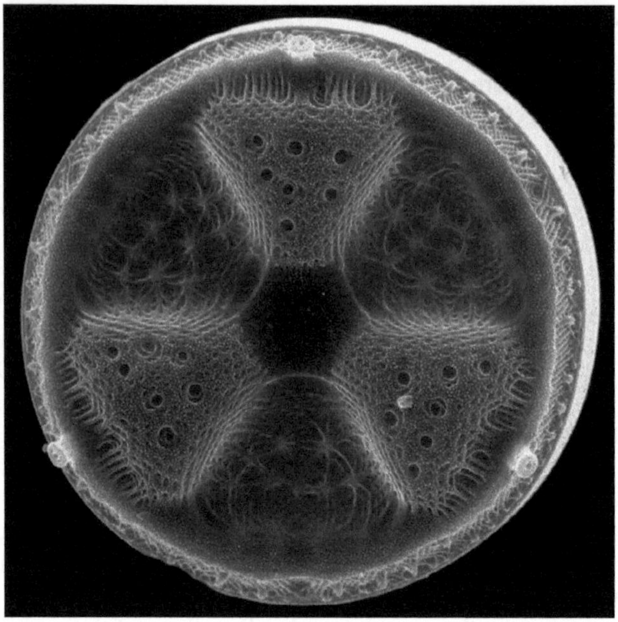

Fig. 1.1 *Actinoptychus senarius*, a diatom, from the Golfo de Nicoya, Costa Rica. Photo: Paul E. Hargraves

Table 1.1 Summary of phytoplankton species reported for each coast of Costa Rica

	Pacific	Caribbean		Pacific	Caribbean
Bacillariophyceae	151	46	Gymnodiniales	14	0
Coscinodiscales	40	6	Noctilucales	2	0
Rhizosoleniales	14	6	Peridiniales	30	0
Bidduphiales	56 plus 2 subspecies	14	**Cyanobacteria**	4 genera, 2 species	3 genera, 2 species
Fragilariales	11 plus 1 subspecies	2	**Ebridea**	2	0
Bacillariales	27	18	**Ciliata**	1	0
Dinophyceae	96	3	Prymnesiophyceae	1	0
Prorocentrales	11	1	**Raphydophyceae**	1	0
Dinophysiales	10	1	**Dictyochophyceae**	2	1
Gonyaulacales	29	1	**TOTAL N° OF TAXA**	**258**	**53**
Species reported only from the Pacific coast		215	Species reported only from the Caribbean coast		10
Number of species occurring along both coasts					43
TOTAL NUMBER OF SPECIES					**268**

Discussion

None of the above mentioned publications considered coccolithophorids, although these carbonate-secreting species are known to be diverse in other tropical coastal areas. When any local area is examined in detail for microalgae, species richness increases dramatically. In Florida, for example, an intensive survey of protists by one of us (PEH) tripled the number of reported species (www.sms.si.edu/irlspec/Species_Rpts.htm) based on a 3-month survey. Likewise, entire taxonomic groups of microalgae (Prymnesiophyceae, Cryptophyceae, Prasinophyceae, and several other classes) have never been examined in Costa Rican waters, and only occasional anecdotal observations are available. In other temperate and tropical coastal regions these (mostly) nanoplankton-sized organisms frequently make significant contributions to primary productivity and can form harmful algal blooms. They are neglected mostly because they are destroyed in common preservatives, their study requires electron microscopy, molecular techniques or both, and competent taxonomists for these difficult groups are scarce. Most species listed in this part are either easily preservable, identifiable by light microscopy, or both.

Phytoplankton so far reported for the Caribbean and Pacific coasts of Costa Rica can be defined as basically neritic and cosmopolitan. Communities of microplankton-sized cells (i.e., 20–200 µm) were dominated by diatoms, and *Skeletonema costatum* was repeatedly identified as an abundant and ubiquitous species (although this name may represent a complex of several species; Sarno *et al.* 2007, J. Phycol. 43:156–170). The species-rich genus *Chaetoceros* played a prominent role in determining the diversity of the Costa Rican flora (Víquez 1983; Brugnoli 1998; Vega 1999; Brugnoli Olivera & Morales Ramírez 2001).

The phytoplankton in the Golfo de Nicoya shared around 50% of species with the Gulf of Panama, the Gulf of California, the Gulf of Guayaquil in Ecuador, and the summer flora of Narragansett Bay in the east coast of the United States (Hargraves & Víquez 1985). Although this suggests a high degree of cosmopolitan species, it is also true that five centuries of interoceanic ship travel makes it difficult to decide which species are truly cosmopolitan, and which are invasive species on a century-long time scale. In other words, many cosmopolitan species could be more accurately described as cryptogenic.

When compared to the floras described for other parts of the world it is evident that much more work is needed on both Costa Rican coasts. As an example, here we report only 13 species of the genus *Ceratium* for the Pacific coast and two species for the Caribbean; however, in the Gulf of California 83 taxa have been described, close to the total number of species reported worldwide (120 spp.) (Cortés-Altamirano & Nuñez 2000).

Knowledge of phytoplankton species richness for the Caribbean coast of Costa Rica (Species List 1.1: 53 spp.) is rather incomplete, probably because the single study (Vargas 1991) was carried out at the mouth of two estuaries with strong freshwater influence, thus excluding many oceanic stenohaline species. It is also noteworthy that for some of the species listed, few or no references from other tropical areas were found, whereas other taxa were ubiquitous and common according to a survey of published records from other areas. In Venezuela, Díaz-Ramos (2000) compiled a total of 461 species (275 species of diatoms, 162 species of dinoflagellates, and 24 species of coccolithophores) based on 40 years of studies by several authors.

Since harmful algal blooms have an impact on public health and economy of Costa Rica (Mata *et al.* 1990; Morales *et al.* 2001), it is necessary to monitor more closely the occurrence of those species known to be toxic to both humans and marine animals of economic importance. Moreover, it has been estimated that hundreds or thousands of species become extinct before they are discovered (which is a non-verifiable statement). It is likely that extinctions and biodiversity loss also take place at some level in coastal marine environments. Before evaluations of such extinction events can be made, it is necessary to know the number of endemic species in marine areas of Costa Rica. Widespread species may be in danger of local extinction, but not necessarily global eradication; in contrast, endemic species are much more susceptible. Given the common occurrence of allelopathy (i.e., biochemical warfare) amongst microscopic organisms, it is likely that potentially valuable marine pharmaceuticals may be found in phytoplankton species, waiting to be discovered. The first step is the establishment of biotic inventories, and there is much to be done. This is contribution # 1695 from Harbor Branch Oceanographic Institution of Florida Atlantic University, and contribution #728 from Smithsonian Marine Station/Ft Pierce, FL. PEH acknowledges support from both institutions.

Specialists

Currently Working with Phytoplankton of Costa Rica

Emilia Calvo V.: Estación de Biología Marina, Universidad Nacional, Puntarenas, Costa Rica – Red tide dinoflagellates. ecalvo@una.ac.cr

Paul E. Hargraves: Graduate School of Oceanography, University of Rhode Island, Kingston, Rhode Island, USA and Harbor Branch Oceanographic Institution of Florida Atlantic University, Ft Pierce, FL 34946 USA – Taxonomy of phytoplankton. pharg@gso.uri.edu

Maribelle Vargas M.: Centro de Investigación en Estructuras Microscópicas, Universidad de Costa Rica, Costa Rica – Red tide organisms. vmontero@cariari.ucr.ac.cr

Roxana Víquez: Escuela de Ciencias Biológicas, Universidad Nacional, Costa Rica – Taxonomy of phytoplankton. rviquezm@una.ac.cr

References

1. Aké-Castillo JA, Hernández-Becerril DU, Meave del Castillo ME (1999) Species of the genus *Thalassiosira* (Bacillariophyceae) from the Gulf of Tehuantepec, México. Bot Mar 42:487–503

Allen WE (1925) Statistical studies of surface catches of marine diatoms and dinoflagellates made by yatch Ohio in tropical waters in 1924. Trans Am Microsc Soc 44:24–30

Allen WE (1939) Surface distribution of marine plankton diatoms in the Panamá region in 1933. Bull Scripps Inst Ocean 4(7):181–196

2. Allen WE (1943) Red water in La Jolla Bay in 1942. Trans Amer Microsc Soc 62:262–264
3. Avaria S (1965) Diatomeas y dinoflagelados de la Bahía de Valparaíso. Rev Biol Mar 12:60–119
4. Balech E (1985) A revision of *Pyrodinium bahamense* Plate (Dinoflagellata) Rev. Paleobot Palynol 45:17–34
5. Balech E (1988) Los dinoflagelados del Atlántico sudoccidental. Publ. Esp. Inst. Esp. Oceanogr., Madrid N° 1:310 p
6. Balech E (2002) Dinoflagelados tecados tóxicos del cono sur americano. 2002. In: Sar EA, Ferrario ME, Reguera B (eds.). Floraciones Algales Nocivas en el Cono Sur Americano. Instituto Español de Oceanografía, Vigo, España, pp. 123–144
7. Balkis N (2004) Tintinnids (Protozoa: Ciliophora) of the Büyükcekmece Bay in the Sea of Marmara. Sci Mar 68:33–44
8. Barnard R, Batten S, Beaugrand G, Buckland C, Conway DVP, Edwards M, Finlayson J, Gregory LW, Halliday NC, John AWG, Johns DG, Johnson AD, Jonas TD, Lindley JA, Nyman J, Pritchard P, Reid PC, Richardson AJ, Saxby RE, Sidey J, Smith MA, Stevens DP, Taylor CM, Tranter PRG, Walne AW, Wootton M, Wotton COM, Wright JC (2004) Continuous plankton records: plankton Atlas of the North Atlantic Ocean (1958–1999). II. Biogeographical charts. Mar Ecol Prog Ser (Suppl 2004):11–75
9. Botes L (2003) Phytoplankton Identification Catalogue- Saldanha Bay, South Africa, April 2001. GloBallast Monogr. Ser, IMO, London N° 7:77
10. Braarud T, Foyn B, Hasle GR (1958) The marine and fresh water phytoplankton of the Dramsfjord and the adjacent part of Oslofjord March-December 1951. Det Norske Videnskaps-Akademi I Oslo. Oslo Kommisjon hos H. Aschehoug & Co. 102 p
11. Bradbury PC, Westefall JA, Townsend JW (1983) Ultrastructure of the dinoflagellate *Polykrikos*. II. The nucleous and its connections to the flagellar apparatus. J Utrast Res 85:24–32

Brugnoli E (1998) Factores abióticos y bióticos que regulan la estructura y dinámica de la comunidad fitoplanctónica en la zona de Punta Morales, Golfo de Nicoya. M.Sc. thesis, University of Costa Rica, San Pedro, 132 p

Brugnoli Olivera E, Morales Ramírez Á (2001) La comunidad fitoplánctica de Punta Morales, Golfo de Nicoya, Costa Rica. Rev Biol Trop 49(Suppl 2): 11–17

12. Burns DA, Mitchell JS (1982) Dinoflagellates of the genus *Dinophysis* Ehrenberg from New Zealand coastal waters. New Zealand J Mar FreshwRes 16:289–298
13. Cembella A (1989) Occurrence of okadaic acid, a major diarreic shellfish toxin in natural population of *Dinophysis* spp. from the eastern coast of North America. J Appl Phycol 1:307–310
14. Cortés-Altamirano R (1984) Mareas rojas producidas por el ciliado *Mesodinium rubrum* (Lohmann) en el área litoral de Mazatlán, Sinaloa, México. Biótica 9:259–269
15. Cortés-Altamirano R (1986) Serie coleccionable Plancton: *Skeletonema costatum* (Grev.) Cleve y *Skeletonema tropicum* Cleve. Rev Cienc Mar, UAS 8:22–23
16. Cortés-Altamirano R, Nuñez-Pastén A (2000) Distribución y abundancia anual de *Ceratium dens* (Peridinales: Ceratiaceae) en el Golfo de California, México. Rev Biol Trop 48:305–311

17. Curl H (1953) The phytoplankton of Apalachee Bay and the northeastern Gulf of México. Oceanogr Inst, Florida State Univ, Contr 102:277–320
18. Cupp EE (1934) Analysis of marine plankton diatom collections taken from the canal zone to California during March, 1933. Trans Amer Microscop Soc 53:22–29
19. Cupp EE (1943) Marine plankton diatoms of the west coast of North America. Bull Scripps Inst Oceanogr 5:1–238
20. Delgado M, Fortuño JM (1991) Atlas del fitoplancton del mar Mediterráneo. Sci Mar 55(Suppl 1):1–133
21. Díaz-Ramos JR (2000) Índice de la microflora marina de Venezuela: diatomeas, dinoflagelados y cocolitofóridos. Rev Biol Trop 48:897–918
22. Dodge JD (1975) The Prorocentrales (Dinophyceae). II. Revision of the taxonomy within the genus *Prorocentrum*. Bot J Linn Soc 71:103–125
23. Dodge JD (1982) Marine dinoflagellates of the British Isles. Her Majesty's Stationary Office. London, 303 p
24. Florida Fish and Wildlife Research Institute. Harmful Algal Blooms and Red Tides - HAB Species. *Heterocapsa niei*. (Downloaded August 1, 2004). http://www.floridamarine.org/gallery/image_details.asp?id = 3977
25. Fukuyo Y, Takano H,Chihara M, Matsuoka K (eds.) (1990) Red tide organisms in Japan –an illustrated Taxonomic Guide. Uchida Rokakuho, Tokyo. 430 p
26. Gómez F, Gorsky G (2003) Annual microplankton cycles in Villefranche Bay, Ligurian Sea, NW Mediterranean. J Plankton Res 25:323–339
27. Gómez-Aguirre S (2003) Eventos de hemotalasia en México: el caso observado en la Bahía de Banderas (abril, 2001). Rev Espec Cienc Quím-Biol 6:5–10
28. Guillard RRL, Hellebust JA (1971) Growth and the production of extracellular substances by two strains of *Phaeocystis poucheti*. J Phycol 7:330–338
29. Hallegraeff GM (1986) Taxonomy and morphology of the marine plankton diatoms *Thalassionema* and *Thalassiothrix*. Diatom Res 1:57–80
30. Hallegraeff GM (1991) Aquaculturist guide to harmful Australian microalgae. CSIRO Division Fisheries, Tasmania, Australia
31. Hallegraeff GM, Reid DD (1986) In phytoplankton species successions and their hydrological environment at a coastal station off Sydney. Aust J Mar Freshwat Res 37:361–377
32. Hargraves PE (1976) Studies on marine plankton diatoms. III The structure and clasification of *Gossleriella tropica*. J Phycol 12:285–291
33. Hargraves PE (2002) Diatoms of the Indian River Lagoon, Florida: An annotated account. Florida Scient 65:225–244
34. Hargraves PE, Víquez R (1980) *Spirulina subsalsa* Oersted en Costa Rica. Estructura y posible importancia comercial. Rev Biol Trop 22:304–307
35. Hargraves PE, Víquez R (1985) Spatial and temporal distribution of phytoplankton in Gulf of Nicoya, Costa Rica. Bull Mar Sci 37:577–585
36. Hasle GR (1960) Phytoplankton and ciliate species from the tropical Pacific. Skrifter utgitt av Det Norske Videnskaps-Akademi i Oslo I. Matematisk-Naturvidenskapelig Klasse 2:1–150
37. Hasle GR (1975) Some living marine species of the diatoms family Rhizosoleniaceae. Beiheft Nova Hedwigia 53:99–140
38. Hasle GR (1980) Examination of *Thalassiosira* type material: *T. minima* and *T. delicatula* (Bacillariophyceae). Norw J Bot 27:167–173
39. Hasle GR (1983) The marine, planktonic diatoms *Thalassiosira oceanica* sp. nov. and *T. partheneia*. J Phycol 19:220–229
40. Hasle GR (1993) Nomenclatural notes on marine planktonic diatoms. The family Bacillariaceae. Beiheft Nova Hedwigia 106:1036–1039
41. Hasle GR, Lange CB (1992) Morphology and distribution of *Coscinodiscus* species from the Oslofjord, Norway, and the Skagerrak, North Atlantic. Diatom Res 7:37–68
42. Hasle GR, Sims P (1985) The morphology of the diatom resting spores *Syringidium bicorne* and *Syringidium simplex*. Brit Phycol J 20:219–225

43. Hasle GR, Syvertsen EE (1980) The diatom genus *Cerataulina*: Morphology and taxonomy. Bacillaria 3:79–113
44. Hasle GR, Syvertsen EE (1996) Marine diatoms. *In*: Tomas CR (ed) Identifying Marine Diatoms and Dinoflagellates. Academic Press Inc., San Diego Pp. 5–385
45. Heiden H, Kolbe RW (1928) Die marinen Diatomeen der Deutschen Süd Polar Expedition 1901–3. Deutschen Süd Polar Exped. 1901–1903, 8:447–715
46. Hermosilla J (1973) Contribución al conocimiento sistemático de los dinoflagelados de la bahía de Concepción, Chile. Inst. Biol., Univ. Concepción, Chile. 149 p
47. Hernández-Becerril DU (1985) Dinoflagelados en el fitoplancton del Puerto de El Sauzal, Baja California. Cienc Mar (Méx) 11:65–91
48. Hernández-Becerril DU (1988) Especies de fitoplancton tropical del Pacífico Mexicano. II. Dinoflagelados y cianobactericas. Rev Lat Amer Microbiol 30:187–196
49. Hernández-Becerril DU (1991) *Protoperidinium* (Dinophyceae) species of the Gulf the California and of the Gulf of Baja California. Anales Inst Cienc Mar Limnol 18:77–88
50. Hernández-Becerril DU (1992) Nota sobre la presencia de la diatomea *Asteromphalus cleveanus* Grunow en aguas de Baja California y su relación con *A. flavellatus* (Brev.) Grev. Acta Bot Mex 17:39–44
51. Hernández-Becerril DU (1992) *Dinophysis taylorii sp. nov.* y otros *Dinophysis* de Baja California, México (Dinophyceae). Rev Biol Trop 40:101–109
52. Hernández-Becerril DU, Navarro N (1996) Thecate dinoflagellates (Dinophyceae) from Bahía Fosforescente, Puerto Rico. Rev Biol Trop 44:465–475
53. Hernández-Becerril DU, Cortés-Altamirano, R, Alonso R (2000) The dinoflagellate genus *Prorocentrum* along the coasts of the Mexican Pacific. Hydrobiol. 418:111–121
54. Howell JF (1953) *Gonyaulax monilata*, sp. nov., the causative dinoflagellate of red tide of Florida in August- September 1951. Trans Am Microsc Soc 82:153–156
55. Hulburt EM (1957) The taxonomy of unarmored Dinophyceae of shallow embayments on Cape Cod, Massachusetts. Biol Bull 112:196–219
56. Jiménez R, Intriago P (1987) Observations on blooms of *Mesodinium rubrum* in the upwelling area off Ecuador. Oceanologica Acta Proc Int Symp. Equatorial Vertical Motion, Paris, 6–10 May 1985, pp 145–154
57. Karentz D, Spero H (1995) Response of a natural population to ambient fluctuations of UVB radiation caused by Antartic ozone depletion. J Plankt Res 17:1771–1789
58. Kifle D, Purdie DA (1993) The seasonal abundance of the phototrophic ciliate *Mesodinium rubrum* in Southhampton water, England. J Plankt Res 17:823–833
59. Kofoid CA, Swezy O (1921) The free living unarmored dinoflagellates. Mem Univ California 5:1–562
60. Koray T, Gokpinar S, Yurga L, Turkoglu M, Polat S (7th Apr. 1999): Microplankton species of Turkish Seas. http://bornova.ege.edu.tr/ korayt/p lankweb/chklists.html (Downloaded August 1, 2004)
61. La Barbera-Sánchez A (2000) Informe Técnico Instituto Nacional de Investigaciones Agricolas (INIA), Venezuela, 38 p
62. Larsen JL (2002) Dinoflagelados atecados potencialmente toxígenos en el cono sur americano. *In*: Sar EA, Ferrario ME, Reguera B (eds) Floraciones Algales Nocivas en el Cono Sur Americano Inst Esp Oceanogr, pp 145–154
63. Larsen JL, Moestrup O (1989) Guide to toxic and potentially toxic marine algae. The Fish Inspection Service, Ministery of Fisheries. Copenhagen, 59 p
64. Lassus P (1988) Plankton toxique et plankton d'eaux rouges sur le cotes européenes. IFREMER, France, 111 p
65. Lebour MV (1930) The Planktonic Diatoms of Northern Seas. Otto Koeltz Science Publ., Königstein, Germany, 243 p
66. Licea S, Moreno JL, Santoyo H, Figueroa G (1995) Dinoflageladas del Golfo de California. Univ. Autón. Baja California Sur, México, 165 p
67. Loeblich AR (1968) A new marine dinoflagellate genus *Cachonima* in axenic culture from the Salton Sea, California, with remarks on the genus *Peridinium*. Proc Biol Soc Wash 81:91–96

68. Loeblich AR, III, Wight WW, Darley WM (1968) A unique colonial marine centric diatom *Coenobiodiscus muriformis* gen. et sp. nov. J Phycol 4:23–29
69. Loeblich A, Sherley JL, Schimidt RJ (1979) The correct possition of flagellar insertion in *Prorocentrum* and description of *Prorocentrum rhathymum* sp. nov Pyrrophyta. J Plankt Res 1:113–120
70. Maclean JL (1989) An overview of *Pyrodinium* red tides in the western Pacific. *In*: Hallegraeff GM, Maclean JL (eds) Biology, Epidemiology and Management of *Pyrodinium* Red Tides. ICLARM Conference Proc 21, Fish. Dept., Ministry of Development, Brunei Darussalam, and ICLARM, Manila, Philippines, pp 1–8
71. MacKenzie L, Sims I, Beuzenberg V, Gillespie P (2002) Mass accumulation of mucilage caused by dinoflagellates polysaccharide exudates in Tasmanian Bay, New Zealand. Harmful Algae 1:69–83
72. Marshall HG (1980) Seasonal phytoplankton composition in the lower Chesapeake Bay and Old Plantation Creek, Cape Charles, Virginia. Estuaries 3:207–216
73. Marshall HG (1981) Occurrence of blue green algae (Cyanophyta) in the phytoplankton of the southeastern coast of the United States. J Plankt Res 3:163–166

Mata L, Abarca G, Maranghello L, Víquez R (1990) Intoxicación paralítica por mariscos (IPM) por *Spondylus calcifer* contaminado con *Pyrodinium bahamense*, Costa Rica, 1989–1990. Rev Biol Trop 38:129–136

74. Montagnes DJS, Lynn DH (1989) The annual cycle of *Mesodinium rubrum* in the waters surrounding the Isles of Shoals, Gulf of Maine. J Plankt Res 11:193–201

Morales A, Víquez R, Rodríguez K, Vargas M (2001) Marea roja producida por *Lingulodinium polyedrum* (Peridiniales, Dinophyceae) en Bahía Culebra, Golfo de Papagayo, Costa Rica. Rev Biol Trop 40(Suppl 2):19–23

75. Moreno JL, Licea S, Santoyo H (1996) Diatomeas del Golfo de California. Univ. Autón. Baja California Sur, México, 273 p
76. Muñoz P, Avaria S (1983) *Scripsiella trochoidea* (Stein) Loeblich III, nuevo organismo causante de marea roja en la bahía de Valparaíso, Chile. Rev Biol Mar Valparaiso 19:63–78
77. Navarro JN (1981a) A survey of the marine diatoms of Puerto Rico. I Suborders Coscinodicineae and Rhizosoleniineae. Bot Mar 24:427–439
78. Navarro JN (1981b) A survey of the marine diatoms of Puerto Rico. II. Suborder Biddulphiineae: Families Biddulphiaceae, Lithodesmiaceae and Eupodiscaceae. Bot Mar 24:615–630
79. Navarro JN (1982a) A survey of the marine diatoms of Puerto Rico. III. Suborder Biddulphiineae: Family Chaetoceraceae. Bot Mar 25:305–319
80. Navarro JN (1982b) A survey of the marine diatoms of Puerto Rico. IV. Suborder Araphidineae: Families Diatomaceae and Protoraphidaceae. Bot Mar 25:247–263
81. Navarro JN (1982c) A survey of the marine diatoms of Puerto Rico. V. Suborder Raphidineae: FamiliesAchnantaceae and Naviculaceae. Bot Mar 25:321–338
82. Navarro JN (1983) A survey of the marine diatoms of Puerto Rico. VI. Suborder Raphidineae: Family Naviculaceae (Genera Haslea, Mastogloia and Navicula). Bot Mar 26:119–136
83. Norris DA (1969) Thecal morphology of *Ornithocercus magnificus* (Dinoflagellata) with notes on related species. Bull Mar Sci 19:175–193
84. Norris DA, Berner L (1970) Thecal morphology of selected species of *Dinophysis* (Dinoflagellata) from the Gulf of Mexico. Contr Mar Sci 15:145–192

Raymond JEG (1980) Plankton and Productivity in the Oceans. Vol. 1 - Phytoplankton, 2nd ed. Pergamon Press, London, 489 p

85. Rines JEB, Hargraves PE (1988) The *Chaetoceros* Ehrenberg (Bacillariophyceae) flora of Narragansett Bay, Rhode Island, U.S.A. Bibliotheca Phycologica 79:196
86. Rivera P (1985) Las especies del género *Nitzschia* Hassall, sección *Pseudo-nitzschia* (Bacillariophyceae) en las aguas marinas chilenas. Gayana Bot 42:9–40
87. Round FE,Crawford RM, Mann DG (1992). The Diatoms. Biology, Morphology of the Genera. Cambridge University Press, Cambridge, 747 p
88. Saunders RP, Glenn DA (1969) Diatoms. Mem Hourglass Cruises 1:1–119

Silva Benavides AM (1986) Productividad primaria, biomasa del fitoplancton y la relación con parámetros físico-químicos en el arrecife coralino del Parque Nacional Cahuita. Lic Biology, Universidad de Costa Rica, San Pedro, 72 p

89. Schmid AM (1990) Intraclonal variation in the valve structure of *Coscinodiscus wailesii* Gran et Angst. Beiheft Nova Hedwigia 100:101–119
90. Simonsen R (1974) The diatom plankton of the Indian Ocean expedition of R/V "Meteor" 1964–1965. "Meteor" Forsch-Ergeb D (19):1–107
91. Sims PA (1996) An atlas of the British diatoms. Biopress Ltd., Bristol, England, 601 p
92. Schiller J (1933) Dinoflagellatae (Peridineae) in monographischer Behandlung. I Teil, Lieferung 3. "Dr. L. Rabenhorst's Kryptogamen-Flora von Deutschland, Österreich und der Schweiz". Akad. Verlagsgesell., Leipzig, 617 p
93. Sotto FB, Aruga Y, Yamaguchi Y (1995) Vertical distribution of *Mesodinium rubrum* in response to tidal motion in a shallow tropical embayment in the Philippines. *In*: Lassus P, Arzul G, Erard-Le Denn E, Gentien P, Marcaillou-Le Baut C (eds) Harmful Marine Algal Blooms. Lavoisier, Paris, pp 237–242
94. Steidinger KA (1968) The genus *Gonyaulax* in Florida waters, 1. Morphology and thecal development in *Gonyaulax polygramma* Stein, 1883. Florida Bd. Conserv. Mar. Lab. Leaf. Ser. 1. Phytoplankton, pt. 1, No. 4:1–5
95. Steidinger KA, Tangen K (1996) Dinoflagellates. *In*: Tomas CR (ed) Identifying Marine Diatoms and Dinoflagellates. Academic Press, New York, pp 387–598
96. Subrahmanyan R (1946) A systematic account of the marine plankton diatoms of the Madras coast. Proc Indian Acad Sci 24(4, Sec B):85–197
97. Uchida T, Yamaguchi M, Matsuyama Y, Honjo T (1995) The red-tide dinoflagellate *Heterocapsa* sp. kills *Gyrodinium instriatum* by cell contact. Mar Ecol Prog Ser 118:301–303
98. Vargas W (1991) Abundancia y distribución espacial-temporal de fitoplancton en los estuarios Vizcaya y Estero Negro, Limón, Costa Rica. Esc Cienc Biol, Univ Nacional, Heredia, Costa Rica, 59 p
99. Vargas M (2001) Taxonomía y ultraestructura de algunas especies de dinoflagelados del Golfo de Nicoya, Costa Rica. Lic Biol, Esc Cienc Biol, Univ Nacional, Heredia, Costa Rica, 83 p

Vargas-Montero M, Freer E (2004a) Presencia de los dinoflagelados *Ceratium dens, C. fusus* y *C. furca* (Gonyaulacales: Ceratiaceae) en el Golfo de Nicoya, Costa Rica. Rev Biol Trop 52(Suppl 1):115–120

Vargas-Montero M, Freer E (2004b) Proliferaciones algales nocivas de cianobacterias (Oscillatoriaceae) y dinoflagelados (Gymnodiniaceae) en el Golfo de Nicoya, Costa Rica. Rev Biol Trop 52(Suppl 1):121–125

Vargas-Montero M, Freer E (2004c) Proliferaciones algales de la diatomea toxigénica *Pseudo-Nitzschia* (Bacillariophyceae) en el Golfo de Nicoya, Costa Rica. Rev Biol Trop 52(Suppl 1):127–132

100. Vega H (1999) Ciclo anual del fitoplancton en la parte norte del Golfo de Nicoya, Costa Rica. Lic. Biol., Esc. Cienc. Biol., Univ. Nacional, Heredia, Costa Rica, 101 p
101. Venrick EL (1992) Phytoplankton species structure in the central North Pacific: Is the edge like the center? J Plankt Res 14:665–680
102. Villarreal TA, Frixel GA (1983) The genus *Actinocyclus* (Bacillariophyceae): frustule morphology of *A. sagittulus* sp. nov. and two related species. J Phycol 19:452–466
103. Víquez R (1983) Occurrence and distribution of planktonic diatoms in the Gulf of Nicoya, Costa Rica. M.Sc. Thesis. University of Rhode Island, USA, 89 p
104. Víquez R, Hargraves PE (1995) Annual cycle of potentially toxic dinoflagellates in the Gulf of Nicoya, Costa Rica. Bull Mar Sci 57:467–475
105. Walker LM, Steidinger KA (1979) Sexual reproduction in the toxic dinoflagellate *Gonyaulax monilata*. J Phycol 15:312–315
106. Whedon WF, Kofoid CA (1936) Dinoflagellata of the San Francisco region I. On the skeletal morphology of two new species, *Gonyaulax catenella* and *Gonyaulax acatenella*. Univ California Publ Zool 41:25–34

107. Wilkerson FP, Grunseich G (1990) Formation of bloom by the symbiotic ciliate *Mesodinium rubrum*: the significance of nitrogen uptake. J Plankt Res 12:973–989
108. Wood EJF (1954) Dinoflagellates in the Australian region. Aust J Mar Freshwat Res 5:117–351
109. Wood EJF (1968) Dinoflagellates of the Caribbean Sea and Adjacent Areas. Univ. Miami Press, Coral Gables, FL, 83 p

Part 2
Marine Benthic Algae

Andrea Bernecker

Caulerpa sertularioides from the Caribbean coast of Costa Rica (Photo: Andrea Bernecker)

Abstract This paper represents a synthesis of all marine benthic algae and Cyanobacteria species reported from the Pacific and Caribbean coasts of Costa Rica. Worldwide and Central American distributions are included, as well as data regarding the habitat and substrate on which these species were found. In total, 396 species of marine benthic algae, and 24 species of Cyanobacteria are reported from Costa Rica. There are 84 species of algae belonging to Chlorophyta,

A. Bernecker
Escuela de Biología, Universidad de Costa Rica, San Pedro, 11501-2060 San José, Costa Rica
e-mail: aberneck@biologia.ucr.ac.cr

I.S. Wehrtmann, J. Cortés (eds.) *Marine Biodiversity of Costa Rica, Central America*,

51 to Phaeophyceae, 260 to Rhodophyta, and 1 species of Xanthophyceae. Six species of algae are endemic. There are 287 species of algae and 10 species of Cyanobacteria reported from the Caribbean, and 156 species of algae and 19 species of Cyanobacteria from the Pacific. Forty seven species of algae and five species of Cyanobacteria occur on both the Caribbean and Pacific coasts.

Introduction

Our knowledge about the marine benthic algae of Costa Rica is based upon relatively few publications. Soto (1983) published a historical review of phycological activities in Costa Rica. The first reports of marine algae from the Caribbean coast were made by Taylor (1933, 1942, 1960) followed by Dawson (1962) and Wellington (1973, 1974). Kemperman & Stegenga (1983) and Stegenga & Kemperman (1983) studied individual species and groups. Two enlarged and updated checklists of the Caribbean were subsequently provided by Soto & Ballantine (1986) and Kemperman & Stegenga (1986). These checklists were both based on literature data and recent collections by the authors, and a total of 262 and 261 species were reported, respectively. Several species were mentioned only in one of the two publications, so the real number of species hitherto known from Costa Rica was slightly higher.

Taylor (1945) made the first reports of benthic marine algae species from the Pacific coast. These initial reports were followed by the more comprehensive work done by Dawson (1957, 1960, 1962). A total of about 160 species have been reported from the Pacific coast of Costa Rica of which 45 species are also present in the Caribbean.

After 1986, research on marine benthic algae in Costa Rica has been scarce. A checklist reporting 54 species was compiled for the Parque Nacional Marino Ballena management plan by Soto & Bermúdez (1990), but it was never published. Studies on Costa Rican Gelidiales of the Caribbean were carried out recently by Thomas & Freshwater (2001), and an inventory of algae associated to mangrove roots in the Pacific coast was provided by Tejada Rivas (2002).

The present paper is an up-to-date account of the species of marine benthic algae and Cyanobacteria reported from both the Pacific and Caribbean coasts of Costa Rica; worldwide and Central American distributions are included, as well as data regarding the habitat and substrate on which these species were found.

Materials and Methods

The checklist of marine benthic algae from Costa Rica provided in this paper is based only on consulted literature. There is no information included regarding recent collections or herbarium material. Numerous taxonomic and nomenclatural changes have been made for the species names previously reported. The names included in this list have been updated following Wynne (1998) and Guiry & Dhonncha (2004). The order of higher taxa follows Wynne (1998) with the species listed alphabetically. World distribution data were obtained from AlgaeBase, a database established by Guiry & Dhonncha (2004). This database is updated daily, so the worldwide distribution data may not be complete, and additional reports from other regions may be obtained in the future. Distribution data within Central America were obtained from the following sources: Panama (Earle 1972; Wysor & De Clerck 2003; Wysor & Kooistra 2003; Wysor 2004), Nicaragua (Dawson 1962; Phillips *et al.* 1982), El Salvador (Dawson 1961), Honduras (Ogden 1998; Wysor & De Clerck 2003; Wysor & Kooistra 2003), Guatemala (Bird & McIntosh 1979), and Belize (Norris & Bucher 1982; Littler *et al.* 1995, 2000; Littler & Littler 1997; Wysor & De Clerck 2003; Wysor & Kooistra 2003; Dhonncha, 2003, personal communication). Details about the habitat and substrate on which these species were found were taken mainly from Humm & Wicks (1980), Littler & Littler (2000), and from publications where individual species were reported (Dawson 1949, 1957, 1960, 1962; Hollenberg 1961; Bird & McIntosh 1979; Wysor & De Clerck 2003; Wysor & Kooistra 2003; Wysor 2004).

Results and Discussion

The marine benthic flora reported from Costa Rica includes in total 396 species of marine benthic algae, and 24 species of Cyanobacteria, of which 84 belong to Chlorophyta, 51 to Phaeophyceae, 260 to Rhodophyta, and 1 species of Xanthophyceae (Species Lists 2.1 and 2.2 are included on the CD-Rom). There are 287 species of algae and 10 species of Cyanobacteria reported from the Caribbean coast (Species List 2.1 is included on the CD-Rom), and 156 species of algae and 19 species of Cyanobacteria from the Pacific coast (Species List 2.2 is included on the CD-Rom). Forty seven species of algae and five species of Cyanobacteria occur on both the Caribbean and Pacific coasts; six species of algae are endemic.

Diversity is much higher in the Caribbean where extensive coral reefs are present especially along the southern part of the coast (Figs. 2.1 and 2.2). Higher species numbers in the Caribbean may also be due to the fact that more investigations have been conducted in this area. It is expected that species numbers reported from the Pacific coast will increase with future collections and investigations in the area.

More species of benthic marine algae have been reported from Costa Rica than from any other Central American country (Table 2.1). This is true not only for the total number of species, but also for species numbers reported from the Caribbean and the Pacific coasts separately. Some identifications of Costa Rican algae remain to be confirmed, thus real species numbers may change slightly. The number of species reported from Costa Rica is much higher than that reported from Panama and Belize, the countries with the next highest numbers. New studies of the flora

Fig. 2.1 *Caulerpa racemosa* from the Caribbean coast of Costa Rica (Photo: Andrea Bernecker)

Fig. 2.2 A red algae (*Caloglossa leprieurii*) which has been reported from both coasts of Costa Rica (Photo: Andrea Bernecker)

Table 2.1 Species numbers of marine benthic algae and Cyanobacteria reported from all Central American countries

	Chlorophyta	Hetero-kontophyta (Phaeophyceae, Xanthophyceae)	Rhodophyta	Cyanobacteria	Total	Caribbean	Pacific	Both coasts
Costa Rica	84	52	260	24	396 (+24 cyanobacteria)	287 (+10 cyanobacteria)	156 (+19 cyanobacteria)	47 (+5 cyanobacteria)
Panama	107	43	103	7	253 (+7 cyanobacteria)	175 (+3 cyanobacteria)	103 (+6 cyanobacteria)	25 (+2 cyanobacteria)
Belize	72	23	79	13?	174	174 (+13)	–	–
Honduras	38	19	58	5	115 (+5 cyanobacteria)	115 (+5 cyanobacteria)	–	–
Nicaragua	34	17	46	2	97 (+2 cyanobacteria)	97 (+2 cyanobacteria)	?	?
El Salvador	18	8	67	?	93	–	93	–
Guatemala	11	1	23	11	35 (+11 cyanobacteria)	27 (+4 cyanobacteria)	10 (+11 cyanobacteria)	2

of Panama, however, indicate that the number of species may be much higher than previously reported. New collections of green and brown algae revealed the presence of more than 250 species, and this number does not include data from new collections of red algae (Wysor & De Clerck 2003; Wysor & Kooistra 2003; Wysor 2004; Wysor, 2003, personal communication). It is likely that the higher number of species reported from Costa Rica is due in part to the fact that more studies on the marine algal flora have been conducted in this country. It is expected that species numbers for Central America will increase with future investigations.

Specialists

ANDREA BERNECKER: Escuela de Biología, Universidad de Costa Rica, San José, Costa Rica – Costa Rica. aberneck@biologia.ucr.ac.cr

D. WILSON FRESHWATER: Center for Marine Science, 1 Marvin Lane, Wilmington, North Carolina 28409, U.S.A. – Costa Rican Gelidiales (Rhodophyta). freshwaterw@uncw.edu

BRIAN WYSOR: Department of Biology, The University of Louisiana at Lafayette, U.S.A. – Benthic marine algae of Panama (Pacific and Caribbean). bwysor@bigelow.org

MARK M. LITTLER: National Museum of Natural History, Smithsonian Institution, Washington, U.S.A. – Benthic marine algae of the Caribbean. http://www.nmnh.si.edu/botany/projects/algae/Staff-ml.htm

DIANE S. LITTLER: National Museum of Natural History, Smithsonian Institution, Washington, U.S.A. – Benthic marine algae of the Caribbean. http://www.nmnh.si.edu/botany/projects/algae/Staff-dl.htm

MICHAEL J. WYNNE: Department of Ecology and Evolutionary Biology and Herbarium, University of Michigan, Ann Arbor, Michigan, USA – Benthic marine algae of the Caribbean. http://www.biology.lsa.umich.edu/~mwynne/

Collections

The main collections of specimens from Costa Rican marine benthic algae are housed in the following herbaria:

HERBARIO DE LA UNIVERSIDAD DE COSTA RICA, SAN JOSÉ, COSTA RICA (USJ): mainly collected by A. Bernecker, E.Y. Dawson and R. Soto

HERBARIO NACIONAL DE COSTA RICA, SAN JOSÉ, COSTA RICA (CR): mainly collected by E.Y. Dawson, Th.C.M. Kemperman & H. Stegenga, R. Soto, J. Valerio and G.M. Wellington

HERBARIUM VRIJE UNIVERSITEIT, AMSTERDAM, NETHERLANDS (AVU): mainly collected by Th.C.M. Kemperman & H. Stegenga

United States National Herbarium, Smithsonian Institution, Washington, USA (US): mainly collected by C.W. Dodge, A.S. Oersted and J. Valerio

University of Southern California, Los Angeles, USA (AHFA): mainly collected by E.Y. Dawson

University of California, Berkeley, USA (UC): mainly collected by E.Y. DawsonCalifornia Academy of Sciences, San Francisco, USA (CAS): mainly collected by E.Y. Dawson

Acknowledgments I am grateful to Brian Wysor (Bigelow Laboratory for Ocean Sciences, Maine, USA) and Michael Wynne (University of Michigan, USA) for kindly providing me with significant literature. They were also very helpful with taxonomic questions, and getting me in touch with other phycologists. I would also like to thank Jorge Cortés and Ingo Wehrtmann (Universidad de Costa Rica) for the invitation to participate in this book about Costa Rican marine biodiversity. Barrett Brooks and Bob Sims (Smithsonian Institution) helped me to obtain the literature from Diane and Mark Littler, and James Norris. I would also like to thank Mike Guiry (National University of Ireland), and Eilís Nic Dhonncha (AlgaeBase Centre, Galway, Ireland) for the exchange we had in taxonomic questions, and the information and literature they sent to me. I am grateful to Gerhard Gottsberger (University of Ulm, Germany) for the access to the library at his university, and to Cindy Fernández (Universidad de Costa Rica) for providing me with copies of some of the literature needed for this work. I wish to express my gratitude also to Wilson Freshwater (University of North Carolina, Wilmington, USA), Olivier De Clerck (Ghent University, Belgium), the late German Bula-Meyer (Laboratorio de Biología, Santa Marta, Colombia), Frederico Gurgel (University of Louisiana, Lafayette, USA), and Christine Maggs (Queen's University, Belfast, Ireland) for literature and helpful remarks, and to Daniella Baltodano and Karla Barquero for their help in revising distribution data in AlgaeBase. Finally, I appreciated the valuable recommendations of two anonymous reviewers of this manuscript. This work was supported by the University of Costa Rica (Vicerrectoría de Investigación, Project No. 111-A3-050).

References

1. Bird KT, McIntosh RP (1979) Notes on the marine algae of Guatemala. Rev Biol Trop 27:163–169
2. Dawson EY (1949) Studies of the Northeast Pacific Gracilariaceae. Allan Hancock Found Publ Occas Pap 7:1–105
3. Dawson EY (1957) Marine algae from the Pacific Costa Rican gulfs. LA County Mus Contr Sci 15:1–28
4. Dawson EY (1960) New records of marine algae from Pacific Mexico and Central America. Pac Nat 1 (20):31–52
5. Dawson EY (1961) Plantas marinas de la zona de las mareas de El Salvador. Pac Nat 2 (8):388–461
6. Dawson EY (1962) Additions to the marine flora of Costa Rica and Nicaragua. Pac. Nat. 3 (13):375–395
7. Earle SA (1972) A review of the marine plants of Panama. Bull Biol Soc Wash 2:69–87
8. Guiry MD, Nic Dhonncha E 2004) AlgaeBase version 2.1. World-wide electronic publication. National University, Ireland, Galway. http://www.algaebase.org (searched April–May 2004; March–July 2005)
9. Hörning I, Schnetter R (1988) Notes on *Dictyota dichotoma, D. menstrualis, D. inidica and D. pulchella* sp. nov. (Phaeophyta). Phyton (Austria) 28:277–291
10. Hollenberg GJ (1961) Marine red algae of Pacific Mexico Part 5. The genus Polysiphonia Pac Nat 2 (6):345–375

11. Humm HJ, Wicks SR (1980) Introduction and guide to the marine bluegreen algae. Wiley, New York, 194 p
12. Kemperman ThCM, Stegenga H (1983) A new *Caulerpa* species (Caulerpaceae, Chlorophyta) from the Caribbean side of Costa Rica, C.A. Acta Bot Neerl 32:271–275
13. Kemperman ThCM, Stegenga H (1986) The marine benthic algae of the Atlantic side of Costa Rica. An annotated, updated and enlarged checklist. Brenesia 25–26:99–122
14. Kraft GT, Wynne MJ (1996) Delineation of the genera *Struvea* Sonder and *Phyllodictyon* J.E. Gray (Cladophorales, Chlorophyta). Phyc Res 44:129–143
15. Littler DS, Littler MM (1997) An illustrated marine flora of the Pelican Cays, Belize. Bull Biol Soc Wash 9:1–449
16. Littler DS, Littler MM (2000) Caribbean reef plants: an identification guide to the reef plants of the Caribbean, Bahamas, Florida and Gulf of Mexico. OffShore Graphics, Inc., Washington, DC, 542 p
17. Littler DS, Littler MM, Brooks BL (1995) Marine algae and seagrasses from the Tobacco Range fracture zone, Belize, C.A. Atoll Res Bull 429:1–43
18. Littler DS, Littler MM, Brooks BL (2000) Checklist of marine algae and seagrasses from the ponds of the Pelican Cays, Belize. Atoll Res Bull 474:151–206
19. Norris JN, Bucher KE (1982) Marine algae and seagrasses from Carrie Bow Cay, Belize. Smithson Contr Mar Sci 12:167–223
19a Norris JN (1985). Studies on *Gracilaria* Grev. (Gracilariaceae, Rhodophyta) from the Gulf of California, Mexico. *In*: Abbott IA, Norris JN (eds) Taxonomy of Economic Seaweeds with Reference to some Pacific and Caribbean Species, Vol. 9. California Sea Grants College Program, pp 123–135
20. Ogden NB (1998) Checklist of marine benthic algae in the Cayos Cochinos Archipelago, Honduras. Rev Biol Trop 46(Suppl 4):81–87
21. Phillips RC, Vadas RL, Ogden NB (1982) The marine algae and seagrasses of the Miskito Bank, Nicaragua. Aquat Bot 13:187–195
21a. Setchell WA (1937) The Templeton Crocker Expedition of the California Academy of Sciences, 1932. No. 34: Reports on the *Sargassums*. Proc Calif Acad Sci 4th Ser 22:127–158
21b. Silva PC, Basson PW, Moe RL (1996) Catalogue of the benthic marine algae of the Indian Ocean. Univ Cal Publ Botany 79:1–1259
22. Soto R (1983) Nuevos informes para la flora bentónica marina de Costa Rica. Brenesia 21:365–370
23. Soto R, Ballantine DL (1986) La flora bentónica marina del Caribe de Costa Rica (notas preliminares). Brenesia 25–26:123–126
24. Soto R, Bermudez M (1990) Parque Nacional Ballena. Plan de Desarrollo. MINAE, San José, 125 p (unpublished)
25. Stegenga H, Kemperman Th CM (1983) Acrochaetiaceae (Rhodophyta) new to the Costa Rican Atlantic flora. Brenesia 21:67–91
Taylor WR (1933) Notes on algae from the tropical Atlantic Ocean II. Pap Mich Acad Sci, Arts Lett 17:395–407
Taylor WR (1942) Caribbean marine algae of the Allan Hancock Expedition, 1939. Rep Allan Hancock Atlantic Exp Rep 2:1–193
26. Taylor WR (1945) Pacific marine algae of the Allan Hancock Expeditions to the Galapagos Islands. Allan Hancock Pacific Exp 12:1–528
27. Taylor WR (1960) Marine algae of the eastern tropical and subtropical coasts of the Americas. University of Michigan Press, Ann Arbor, MI, 870 p
28. Tejada Rivas OL (2002) Macroalgas asociadas a raíces de mangle: costa Pacífica de Costa Rica. M.Sc. thesis, University of Costa Rica, San José, Costa Rica
29. Thomas DT, Freshwater DW (2001) Species of Costa Rican Gelidiales (Rhodophyta): four Caribbean taxa including *Pterocladiella beachii* sp. nov. Phycologia 40:340–350
30. Wellington GM (1973) Additions to the Atlantic benthic flora of Costa Rica. Brenesia 2:17–20

31. Wellington GM (1974) The benthic flora of Punta Cahuita: annotated list of species with additions to the Costa Rican Atlantic flora. Brenesia 3:19–30
32. Wynne MJ (1998) A checklist of benthic marine algae of the tropical and subtropical western Atlantic: first revision. Beiheft zur Nova Hedwigia 116, 155 p
33. Wysor B (2004) An annotated list of marine Chlorophyta from the Pacific coast of the Republic of Panama with a comparison to Caribbean Panama species. Nova Hedwigia 78:209–241
34. Wysor B, De Clerck O (2003) An updated and annotated list of marine brown algae (Phaeophyceae) of the Caribbean coast of the Republic of Panama. Bot Mar 46:151–160
35. Wysor B, Kooistra WHCF (2003) An annotated list of marine Chlorophyta from the Caribbean coast of the Republic of Panama. Nova Hedwigia 77:487–523

Part 3
Seagrasses

Jorge Cortés and Eva Salas

Seagrass bed in Puerto Vargas, Caribbean coast of Costa Rica. (Photo: Ingo S. Wehrtmann)

Abstract Four species of seagrasses have been reported for the Caribbean coast of Costa Rica: *Thalassia testudinum, Syringodium filiforme, Halophila decipens*, and *Halodule wrightii*, which are all common throughout the Caribbean. Only two species have been reported from the Pacific coast, *Ruppia maritima* and

J. Cortés and E. Salas
Centro de Investigación en Ciencias del Mar y Limnología (CIMAR), and Escuela de Biología, Universidad de Costa Rica, San Pedro, San José 11501-2060, Costa Rica.
e-mail: jorge.cortes@ucr.ac.cr; evamaria_cr@yahoo.com

I.S. Wehrtmann, J. Cortés (eds.) *Marine Biodiversity of Costa Rica, Central America*,

Halophila baillonis. There are few studies on the seagrasses of Costa Rica, and even less for other Central American countries except Guatemala.

Introduction

Seagrasses are monocotylydoneous flowering plants (Division: Tracheophyta) that live completely submerged in marine environments. There are considerable taxonomic discrepancies over the nomenclature and relationship of seagrasses, and hence no definite number of species. Phillips & Meñez (1988) consider that there are close to 50 species of seagrasses found in most shallow coastal areas of the world, while Spalding *et al.* (2003) raise the number to 59. Some species form extensive beds that are highly productive and rich in biodiversity (CARICOMP 1997; Kjerfve 1998).

There are few works on the seagrasses of Costa Rica. Three publications mentioned the species present in Costa Rica (Dawson 1962; Gómez 1984; Davidse *et al.* 1994). Cortés & Guzmán (1985) indicated that seagrasses were present in the lagoons of coral reef systems of the Caribbean, and Cortés *et al.* (1992), reported the massive death of seagrasses and reef organisms that resulted from the Limón earthquake in April 1991. This event uplifted the southern section of the Caribbean coast of Costa Rica between 0.5 and 1.9 m.

Only two studies on seagrass ecology in Costa Rica have been published. The first concerned biomass, productivity, and density of *Thalassia testudinum* at three sites in Cahuita National Park, Caribbean coast by Paynter *et al.* (2001). They found that the site with the highest biomass and productivity had intermediate salinity, temperature, and sediment grain size compared to other sites studied in Cahuita. The second study reported disappearance of several seagrass beds on the Pacific coast of Costa Rica after a strong storm uprooted the seagrasses and all its associated fauna (Cortés 2001). The only area in Central America where the ecology of seagrasses has been studied is in Guatemala. Arrivillaga & Baltz (1999), and Arrivillaga (2000) studied the use of seagrass beds by fishes and decapod crustaceans in Guatemala Caribbean coast.

Six species have been reported for the Caribbean coast of Costa Rica, four of them: *T. testudinum, S. filiforme, H. decipens*, and *H. wrightii* are all common throughout the Caribbean; the most abundant and widespread being *T. testudinum*, followed by *S. filiforme*. Dawson (1962) reports two more species, apart from *Thalassia* (written *Thallassia*) and *Syringodium: H. baillonis* and *R. maritima*. The first is probably *H. wrightii* and the second may have been present but it has not been seen in the last 40 years. In some regions the seagrasses cover extensive areas, for example in the Nicaragua Rise (Phillips *et al.* 1982). The species *H. decipens* and *H. wrightii* have been observed in deeper waters of reef fronts at several sites on the coast. Even though *H. wrightii* also forms frequently narrow fringes in shallow areas along the beaches it has only been found in the deeper areas in Costa Rica.

From the Pacific coast of Costa Rica only two species have been reported: *R. maritima* and *H. baillonii* (Gómez 1984; Cortés 2001). But more species may be present on the Pacific coast, since Hartog (1960) mentioned the presence of *Diplanthera ciliata* and *Halophila beaudettei* (reported as *Diplanthera*) in Panama, and *D. dawsonii* in Nicaragua, also, *H. beaudettei* and *Naja marina* have been reported from neighboring countries (Davidse *et al.* 1994).

Here we present a list of species of seagrasses from the Caribbean (Species List 3.1 is included on the CD-Rom) and Pacific coast (Species List 3.2 is included on the CD-Rom) of Costa Rica based on published reports and herbarium collections. Other data are included, such as world distribution, presence in other Central American countries, habitat, and depth range in Costa Rica. There are few ecological studies on the seagrasses of Costa Rica, and less for other Central American countries. From some countries there is no information of the species present much less on ecological aspects, with the exception of work done in Guatemala over a decade ago. There is practically no information on the conservation status of seagrasses in the region.

Specialists

ALEJANDRO ARRIVILLAGA: Universidad del Valle, Guatemala – Caribbean. aarrivillaga@tnc.org

VANESSA NIELSEN: CIMAR, Universidad de Costa Rica. – Caribbean. vnielsen@cariari.ucr.ac.cr

BRIGITTA I. VAN TUSSENBROEK: Instituto de Ciencias del Mar y Limnología, UNAM, Puerto Morelos, Yucatán, México – Caribbean. vantuss@mar.icmyl.unam.mx

Collections

There are collections of Costa Rican seagrasses in the herbaria of the University of Costa Rica, National Museum of Costa Rica, and in the Milwaukee Museum.

Recommendations

More work is needed on both coasts of Costa Rica and in the other Central American countries on the species richness, distribution, conservation status, and ecology of the seagrasses.

References

Arrivillaga A (2000) Ecology of seagrass fishes and macroinvertebrates on Guatemala's Atlantic coast. Ph.D. dissertation, Lousiana State University, Baton Rouge, Louisiana

Arrivillaga A, Baltz DM (1999) Comparison of fishes and macroinvertebrates on seagrass and bare-sand sites of Guatemala's Atlantic coast. Bull Mar Sci 65:301–319

CARICOMP (1997) Variation in ecological parameters of *Thalassia testudinum* across the CARICOMP network. Proc 8th Int Coral Reef Symp, Panamá 1:663-668

1. Cortés J (2001) Requiem for an eastern Pacific seagrass bed, Bahía Culebra, Costa Rica. Rev. Biol. Trop. 49 (Suppl 2):273–278
2. Cortés J, Guzmán HM (1985) Arrecifes coralinos de la costa Atlántica de Costa Rica. Brenesia 23:275–292

Cortés J, Soto R, Jiménez C, Astorga A (1992) Death of intertidal and coral reef organisms as a result of a 7.5 earthquake. Proc 7th Int Coral Symp, Guam 1:235–240

3. Davidse G, Sousa S M, Chater AO (eds) (1994) Flora Mesoamericana: Alismataceae a Cyperaceae, vol. 6. Institute of Biology, Universidad Nacional Autónoma de México (UNAM), México, D.F. 543 p
4. Dawson EY (1962) Additions to the marine flora of Costa Rica and Nicaragua. Pac. Nat. 3:375–395
5. Fonseca AC, Arrivillaga A (2003) Coral reefs of Guatemala. *In*: Cortés J (ed) Latin American Coral Reefs. Elsevier Science, Amsterdam, pp 159–169
6. Gómez LD (1984) Las Plantas Acuáticas y Anfibias de Costa Rica y Centroamérica: 1. Lilliopsida. UNED, San José, 430 p

Hartog C den. (1960) New sea grasses from Pacific Central America. Pac Nat 1(15):1–8

Kjerfve B (ed) (1998) CARICOMP - Caribbean Coral Reef, Seagrass, and Mangrove Sites. UNESCO, Paris, 345 p

7. Ogden JC, Ogden NB (1998) Reconnaissance survey of the coral reefs and associated ecosystems of Cayos Cochinos, Honduras. Rev Biol Trop 46(Suppl 4):67–74
8. Paynter CK, Cortés J, Engels M (2001) Biomass, productivity and density of the seagrass *Thalassia testudinum* at three sites in Cahuita National Park, Costa Rica. Rev Biol Trop 49(Suppl 2):265–272
9. Phillips RC, Meñez EG (1988) Seagrasses. Smithson Contr Mar Sci 34:1–104
10. Phillips RC, Vadas RL, Ogden JC (1982) The marine algae and seagrasses of the Miskito Bank, Nicaragua. Aquat Bot 13:187–195
11. Ryan J, Zapata Y (2003) Nicaragua's coral reefs: status, health and management strategies. *In*: Cortés J (ed) Latin American Coral Reefs. Elsevier Science, Amsterdam, pp. 203–222
12. Salas S USJ-741 (University of Costa Rica Herbarium)
13. Soto R USJ-871 (University of Costa Rica Herbarium)

Spalding M, Taylor M, Ravilious C, Short F, Grenn E (2003) The distribution and status of seagrasses. *In*: Green EP, Short FT (eds) World Atlas of Seagrasses. University of California Press, Berkeley, pp 5–26

Part 4
Mangroves

Ana Margarita Silva-Benavides

Mangroves in Térraba, southern Pacific of Costa Rica (Photo: Ingo S. Wehrtmann)

Abstract Costa Rica has an estimated mangrove area of 412 km^2, and 99% of that is on the Pacific coast, which is longer and has a more complex geomorphology than the Caribbean coast. The Pacific coast includes many estuaries, gulfs,

A.M. Silva-Benavides
Escuela de Biología. Universidad de Costa Rica. 11501-2060 San José, Costa Rica, Centro de Investigación en Ciencias del Mar y Limnología (CIMAR), Universidad de Costa Rica, San José, Costa Rica.
e-mail:msilva@biología.ucr.ac.cr

and embayments that provide optimal habitats for mangrove development, and it has the highest diversity and greatest structural complexity of mangrove forests. Along the northern Pacific coast, mangrove forests are less structured, with a canopy approximately 1–1.5 m high due to the long dry season and low rainfall. The central Pacific coast shows a transition zone where the forests are higher with trees reaching 30 m as a result of the increase in precipitation and the reduction of the dry season in this zone. The southern Pacific coast (Térraba-Sierpe and Golfo Dulce) presents the most complex mangrove systems, with trees reaching over 40 m in height, partially due to the precipitation of the zone and high levels of freshwater input throughout the year. The mangrove forests of the Caribbean are located on the southern part of the coast, with *Rhizophora mangle* as the dominant species. The mangrove species in Costa Rica are included in four families: Rhizophoraceae (*R. mangle, R. racemosa, R. harrisonii*), Combretaceae (*Laguncularia racemosa, Conocarpus erecta*), Avicenniaceae (*Avicennia germinans, A. bicolor*) and Pellicieraceae (*Pelliciera rhizophorae*). Other common associated species are *Mora oleifera, Acrostichum aureum*, and *Hibiscus tiliaceus*.

Introduction

Costa Rica has mangrove forests along both the Pacific and Caribbean coasts. The estimated coverage of mangroves along the Pacific coast is 41,290 ha, representing 99% of all the mangrove area of the country, while in the Caribbean they cover about 250 ha (Polanía 1993; Coll *et al.* 2001).

After the consolidation of Central America as an isthmus, species of *Rhizophora, Avicennia, Pelliciera*, and *Laguncularia* were common on both coasts. Following a regional climate change to a more dry seasonal weather, the floristic composition of the two coasts changed (Jiménez 1999). Three groups of mangrove vegetation appeared: one group of dry seasonal climates, represented by *Avicennia bicolor*; a second group restricted to rainy climates represented by *Pelliciera rhizophorae*; and a third group composed by *Rhizophora mangle* (Fig. 4.1) and *R. racemosa*, which can be found in both dry and rainy climates (Jiménez 1999). Forests from the southern humid regions have the highest floristic diversity and structural development with *P. rhizophorae* and *R. racemosa* as the dominant species.

Since pre-Columbian times indigenous people were linked to the mangrove forest resources, using fish as a common protein source. More recently, the extraction of cockles from the mangroves is the main resource directly harvested by local communities. Additionally, in Costa Rica as well as in other tropical countries, aquaculture activities have caused a great impact on the mangrove forests, especially on the northern and central Pacific coast, where they have led to the degradation of most of the forests. In Costa Rica, the mangrove ecosystem plays an important role in the socioeconomic livelihood of the coastal communities (Jiménez 1994a).

Fig. 4.1 The common mangrove species *Rhizophora mangle*, present on both coasts of Costa Rica (Photo: Ingo Wehrtmann)

Distribution and Characteristics of the Mangrove Forests

The Pacific coast of Costa Rica is about 1,250 km long and has bays, inlets, estuaries, islands, and gulfs. In contrast, the Caribbean coast is only 212 km long and lacks the geomorphologic features of the Pacific (Flores 1995). The tidal range is an important factor, which affects the distribution and physiognomy of the mangrove forests of the Pacific and Caribbean coasts of Costa Rica. The tides on the Pacific are semidiurnal, with a mean amplitude of 2.8 m (Lizano 1997), while in the Caribbean, the mean amplitude is just 0.3 m.

CARIBBEAN COAST. Mangroves on the Caribbean (Moín and the Refugio Nacional de Vida Silvestre Gandoca-Manzanillo) are located in the southern part of the coast (Cortés *et al.* 2001). The climate of the Caribbean zone is characterized by high year-round precipitation, with an average of 3,531 mm per year (Cortés 1998). The Gandoca mangrove is the largest of the Caribbean coast of Costa Rica (~12.5 ha), and this area is now protected by law, having been set aside within the Refugio Nacional de Vida Silvestre Gandoca-Manzanillo. Dominant species are *R. mangle, Laguncularia racemosa, Conocarpus erecta* and *Avicennia germinans*; associated species, which are seasonally submerged, are the following: *Hibiscus tiliaceus, Mora oleifera, Annona glabra, Montricardia arborescens, Hippomane mancinella, Tabebuia palustris, Pavonia spicata, Carapa guianensis, Acrostichum aureum, Typha latifolia, Raphia taedigera, Manicaria saccifera*, and *Muellera frutescens* (Coll *et al.* 2001). These plants form the characteristic vegetation bordering the outermost section of the mangrove forest (Tomlinson 1986; Jiménez 1994a; Coll *et al.* 2001). They are considered minor or associated components of the mangroves, and they are not found directly in the mangrove habitat because they are unable to tolerate high salt levels (Jiménez 1994a).

Pacific coast. The Pacific mangrove forests of Costa Rica are distributed along the entire coast, and they can be grouped into three major sections: northern, central, and southern Pacific (Jiménez 1994a).

Northern Pacific: This zone comprises the area from the Nicaraguan border to north of Tivives, and includes the Golfo de Nicoya with more than 4,100 ha of mangrove forests. Due to the fishery and salt industries, these mangroves represent one of the most altered coastal areas of the country. The north Pacific coast has a well-marked seasonal climate with a dry tropical climate during 4–6 months of the year and a rainy season during the rest of the year (December–April); the mean annual rainfall is relatively low (maximum of 2,000 mm), and the average temperature is 27.5 °C (Jiménez 1994a)

Central Pacific: The central Pacific coast extends from the eastern end of Golfo de Nicoya to north of the Térraba-Sierpe mangrove complex. It is considered as a transition area, with a less-pronounced dry season and an average precipitation between 2,500 and 3,000 mm per year. The mean annual temperature is 26.5 °C with a relative humidity of 74%. This climate favors a more structurally developed mangrove forest than the ones along the northern coast (Jiménez 1994a).

Southern Pacific: This zone covers the area from Térraba-Sierpe to the border with Panama. The area has considerably higher rainfalls over the year than the above-mentioned zones and is comparable to that of the Caribbean coast (Valerio 1999). The mean annual precipitation of the southern zone is between 4,500 and 5,500 mm; the average temperature is 26.5 °C. There are no dry months, and the relative humidity is higher than 90%. The high input of freshwater throughout the year supports a well-structured mangrove forest (Polanía 1993).

Térraba-Sierpe mangroves: These mangroves are located at the estuary of two large rivers, Río Grande de Térraba and Río Sierpe. The area represents approximately 40% (16,700 ha) of the total mangrove forest of Costa Rica (Lahmann 1993) and is considered as the less-altered mangroves of the country; however, human pressure has increased due to economic growth. These forests are the most structurally complex mangroves of the country, mainly due to the size of *R. racemosa, R. mangle*, and *A. germinans*, some of which can reach a height of up to 40 m.

Golfo Dulce mangroves: There are mangrove swamps in several coastal estuaries (Rincón, Golfito, and Coto), dominated by a mixture of *R. mangle, R. racemosa*, and *P. rhizophorae*. The local communities depend mainly on the extraction of fishes and bivalves (Silva & Chávez 2001).

Floristic Characteristics

The basic floristic composition of the Costa Rican mangroves is represented by eight species from four families: Rhizophoraceae, Pellicieraceae, Avicenniaceae, and Combretaceae. Four species have been reported from the Caribbean (Species List 4.1

are included on the CD-Rom), all of them also present on the Pacific coast, and eight from the Pacific coast of Costa Rica (Species List 4.2 is included on the CD-Rom). Rhizophoraceae is represented by *R. mangle, R. racemosa*, the most abundant species, and *R. harrisonii*. Although these trees share anatomical and physiological characteristics (Bodero 1993), they differ in heights, diameters, and flowering periods (Jiménez 1987, 1994a). *R. mangle* grows especially on the sea front, along muddy beaches and is most abundant in soils with salinity near or less than 35 psu; however, it is possible to find it in salinities higher than 50 psu. Under these conditions, the height, diameter, basal area and bark volume of these species tend to decrease, reaching a height of only 1–1.5 m (Jiménez 1994b). Flowers are produced year-round, with peaks in May, September, and October (Jiménez 1994a). *R. racemosa* inhabits the areas of deeper inundation, mainly on convex river banks. The trees grow straight and can reach heights of around 40 m. The species blooms between February and April (Jiménez 1994a).

Pellicieraceae is represented by *P. rhizophorae*. The presence of the species is restricted to the central and southern part of the country, where it occupies the intertidal zone, typically sheltered sites such as estuarine banks or protected beaches.

L. racemosa and *C. erecta* are the two representatives of Combretaceae. The first species is adapted to low salinity, grows in higher areas of the shoreline, and bears fruits mainly from June to December. The second species grows in lower salinity zones (<10 psu), and is generally found on sandy soils in the peripheral zones towards land.

Avicenniaceae is represented in Costa Rica by the following two species: *A. germinans* blooms between January and May, grows typically on sandy soil, and can be found closer to the edge of the water due to its considerable tolerance to high salinities (Jiménez 1994a); *A. bicolor* is less abundant than *A. germinans*, limited to areas with seepage and runoff, and attains basal areas of 40 m^2/ha (Jiménez-Ramón 1990), and bears flowers from December to February.

Plants associated with the nuclear vegetation of the mangroves of both the Caribbean and Pacific coasts are included in Species Lists 4.1 and 4.2 (both on the CD-Rom).

Research on Mangroves

Concerning floristic aspects of mangroves, Soto and Jiménez (1982) found that the salinity gradients in the soil affect the floristic composition, zonation, structural, and dynamical characteristics of the vegetation. Two floristic groups have been described for Central American mangrove forests corresponding to dry and rainy climate (Jiménez 1999). These mangrove communities were characterized according to structural and diversity development, where rain and runoff were the main factors controlling the floristic aspects of mangrove vegetation.

Stands of *A. bicolor, A. germinans, R. harrisonii*, and *L. racemosa* were studied by Jiménez (1987, 1988a, b, 1990), and Jiménez & Soto (1985) to characterize the structure and function of a mangrove forest from the Pacific of Costa Rica; his results showed that growth, mortality, and reproductive phenology were related to freshwater input. Jiménez-Ramón & Sauter (1991) studied the zonation patterns of *A. bicolor* and *R. racemosa* in the same forest. The authors concluded that species zonation was related to depth of tidal inundation and the ability of propagules to become established. Regarding *R. racemosa*, Jiménez (1988c) showed that the density of these plants depended on both the season and the seedling crop size.

Pizarro (1994), studying the mangroves of Golfo de Nicoya, Tivives, Quepos, and Térraba-Sierpe, provided information on phenology, regeneration, and structure of each nuclear mangrove species (*R. mangle, R. racemosa, A. bicolor, A. germinans*, and *L. racemosa*). To analyze the dynamics of nutrient concentrations, Soto (1992) studied mature and senescent leaves of several mangrove and non-mangrove species in the Pacific coast of Costa Rica. He reported higher concentration of nutrients in mature than in senescent leaves, as well as differences among species.

Mainardi-Grellet (1996) presented information on the distribution and composition of mangrove species from the Térraba-Sierpe mangrove forest. Lizano *et al.* (2001) used satellite images to study the mangrove distribution patterns in the Sierpe-Térraba mangroves; they confirmed the higher structural development and diversity of species, probably related to the high input of freshwater in this area.

Silva (2005) studied a mangrove forest in Golfito, Golfo Dulce, where *R. mangle* and *R. racemosa* were the dominant species. Close to the tidal channel, where the salinity can vary from 18 to 34 psu, trees with a height of 5–10 m were common, and *R. mangle* was the dominant species. In the innermost areas of the forest, where salinity ranged from 0 to 30 psu, the structural development was high, especially in *R. racemosa* and *A. germinans*, with an average height of 30–40 m. Stands of *L. racemosa, P. rhizophorae, A. aureum*, and *M. oleifera* were located in the landward edge.

The Playa Blanca mangrove showed a relationship between forest structure and position along the intertidal gradient (M. Silva-Benavides, unpublished data, 2007). In these mangroves, *P. rhizophorae* was common on the seaward side, where the soil was sandy. *Rhizophora racemosa* dominated the upper intertidal zones, which were composed by clay substrate. The fern *A. aureum* was the most widespread species among the associated flora on more elevated sites and around dry and less saline areas inside the mangrove. The effect of hydrology and soil conditions across the intertidal gradient reflected the importance of the edaphic factors on the mangrove distribution (Delgado-Sánchez *et al.* 2001).

Delgado-Sánchez *et al.* (1999) monitored over a 6-year period the density, growth, mortality, and transition rates of seedlings, saplings, and trees of *A. bicolor* along the Pacific coast of Costa Rica. They also measured the establishment of *L. racemosa, A. germinans*, and *A. bicolor* propagules along intertidal gradients on sand bars and islands in the Río Tempisque estuary, Golfo de Nicoya, Costa Rica (Delgado-Sánchez *et al.* 2001).

Several authors reported on the structure of the Caribbean mangrove forest in Laguna Gandoca, Limón (Cortés 1998; Coll *et al.* 2001; Cortés *et al.* 2001). They identified five plant associations: mangroves, palm swamp and ferns, mixed palm trees, very humid tropical rain forest, and tropical beach vegetation. *R. mangle* was the dominant species, followed by *A. germinans, L. racemosa*, and *C. erecta*. Jiménez (1984) proposed a hypothesis for the distribution of *P. rhizophorae* along the Caribbean coast: at the beginning of the Miocene *P. rhizophorae* was a common species, but during the Pliocene its distribution was reduced. This distributional change was related to climate variations and its influence on soil salinities (Jiménez 1984, 1994a). At present, *P. rhizophorae* is restricted to the Pacific side of Costa Rica (Graham 1995).

Acknowledgments I gratefully acknowledge Carlos Morales and Ingo Wehrtmann for assistance on this paper.

References

Bodero A (1993) Mangrove ecosystems of Ecuador. *In*: Lacerda LD (ed) Conservation and Sustainable Utilization of Mangrove Forests in Latin America and Africa regions. The International Society For Mangrove Ecosystems, Okinawa, Japan, pp 55–75

1. Coll M, Fonseca A, Cortés J (2001) El manglar y otras asociaciones vegetales de la laguna de Gandoca, Limón, Costa Rica. Rev Biol Trop 49(Suppl 2):321–329

Cortés J (1998) Cahuita and Laguna Gandoca, Costa Rica. *In*: Kjerfve J (ed) CARICOMP: Caribbean coral reef, seagrass and mangrove sites. Unesco, París, pp 107–113

Cortés J, Fonseca AC, Coll M (2001) Descripción del manglar del Refugio Nacional de Vida Silvestre Gandoca-Manzanillo. Limón, Costa Rica. *In*: Pizarro F, Córdoba R, Gómez C (eds) Humedales de Centroamérica: síntesis de 27 estudios e iniciativas sobre educación, investigación, manejo y conservación de humedales y zonas costeras. UICN/ORMA, San José, Costa Rica, pp 90–91

Delgado-Sánchez P, Jiménez JA, Justic D (1999) Population dynamics of mangrove *Avicennia bicolor* on the Pacific coast of Costa Rica. Wetlands Ecol Manag 7:113–120

Delgado-Sánchez P, Hensel PF, Jiménez-Ramón JA, Day JA (2001) The importance of propagule establishment and physical factors in mangrove distributional patterns in a Costa Rican estuary. Aquat Bot 71:157–178

2. Field C (1995) Journey amongst mangroves. International Society for Mangrove Ecosystems, Okinawa, Japan, 138 p

Flores E (1995) Geografía de Costa Rica. Ed. EUNED, San José, Costa Rica, 369 p

Graham A (1995) Diversification of Gulf/Caribbean mangrove communities through Cenozoic time. Biotropica 27:20–27

Jiménez JA (1984) A hypothesis to explain the restricted distribution of the mangrove *Pelliciera rhizophorae* Tr & Pl Biotropica 16:304–308

Jiménez JA (1987) A clarification on the existence of *Rhizophora* species along the Pacific coast of Central America. Brenesia 28:25–32

Jiménez JA (1988a) Dynamics and dispersion patterns of two mangrove populations on the Pacific coast of Costa Rica. Ph.D. dissertation, University of Miami, Florida, 176 p

3. Jiménez JA (1988b) Floral and fruiting phenology of trees in a mangrove forest on the dry Pacific coast of Costa Rica. Brenesia 29:33–50

Jiménez JA (1988c) The dynamics of *Rhizophora racemosa* forests on the Pacific coast of Costa Rica. Brenesia 30:1–12

Jiménez JA (1994a) Los manglares del Pacífico centroamericano. Ed. Fundación UNA, Heredia, Costa Rica, 336 p

Jiménez JA (1994b) Bosques de manglares en la costa Pacífica de América Central. Rev Forest Centroamer 9:13–17

Jiménez JA (1999) Ambiente, distribución y características estructurales en los manglares del Pacífico de Centro América: contrastes climáticas. *In*: Yáñez-Arancibia A, Lara-Dominguez AL (eds) Ecosistemas de manglar en América Tropical. Instituto de Ecología, México, pp 51–70

Jiménez JA, Soto R (1985) Patrones regionales en la estructura y composición florística de los manglares de la costa Pacífica de Costa Rica. Rev Biol Trop 33:25–37

Jiménez-Ramón JA (1990) The structure and function of dry weather mangroves on the Pacific coast of Central America, with emphasis on *Avicennia bicolor* forests. Estuaries 13:182–192

Jiménez-Ramón JA, Sauter K (1991) Structure and dynamics of mangrove forests along a flooding gradient. Estuaries 14:49–56

Lahmann EJ (1993) The mangrove forests of Sierpe, Costa Rica. *In*: Towards the Wise Use of Wetlands. Ramsar Convention Bureau, Wise Use Project, Gland, pp 82–88

Lizano O (1997) Las mareas extraordinarias de 1997 en la costa del Pacífico de Costa Rica. Top Meteor Oceanogr 4:169–179

Lizano O, Amador J, Soto R (2001) Caracterización de manglares de Centroamérica con sensores remotos. Rev Biol Trop 49(Suppl 2):331–340

Mainardi-Grellet V (1996) El manglar de Térraba-Sierpe en Costa Rica. Ser Téc Inf Tec./CATIE, Turrialba, Costa Rica No. 284, pp 1–91

Pizarro F (1994) Diagnóstico de los manglares de la costa Pacífica de Costa Rica. *In*: Angulo JF, Suman H (eds) El ecosistema de manglar en América Latina y la Cuenca del Caribe: su manejo y conservación. University of Miami, Florida, USA, pp 34–63

Polanía J (1993) Mangroves of Costa Rica. *In*: Lacerda LD (ed) Conservation and Sustainable Utilization of Mangrove Forests in Latin America and Africa regions. International Society for Mangrove Ecosystems, Okinawa, Japan, pp 129–138

4. Silva M (2005) Manglares de Golfito. *In*: Lobo J, Bolaños F (eds) Historia natural de Golfito. Ed. INBio, Heredia, Costa Rica, pp 55–65

Silva M, Chávez M (2001) Cuantificación sociobiológica del manglar de Purruja, Golfito. Recomendaciones para su manejo. Puntarenas, Costa Rica. *In*: Pizarro F, Córdoba R, C Gómez (eds) Humedales de Centroamérica: síntesis de 27 estudios e iniciativas sobre educación, investigación, manejo y conservación de humedales y zonas costeras. UICN/ORMA, San José, Costa Rica, pp 82–87

Soto R (1992) Nutrient concentration and retranslocation in coastal vegetation and mangroves from the Pacific coast of Costa Rica. Brenesia 37:33–50

Soto R, Jiménez J (1982) Análisis fisonómico estructural del manglar de Puerto Soley, La Cruz, Guanacaste, Costa Rica. Rev Biol Trop 30:161–168

5. Tomlinson PB (1986) The botany of mangroves. Cambridge University Press, Cambridge, 405 p

Valerio C (1999) Costa Rica, ambiente y biodiversidad. Ed. INBio, Heredia, Costa Rica, 139 p

Part 5
Foraminifera

Jorge Cortés, Claudia Mora-Baumgartner, and Vanessa Nielsen

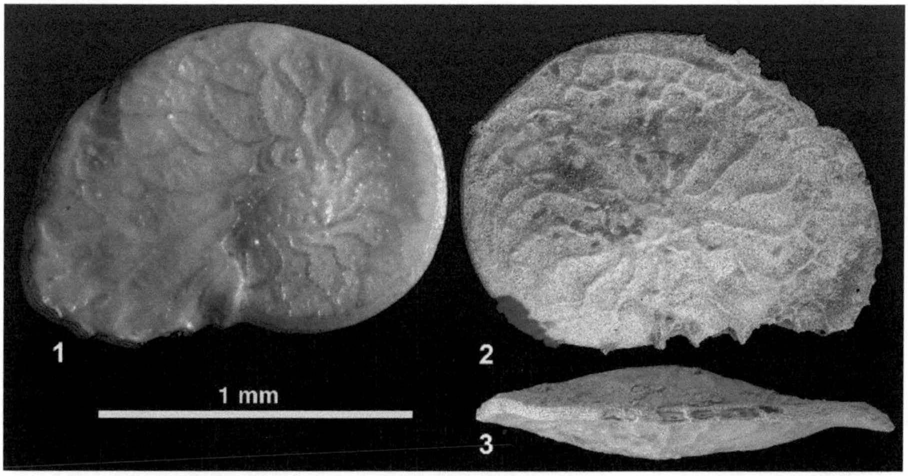

Recent larger Foraminifera *Heterostegina depressa* d'Orbigny from Puerto Vargas coral reef (Caribbean coast, Costa Rica). (Photo: Claudia Mora-Baumgartner)

Abstract Eighty-four species of benthic and one species of planktonic Foraminifera, classified under 40 genera and 34 families reported for Costa Rica are listed in this paper. These lists are based on literature data and ongoing studies. All (except for four species from the Caribbean) are reports from the Pacific Ocean, and most are from offshore or have no specific indication of where in Costa Rica the Foraminifera

J. Cortés and V. Nielsen
Centro de Investigación en Ciencias del Mar y Limnología (CIMAR), and Escuela de Biología, Universidad de Costa Rica, San Pedro, San José 11501-2060, Costa Rica.
e-mail: jorge.cortes@ucr.ac.cr; vnielsen@cariari.ucr.ac.cr

C. Mora-Baumgartner
Institut de Géologie et Paléontologie, Université de Lausanne, BFSH-2, CH-1015 Lausanne, Switzerland.
e-mail: Claudia.Baumgartner@igp.unil.ch

I.S. Wehrtmann, J. Cortés (eds.) *Marine Biodiversity of Costa Rica, Central America*,

were collected. Of the other Central American countries there is little information except from Panama. More research is needed on Foraminifera, since they may be a predominant group in some areas and ecosystems, for example the meiofauna of Caño Island, and much more research is need on planktonic Foraminifera.

Introduction

Foraminifera are benthic or planktonic protists of the Phyllum Granuloreticulosa. Over 40,000 species have been described, most of them fossil. Foraminiferans are found from the equator to the poles and most are benthic (Brusca & Brusca 2003). The test or skeletons, made of calcium carbonate, are excellent environmental recorders and the group date back to the Lower Cambrian. They are useful as stratigraphic and bio-indicators, and for the reconstruction of paleoenvironments (Cassell & Sen Gupta 1989; Collins 1996; Elderfield & Ganssen 2000; Hallock *et al.* 2003).

Benthic Foraminifera. Most of the work on microbenthic Foraminifera collected in Costa Rica (Fig. 5.1) was done between the 1930s and 1950s (Natland 1938; Cushman & McCulloch 1939, 1940, 1942, 1948, 1950; Bandy & Arnal

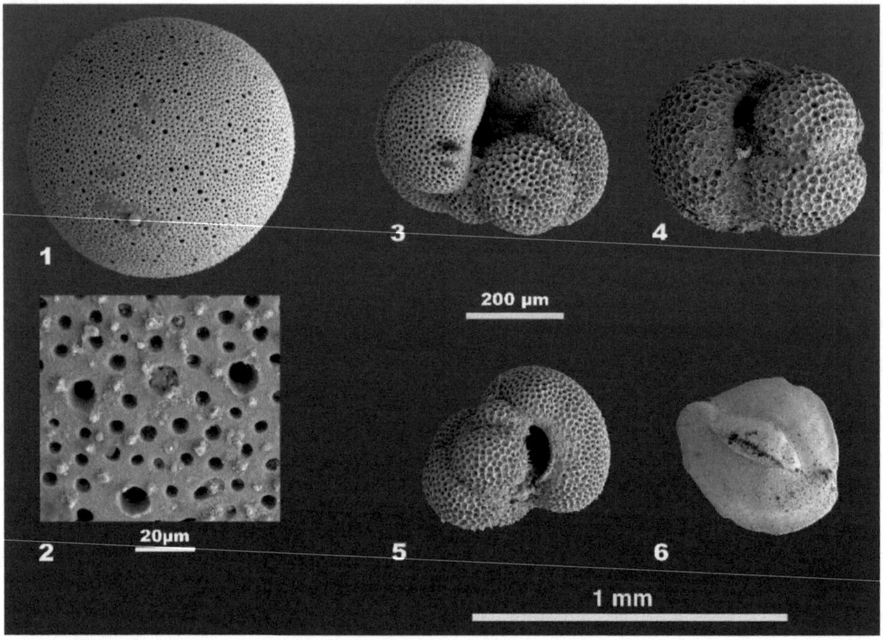

Fig. 5.1 A selection of Foraminifera from Costa Rica. (1) *Orbulina universa*; (2) *Orbulina universa*, detail of test surface; (3) *Globigerinoides sacculifer*; (4) *Globigerinoides rubber*; (5) *Globigerinoides* sp.; (6) *Quinqueloculina lamarckiana*. Photos: Claudia Mora-Baumgartner

1957). M. L. Natland (1938) reported one species, *Planulina limbata* collected off Costa Rica, in his paper on Foraminifera from off the west coast of North America. The Joseph Augustine Cushman and Irene Agnes McCulloch papers are reports based on the collections done during the Allan Hancock expeditions to the eastern Pacific between 1931 and 1941. Finally, Bandy & Arnal (1957) reported many species in their paper on recent Foraminifera off the west coast of Central America. Several papers mentioned the presence of microbenthic Foraminifera (no identification to species were provided) in the meiofauna of a mudflat in Punta Morales, Gulf of Nicoya (De la Cruz & Vargas 1986, 1987; Vargas 1988), where they constitute between 3.9% and 6.1% of the individuals. At Caño Island, the microbenthic Foraminifera are the predominant group in the meiofauna, 21.2% of all individuals collected (Guzmán *et al.* 1987). Finally, in a thesis, the presence of *Bolivina seminuda* is reported from Golfo Dulce (Beese 1995). Virtually no information is presently available on larger benthic Foraminifera. We have recently observed *Heterostegina depressa* and *Amphistegina* spp. in the sediments samples from Puerto Vargas, Cahuita National Park, on the Caribbean coast.

Planktonic Foraminifera. Little is known of the distribution of planktonic Foraminifera in coastal waters of Central America. However, in the zooplankton at the coral reef in Cahuita National Park, planktonic Foraminifera are the second must abundant group with 1% to 34% of the individuals (Morales & Murillo 1996). In a thesis, the presence of *Globigerinoides sacculifer* was reported from Golfo Dulce (Beese 1995). So far there are no other reports of foraminiferan species from the Caribbean except for the planktonic Foraminifera from Cahuita (Morales & Murillo 1996).

Remarks on the list. The list of species presented here is classified at family level after the classification proposed in the European Register of Marine Species (http://erms.biol.soton.ac.uk/cgi-bin) and Systema Naturae (http://sn2000.taxonomy.nl). Where possible, the generic assignments of the species cited in the early work have been actualized using the references cited above and, for species not listed therein, the work by Loeblich & Tappan (1987) was consulted. The names used in the cited literature are given in parentheses as (in ref:). These lists contain four species of Foraminifera from the Caribbean (Species List 5.1 is included on the CD-Rom) and 80 from the Pacific Ocean (Species List 5.2 is included on the CD-Rom). Eighty-four species are placed in alphabetical order in 40 genera and 34 families. Most are from the Pacific, with the exception of four species from the Caribbean, and mainly from offshore or have no specific indication of where in Costa Rica they were collected. Other sampled sites are Salinas, Santa Elena and Culebra bays, Golfo Dulce, and Isla del Coco.

Of the other Central American countries there is little information except from Panama. Species are listed from the Pacific of Panama collected by the Allan Hancock expeditions (see the Cushman and McCulloch papers) and from the Caribbean (Culver & Buzas 1982). Much more research is needed on the Foraminifera, since in some systems, for example the meiofauna of Caño Island (Guzmán *et al.* 1987) it is the predominant group. Many species are encountered in sand samples from the shallow subtidal of both the Caribbean and Pacific coasts (personal observation) but have not been worked on systematically.

Acknowledgments We appreciate the support with the literature by K. Hale and H. Albertson, RSMAS Library, University of Miami. Funding to study the marine biodiversity of Costa Rica was provided by the Vicerrectoría de Investigación of the Universidad de Costa Rica (Projects 808-92-237; 808-96-601) and by CONICIT (Project 90-326-BID). We appreciate the review by R. Petersen.

Specialists

Claudia Mora-Baumgartner: Institut de Géologie et Paléontologie, Université de Lausanne, Claudia.Baumgartner@igp.unil.ch – Recent and fossil, worldwide

Elspeth Urquart: Rosenstiel School of Marine and Atmospheric Sciences, University of Miami, eurquhar@rsmas.miami.edu – Worldwide

Collections: There are small collections of Costa Rican Foraminifera in Escuela Centroamericana de Geología, Universidad de Costa Rica; and at the Institut de Géologie et Paléontologie, Université de Lausanne, Switzerland

Recommendations: More work is needed on both coasts of Costa Rica, especially on the Caribbean coast, and in the other Central American countries on the species richness, distribution, and ecology of the Foraminifera.

References

1. Bandy OL, Arnal RE (1957) Distribution of recent Foraminifera off the west coast of Central America. Bull Am Assoc Petrol Geol 41:2037–2053
2. Beese D (1995) (unpublished). Zur Rekonstruktion des Paläoklimas im Golfo Dulce, Costa Rica anhand der stabilen Isotopen-verhältnisse in Foraminiferen- und Pteropodenschalen. Diploma. Teil 2 (Lab.) Fachber. Geow., Universität Bremen, 41 p

Brusca RC, Brusca GJ (2003) Invertebrates, 2nd ed. Sinauer Assoc, Sunderland, MA, p. 936

Cassell DT, Sen Gupta BK (1989) Pliocene Foraminifera and environments, Limon, Costa Rica. J Paleont 63:146–157

Collins LS (1996) Environmental changes in Caribbean shallow waters relative to the closing Tropical American Seaway. *In*: Jackson JBC, Budd AF, Coates AG (eds) Evolution and Environment in Tropical America. University of Chicago Press, Chicago, pp 130–16

Culver SJ, Buzas MA (1982) Distribution of recent benthic Foraminifera in the Caribbean region. Smithsonian Contr Mar Sci 14:1–382

3. Cushman JA, McCulloch I (1939) A report on some arenaceous Foraminifera. Allan Hancock Pac Exp 6(1):1–113
4. Cushman JA, McCulloch I (1940) Some Nonionidae in the collections of the Allan Hancock Foundation. Allan Hancock Pac Exp 6(3):145–178
5. Cushman JA, McCulloch I (1942) Some Virgulininae in the collections of the Allan Hancock Foundation. Allan Hancock Pac Exp 6(4):179–230
6. Cushman JA, McCulloch I (1948) The species of *Bulimina* and related genera in the collections of the Allan Hancock Foundation. Allan Hancock Pac Exp 6(5):231–294
7. Cushman JA, McCulloch I (1950) Some Lagenidae in the collections of the Allan Hancock Foundation. Allan Hancock Pac Exp 6(6):295–364

De la Cruz EM, Vargas JA (1986) Estudio preliminar de la meiofauna de la playa fangosa de Punta Morales, Golfo de Nicoya, Costa Rica. Brenesia 25–26:89–97

De la Cruz E, Vargas JA (1987) Abundancia y distribución vertical de la meiofauna en la playa fangosa de Punta Morales, Golfo de Nicoya, Costa Rica. Rev Biol Trop 35:363–367
Elderfield H, Ganssen G (2000) Past temperature and $\delta^{18}O$ of surface ocean waters inferred from foraminiferal Mg/Ca ratios. Nature 405:442–445
Guzmán HM, Obando VL, Cortés J (1987) Meiofauna associated with a Pacific coral reef in Costa Rica. Coral Reefs 6:107–112
Hallock P, Lidz BH, Cockey-Burkhard EM, Donnelly KB (2003) Foraminifera as bioindicator in coral reef assessment and monitoring: the foram index. Environ Monit Asses 81:221–238
Loeblich AR, Tappan H (1987) Foraminiferal Genera and Their Classification. Van Nostrand Reinhold, New York, 970 p
Morales R, A, Murillo MM (1996) Distribution, abundance and composition of coral reef zooplankton, Cahuita National Park, Limon, Costa Rica. Rev Biol Trop 44:619–630
8. Natland ML (1938) New species of Foraminifera from off the west coast of North America and from the Later Tertiary of the Los Angeles Basin. Bull Scripps Inst Oceanogr Tech Ser 4(5):137–164
Vargas JA (1988) A survey of the meiofauna of an eastern Pacific intertidal mudflat. Rev Biol Trop 36:541–544

Part 6
Sponges

Jorge Cortés, Noam Van Der Hal, and R.W.M. Van Soest

Ailochroia crassa, one of the sponge species recorded for the coral reef in Cahuita, Caribbean Costa Rica, (Photo: Jorge Cortés).

J. Cortés
Centro de Investigación en Ciencias del Mar y Limnología (CIMAR) and Escuela de Biología,Universidad de Costa Rica, 11501-2060 San José, Costa Rica.
e-mail: jorge.cortes@ucr.ac.cr

N. Van Der Hal and R.W.M. Van Soest
Zoological Museum, University of Amsterdam, P.O. Box 94766, 1090 GT, Amsterdam, The Netherlands.
e-mails: nvanderhal@yahoo.com; R.W.M.vansoest@uva.nl

I.S. Wehrtmann, J. Cortés (eds.) *Marine Biodiversity of Costa Rica, Central America*,

Abstract A total of 127 species of sponges distributed in two classes, 14 orders, 42 families, and 72 genera are reported for Costa Rica in this part. Sixty-five species from the Caribbean coast are included here, belonging to 1 class, 10 orders, 29 families, and 45 genera; and 62 species in 2 classes, 13 orders, 31 families, and 44 genera from the Pacific. There are no species in common between both sides of Costa Rica. Twenty-nine species are new records for the Caribbean and 53 species for the Pacific of Costa Rica. Four species, *Stelletta pudica, Amorphinopsis atlantica, Axinyssa lithophaga*, and *Hymeniacidon caerulea*, may be new records for the Caribbean Sea. There are probably many more species, since only a few sites have been studied, and several habitats have not been sampled in the Caribbean or Pacific of Costa Rica, for example, caves, mangroves, or deep waters. There are few studies from other Central American countries, with the exception of Belize and Panama.

Introduction

Sponges (Phylum Porifera) represent an important group of aquatic organisms, found all the way from freshwater to the deep oceanic environments and from the tropics to the poles. There are more than 7,000 valid species of sponges in the world (Hooper & Van Soest 2002, 9,000 species in Bergquist 2001). And in the Caribbean, there are probably more than 325 species reported by Hartman (1977) (see van Soest 1994, 640 species), which included only species from areas shallower than 120 m. Sponges are among the most diverse and abundant components of the fauna on tropical benthic ecosystems, such as coral reefs and mangroves (Rützler *et al.* 2000; Díaz & Rützler 2001). There are only four papers and one thesis (Loaiza 1989) on marine sponges of Costa Rica. The first and only paper with Pacific species includes the species collected around Isla del Coco during the cruise of the Albatross in 1891 (Wilson 1904). The second paper contains a list and description of coral boring sponges from the Caribbean coast of Costa Rica (Risk *et al.* 1980), the third paper is the publication of the thesis by Betty Loaiza Coronado (Loaiza 1991), in which she described the sponges from two sites on the Caribbean coast, Isla Uvita and Puerto Vargas. The last paper is a compendium of all that could be found in the literature and in the collection of the Museo de Zoología, Universidad de Costa Rica on sponges from Costa Rica (Cortés 1996).

A total of 127 species of sponges from Costa Rica are reported in this part. They comprise 2 classes, 14 orders, 42 families, and 72 genera. Sixty-five species from the Caribbean coast (Species List 6.1 is included on the CD-Rom) are presented here, comprising 1 class, 2 subclasses, 10 orders, 29 families, and 45 genera; and 62 species from the Pacific (Species List 6.2 is included on the CD-Rom) in 2 classes, 13 orders, 31 families, and 44 genera. There are no species in common between both sides. Four species, *Stelletta pudica, Amorphinopsis atlantica, Axinyssa lithophaga*, and *Hymeniacidon caerulea*, may be new records for the Caribbean Sea. There are probably more species in Costa Rica than the 127 species reported here, since only a few sites have been studied, and several habitats have not been sampled in the Caribbean or Pacific of Costa Rica, for example, caves, mangroves, or deep waters. Recently, one of us (Noam van der Hal) collected sponges from the southern Caribbean coast and from one area of the north Pacific of Costa Rica, increasing the number of new reports from the country: 28 for the Caribbean and 47 species for the Pacific (Species Lists 6.1 and 6.2 are included on the CD-Rom).

Comparing with other Caribbean localities nearby, the number of species reported here for Costa Rica (65) is lower, but it may be due to the difference in area, habitats, and intensity of research in the other countries: 89 species from Colombia (Zea 1987), more than 120 species from Panamá (Guzmán 2003; Díaz 2005), and over 180 species from Belize (Rützler *et al.* 2000). Recent surveys on sponges in three islands of Bocas del Toro (Panamá) added 41 species to the list known from this area (Díaz 2005). Most of the species that have been reported from the Caribbean of Costa Rica are among the common species of the Caribbean Sea. But, in the Caribbean an important number of new species or taxa are found when poorly studied regions or habitats are explored (Díaz 2005). Even in well-studied areas, for example in Jamaica, a classically well-studied sponge fauna, five and nine new sponge species were described recently from shallow and deep reefs, respectively (Lehnert & van Soest 1998, 1999). A study on the mangrove sponge epibiont community in Belize reveals 149 taxa, of which 49 were considered undescribed species or subspecies (Rützler *et al.* 2000). Therefore, it is expected that future studies on the Costa Rican sponge fauna will reveal more species than the ones indicated here, with a portion of new species and subspecies.

Fig. 6.1 An unidentified sponge from Pacific Costa Rica covered with ophiuroids. (Photo: Jorge Cortés)

Sponges of the eastern tropical Pacific have been studied much less than the Caribbean (Fig. 6.1). Sponges of the region were studied by Wilson (1904), Laubenfels (1935, 1936, 1939), Dickinson (1945), and more recently in Mexico by G. Green, P. Gómez and J.L. Carballo (Green & Gómez 1986; Green *et al.* 1986; Gómez & Bakus 1992; Gómez 1998; Carballo & Cruz-Barraza 2005; Carballo *et al.* 2003, 2004a, b, 2006; Cruz-Barraza & Carballo 2005), and Galápagos (Desqueyroux-Faúndez & van Soest 1997). There are no published records of sponges from the Pacific of any Central American country except for one report from the Pacific coast of Honduras (*Geodia dysoni* Bowerbank, 1873, probably *Isops dysoni*), and several reports from Costa Rica (Wilson 1904) and Panama (Wilson 1904; Laubenfels 1936; Wulff 1996; Boury-Esnault *et al.* 1999; Díaz *et al.* 2005).

Specialists

José Luis Carballo Cenizo: Instituto de Ciencias del Mar y Limnología, UNAM, Mazatlán, México. – Eastern Pacific. carballo@ola.icmyl.unam.mx

María Cristina Díaz: Museo Marino, Isla Margarita, Venezuela. – Caribbean. Crisdiaz@ix.netcom.com

Patricia Gómez: Instituto de Ciencias del Mar y Limnología, UNAM, México. – Caribbean and eastern Pacific

Klaus Rützler: National Museum of Natural History, Smithsonian Institution, Washington, DC. – Caribbean. ruetzler.klaus@nmnh.si.edu

Rob W. M. van Soest: Zoological Museum, University of Amsterdam, The Netherlands. – World-wide. R.W.M.vanSoest@uva.nl

Sven Zea: INVEMAR, Santa Marta, Colombia. – Caribbean. szea@invemar.org.co

Collections

There are collections of Costa Rican sponges in the Museo de Zoología at the Universidad de Costa Rica (Caribbean and Pacific), at the National Museum of Natural History, Smithsonian Institution, Washington, DC (Pacific species), and at the Zoological Museum of the University of Amsterdam (Caribbean and Pacific species).

Recommendations

Much more work is needed on both coasts of Costa Rica, especially on the Pacific coast, as well as in the rest of the Central American countries. Information is needed on the species present, distribution, conservation status, and ecology of the sponges. Biologists specialized on the taxonomy of the group is necessary to obtain a better grasp on the diversity and ecologic importance of this conspicuous benthic group in Costa Rica's waters.

Acknowledgments This compilation was possible thanks to the supply of literature by Kay Hale (previously at the RSMAS Library, University of Miami Library), and Carla Dietrich, Klaus Rützler and Allen Collins at the National Museum of Natural History, Smithsonian Institution. This part benefited greatly from the review by José Luis Carballo Cenizo, UNAM, and comments by two anonymous reviewers. Part of this research was done while holding a Fulbright Scholar Fellowship at the NMNH, Washington, DC. This is a contribution of CIMAR and the Museo de Zoología, Escuela de Biología, both at the Universidad de Costa Rica.

References

Bergquist PR (2001) The Porifera. *In*: Anderson DT (ed) Invertebrate Zoology, 2nd ed. Oxford University Press, Australia, pp 10–27

1. Boury-Esnault N, Klautau M, Bezac C, Wulff J, Solé-Cava AM (1999) Comparative study of putative conspecific sponge populations from both sides of the Isthmus of Panama. J Mar Biol Assoc UK 79:39–50

Carballo JL, Cruz-Barraza JA (2005) *Cliona microstrongylata*, a new species of boring sponge from the Sea of Cortés (Pacific Ocean, México). Cah Biol Mar 46:379–387

2. Carballo JL, Gómez P, Cruz-Barraza JA, Flores-Sánchez DM (2003) Sponges of the Family Chondrillidae (Porifera: Demospongiae) from the Pacific coast of Mexico, with the description of three new species. Proc Biol Soc Wash 116:515–527

Carballo JL, Cruz-Barraza JA, Gómez P (2004a) Taxonomy and description of clionid sponges (Hadromerida, Clionaidae) from the Pacific Ocean of Mexico. Zool J Linnean Soc 141:353–397

Carballo JL, Yáñez B, Nava H (2004b) Persistence of the sponge *Suberites aurantiaca* (Duchassaing and Michelotti, 1864) in an estuarine ecosystem (Pacific coast, Mexico). Bull Mar Sci 75:487–497

3. Carballo JL, Zubía E, Ortega MJ (2006) Biological and chemical characterizations of three new species of *Dysidea* (Porifera: Demospongiae) from the Pacific Mexican coast. Bioch Syst Ecol 34:498–508
4. Cortés J (1996) Biodiversidad marina de Costa Rica: Filo Porifera. Rev Biol Trop 44:911–914

Cruz-Barraza JA, Carballo JL (2005) First record of *Plakortis* Schulze (Porifera: Homosclerophorida) from the northeast Pacific coast, with the description of *Plakortis albicans* sp. nov. Zootaxa 868:1–12

Desqueyroux-Faúndez R, van Soest RWM (1997) Shallow waters Demosponges of the Galápagos Islands. Rev Suisse Zool 104:379–467

5. Díaz MC (2005) Common sponges from shallow marine habitats from Bocas del Toro region, Panama. Carib J Sci 41:465–475

Díaz MC, Rützler K. (2001) Sponges: an essential component of Caribbean coral reefs. Bull Mar Sci 69:535–546

6. Díaz MC, van Soest RWM, Rützler K, Guzman HM (2005) *Aplysina chiriquiensis*, a new pedunculate sponge from the Gulf of Chiriquí, Panamá, eastern Pacific (Aplysinidae, Verongida). Zootaxa 1012:1–12

Dickinson MG (1945) Sponges of the Gulf of California. Allan Hancock Pac Exp 11:1–251

7. Fonseca E, AC, A Arrivillaga (2003) Coral reefs of Guatemala. *In*: Cortés J (ed) Latin American Coral Reefs. Elsevier Science, Amsterdam, pp 159–170

Gómez P (1998) First record and new species of *Gastrophanella* (Porifera: Demospongiae: Lithistida) from the central East Pacific. Proc Biol Soc Wash 111:774–780

Gómez P, Bakus GJ (1992) *Aplysina gerardogreeni* and *Aplysina aztecus* (Porifera: Demospongiae), new species from the Mexican Pacific. An Inst Cienc Mar Limnol, UNAM 19:175–180
Green G, Gómez P (1986) Estudio taxonómico de las esponjas de la Bahía de Mazatlán, Sinaloa, México. An Inst Cienc Mar Limnol, UNAM 13:273–300
8. Green G, Fuentes-Velázquez LE, Gómez-López P (1986) Nuevos registros de porífera del arrecife La Blanquilla, Veracruz, México. An Inst Cienc Mar Limnol, UNAM 13:127–146
9. Guzmán HM (2003) Caribbean coral reefs of Panama: present status and future perspectives. *In*: Cortés J (ed) Latin American Coral Reefs. Elsevier Science, Amsterdam, pp 241–274
Hartman WD (1977) Sponges as reef builders and shapers. Stud Geol, Am Assoc Petrol Geol 4:127–134
10. Hechtel GJ (1965) A systematic study of the Demospongiae of Port Royal, Jamaica. Peabody Mus Nat Hist Bull 20:1–103
11. Hooper JNA, van Soest RWM (2002) Systema Porifera: a guide to the classification of sponges. Kluwer/Plenum, New York, 1725 p
12. Laubenfels MW de (1934) New sponges from the Puerto Rican deep. Smithsonian Miscell Coll, Washington 91(17):1–28
Laubenfels MW de (1935) Some sponges of Lower California (Mexico). Am Mus Novitates 779:1–14
13. Laubenfels MW de (1936) A comparison of the shallow-water sponges near the Pacific end of the Panama canal with those of the Caribbean end. Proc US Natl Mus 83:441–466
Laubenfels MW de (1939) Sponges collected on the Presidential cruise of 1938. Smithson Misc Coll 98:1–7
Lehnert H, van Soest RWM (1998) Shallow water sponges from Jamaica. Beaufortia 48:71–103
Lehnert H, van Soest RWM (1999) Deep water sponges from Jamaica. Beaufortia 49:141–169
Loaiza B (1989) Generalidades del Phylum Porifera y bases para su identificación con sinopsis de algunas de ellas, en Limón, Costa Rica. Tesis de Licenciatura, Escuela de Ciencias Biológicas, Universidad Nacional, Heredia, 173 p
14. Loaiza B (1991) Estudio taxonómico de las esponjas del Parque Nacional Cahuita, sector Puerto Vargas e Isla Uvita, Limón, Costa Rica. Brenesia 36:21–62
15. Risk MJ, Murillo MM, Cortés J (1980) Observaciones biológicas preliminares sobre el arrecife coralino en el Parque Nacional Cahuita, Costa Rica. Rev Biol Trop 28:361–382
16. Rützler K, Díaz MC, van Soest RWM, Zea S, Smith KP, Alvarez B, Wulff J (2000) Diversity of sponge fauna in mangroves ponds, Pelican Cays, Belize. Atoll Res Bull 476:231–248
17. Van Soest RWM (1981) A checklist of the Curaçao Sponges (Porifera: Demospongiae) including a pictorial key to the more common reef forms. Versl Techn Gegv, Inst Taxon Zool, Universiteit, Amsterdam 31:1–39
Van Soest RWM (1994) Demosponge distribution patterns. *In*: van Soest RWM, van Kempen TMG, Braekman J-C (eds) Sponges in Time and Space. Balkema, Rotterdam, pp 213–223
18. Wiedenmayer F (1977) Shallow-water sponges of the western Bahamas. Experientia Suppl. 28:1–287
19. Wilson HV (1904) Reports on an exploration off the west coasts of Mexico, Central and South America, and off the Galapagos Islands, in charge of Alexander Agassiz, by the US Fish Commission Steamer Albatross, during 1891, Lieut. Commander Z.L. Tanner U.S.N., commanding. XXX. The sponges Mem Mus Comp Zoöl Harvard Coll XXX:1–164
20. World Porifera Database: www.vliz.be/vmdcdata/porifera
Wulff JL (1996) Do the same sponge species live on both the Caribbean and eastern Pacific sides of the Isthmus of Panama? Bull Inst Royal Sci Natur Belgique 66(Suppl):165–173
21. Zea S (1987) Esponjas del Caribe Colombiano. Catálogo Científico, Bogotá, Colombia, 286 p

Part 7
Hydrozoa, Scyphozoa, and Cubozoa (Medusozoa)

Karina Rodríguez-Sáenz and Lourdes Segura-Puertas†

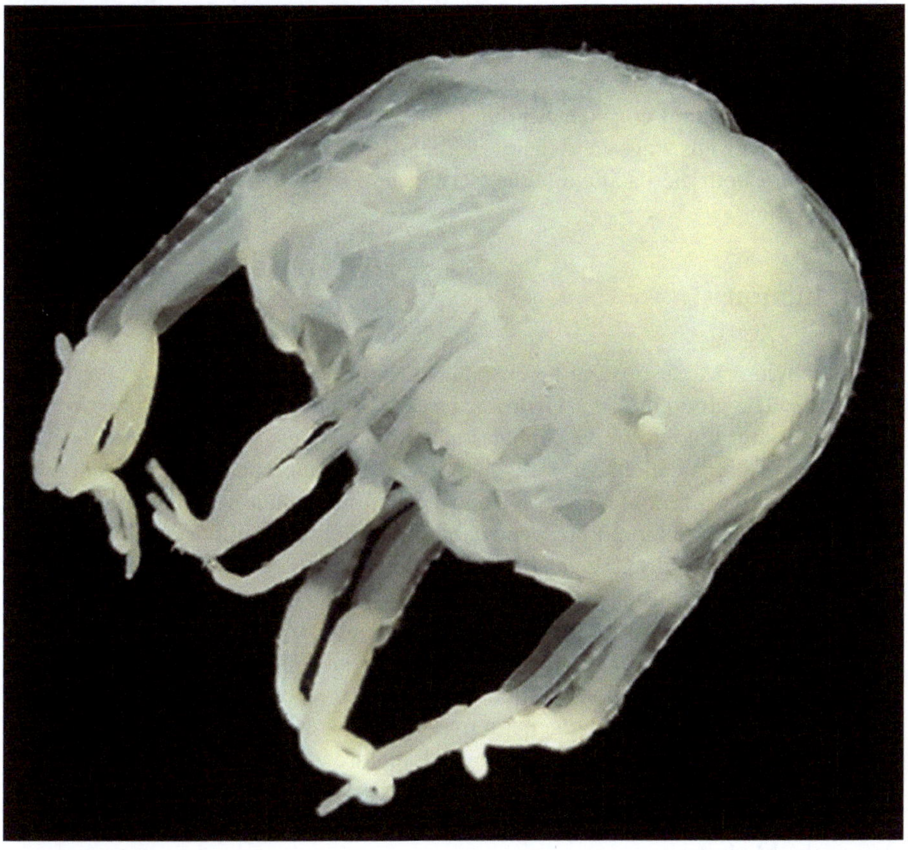

Cubozoa (*Tripedalia cystophora*) collected in Punta Morales, Puntarenas, Pacific Costa Rica (Photo: Ingo S. Wehrtmann)

K. Rodríguez-Sáenz
Sistema de Estudios de Posgrado en Biología, Universidad de Costa Rica. 2060 San Pedro, San José, Costa Rica.
e-mail: karodrig@cariari.ucr.ac.cr
Present address: Habitat Crítico: Environmental Consulting, Apdo. 389-1300, San José, Costa Rica.
e-mail: karinarodriguez@racsa.co.cr

L. Segura-Puertas† (deceased)
Universidad Nacional Autónoma de México, Instituto de Ciencias del Mar y Limnología, Unidad Académica Puerto Morelos. P. O. Box 1152, 77501 Cancún, Q. Roo, México.
e-mail: lsegura@mar.icmyl.unam.mx

I.S. Wehrtmann, J. Cortés (eds.) *Marine Biodiversity of Costa Rica, Central America*,

Abstract This paper presents a checklist of 84 medusozoan species recorded in Costa Rican waters, including the Costa Rican Dome, Golfo de Papagayo, Golfo de Nicoya, Golfo Dulce, Isla del Coco, and the Caribbean coast. The species comprise 3 classes, 2 subclasses, 40 families, and 63 genera. Among Hydrozoa, 21 species are new records to Costa Rican waters, and *Tripedalia cystophora* is a new record of Cubozoa to the Eastern Tropical Pacific ocean. Forty-four species (57%) of Hydrozoa were recorded in the Golfo de Papagayo, a coastal upwelling region in the northern Pacific of Costa Rica. Few species collected in the Costa Rica's Caribbean waters are also present in the country's Pacific coast. Additional surveys are needed to better understand Costa Rican marine gelatinous plankton diversity and ecology. Further studies should consider benthic habitats, coastal lagoons, and deep waters. Analysis of hydroids stages and their released medusa will also increase the number of species for both the Pacific and the Caribbean coasts.

Introduction

The subphylum Medusozoa is a conspicuous group of the most simply organized metazoans. The group includes Hydromedusae, Scyphomedusae, and Cubomedusae. Medusozoa are diploblastic, acelomate radiata metazoans, having their cells at the tissue level of organization. They are distinguished by alternations of stages between a benthic, asexually reproducing polyp and a planktonic sexually reproducing medusa that differ radically in morphology and ecology. The polyp is solitary or colonial, and usually consists of morphologically similar units (Goy & Toulemont 1997). The medusa is planktonic, usually free-swimming and solitary. They typically have a ciliated planula larva. Nearly ubiquitous within the subphylum are stinging cells or nematocysts, which are employed for prey capture and for defense.

The phylogenetic position of Medusozoa is a controversial issue. Bouillon & Boero (2000) provided a phylogenetic hypothesis for Hydrozoa based on embryological, developmental, and morphological features. Recently, studies based on molecular sequences of the small subunit of the ribosome and cladistic analysis (Marques & Collins 2004) have provided information on the phylogenetic relationship of the Medusozoa.

Hydromedusae are the most highly diversified class of the subphylum Medusozoa, exhibiting the greatest variation in life cycles, and the polyp or medusa stages are entirely lacking in some species. Most species are dioecious. The hydromedusae have a velum (except *Obelia*), their mesoglea is acellular, nematocysts are usually restricted to the ectoderm (Bouillon 1999), and few hydromedusae sting.

The Scyphozoa and Cubozoa classes include most of the larger solitary marine invertebrates of the world, even though, according to Mianzan & Cornelius (1999), for most regions of the world there is no reliable taxonomic account of them, and little is known about their role in marine ecosystems. The Scyphozoa includes approximately 200 exclusively marine species, occurring in both pelagic and benthic regions. Cubozoa constitutes a particular group of about 15 species. They occur in

Fig. 7.1 An unidentified jelly fish (Scyphozoa) observed at the north Pacific of Costa Rica (Photo: Jorge Cortés)

warm shallow waters in the tropical and subtropical regions. Some species have potent toxins and are dangerous to humans (Mianzan & Cornelius 1999).

Planktonic cnidarians are widespread in the oceans, both in vertical and horizontal distribution, but with ecological preferences for different water masses. They are useful indicators of water masses and changes in marine ecosystems. Thus, the study of their population fluctuations can provide valuable information on the conditions and health of marine environments, particularly for fisheries, since many species prey on different life stages of fish and other fishery resources (CIESM 2001).

In this paper, we present a preliminary list of Hydrozoa (for Siphonophora: see Part 8 in Chapter IV), Scyphozoa (Fig. 7.1), and Cubozoa from Costa Rican waters.

The higher taxa follow the phylogenetic arrangement proposed by Marques & Collins (2004). Families, genera, and species are arranged alphabetically. Information of species collected as part of international cruises and national efforts in marine studies are also included.

Taxonomy

Several morphological characters are important for the identification of medusae: the general body shape and size, the consistency of the mesoglea, the length and the final section of the manubrium, the shape of the mouth, the number, kind, and position of tentacles, types of nematocysts, the number and position of radial canals, and the presence of distinct sensory structures. Many of these gelatinous organisms are small and have a fragile, transparent tissue, being difficult to see in the water. In contrast, a number of scyphomedusae are conspicuous. Their tissue is composed of 95% or more of water, and their bodies are usually delicate and easily damaged by traditional trawl nets used for capture, which makes identification even harder. Recent research shows the success of manual collection by SCUBA diving, submersibles, and the use of underwater video (Wrobel & Mills 1998).

Although the present contribution focuses on the medusae stage only, it is relevant to emphasize the importance of studying both hydroid and medusa stages. Taxonomic studies including the complete life cycles are important to document the number of medusozoa species in Costa Rica.

Distribution and diversity in Costa Rica: The checklist includes species collected in Costa Rican waters only. Studies on the medusozoan fauna in Costa Rica are scarce; several of these studies were part of scientific expeditions along the Eastern Pacific Ocean ("Eastern Pacific Zaca") published by Bigelow (1940) and in the Caribbean Sea ("Thomas Washington") (Alvariño 1972). Recently, Nowaczyk (1998) and Rodríguez (2005) present information on the vertical distribution and seasonal variations of planktonic cnidarians in Costa Rican coastal waters, particularly in Golfo Dulce and Golfo de Papagayo.

In total, 83 species of Medusozoa have been recorded in Costa Rican waters (Species Lists 7.1 and 7.2, both included in the CD-Rom), comprising 3 classes, 2 subclasses, 40 families, and 63 genera. The greatest species richness was observed in Hydrozoa, represented by 76 species (90.4%), followed by five species of Scyphozoa (6.0%) and three of Cubozoa (3.6%). Among the Hydrozoa, the Costa Rican fauna is rather poor in Limnomedusae (one family), Narcomedusae, and Trachymedusae (three families each) compared to Anthomedusae (17 families) and Leptomedusae (10 families).

Here we include 21 new records of hydrozoans from Costa Rican waters. *Tripedalia cystophora* is a new record of Cubozoa for the Eastern Tropical Pacific region. All specimens have been collected from the Pacific coast. Forty-four species (57%) of Hydrozoa have been recorded in the Golfo de Papagayo, a coastal upwelling region in the northern Pacific of Costa Rica (Rodríguez 2006).

Specimens of the hydromedusae *Aegina citrea* and *Solmundella bitentaculata* were collected by Alvariño (1972) from Costa Rican Caribbean waters, while a few specimens of *Liriope tetraphylla* and *Aglaura hemistoma* from the same area were collected (K. Rodríguez, unpublished data 2001).

Studies in the Caribbean region show a high diversity of hydromedusae (Human & Deloach 2002), but only four species are reported for the Caribbean coast of Costa Rica. Furthermore, Larson (1982) collected 62 species of hydromedusae, 4 species of cubomedusae, and 5 species of scyphozoans in different reef areas of Carrie Bow Cay (Belize). Recently, Segura-Puertas *et al.* (2003) listed 88 species from the Mexican Caribbean Sea. These data show that there are large gaps in the knowledge of medusozoan fauna in the Caribbean coast of Costa Rica, and clearly indicate the need for more research.

Most species listed here (except *A. citrea*) have been recorded from the Pacific coast (Species List 7.2 is included on the CD-Rom), with only four species for the Costa Rica Caribbean Sea (4.8%) (Species List 7.1 is included on the CD-Rom). Despite previous research, additional work in both oceanic and coastal waters is necessary, including estuarine ecosystems, where studies on Medusozoa have not been done.

Specialists

Ángeles Alvariño: Emeritus research biologist, Southwest Fisheries Science Center, NOAA, P.O. Box 271, La Jolla, California 92038, USA – Taxonomy and biogeography of Medusozoa in the world's oceans

Mary N. Arai: Department of Biological Sciences, University of Calgary, Calgary, Alberta, Canada – Biology, ecology, systematics, and physiology of worldwide Scyphozoa. arai@acs.ucalgary.ca

Ferdinando Boero: Dipartamento di Biologia, Staziones di Biologia Marina, Universita di Lecce, 73 100 Lecce, Italy – Systematics, ecology, and distribution of worldwide Hydrozoa. boero@unile.it

Jean Bouillon: Laboratoire de Zoologie, Université Libre de Bruxelles, 50 av. F.D. Roosvelt, 1050 Bruxelles, Belgium – Systematics, ecology, distribution of worldwide Hydromedusa

Ronald Larson: Department of Biology, University of Victoria, British Columbia, Canada V8W 2Y2 – Worldwide; ecology of jellyfish

Claudia Mills: Friday Harbor Laboratories. University of Washington, Washington, DC Ecology, physiology of worldwide medusae

†Francesc Pagès: Insitut de Ciències del Mar (CSIC), Passeig Nacional s/n, 08039 Barcelona, Spain – Ecology and taxonomy of gelatinous zooplankton (Hydromedusa and Siphonophora)

Sergio Palma: Escuela de Ciencias del Mar, Universidad Católica de Valparaíso, Chile. Casilla 1020, Valparaíso, Chile – Gelatinous zooplankton; southeastern Pacific – spalma@ucv.ac.ch

Karina Rodríguez: Hábitat Crítico, Environmental Consulting, Apdo. 389-1300 San José, Costa Rica – Ecology and taxonomy of eastern tropical Pacific Hydromedusa – karinarodriguez@racsa.co.cr

†Lourdes Segura-Puertas: Universidad Nacional Autónoma de México, Instituto de Ciencias del Mar y Limnología, Unidad Académica Puerto Morelos, P.O. Box 1152, 77501 Cancún, Q. Roo, México – Taxonomy, ecology, and toxicology of Medusozoa

Collections

Most specimens reported here are belong to a collection that is kept by K. Rodríguez as part of her Master thesis research and stored at CIMAR, Universidad de Costa Rica. A few specimens collected in previous projects are deposited at the Museo de Zoología, Universidad de Costa Rica. Zooplankton samples collected at the Costa

Rica Dome are kept in the Laboratorio de Invertebrados, Facultad de Ciencias, Universidad Nacional Autónoma de México, México, D.F.

Future Research

Additional surveys are needed to increase our knowledge on the diversity and distribution of Medusozoa in Costa Rica. Future research should consider benthic habitats, coastal lagoons, deep waters and the Caribbean region. Life cycles, feeding methods, and growth rates are also areas of interest for research. Further analysis of hydroid stages and their laboratory-released medusae will also increase the number of species of Costa Rica's Caribbean and Pacific waters. Additional information about seasonal changes in species composition, predatory impact on fisheries, pelagic ecosystem dynamics, and reproductive trends in tropical species should be part of a new phase of gelatinous plankton research in Costa Rica.

References

Alvariño A (1972) Zooplancton del Caribe, Golfo de México y regiones adyacentes del Pacífico. Mem. IV Congr. Nac. Oceanog., México, pp 223–247

1. Alvariño A (1976) El zooplancton de la región de la América Central en el Pacífico. Mem. III Simp. Latinoamer. Oceanogr. Biol., San Salvador, El Salvador, pp 47
2. Bigelow HB (1940) Eastern Pacific Expeditions of the New York Zoological Society. XX. Medusae of the Templeton Crocker and Eastern Pacific *Zaca* Expeditions, 1936–1938. Zoologica XXV:281–321
3. Bouillon J (1999) Hydromedusae. *In*: Boltovskoy D (ed) South Atlantic Zooplankton, Backhuys, Leiden, The Netherlands, pp 385–465

Bouillon J, Boero F (2000) Phylogeny and classification of Hydroidomedusae. Thalassia Salentina 24:1–296

4. CIESM (2001) Gelatinous zooplankton outbreaks: theory and practice. CIESM workshop 14, Monaco. 112 p. www.ciesm.org/publications/Naples01.pdf
5. Cortés J (1996–1997) Biodiversidad marina de Costa Rica: filo Cnidaria. Rev Biol Trop 44(3)/45(1):323–334

Goy J, Toulemont A (1997) Méduses. Musée Océanographique de Monaco. Abysses 5:1–159

Human P, Deloach N (2002) Reef creature identification: Florida Caribbean Bahamas, 2nd ed. New World, Jacksonville, FL, 420 p

6. Kramp PL (1968) The hydromedusae of the Pacific and Indian Oceans. The Carlsberg Foundation's Oceanographical Expedition around the world 1928–1930 and previous Dana Expeditions. Andr. Fred. Host. Sections II & III, 196 p

Larson RJ (1982) Medusae (Cnidaria) from Carrie Bow Cay, Belize. *In*: Rützler K, Macintyre IG (eds) The Atlantic Barrier Reef Ecosystem at Carrie Bow Cay Belize. Vol. I: structure and communities. Smithsonian Institution Press, Washington, DC, pp 253–258

Marques AC, Collins AG (2004) Cladistic analysis of Medusozoa and cnidarian evolution. Invertebr Biol 123:23–42

7. Mayer A (1910) Medusae of the World. Vol III: Scyphozoa. Carnegie Institute, Washington, DC, 735 p

Mianzan H, Cornelius P (1999) Cubomedusae and Scyphomedusae. *In*: Boltovskoy D (ed) South Atlantic Zooplankton. Backhuys, Leiden, The Netherlands, pp 513–559

8. Nowaczyk J (1998) Gelatinöses Zooplankton im Golfo Dulce (Costa Rica) während des Überganges von der Regen- zur Trockenzeit 1997–1998. M.Sc. thesis, University of Bremen, Germany, 82 p
9. Pagés F, Gili JM, Bouillon J (1992) Planktonic cnidarians of the Benguela Current. Sci Mar Mar 56(Suppl 1):1–144
10. Rodríguez K (2005) Distribución espacial y temporal de la biomasa, composición y abundancia del zooplancton, con énfasis en hidromedusas de Bahía Culebra durante La Niña 1999 y el 2000. M.Sc. thesis, University of Costa Rica, San José, Costa Rica, 156 p
11. Segura-Puertas L (1984) Morfología, sistemática y zoogeografía de las medusas (Cnidaria: Hydrozoa y Scyphozoa) del Pacífico Tropical Oriental. An Inst Cienc Mar Limnol UNAM 8:1–320
12. Segura-Puertas L (1991) Medusas (Cnidaria: Hydrozoa y Scyphozoa) de la región del Domo de Costa Rica. Rev Biol Trop 39:159–163

Segura-Puertas L, Suárez-Morales E, Celis L (2003) A checklist of the Medusae (Hydrozoa, Scyphozoa and Cubozoa) of Mexico. Zootaxa 194:1–15

Wrobel D, Mills C (1998) Pacific coast pelagic invertebrates: a guide to the common gelatinous animals. Monterey Bay Aquarium/Sea Challengers, California, 112 p

Part 8
Siphonophores

Karina Rodríguez-Sáenz and Rebeca Gasca

Physophora hydrostatica, a species from Pacific Costa Rica. (Photo: Steven H.D. Haddock)

Abstract Thirty-seven species of siphonophores are listed for the Costa Rican coastal and oceanic waters. All species were recorded from the Pacific coast, seven of them have been also collected off the Caribbean coast. These figures represent 10% of the 70 species known from the Caribbean Sea, and 42% of the 87 species known from the eastern tropical Pacific. Overall, the species recorded in Costa Rican waters represent close to 37% of the nearly 100 species known to be distributed in the tropical-equatorial belt. The relatively low biodiversity of siphonophores in the Caribbean coast of Costa Rica is clearly a result of the scarcity of research in this region. All the species currently known from Costa Rica are epipelagic forms living between the surface and 200 m, and have been recorded previously in the corresponding oceanographic regions of the Atlantic and Pacific oceans. Additional work in both oceanic and coastal waters is necessary, including coastal and estuarine ecosystems and also deep waters (>200 m), which most probably harbor a diverse siphonophore fauna.

Introduction

The Subclass Siphonophora is a peculiar group of marine cnidarians (Class: Hydrozoa); it is part of the commonly known "gelatinous zooplankton," which includes hydromedusae, scyphomedusae, ctenophores, salps, and appendicularians among other taxa. Most of the species of siphonophores are fully planktonic forms. The Portuguese Man-of-War (*Physalia physalis*), floating at the surface of the water, and a small group of benthic forms (Family: Rhodaliidae) (Pugh 1999b) are two noteworthy exceptions.

The siphonophores are efficient predators distributed in the entire world ocean from the surface to thousands of meters deep. Most of them inhabit oceanic waters, but a few species can live in neritic areas, and are often present in high densities in both neritic and oceanic waters. They are all carnivorous forms, capturing their prey using poisonous nematocysts, and most species are assignable to a passive feeding behavior. The use of luminescent lures to attract preys has been discovered recently (Haddock *et al.* 2005a). Because of their predating efficiency, siphonophores can cause major impacts on the structure and dynamics of the zooplankton communities

K. Rodríguez-Sáenz
Habitat Crítico S.A., Environmental Consulting. P.O. Box 389-1300, San José, Costa Rica
e-mails: karinarodriguez@racsa.co.cr; habitat@ice.co.cr

R. Gasca
El Colegio de la Frontera Sur (ECOSUR)-Chetumal. A.P. 424. Chetumal, Quintana Roo 77000, México
e-mail: rgasca@ecosur.mx

(Pagés *et al.* 1992). Recent studies have shown how pollution caused by human activities may have an impact on gelatinous zooplankton, including the population dynamics of siphonophores. These changes can eventually impact other components of the planktonic ecosystems (CIESM 2001).

We present a checklist of the siphonophores collected in coastal and oceanic environments of Costa Rica's Caribbean Sea and Pacific Ocean (Species Lists 8.1 and 8.2 are included on the CD-Rom). The lists include scientific results of international expeditions such as that of the "Albatross" in the eastern tropical Pacific during 1904 and 1905, and the "BONACCA" in the Western Caribbean Sea in 1963 (Alvariño 1972); they covered large areas of these regions, including a few stations in Costa Rican waters. Recent surveys of siphonophores in Costa Rican waters have been done and included basic aspects on the composition and distribution of this group (Ramírez 1988; Gasca & Suárez-Morales 1992; Nowaczyk 1998, Rodríguez 2001 unpublished data).

Taxonomic status/problems: The classic and best account on siphonophore taxonomy and morphology was written by Totton (1965); however, the systematics of the group has changed since then (Bouillon *et al.* 1992). Many species, mainly deepwater forms, have been described recently from different regions of the world (Pugh 1992a, b, c, 1995, 1999a, 2001, 2002, 2004; Pugh & Pagès 1995, 1997). Valuable publications on the biology and reproduction of the siphonophores are those by Mackie *et al.* (1987) and Carré & Carré (1993).

Siphonophores have been grouped into three orders, based on the presence of an apical pneumatophore (Cystonectae and Physonectae; absent in Calycophorae), and on the presence (Physonectae and Calycophorae) or absence (Cystonectae) of nectophores. The structure and life cycle of this group are complex. They have been designated as "colonies" formed by zooids, medusoids, and polypoids, each with different forms and functions. The complexity of siphonophores is not restricted to their morphological and physiological diversity; it extends also to the specific terminology used to define each of these variable structures. A recent revision of the morphological nomenclature in siphonophores is given in Haddock *et al.* (2005b). Nectophores and bracts are of particular importance for taxonomic identification.

Siphonophores are fragile, thus making it difficult to study them both taxonomically and ecologically. The use of traditional trawl nets may produce serious damages to the organisms and consequently complicate their identification and/or quantification. In recent years, blue water diving, submersibles, and remotely operated vehicles (ROVs) have solved most of this problem. It is now possible to obtain complete organisms and images of siphonophores, thus facilitating the identification of species and providing valuable information about their biology, ecology, and behavior. However, this kind of equipment is expensive (especially submersibles), and for developing countries it is almost impossible to acquire and maintain them. Due to this situation, most of the information on deepwater siphonophores is concentrated in a few sites around the world; hence, epipelagic trawls are, historically and currently, the most important source of the available information on the group. Moreover, the task of fixation-preservation of an intact organism is difficult. The slow addition of magnesium chloride to the seawater while the organism is swimming helps to create a condition for relaxation previous to the fixation and preservation in formalin.

Their phylogenetic relations are still obscure and under investigation. There are few specialists working in the field, and some of them are listed below.

Species richness. The number of nominal species of siphonophores known to date is about 170 (worldwide), with many others discovered but yet undescribed. There are surveys on the general diversity of siphonophores in different regions of the Atlantic Ocean, such as the works by Pugh (1999b) in the southwestern Atlantic (96 species), by Stepanjants (1975) and by Suárez & Gasca (1991) in the Caribbean Sea and Gulf of Mexico (62 species), and by Pugh & Gasca (in press) for the Gulf of Mexico (80 species). In the Central Pacific Ocean, between 10°N and 20°S, Stepanjants (1977) recorded 55 species. Suárez & Gasca (1991) recorded 48 species from Mexican Pacific waters.

Overall, the diversity of siphonophores known in Costa Rica represents close to 20% of the known diversity of the group worldwide. In regional terms, the number of species estimated to inhabit the Pacific Ocean adjacent to the Costa Rican waters is *ca*. 90 species (Alvariño 1972, 1974, Stepanjants 1977, Gasca 2002), representing 42% of the species reported from this area. Roughly seven species have been reported from the Caribbean Sea of Costa Rica, which represents only 10% of the 70 species known from the Caribbean region (Alvariño 1972, 1974, Owre & Foyo 1972, Stepanjants 1975, Michel & Foyo 1976, Gasca 2002). These data show the need to increase surveys of the zooplankton community in both coasts of Costa Rica, but particularly in the Caribbean waters.

Specialists

PHIL R. PUGH: National Oceanography Centre, Southampton S014 3ZH, United Kingdom – Taxonomy. prp@noc.soton.ac.uk

CLAUDE CARRE and DANIELE CARRE: Station Zoologique, Faculté des Sciences de Paris, 06 Villefranche-sur-mer. France.

CASEY W. DUNN: Brown University Providence, 80 Waterman Street Box G-W, RI 02912 USA – casey_dunn@broun.edu

REBECA GASCA: El Colegio de la Frontera Sur (ECOSUR)-Chetumal. A.P. 424. Chetumal, Quintana Roo 77000. México – Taxonomy, geographic distribution. rgasca@ecosur.mx

Collections

Specimens of most of the species reported herein are deposited in a collection kept by K. Rodríguez as part of her Master thesis research. This material is held at CIMAR, Universidad de Costa Rica, San José. A few specimens, collected in previous projects, are deposited in the Museo de Zoología of the Universidad de Costa Rica.

Conclusions

One of the reasons for the relatively low biodiversity found in the Caribbean coast of Costa Rica with respect to the Pacific coast is the scarcity of research in those waters. Additional collections of gelatinous zooplankton in both oceanic and coastal waters are necessary, including estuaries, but emphasizing the neritic and oceanic areas. The same recommendation is valid for the rest of Central America. Through understanding the ecology and distribution of siphonophores as an important group of predators, it will be easier to evaluate the processes determining the dynamics of the marine planktonic biota of Costa Rica, including the ecology and development of local fisheries.

References

1. Alvariño A (1972) Zooplancton del Caribe, Golfo de México y regiones adyacentes del Pacífico. Mem. IV Congr. Nac. Oceanogr. México, pp 223–247

Alvariño A (1974) Distribution of siphonophores in the regions adjacent to the Suez and Panama Canals. Fish Bull 22:41–76

Bouillon J, Boero F Cicogna F Gili JM Hughes RH (1992) Non-siphonophoran Hydrozoa: what are we talking about? Sci Mar 56:279–284

CIESM (2001) Gelatinous zooplankton outbreaks: theory and practice. CIESM workshop ser. No. 14, 112 p. Monaco. www.ciesm.org/publications/Naples01.pdf

Carré C, Carré D (1993) Ordre des Siphonophores. *In*: Doumenc D (ed) Traité de Zoologie. Anatomie, Systématique, Biologie. Tome III. Fascicule 2. Cnidaires. Cténaires. Masson, Paris, pp. 523–596

2. Cortés J (1996–1997) Biodiversidad marina de Costa Rica: Filo Cnidaria. Rev Biol Trop 44/45:323–334

Gasca R (2002) Lista faunística y bibliografía comentada de los sifonóforos (Cnidaria) de México. An Inst Biol UNAM 73:123–143

3. Gasca R Suárez E (1992) Sifonóforos (Cnidaria: Siphonophora) del Domo de Costa Rica. Rev Biol Trop 40:125–130

Haddock SHD, Dunn CW, Pugh PR, Schnitzler CE (2005a) Bioluminescent and red-fluorescent lures in a deep-sea Siphonophore. Science 309:263.

Haddock SHD, Dunn CW, Pugh PR (2005b) A re-examination of siphonophore terminology and morphology, applied to the description of two new prayinae species with remarkable bio-optical properties. J Mar Biol Ass U.K. 85:1–13.

4. Kirkpatrick P Pugh PR (1984) Siphonophores and Velellids: Keys and notes for the identification of the species. Linn Soc London Estuar Brack.-Water Sci Assoc, London, 154 p

Mackie GO, Pugh PR, Purcell JE (1987) Siphonophore biology. Adv Mar Biol 24:97–262

Michel HB, Foyo M (1976) Caribbean Zooplankton. Part. I. Siphonophora, Heteropoda, Copepoda, Euphausiacea, Chaetognatha and Salpidae. Off. Naval Res., Dept. Navy, Washington, DC, 549 p

5. Nowaczyk J (1998) Gelatinöses Zooplankton im Golfo Dulce (Costa Rica), während des Überganges von der Regen zur Trockenzeit 1997–1998. M.Sc. thesis, Univ. Bremen, Bremen, 82 p

Owre HBM, Foyo M (1972) Studies on Caribbean zooplankton. Description of the program and results of the first cruise. Bull Mar Sci 22:483–521

6. Pagès F, Gili JM Bouillon J (1992). Planktonic cnidarians of the Benguela current. Sci Mar 56(Suppl 1):144

Pugh PR (1992a) The status of the genus *Prayoides* (Siphonophora: Prayidae). J Mar Biol Assoc U.K. 72:895–909
Pugh PR (1992b) *Desmophyes haematogaster*, a new species of prayine siphonophore (Calycophorae, Prayidae). Bull Mar Sci 50:89–96
Pugh PR (1992c) A revision of the sub-family Nectopyramidinae (Siphonophora, Prayidae). Phil Trans Royal Soc Lond 335:281–322
Pugh PR (1995) *Clausophyes tropica* (Siphonophorae, Calicophora) a new siphonophore species from the tropical Atlantic. Bull Mar Sci 57:453–459
Pugh PR (1999a) A review of the genus *Bargmannia* Totton, 1954 (Siphonophorae, Physonectae, Pyrostephidae). Bull Nat Hist Mus Lond (Zool.) 65:51–72
Pugh PR (1999b) Siphonophorae. *In*: Boltovskoy D (ed) South Atlantic Zooplankton. Backhuys, Leiden, The Netherlands, pp 467–511
Pugh PR (2001) A review of the genus *Erenna* Bedot, 1904 (Siphonophora, Physonectae). Bull Nat Hist Mus Lond (Zool.) 67:169–182
Pugh PR (2002) A new species of *Rosacea* (Siphonophora: Calycophorae: Prayidae) from the Gulf of Oman. J Mar Biol Ass U.K. 82:171–172
Pugh PR (2004) A new species of *Physophora* (Siphonophora: Physonectae: Physophoridae) from the North Atlantic, with comments on related species. Syst Biodiv 2:251–270
Pugh PR, Gasca R (2008) Siphonophora (Cnidaria: Hydrozoa) of the Gulf of Mexico. In: Felder DL, Camp DA (eds) Gulf of Mexico Origin, Waters and Biota. Biodivesity, vol 1 Texas A&M University Press, College Station, Texas (in press)
Pugh PR Pagès F (1995) Is *Lensia reticulata* a diphyinae species (Siphonophorae, Calycophorae, Diphyidae)? A redescription. Sci Mar 52:181–192
Pugh PR, Pagès F (1997) A re-description of *Lensia asymmetrica* Stepanjants, 1970 (Siphonophorae, Diphyidae). Sci Mar 61:153–161
Ramírez AL (1988) Sistemática y distribución de los sifonóforos del Domo de Costa Rica. Thesis, Fac. Ciencias, UNAM, México, 207 p
7. Rodríguez K (2001) Relación de eventos oceanográficos con la abundancia y composición del zooplancton, con énfasis en el zooplancton gelatinoso de la Isla del Coco. Consulting report for SINERGIA 66. Unpublished data, 11 p
8. Rodríguez K (2005) Distribución espacial y temporal de la biomasa, composición y abundancia del zooplancton, con énfasis en hidromedusas de Bahía Culebra durante La Niña 1999 y el 2000. M.Sc. thesis, University of Costa Rica, San José, Costa Rica, 156 p
Stepanjants S (1975) Species composition and distributional pattern of Siphonophora of the Caribbean, Gulf of Mexico and adjacent waters of the Atlantic. Trans P.P. Shirshov Inst Oceanol 100:96–126
Stepanjants S (1977) Siphonophora of the central part of the Pacific Ocean. *In*: Explorations of the Fauna of the Seas XX (XXVIII). Marine Plankton. (Systematics and Faunistics), pp 54–81
9. Suárez E, Gasca R (1991) Sifonóforos de México. Biología y Ecología. Centro de Investigaciones de Quintana Roo (CIQRO). Chetumal, Q. Roo, México. 178 p
Totton AK (1965) A synopsis of the Siphonophora. Trust Brit Mus (Nat Hist), Lond, 230 p

Part 9
Zoanthids, Sea Anemones and Corallimorpharians

Jorge Cortés

The sea anemone *Stichodactyla helianthus* from Cahuita, Caribbean coast of Costa Rica. (Photo: Andrea Bernecker)

J. Cortés
Centro de Investigación en Ciencias del Mar y Limnología (CIMAR) and Escuela de Biología, Universidad de Costa Rica, 11501-2060 San José, Costa Rica.
e-mail: jorge.cortes@ucr.ac.cr

Abstract Five species of zoanthids, six of actiniarians and two of corallimorpharians have been recognized from the Caribbean coast, and one species of sea anemone from the Pacific of Costa Rica. The Caribbean species are common throughout the Caribbean Sea. Many more species have been observed on both coasts, but have not been identified.

Introduction

Members of three orders of anthozoans (Phylum: Cnidaria), Zoanthiniaria, Actinaria, and Corallimorpharia are commonly called sea anemones, but only the Actinaria are anemones in the strict sense (www.kgs.ukans.edu/Hexacoral/Biodata/index.html). There are about 250 species of zoanthids, 900 species of actiniarians, and 50 species of corallimorpharians in the world (http://phylogeny.arizona.edu/tree/eukaryotes/animals/cnidaria/cnidaria.html).

There is little information on zoanthids and corallimorpharians. Even in specialized web pages like: www.kgs.ukans.edu/Hexacoral/Biodata/index.html, there is no information on the taxonomic status of several species of zoanthids. Because of the problem of preserving these soft-body organisms there are few specimens in museum collections.

In this part I present a list of species that I have observed and identified in the field, using Colin (1988) and Humann (1992). Five species of zoanthids (e.g., *Zoanthus sociatus*, Fig. 9.1), six of actiniarians and two of corallimorpharians have been recognized from the Caribbean coast and one species of sea anemone from the Pacific of Costa Rica (Species Lists 9.1 and 9.2 are included on the CD-Rom). Many more species have been observed on both coasts but have not been identified, stressing the need for more taxonomic work on these groups.

Fig. 9.1 *Zoanthus sociatus* from Cahuita, Caribbean coast of Costa Rica. (Photo: Jorge Cortés)

Specialists

FABIÁN H. ACUÑA: Laboratorio de Cnidarios, Universidad Nacional de Mar del Plata, Argentina – Atlantic Actiniarians. facuna@mdp.edu.ar

DAPHEN FAUTIN: Division of Biological Sciences, University of Kansas, Lawrence – Worldwide. fautin@ukans.edu

RICHARD MARISCAL: Florida State University, Tallahassee,Florida – Worldwide.

JOHN RYLAND: School of Biological Sciences, University of Wales Swansea, Swansea, Wales, UK – Worldwide. J.S.Ryland@Swansea.ac.uk

Collections

There is a small collection of Costa Rican zoanthids and anemones in the Museo de Zoología of the Universidad de Costa Rica.

Recommendations

Much more work is needed on these groups, not only in Costa Rica, but worldwide. Dr. Daphne Fautin recommends that good underwater photographs, from several angles, should be taken of an organism before collecting it, because once preserved, even after they have been relaxed, they lose their original shape and color.

Acknowledgments I thank Daphne Fautin and John Ryland for their help and review of the manuscript.

References

Colin PI (1988) Caribbean reef invertebrates and plants. T.F.H. Publications, Neptune City, NJ, 512 p

1. Cortés J (1996–1997) Biodiversidad marina de Costa Rica: Filo Cnidaria. Rev Biol Trop 44(3)/45(1):323–334
2. Cortés J (1979–2002) Personal observations
3. Fonseca E, Arrivillaga AC, Arrivillaga A (2003) Coral reefs of Guatemala. *In*: Cortés J (ed) Latin American Coral Reefs. Elsevier Science, Amsterdam, pp 159–170

Humann P (1992) Reef Creatures Identification: Florida, Caribbean, Bahamas. New World Publications, Jacksonville, FL, 328 p

Part 10
Octocorals

Odalisca Breedy

Leptogorgia alba with extended polyps from Bahia Culebra, northern Pacific of Costa Rica. (Photo: Jaime Nivia)

Abstract Preliminary studies on diversity of the octocoral fauna in shallow waters (down to 55 m depth) of Costa Rica have brought to light 56 species of octocorals, 26 species in the Caribbean, and 30 species in the Pacific. According to field observations, 50% of the species of octocorals from shallow waters has been surveyed, and nothing is known about the deepwater fauna.

O. Breedy
Museo de Zoología, Escuela de Biología, Universidad de Costa Rica, 11501-2060 San José, Costa Rica.
e-mail: odalisca@racsa.co.cr

I.S. Wehrtmann, J. Cortés (eds.) *Marine Biodiversity of Costa Rica, Central America*,

Introduction

The octocorals are a widespread group of marine animals. They occur from littoral waters down to the deep-sea abysses and from Arctic to Antarctic oceans. They are found in variable densities in almost all the marine habitats.

Octocorals are commonly found along the coasts of Costa Rica, ranging widely in distribution and depth. The first reports on octocorals in Costa Rica came from explorations to the west tropical coasts of America, based on collections by F.H. Bradley for the Museum of Yale College (now known as Yale Peabody Museum, Yale University) and J.A. McNeil during the 1860s (Verrill 1868). As a result six species of octocorals for the Pacific coast were reported by Verrill (1868). Deichmann (1941) reported a sea pen from Isla del Coco, and Cortés (1996–1997) reported another species from the Golfo de Nicoya found in the Museo de Zoología, Universidad de Costa Rica (UCR).

Wellington (1974) published initial reports on the Costa Rica Caribbean octocorals. He reported four species in a general study of the Cahuita coral reef. Guzmán and Cortés (1985) identified 23 species living along the Caribbean coast up to 16m in depth. Later, Guzmán & Jiménez (1989) added another species to the list, and Cortés & Jiménez (2003) reported the occurrence of another species in deeper waters. Additionally, Sánchez (2001) described a new species found at one site in the Caribbean coast. So far, 26 species have been reported living in the shallow Caribbean waters up to 35m in depth.

Since the year 2000, cooperative research initiatives between the Centro de Investigación en Ciencias del Mar y Limnología (CIMAR, UCR) and the Smithsonian Tropical Research Institute (STRI, Panama) have began with the goal of evaluating and studying the biodiversity and community structure of octocorals in the Pacific. Scuba-diving surveys on reefs, rocky outcrops along the coasts, islets, and adjacent islands up to 35m in depth have been conducted. Samples from 60 to 70m have also been acquired by dredging as a result of cooperative programs with the University of Bremen (R/V *Victor Hensen* 1995), and STRI (R/V *Urracá* 2005). As a consequence, a representative collection of octocoralian fauna was obtained, deposited in the Museo de Zoología, Universidad de Costa Rica, and is under study; new species have been described and geographical ranges have been expanded. After more than a century of neglect in taxonomic studies in the shallow waters of the eastern Pacific, Breedy (2001) described a new species, and then, Breedy & Guzmán (2003), described ten species as new and reported seven more. Thus far, the species account is 30 for the shallow waters of the Pacific of Costa Rica (Species List 10.2 is included on the CD-Rom), but more species have been observed and collected, among them several species of the genera *Muricea, Heterogorgia*, and *Ellisella*. These genera are under study and will definitively increase the current account.

Structure and Systematics. Octocorals are colonial coelenterates characterized by polyps bearing eight pinnate tentacles. Polyps arise from a basal common tissue called coenenchyme. Embedded in the coenenchyme are calcareous skeletal elements called sclerites. The sclerites stiffen parts of the polyp body, fill the coenenchyme, and provide support for the octocoral colony (Grasshoff 2001).

Octocoral taxonomy is currently based on morphology of the colonies, structure of the axes, the sclerites, and color of colonies and sclerites. Based on Bayer (1981), Fabricius & Alderslade (2000) presented a systematic approach for the higher categories (e.g., superorders, orders, suborders), which is adopted in this part.

Diversity of Octocorals

Basically the subclass Octocorallia has three distinctly separated orders: Helioporacea (blue coral), Pennatulacea (sea pens), and Alcyonacea (soft corals and gorgonians).

The first order is the only one that contains a species (*Heliopora coerulea*, blue coral) that forms massive aragonite skeletons. This species contains zooxanthellae (symbiotic algae in the polyp tissue), like hard corals (Scleractinia), and is found in the fossil record from 100 million years ago (Fabricius & Alderslade 2000). Blue coral is found at Marshall Islands, Gilbert Islands, Samoa, and south Japan (Colin & Arneson 1995). Other species of this order are found deep in the Atlantic Ocean and from Madagascar (Bayer 1992). They are not found in the eastern Pacific or in the Caribbean.

Pennatulacean colonies are formed by a large polyp called the oozoid, on the wall of which the coenenchyme spreads with numerous small (secondary) polyps; the large primary polyp may be additionally supported by a horny axis. Part of the oozoids forms the colony peduncle that anchors the colony in sand or soft substrates. The other part of the polyps forms the rachis, which bears other kinds of polyps: autozooids and siphonozooids. In some species the emergent part looks like a feather (sea pens) (Williams 1990; Fabricius & Alderslade 2000). Only two species of sea pens have been recorded for Costa Rica, but some others have been collected.

The order Alcyonacea includes the soft corals and the gorgonians. Soft corals form fleshy colonies characterized by having polyps aggregated or concentrated into polyparies. An internal medulla or axis is absent and the polyps are embedded into a soft coenenchymal tissue, which may or may not contain sclerites (Williams 1992). Only the family Clavulariidae from the Stolonifera group, with one species identified, represents the soft corals in Costa Rica.

Gorgonians include sea rods, sea whips, sea candelabra, sea sausages, sea feather plumes, and sea fans. The most common octocorals in Costa Rican waters are the gorgonians. They present diverse growth forms: incrusting colonies, upright fans and bushes with slender branches, or simple whips. Gorgonian colonies have a central scleroproteinous axial structure (gorgonin). A layer of coenenchyme with sclerites and polyps (Bayer 1961) surrounds it. Four families from this group are found in the shallow waters of Costa Rica: Anthothelidae, Ellisellidae, Plexauridae, and Gorgoniidae.

The Caribbean species reported for Costa Rica (Species List 10.1 is included on the CD-Rom) are common in other regions: Swan Islands, Honduras (Tortora &

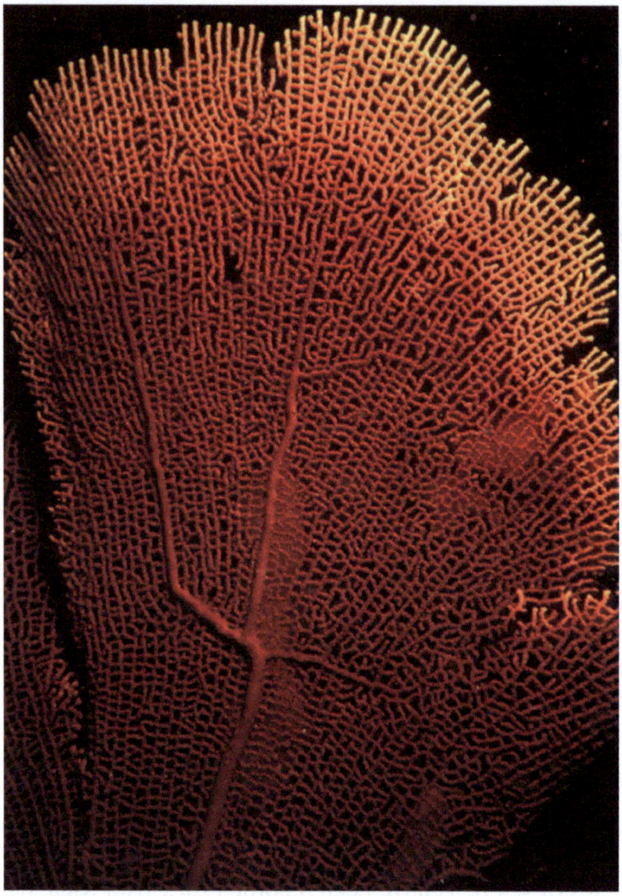

Fig. 10.1 *Pacifigorgia irene*, a common species at Isla del Caño, Pacific Costa Rica. (Photo: Jorge Cortés)

Keith 1980), Carrie Bow, Belize (Muzik 1982), Cayos Cochinos, Honduras (Guzmán 1998), and Bocas del Toro, Panama (Guzmán & Guevara 1998). There is no published information on the Caribbean fauna of the other Central American countries.

The Pacific species found in Costa Rica (Species List 10.2 is included on the CD-Rom) include 28 alcyonaceans and two pennantulaceans. Fourteen species belong to the genus *Pacifigorgia*, eleven to the genus *Leptogorgia* (Fig. 10.1) which have been properly studied and collected; the others are in need of taxonomic revisions and collection. With the exception of five *Pacifigorgia*, and *Leptogorgia exigua*, the other species are also reported for Panama. Four species for Nicaragua, and three for El Salvador (Verrill 1868, 1870; Hickson 1928; Breedy & Guzmán 2003, 2005). However, studies on this fauna in Central America are limited, and these numbers are expected to increase. Extensive surveys are needed before attempting to account octocoral species richness and understand the geographic distribution, not only in Central America, but also in the tropical American waters.

Specialists

PHIL ALDERSLADE: Museum and Art Gallery of the Northern Territory, Australia – Worldwide. phil.alderslade@nt.gov.au

CLOVIS CASTRO: Museu Nacional do Brasil, Brazil – Western Atlantic. cbmnjr@omega.Inee.br

STEPHEN CAIRNS: Natural History Museum, Smithsonian Institution, USA – Worldwide. cairns.stephen@ si.edu

HÉCTOR GUZMÁN: Smithsonian Tropical Research Institute, Panama – Caribbean and eastern Pacific. guzmanh@si.edu

LEEN VAN OFWAGEN: National Museum of Natural History, The Netherlands – Worldwide. ofwegen@naturalis.nnm.nl

JUAN ARMANDO SÁNCHEZ: Universidad de Los Andes, Colombia – Caribbean and New Zealand. juansanc@uniandes.edu.co

GARY WILLIAMS: California Academy of Sciences, USA – Worldwide. gwilliams@CalAcademy.org

ODALISCA BREEDY: Museo de Zoología, Escuela de Biología, Universidad de Costa Rica, 2060 San José, Costa Rica – Caribbean and eastern Pacific. odalisca@racsa.co.cr

Collections

The main collections containing type material of Costa Rican species are housed in the following museums:

- Museo de Zoología, Universidad de Costa Rica (UCR)
- Yale Peabody Museum of Natural History, New Haven, USA (YPM)
- Museum of Comparative Zoology, Harvard University, Cambridge, USA (MCZ)
- Natural History Museum, Smithsonian Institution, Washington, D.C., USA (USNM)

Recommendations

Undoubtedly, the knowledge of biodiversity is linked to systematics. There are still many groups of organisms and locations for which taxonomic research is unavailable. Many taxonomic questions have not even been addressed, as the validity of species that currently have been recognized, the fundamental idea of species distinctness, the relationship species–environment, and the implications for ecosystem function.

The first steps in the improvement of the knowledge of the octocorallian fauna are necessarily related to systematic studies, which will be achieved through the fulfillment of the following tasks: (1) Taxonomic revisions, (2) genetic studies, (3) extensive surveys focusing on the least explored areas, (4) exploration of the deep zones, (5) preparation of reliable identification guides, (6) training and encouragement of new taxonomists, and (7) reorientation of museums to function as centers of research and inventory.

References

Bayer F (1956) Octocorallia. *In*: Moore RC (ed) Treatise of Invertebrate Paleontology Part F. Coelenterata. Geol Soc Am Univ. Kansas Press/Lawrence, Boulder, CO/Kansas, pp 163–231

Bayer F (1961) The Shallow Water Octocorallia of the West Indian Region. A Manual for Marine Biologists. Martinus Nijhoff, The Hague, 400 p

Bayer FM (1981) Key to the genera of Octocorallia exclusive of Pennatulacea (Coelenterata: Anthozoa), with diagnoses of new taxa. Proc Biol Soc Wash 94:901–947

Bayer FM (1992) The helioporacean octocoral *Epiphaxum*: recent and fossil. A monographic iconography. Stud Trop Oceanogr 15:1–77

1. Breedy O (2001) A new species of *Pacifigorgia* from the eastern Pacific (Coelenterata: Octocorallia: Holaxonia). Bull Biol Soc Wash 10:181–187
2. Breedy O, Guzmán HM (2002) A revision of the genus *Pacifigorgia* (Coelenterata: Octocorallia: Gorgoniidae). Proc Biol Soc Wash 115:782–839
3. Breedy O, Guzman HM (2003) Octocorals from Costa Rica. The genus *Pacifigorgia* (Coelenterata: Octocorallia: Gorgoniidae). Zootaxa 281:1–60
4. Breedy O, Guzman HM (2005) A new species of *Leptogorgia* (Coelenterata: Octocorallia: Gorgoniidae) from the shallow waters of the eastern Pacific. Zootaxa 899:1–11
5. Breedy O, Guzman HM (2007). A revision of the genus *Leptogorgia* Milne Edwards & Haime, 1857 (Coelenterata: Octocorallia: Gorgoniidae) in the eastern Pacific. Zootaxa 1419: 1–90

Colin PL, Arneson C (1995) Tropical Pacific Invertebrates. A Field Guide to the Marine Invertebrates Occurring on Tropical Pacific Coral Reefs, Seagrass Beds and Mangroves. Coral Reef Press, Beverly Hills, CA, 296 p

6. Cortés J, Jiménez CE (2003) Past, present and future of the coral reefs of the Caribbean coast of Costa Rica. *In*: Cortés J (ed) Latin American Coral Reefs. Elsevier Science, Amsterdam, pp 223–239
7. Cortés J (1996–1997) Biodiversidad marina de Costa Rica: Filo Cnidaria. Rev Biol Trop 44(3)/45(1):323–334
8. Deichmann E (1941) Coelenterates collected on the presidential cruise of 1938. Smithson. Miscell Coll 99(10):1–17

Fabricius K, Alderslade P (2000) Soft Corals and Sea Fans: A Comprehensive Guide to the Tropical Shallow Water Genera of the Central-West Pacific, the Indian Ocean and the Red Sea. Science Communication. Aust Inst Mar Sci, Australia, 264 p

9. Fonseca AC, Arrivillaga A (2003) Coral reefs of Guatemala. *In*: Cortés J (ed) Latin American Coral Reefs. Elsevier Science B.V., Amsterdam, pp 159–170

Grasshoff M (2001) Coral reef gorgonians of New Caledonia. L'Institut de Recherche pour le Développment, Paris Coll. Faune et Flore tropicales 38:335 p

Guzmán HM (1998) Diversity of stony, soft and black corals (Anthozoa: Scleractinia, Gorgonacea, Antipatharia; Hydrozoa: Milleporina) at Cayos Cochinos, Bay Islands, Honduras. Rev Biol Trop 46(Suppl 4):75–80

10. Guzmán HM, Cortés J (1985) Organismos de los arrecifes coralinos de Costa Rica VI. Descripción y distribución geográfica de octocoralarios (Cnidaria) de la costa Caribe. Brenesia 24:125–174

11. Guzmán HM, Jiménez CE (1989) *Pterogorgia anceps* (Pallas) (Octocorallia: Gorgoniidae): nuevo informe para la costa caribeña de Costa Rica. Rev Biol Trop 37:231–232

Guzmán HM, Guevara CA (1998) Arrecifes coralinos de Bocas del Toro, Panamá: I. Distribución, estructura y estado de conservación de los arrecifes continentales de la Laguna de Chiriquí y la Bahía Almirante. Rev Biol Trop 46:601–623

12. Hertlein LG (1963) Contribution to the biogeography of Cocos Island, including a bibliography. Proc Calif Acad Sci 32(4th series):219–289

Hickson SJ (1928) The Gorgonacea of Panama Bay together with a description of one species from the Galapagos Islands and one of Trinidad. Vidensk Medd naturhist Forening København 85:325–422

Muzik K (1982) Octocorallia from Carrie Bow Cay, Belize. *In*: Rützler K, Macintyre IG (eds) The Atlantic Barrier Ecosystem at Carrie Bow Cay. I. Structure and Communities. Smithson Inst Press, Washington, DC, pp 309–316

13. Sánchez JA (2001) Systematics of the southwestern Caribbean *Muriceopsis* Aurivillius (Cnidaria: Octocorallia) with the description of a new species. Bull Biol Soc Wash 10:160–180

Tortora LR, Keith DE (1980) Octocorals of the Swan Islands, Honduras. Carib J Sci 16:65–72

14. Verrill AE (1868) Notes on Radiata in the Museum of Yale College, Number 6: Review of the corals and polyps of the West Coast of America, 2nd ed. Trans Connecticut Acad Arts Sci 1:377–422

15. Verrill AE (1870) Notes on Radiata in the Museum of Yale College, Number 6: Review of the corals and polyps of the West Coast of America, 2nd ed. Trans Connecticut Acad Arts Sci 1:503–558

Wellington GM (1974) An ecological description of the marine and associated environments at Monumento Nacional Cahuita, Costa Rica. Subdirección de Parques Nacionales, San José, 81 p

Williams GC (1990) The Pennatulacea of southern Africa (Coelenterata, Anthozoa). Ann S Afr Mus 99:31–119

Williams GC (1992) The Alcyonacea of southern Africa. Stoloniferous octocorals and soft corals (Coelenterata, Anthozoa). Ann S Afr Mus 100:249–358

Part 11
Stony Corals

Jorge Cortés

Tubastrea coccinea from Isla del Coco, Pacific Costa Rica. (Photo: Jorge Cortés)

Abstract On the Caribbean coast of Costa Rica there are 3 species of hydrocorals and 44 species of scleractinian corals: 5 azooxanthellate and 39 zooxanthellate corals, all common Caribbean species. On the Pacific side of Costa Rica, there are 5 species of hydrocorals, all from Isla del Coco, with 2 species endemic to the island, and 46 species of scleractinian corals: 24 zooxanthellate and 22 azooxanthellate. Half of the azooxanthellate corals are from deepwaters around Isla del Coco, and one is endemic to the island. The shallow-water stony coral species composition and richness of the

J. Cortés
Centro de Investigación en Ciencias del Mar y Limnología (CIMAR) and Escuela de Biología, Universidad de Costa Rica, San Pedro, San José 11501-2060, Costa Rica.
e-mail: jorge.cortes@ucr.ac.cr

I.S. Wehrtmann, J. Cortés (eds.) *Marine Biodiversity of Costa Rica, Central America*,

Costa Rican Pacific reefs is comparable to neighboring countries, but the number of species on the Caribbean is lower than in the neighboring countries.

Introduction

The stony or hard corals are simple but extremely important animals because they built coral reef frameworks, the most diverse ecosystems of the oceans. In the reef structure we find representatives of almost every phyla, plus almost any imaginable interaction between them: symbiosis, parasitism, mutualism, etc. (Reaka-Kudla 1997). There are two main groups of stony corals: the hydrocorals (hydrozoans with calcareous skeletons) and the true corals (scleractinians). This last group is divided into species with symbiotic algae (zooxanthellae), which are the main reef-building corals, and the corals without symbiotic algae, the azooxanthellate.

The world is divided in two biogeographic provinces in terms of the corals: the Caribbean-Atlantic province and the Indo-Pacific province. The separation of these two provinces is the Central American Isthmus, of which Costa Rica is a part (Cortés & Jiménez 2003a, b). On the Caribbean coast of Costa Rica we have corals that are the same as in Bermuda, the Gulf of Mexico, Saba, and Brazil, while on the Pacific side some corals are the same as in Hawaii, French Polynesia, the Great Barrier Reef, the Maldives, and even the Red Sea (Veron 2000a, b, c). Nine of the ten genera of corals from Brazil are also found in the Caribbean, but only ten Brazilian species of the 50+ species in the Caribbean are in common. Two and four genera are shared between the eastern Pacific and Brazil and the Caribbean, respectively, but there are no species of zooxanthellate corals in common (Cortés 1986, 2003).

Coral reefs world-wide are being impacted by natural and anthropogenic disturbances (Wilkinson 2000). In Costa Rica, the main impact on the Caribbean has been the excess terrigenous sedimentation (Cortés & Jiménez 2003b) while in the Pacific the main impact has been the El Niño warming events in the past decade (Guzmán & Cortés 2001; Jiménez *et al.* 2001).

Here I list the known species of stony corals from the Caribbean (Species List 11.1 is included on the CD-Rom) and Pacific (Species List 11.2 is included on the CD-Rom) coast of Costa Rica, which are different faunas, since there are no species in common between both sides (Cortés 1986, 2003). Some azooxanthellate corals may be found on both coasts such as *Phyllangia dispersa* which is an eastern Pacific species, but was collected at Cahuita on the Caribbean coast. Another example is *Tethocyathus prahli* that has been collected on the Caribbean coast of Colombia and at Isla del Coco in the Pacific of Costa Rica (Lattig & Cairns 2000). The Costa Rican coral fauna is relatively well known, but even so, new species have been described recently (Glynn 1999; Lattig & Cairns 2000; Glynn *et al.* 2001) and apparently there are several more new species of corals that are being evaluated at the moment.

Three species of hydrocorals and 44 species of scleractinian corals: 5 azooxanthellate and 39 zooxanthellate corals are listed here from the Caribbean of Costa Rica (Species List 11.1). All these species are found in other places in the

Caribbean. Five species of hydrocorals, all from Isla del Coco, including 2 endemics, and 46 species of scleractinian corals: 24 zooxanthellate and 22 azooxanthellate have been reported for the Pacific side of Costa Rica (Species List 11.2). Half of the azooxanthellate corals are from deepwaters around Isla del Coco, and two are endemic to the island. The number of stony corals on the Pacific of Costa Rica is comparable to other countries in the region, but the number of coral species on the Caribbean coast is less than the neighboring countries.

Specialists

S.D. Cairns: NMNH, Washington, D.C. – Hydrocorals and ahermatypic scleractinians – Worldwide. Cairns.Stephen@NMNH.SI.EDU

J. Cortés: CIMAR, Universidad de Costa Rica – Scleractinians, Eastern Pacific and Caribbean. jorge.cortes@ucr.ac.cr

P.W. Glynn: RSMAS, University of Miami – Scleractinians, Eastern Pacific. pwglynn@rsmas.miami.edu

H.M. Guzmán: STRI, Panamá – Scleractinians, Eastern Pacific and Caribbean. guzmanh@si.edu

C.E. Jiménez: ZMT, Universität Bremen – Scleractinians, Eastern Pacific. carlos.jimenez@zmt.uni-bremen.de

H. Reyes-Bonilla: Universidad Autónoma de Baja California Sur, La Paz, México – Scleractinians, Eastern Pacific and Caribbean. bludemos@hotmail.com

Collections

The largest collection of Costa Rican corals is at the Museo de Zoología of the Universidad de Costa Rica. Most of the corals found in Costa Rica are in the collection. Another important depository of Costa Rican corals (all the deepwater species) is the Smithsonian Institution National Museum of Natural History in Washington, DC.

Recommendations

The shallow-water stony coral fauna is probably well known, especially the reef-building corals, but even so, new species were recently described (Glynn 1999; Glynn *et al.* 2001). The deepwater fauna and the shallow-water azooxanthellate corals need more collections and analysis not only in Costa Rica but in the entire Central American region.

References

1. Cairns SD (1982) Stony corals (Cnidaria: Hydrozoa, Scleractinia) of Carrie Bow Cay, Belize. *In*: Rützler K, Macintyre IG (eds) The Atlantic Barrier Reef Ecosystem at Carrie Bow Cay, Belize, I. Structure and Communities. Smithson Institution Press, Washington, pp 271–302
2. Cairns SD (1991) A revision of the ahermatypic Scleractinia of the Galápagos and Cocos Islands. Smithson Contr Zool 504:32
3. Cairns SD (1991) New records of Stylasteridae (Hydrozoa: Hydroida) from the Galápagos and Cocos Islands. Proc Biol Soc Wash 104:209–228
4. Cairns SD (2000) A revision of the shallow-water azooxanthellate Scleractinia of the Western Atlantic. Studies Nat Hist Carib Reg 75:240
5. Cairns SD, Hoeksema BW, van der Land J (1999) Appendix: List of extant stony corals. Atoll Res Bull 459:13–46

Cortés J. (1986) Biogeografía de corales hermatípicos: el istmo centroamericano. Anal Inst Cien Mar Limnol UNAM 13:297–304

6. Cortés J (1992) Nuevos registros de corales (Anthozoa: Scleractinia) para el Caribe de Costa Rica: *Rhizosmilia maculata* y *Meandrina meandrites*. Rev Biol Trop 40:243–244
7. Cortés J (1992) Organismos de los arrecifes coralinos de Costa Rica: V. Descripción y distribución geográfica de hydrocorales (Cnidaria; Hydrozoa: Milleporina & Stylasterina) de la costa Caribe Brenesia 38:45–50
8. Cortés J (1996–1997) Biodiversidad marina de Costa Rica: Filo Cnidaria. Rev Biol Trop 44(3)/45(1):323–334

Cortés J (2003) Coral reefs of the Americas: An introduction to Latin American Coral Reefs. *In*: Cortés J (ed) Latin American Coral Reefs. Elsevier Science B.V., Amsterdam, pp 1–7

9. Cortés J (in preparation) Organismos de los arrecifes coralinos de Costa Rica: VII. Descripción y distribución geográfica de los corales azooxantelados (Anthozoa: Scleractinia) de aguas someras del Caribe y Pacífico
10. Cortés J, Guzmán HM (1985) Organismos de los arrecifes coralinos de Costa Rica. III: Descripción y distribución geográfica de corales escleractinios (Cnidaria: Anthozoa: Scleractinia) de la costa Caribe Brenesia 24:63–124
11. Cortés J, Guzmán HM (1998) Organismos de los arrecifes coralinos de Costa Rica: Descripción, distribución geográfica e historia natural de los corales zooxantelados (Anthozoa: Scleractinia) del Pacífico. Rev Biol Trop 46:55–92

Cortés J, Jiménez CE (2003a) Past, present and future of the coral reefs of the Caribbean coast of Costa Rica. *In*: Cortés J (ed) Latin American Coral Reefs. Elsevier Science, Amsterdam, pp 223–239

Cortés J, Jiménez CE (2003b) Corals and coral reefs of the Pacific of Costa Rica: history, research and status. *In*: Cortés J (ed) Latin American Coral Reefs. Elsevier Science, Amsterdam, pp 361–385

12. Durham JW, Barnard JL (1952) Stony corals of the Eastern Pacific collected by the Velero III and the Velero IV. Allan Hancock Pac Exp 16:1–110
13. Fonseca AC, Arrivillaga A (2003) Coral reefs of Guatemala. *In*: Cortés J (ed) Latin American Coral Reefs. Elsevier Science, Amsterdam, pp 159–169
14. Gibson J, Carter J (2003) The reefs of Belize. *In*: Cortés J (ed) Latin American Coral Reefs. Elsevier Science, Amsterdam, pp 171–202
15. Glynn PW (1999) *Pocillopora inflata*, a new species of scleractinian coral (Cnidaria: Anthozoa) from the tropical eastern Pacific Pac Sci 53:168–180
16. Glynn PW, Maté JL, Stemann TA (2001) *Pavona chiriquensis*, a new species of zooxanthellate coral (Cnidaria: Anthozoa: Agariciidae) from the eastern tropical Pacific. Bull Biol Soc Wash 10:210–225
17. Guzmán HM (1998) Diversity of stony, soft and black corals (Anthozoa: Scleractinia, Gorgonacea, Antipatharia: Hydrozoa: Milleporina) at Cayos Cochinos, Bay Islands, Honduras. Rev Biol Trop 46(Suppl 4):75–80

18. Guzmán HM (2003) Caribbean coral reefs of Panama: present status and future perspectives. *In*: Cortés J (ed) Latin American Coral Reefs. Elsevier Science, Amsterdam, pp 241–274
Guzmán HM, Cortés J (2001) Changes in reef community structure after fifteen years of natural disturbances in the eastern Pacific (Costa Rica). Bull Mar Sci 69:133–149
19. Hoeksema BW (1989) Taxonomy, phylogeny and biogeography of mushroom corals (Scleractinia: Fungiidae). Zool Verh 254:295
Jiménez C, Cortés J, León A, Ruiz E (2001) Coral bleaching and mortality associated with El Niño 1997/98 event in an upwelling environment in the eastern Pacific (Gulf of Papagayo, Costa Rica). Bull Mar Sci 69:151–169
20. Lattig P, Cairns SD (2000) A new species of *Tethocyathus* (Cnidaria: Anthozoa: Scleractinia: Caryophylliidae), a trans-isthmian azooxanthellate species. Proc Biol Soc Wash 113:590–595
21. Maté JL (2003) Corals and coral reefs of the Pacific coast of Panamá. *In*: Cortés J (ed) Latin American Coral Reefs. Elsevier Science, Amsterdam, pp 387–417
Reaka-Kudla ML (1997) The global biodiversity of coral reefs: a comparison with rain forest. *In*: Reaka-Kudla ML, Wilson DE, Wilson EO (eds) Biodiversity II: Understanding and Protecting our Biological Resources. Joseph Henry Press, Washington, DC, pp. 83–108
22. Reyes-Bonilla H, Barraza JE (2003) Corals and associated marine communities from El Salvador. *In*: Cortés J (ed) Latin American Coral Reefs. Elsevier Science, Amsterdam, pp 351–360
23. Ryan J, Zapata Y (2003) Nicaragua's coral reefs: status, health and management strategies. *In*: Cortés J (ed) Latin American Coral Reefs. Elsevier Science, Amsterdam, pp 203–222
24. Veron JEN (2000a) Corals of the World, Vol. 1. AIMS, CRR, Queensland, 463 p
25. Veron JEN (2000b) Corals of the World, Vol. 2. AIMS, CRR, Queensland, 429 p
26. Veron JEN (2000c) Corals of the World, Vol. 3. AIMS, CRR, Queensland, 490 p
Wilkinson C (ed) (2000) Status of Coral Reefs of the World: 2000. AIMS, Cape Ferguson, Australia, 361 p

Part 12
Sipunculans

José A. Vargas and Harlan K. Dean

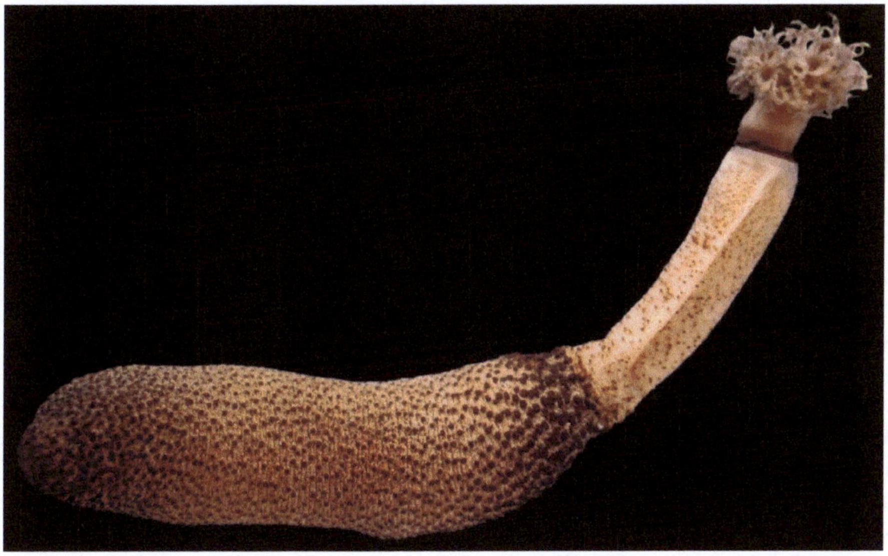

Antillesoma antillarum, a species from the Caribbean coast of Costa Rica (Photo: Jeffrey Sibaja)

Abstract The Phylum Sipuncula (sometimes called peanut worms) includes about 150 species found in marine waters worldwide. Sipunculans have a body divisible into a retractable introvert and a trunk, with the anus located towards the anterior end

J.A. Vargas
Universidad de Costa Rica (UCR), Centro de Investigación en Ciencias del Mar y Limnología (CIMAR), 11501-2060 San José, Costa Rica.
e-mail: javargas@cariari.ucr.ac.cr

H.K. Dean
Museum of Comparative Zoology (MCZ), Harvard University, 26 Oxford St. Cambridge. Massachusetts. 02138. USA.
e-mail: harlandean@comcast.net

I.S. Wehrtmann, J. Cortés (eds.) *Marine Biodiversity of Costa Rica, Central America*,

of the body. The primary literature contains 16 identified species belonging to nine genera of peanut worms found in Costa Rican coastal waters. Most of the collections were conducted after 1975, particularly in the estuary of Golfo de Nicoya (Pacific coast) and the reefs near Cahuita in the Caribbean coast. *Antillesoma antillarum* and *Phascolosoma perlucens*, found on both coasts, are abundant in intertidal rocky shores. The genus *Aspidosiphon* is the most diverse, represented by four species in the Pacific and one species in the Caribbean. *Sipunculus nudus*, a cosmopolitan species in warm and temperate waters, is found in intertidal sand flats in the upper Golfo de Nicoya. Five species (*Aspidosiphon elegans, Apionsoma trichocephalus, Nephasoma pellucidum, Phascolion strombus*, and *Siphonosoma vastum*) are the first records for the Eastern Tropical Pacific region. There are important coastal areas in Costa Rica that are yet to be explored for sipunculans. The coastline southeast of the port of Quepos in the Pacific and the sandy shores north of the port of Limón in the Caribbean are a priority in future surveys, as well as the shallow and deep waters surrounding Isla del Caño and Isla del Coco (Pacific coast).

Introduction

The Phylum Sipuncula includes about 150 species found in marine waters worldwide (Cutler 1994). Due to the pyriform shape of the contracted body of some species (such as *Antillesoma antillarum*), these organisms are called peanut worms but some species, such as *Apionsoma trichocephalus* and *Xenosiphon branchiatus*, have vermiform bodies. The sipunculan body is divisible into a trunk and a retractable introvert. The adult trunk ranges in length from 3 to >400 mm, and the introvert may be one-quarter to ten times the trunk length (Cutler 1994). The anus is located at the anterior (not posterior) end of the trunk, in a dorsal position, a character that easily separates a sipunculan from other marine invertebrates, such as small holothurians. Most of the sipunculans are deposit feeders and consume detritus and fecal material, as well as bacteria, algae, and small invertebrates (Cutler 1994). No species of peanut worms are presently under fishing pressure. However, several species are occasionally used as fish bait or as food for human consumption (Saiz Salinas 1993).

Sipunculans may be found in marine and estuarine habitats, and from the intertidal zone to abyssal depths. Their wide range of habitats is described by Hyman (1959) as: "*they lead a sedentary existence in burrows in sandy, muddy, mucky, gravelly, or shelly bottoms, in clefts and interstices of rocks, in porous lava, in the holdfast tangles of kelp, under beds of eelgrass and other vegetation, among coralline algae, under rocks, among corals, especially in the cavities in rotting coral, in sponges, in empty shells and tubes of other animals, and in almost any protected situation. Definite tubes are never formed although the walls of burrows may be smoothed with secretion.*" In short, sipunculans are found almost anywhere they can find a protective burrow-like habitat, which still allows them to extend their introvert to obtain food.

The phylogenetic position of the phylum is unclear. The ribosomal RNA sequence data, the presence of a molluscan cross in early development, and occurrence of a trocophore larvae point to an ancestral form common to annelids, sipunculans, and molluscs by the earliest Paleozoic (Cutler 1994). Scheltema (1993) has postulated that Sipuncula and Mollusca are sister groups based on comparisons of developmental and larval characteristics between peanut worms and aplacophoran molluscs. An arrangement of the Phylum Sipuncula into two classes, four orders, and six families is a recent effort made by Cutler & Gibbs (1985) and Gibbs & Cutler (1987). In this part (Species Lists 12.1 and 12.2 are included on the CD-Rom) we follow the classification by Cutler (1994).

When collecting sipunculans, relaxing or narcotizing them before fixing will greatly expedite subsequent identification. The use of a few drops of ethanol in seawater, or a solution of 7% magnesium chloride, for narcotizing, gives good results (Saiz Salinas 1993). In addition, placing tropical specimens in a tray with seawater and keeping it in a refrigerator or cooler for a few hours may help to attain a relaxed position before preservation. The tentacles and hooks at the distal tip of the introvert are critical characteristics in most genera and it is desirable that the introvert be fully protruded before fixation (Cutler 1994). For fixation, a solution of 5% formalin in seawater is used, and after 24 h the specimens must be transferred to 70% ethanol for final preservation and storage. The monograph by Cutler (1994) on the sipunculans has updated the earlier review by Stephen & Edmonds (1972) and should be consulted when attempting to identify these worms. Illustrations of the internal anatomy of some of the species listed in this part are available for *Sipunculus nudus* in Andreae (1881), Hyman (1959), and Saiz Salinas (1993); for *X. branchiatus* in Hyman (1959) and Cutler (1994); for *Siphonosoma vastum, A. antillarum*, and *A. pectinatum* in Cutler (1994); for *Phascolion strombus* in Hyman (1959), Saiz Salinas (1993), and Cutler (1994); and for *Aspidosiphon muelleri* in Saiz Salinas (1993) and Cutler (1994).

The first sipunculans reported from Costa Rica were collected in the mid-19th century at the Pacific coast port of Puntarenas (later mistaken by some authors with Punta Arenas, Chile) and described by Grube (1859). Twentieth century collections began in 1975 as part of the field trips of the Invertebrate Zoology course taught at the Universidad de Costa Rica (UCR) by Dr. Manuel M. Murillo, and from 1979 on as part of the research projects sponsored by the Marine Science and Limnology Research Center (CIMAR) of UCR. The first published report of identified sipunculan species from Costa Rica is by Fischer (1981), followed by Pepe (1985), Cutler *et al.* (1992), Dean & Cutler (1998), and Fonseca & Cortés (1998). The latest review by Dean (2001) lists 15 species of peanut worms from Costa Rica, with five species as the first records for the tropical Eastern Pacific Ocean. However, most of the collections come from the estuary of Golfo de Nicoya (intertidal and subtidal) and the Península de Nicoya (intertidal), thus leaving long segments of coastal areas open for further exploration. The sandy beaches and shallow waters of the Caribbean coast from the port of Limón to the border with Nicaragua, and the Pacific coastline southeast of the port of Quepos to the mouth of Golfo Dulce, and the offshore Isla del Coco are to be given priority in future surveys, as these sites might yield new additions to the sipunculan fauna of Costa Rica.

We were recently made aware of the publications by Fischer (1981) and Pepe (1985). Fischer (1981) listed only *Phascolosoma perlucens* and an unidentified species of *Paraspidosiphon* from rocky areas on the Pacific coast. Pepe (1985) listed *A. antillarum* and *P. perlucens* from both the Pacific and Caribbean coasts, *Paraspidosiphon fischeri* from the Pacific coast, as well as five others identified to the genus level: *Phascolosoma* sp. (Pacific and Caribbean), *Paraspidosiphon* sp. 2, sp. 3 (Pacific and Caribbean, respectively), *Lithacrosiphon* sp., and *Themiste* sp. (Caribbean), which brings the total number of identified species to 16, and the number of genera of sipunculans recorded from Costa Rica to nine (Species Lists 12.1 and 12.2 are included on the CD-Rom). Fischer (1990) has also reported that these worms are considered important organisms, which bioerode limestone, non-carbonate sediments, magmatic, and volcanic rocks on the Pacific coast of Costa Rica, but no species were identified and listed.

Of particular interest is the finding of *S. nudus* in sediments from the Golfo de Nicoya (Cutler *et al.* 1992; Dean 2001; Dittmann & Vargas 2001). *S. nudus* is the only member of the genus with a worldwide distribution in warm and temperate habitats (Cutler 1994). Its wide latitudinal range, relatively large size (5–15 cm long), stable taxonomic status, and the availability of information on its anatomy (Andreae 1881; Hyman 1959; Saiz Salinas 1993) and physiology (Pörtner *et al.* 1986; Ruppert & Rice 1995) make this species a good candidate for a cosmopolitan indicator of coastal marine pollution (Carriker 1976; Vargas & Abdullah 1997; Yang & Wang 2002; Spongberg 2006). No species of the genus *Sipunculus* were found in recent collections of sipunculans from intertidal and shallow subtidal habitats around the Caribbean island of Barbados by Cutler & Schulze (2004), but a *Sipunculus* sp. was reported by Schulze (2005) from near by Bocas del Toro, Panama. In the Golfo de Nicoya estuary *S. nudus*, as well as several other species of sipunculans, are found in intertidal environments adjacent to fringing mangroves (Dittmann & Vargas 2001), where human impacts on this estuary are on the rise (Vargas & Mata 2004).

There are no sipunculan species described originally from Costa Rican waters. The name *P. puntarenae* appears in the older literature, such as in Grube (1859) and Hyman (1959). The specific name refers to the port of Puntarenas, but the recent reviews by Cutler & Cutler (1990), and Cutler (1994) consider *P. puntarenae* as difficult to separate morphologically from *P. nigrescens*; thus, these taxa are considered conspecific and their junior synonym is preferred to avoid confusion (Cutler *et al.* 1992).

Based on the reports by Cutler *et al.* (1992), Cutler (1994), Dean & Cutler (1998), Dean (2001), and Schulze (2005), it is important to mention that six of the species of peanut worms reported from Costa Rica (*A. antillarum, P. perlucens, P. nigrescens, A. pectinatum*, and *A. fischeri*, and *A. elegans*) have been found also in coastal waters of neighboring Panama. *S. nudus* and *A. gracilis* have been found also in Guatemala, but the nearest record for *A. muelleri* is from Juan Fernández Island off Chile. The nearest record of *X. branchiatus* is from Ecuador, and the closest report for *S. phalloides* is from the Galápagos Islands. Five species (*A. elegans, A. trichocephalus, Nephasoma pellucidum, P. strombus*, and *S. vastum*) found in Costa Rica are also the first reports for the Eastern Tropical Pacific.

Specialists and Collections

The number of sipunculan specialists is small worldwide in spite of the crucial role of systematics in supporting management and conservation measures, and in assessing pollution impacts in marine ecosystems (Carriker 1976). For taxonomic advice contact Dr. Harlan K. Dean or Dr. José Saiz Salinas (Facultad de Ciencias, Dpto. de Zoología y Dinámica Celular Animal, Universidad del País Vasco, Apdo. 644 P.K. E-48080, Bilbao, España). Future collections of sipunculans should include specimens fixed in 95% ethanol for DNA studies, as well as specimens fixed in formalin for routine morphological study, as recently done by Cutler & Schulze (2004). We hope that this part will encourage young scientists to get involved in the study and conservation of the species included in the Phylum Sipuncula.

Voucher specimens of most of the species mentioned in this part are deposited either at the Museum of Comparative Zoology (Contact: Curator of Invertebrates, Museum of Comparative Zoology, Harvard University, 26 Oxford St. Cambridge, MA 02138, USA), or at the Museo de Zoología of the Universidad de Costa Rica (Contact: Curator of Invertebrates, Museo de Zoología, Escuela de Biología, Universidad de Costa Rica, 2060, Costa Rica; mzucr@biologia.ucr.ac.cr), or at both. Museum catalogue numbers are included in Cutler *et al.* (1992) and Dean (2001).

Acknowledgments This paper has been possible, in part, thanks to the support of the Universidad de Costa Rica, and grant 808-AO-506 Coastal Pollution of Costa Rica from the Costa Rica-United States of America (CR-USA) Foundation for Cooperation to José A. Vargas.

References

1. Andreae J (1881) Zur Anatomie des *Sipunculus nudus* L. Zool Anz 4:477–481
2. Carriker MR (1976) The crucial role of systematics in assessing pollution effects on the biological utilization of estuaries. *In*: Proceedings of the Estuarine Pollution Control and Assessment Conference. US Environmental Protection Agency, Office of Water Planning and Standards, Washington, DC, pp 487–506
3. Cutler EB (1994) The Sipuncula: their systematics, biology, and evolution. Comstock Publishing Association, Ithaca, NY, 453 p
4. Cutler EB, Cutler NJ (1982) A revision of the genus *Siphonosoma* (Sipuncula). Proc Biol Soc Wash 95:748–762
5. Cutler EB, Cutler NJ (1983) An examination of the *Phascolosoma* subgenera *Antillesoma, Rueppellisoma* and *Satonus* (Sipuncula). Zool J Linn Soc 77:175–187
6. Cutler EB, Cutler NJ (1985) A revision of the genera *Sipunculus* and *Xenosiphon* (Sipuncula). Zool J Linn Soc 85:219–246
7. Cutler EB, Cutler NJ (1989) A revision of the genus *Aspidosiphon* (Sipuncula: Aspidosiphonidae). Proc Biol Soc Wash 102:826–865
8. Cutler NJ, Cutler EB (1990) A revision of the subgenus *Phascolosoma* (Sipuncula: *Phacolosoma*). Proc Biol Soc Wash 103:691–730

9. Cutler EB, Gibbs PE (1985) A phylogenetic analysis of higher taxa in the phylum Sipuncula. Syst Zool 34:162–173
10. Cutler EB, Schulze A (2004) Sipuncula from Barbados, including two new for the island plus *Siphonosoma vastum*; first record from the Atlantic Ocean. Bull Mar Sci 74:225–228
11. Cutler NJ, Cutler EB, Vargas JA (1992) Peanut worms (Phylum Sipuncula) from Costa Rica. Rev Biol Trop 40:153–158
12. Dean HK (2001) Marine biodiversity of Costa Rica. The Phyla Sipuncula and Echiura. Rev Biol Trop 49:(Suppl 2):85–90
13. Dean HK, Cutler EB (1998) Range extension of *Nephasoma pellucidum* and new records of *Apionsoma (Edmonsius) pectinatum* (Sipuncula) and *Thalassema steinbecki* (Echiura) from the Pacific of Costa Rica. Rev Biol Trop 46:(Suppl 6):279–280
14. Dittmann S, Vargas JA (2001) Tidal flat benthos compared between Australia and Central America. *In*: Reise K (ed.) Ecological Comparisons of Sedimentary Shores. Ecological Studies, vol 151. Springer, Berlin, pp 275–293
15. Fischer R (1981) Bioerosion of basalt of the Pacific coast of Costa Rica. Senckenberg marit 13:1–41
16. Fischer R (1990) Biogenetic and nonbiogenetically determined morphologies of the Costa Rica Pacific coast. Z Geomorph NF 34:313–321
17. Fonseca AC, Cortés J (1998) Coral borers of the Eastern Pacific: *Aspidosiphon (A.) elegans* (Sipuncula: Aspidosiphonidae) and *Pomatogebia rugosa* (Crustacea: Upogebiidae). Pac Sci 52:170–175
18. Gibbs PE, Cutler EJ (1987) A classification of the phylum Sipuncula. Bull British Mus Nat Hist Zool 52:43–58
19. Grube E (1859) Annulata Oerstediana. Enumeratio Annulatorum quae in itinere der Indiam Occidentalem et American Centralis annis 1845–1848. Vidensk Medd Dansk naturh Forensing København Part 3:(1859):105–120
20. Hyman LH (1959) The invertebrates: volume V. Smaller Coelomate groups. McGraw-Hill, New York, 783 p
21. Pepe PJ (1985) Littoral endolithic fauna of the Central American Isthmus. Rev Biol Trop 33:191–194
22. Pörtner HO, Vogeler S, Grieshaber MK (1986) Recovery from anaerobiosis in the intertidal worm *Sipunculus nudus*. J Exp Biol 122:37–50
23. Ruppert E, Rice ME (1995) Functional organization of dermal coelomic canals in *Sipunculus nudus* (Sipuncula) with a discussion of respiratory designs in sipunculans. Invert Biol 114:51–63
24. Saiz Salinas JI (1993) Sipuncula. *In*: Ramos MA (ed.) Fauna Ibérica, vol 4. Museo Nacional de Ciencias Naturales, CSIC, Madrid, pp 1–200
25. Scheltema AH (1993) Aplacophora as progenetic aculiferans and the coelomate origin of mollusks as their sister taxon of Sipuncula. Biol Bull 184:57–78
26. Schulze A (2005) Sipuncula (Peanut Worms) from Bocas del Toro, Panama. Carib J Sci 41:523–527
27. Spongberg A (2006) PCB concentrations in interidal sipunculan (Phylum Sipuncula) marine worms from the Pacific coast of Costa Rica. Rev Biol Trop 54(Suppl. 1):27–33
28. Stephen AC, Edmonds SJ (1972) The phyla Sipuncula and Echiura. Trustees of the British Museum (Natural History), London, 528 p
29. Vargas JA, Abdullah M (1997) Trace metals in *Sipunculus nudus* (Sipuncula) and *Glottidia audebarti* (Brachiopoda) from a tropical estuarine sand flat. *In*: Abstracts. Third International Ocean Pollution Symposium (3IOPS). Harbor Branch Oceanographic Institution, Florida, pp 76
30. Vargas JA, Mata A (2004) Where the dry forest feed the sea: the Gulf of Nicoya estuary. *In*: Frankie GW, Mata A, Vinson SB (eds) Biodiversity Conservation in Costa Rica. University of California Press, Berkeley, CA, pp 126–135
31. Yan QL & Wang WX (2002) Metal exposure and bioavailability to a marine deposit-feeding Sipuncula, *Sipunculus nudus*. Environ Sci Technol 36:40–47

Part 13
Polychaetes and Echiurans

Harlan K. Dean

Iphione ovata, a polychaete found in Pacific Costa Rica. (Photo: Arthur Anker)

Abstract A total of 724 species of polychaetes have been reported in the primary literature from Central America with 583 from the Pacific coast and 208 from the Caribbean coast, of these 67 species were reported to occur along both coasts. A total of 317 species of polychaetes have been reported from Costa Rica, 317 from the Pacific coast, and 4 of these also from the Caribbean. There has been only a single record of a member of the problematic phylum Echiura from Central America, that from the Pacific coast of Costa Rica. The number of species of polychaetes and echiurans from Central America is lower than expected in relation to surveys from other geographic regions and this is most probably a function of lower sampling effort.

H.K. Dean
Museum of Comparative Zoology (MCZ), Harvard University, 26 Oxford St. Cambridge. Massachusetts. 02138. USA.
e-mail: Harlandean@umb.edu

I.S. Wehrtmann, J. Cortés (eds.) *Marine Biodiversity of Costa Rica, Central America*,

Introduction

The polychaetes are an important taxonomic group of animals in benthic environments, both in a numerical and a biological sense. They are usually the most common phylum of organisms in any bottom sample and, quite often, one or more polychaete species will be the numerical dominants. Many are selective or non-selective deposit feeders and obtain organic nutrients from ingested sediments, often providing structure to the community and modifying particle flux through the construction of burrows and tubes (Carey 1983; Meadows & Tait 1989; Paiva 1993). Other polychaetes are filter feeders, carnivores, or commensals living in association with other organisms and also playing a major role in the trophic structure of the bottom community (Paiva 1993; Muniz & Pires 1999). Polychaetes are also quite important in the secondary production of the benthos, serving as important food sources for many commercially important fish and macro invertebrates (Simenstad & Cailliet 1986; Nonato *et al.* 1990; Amaral *et al.* 1994).

Members of the Class Polychaeta are a morphologically diverse group of metameric organisms perhaps reflecting the wide variety of niches they fill in their communities. They may be found in a wide variety of habitats ranging from muds, sands, algae, rocks, corals, and the plankton. They make up an important part of marine communities from the intertidal to abyssal depths. Currently, there are about 80 recognized families and approximately 16,000 described species within the taxon.

There are believed to be many more species of polychaetes as yet to be described since the polychaetes of only a few regions have been sampled intensively. While the polychaetes of the eastern north Atlantic and the Mediterranean, as well as those of both sides of North America, have been relatively well sampled, those of Central America have not. Collections from both the Caribbean and Pacific coasts of Central America have been made in a relatively unorganized fashion with large areas essentially unexplored. For instance, the polychaetes of the Pacific coast of Costa Rica and Panama have been fairly well sampled as have those of the Caribbean coast of Belize and Panama but the polychaete fauna of other regions of Central America is essentially unknown.

The Echiura has been recognized as a phylum of approximately 135 species, however, recent molecular analysis has indicated that they may belong among the Polychaeta. McHugh (1997) found that the echiurans included in her analysis clustered with several species of polychaetes suggesting that they should be included within the Class Polychaeta. The results of Rouse & Fauchald (1997), based on both molecular and morphological characters, did not support her results and to this point the systematic position of these worms is considered unresolved (Rouse & Pleijel 2001). The single record of the Echiura from Central America (Dean 2001c) is, therefore, included here with the polychaetes as a group of uncertain relationship.

Methodological Background

This list (Species Lists 13.1 and 13.2 are included on the CD-Rom) is an attempt to summarize much of what is known of not only the polychaetes of Costa Rica but also those of all of Central America. It is believed that the restriction of the list to only the polychaetes of Costa Rica would provide a misleading view of the actual fauna. For example, only four species of polychaetes have been reported from the Caribbean side of Costa Rica while 317 have been reported from the Pacific side. This is obviously the result of a difference in sampling effort, not necessarily an indication of differences in the polychaete fauna of the two coasts. While it is obvious that this list still provides an incomplete picture of the Central American fauna, hopefully it can be used to better characterize the polychaete communities of this region as well as point the way for future studies.

The citations listed here are based upon a review of the primary literature up to January 2003 and does not include any citations of academic theses or government reports. All attempts were made to recognize any synonymies, however, these synonymies are not explicitly noted in the lists. For example, Dean (1996b) cited the occurrence of *Ceratonereis crosslandi* but that citation is listed here under *C. singularis* only since Perkins (1980) reviewed the genus and took *C. singularis* out of synonomy with *C. mirabilis*.

Caution must also be advised regarding the validity of many of the species identifications in this list. Some species are probably not valid species as they were poorly described, there is no type material, and they have never been collected again (ex. *Dorvillea bioculata* (Grube 1856) is considered a doubtful species by Hartman (1959)). Additionally, species previously described as cosmopolitan such as *Capitella capitata* and *Synelmis albini* have been found to actually be a complex of distinct species, often easily recognized as such upon closer examination of specimens (Grassle & Grassle 1976; Salazar-Vallejo 2003). Species with widely separated known distributions such as *Decamastus nudus*, reported from the eastern Pacific and Indian Ocean, and *Ancystrosyllis hamata*, reported from the eastern Pacific and the Mediterranean, could also be cases of misidentification. However, recent reports of long-distance transport of polychaetes as a result of human activities (ex. transport of ballast water) could also explain these distributions. Hopefully, despite these dangers, this list of identified polychaete species will act as a guide in the refinement of our knowledge of the polychaete fauna of Central America.

Diversity of Polychaeta and Echiura

There are a total of 724 species of polychaetes reported from Central America with 583 species from the Pacific coast (Fig. 13.1) and 208 from the Caribbean coast. There are 67 species which have been reported to occur on both sides of Central America. There have been 317 species of polychaetes reported from the Pacific coast of Costa Rica and only four from the Caribbean side. Of the 387 species of polychaetes recorded from Panama, 294 were found on the Pacific coast and 93 on the Caribbean coast (a reflection of the relatively greater collecting effort on the Caribbean coast of Panama relative to that of neighboring Costa Rica). Forty-two species were reported from both coasts of Panama. Only 14 polychaete species have been reported from Guatemala, 13 of these from the Pacific coast. There are 119 species reported from Belize, 57 from El Salvador, 12 from Honduras, and only 6 from Nicaragua.

Fig. 13.1 The polychaete *Notopygos ornata*, reported from Pacific Costa Rica. Photo: Leslie Harris

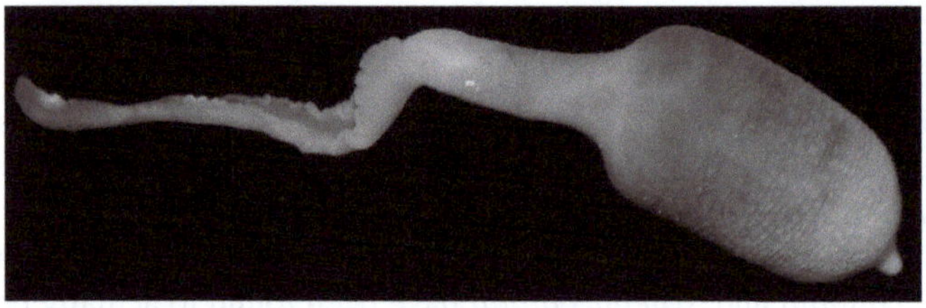

Fig. 13.2 *Thalassema steinbecki*, an echiuran reported from Pacific Costa Rica. Photo: Arthur Anker

The single report of a species of Echiura (Fig. 13.2) from Central America is from the Golfo de Nicoya on the Pacific coast of Costa Rica (Dean 2001c). The single record from this region is a simple reflection of a lack of sampling effort and/or a failure to recognize collected material.

The 724 recognized species of polychaetes reported from Central American waters is a lower number than would be expected for such an environmentally rich area. Salazar-Vallejo *et al.* (1988, 1996b) cited about 1,100 species of polychaetes from the Pacific and Caribbean coasts of Mexico and 1,240 species from the Grand Caribbean region (Bermuda to the central coast of Brazil and South Carolina (USA) to the Gulf of Mexico), but in both cases the authors indicated that these numbers are probably a great underestimate. Costello *et al.* (2001) have also listed 1,848 polychaete species from more temperate European, Mediterranean, and Baltic waters. Comparison of the present list of polychaete species with these numbers makes it apparent that the number of Central American polychaete species has been vastly underestimated most likely as a result of a lack of sample effort. One needs only to cite the results of Grassle (1973), who reported 103 species of polychaetes from a single head of coral in an area of the Great Barrier Reef where previous work had only listed a total of 97 species.

The greater number of polychaete species on the Pacific coast of Central America relative to the Caribbean coast may, indeed, indicate a higher polychaete diversity in the eastern Pacific. Fauchald (1977) also reported the same trend in his study of intertidal polychaetes from both coasts of Panama with a more balanced sampling effort. It is apparent, however, that an appreciable portion of the disparity in polychaete species numbers between the two coasts of Central America is a result of the greater sampling effort from the Pacific coast. Other than work conducted in Panama (Monro 1933a, b; Fauchald 1977) and Belize (Young & Young 1982; Russell 1989a, b, 1991, 1995), few major surveys of the polychaetes have been conducted on the Caribbean side of Central America. In contrast, a great deal of sampling and identification has been conducted on the Pacific shores of Panama (Monro 1928a, b, 1933a, b, Hartman 1939, 1940, 1944a, b, 1947, 1950; Fauchald 1977; Capa *et al.* 2001a, b, c), and Costa Rica (Vargas *et al.* 1985; Vargas 1987, 1988, 1989; Maurer *et al.* 1988, Dean 1996a, b, 1998a, b, 2001a, b).

In summary, while our knowledge of this highly important group of organisms in Central America has grown considerably in the last two decades, much work remains to be done. The polychaetes are often numerical dominants and often are some of the most biologically important species in the sediments. Our understanding of benthic communities and the responses of such communities to environmental degradation cannot occur without a much more extensive understanding of this diverse group in Central America.

Specialists and Collections

Given the uncertainty of some of the identifications of polychaetes from Central America, it is recommended that the original material should be examined whenever possible. What follows are the locations of some of the larger collections of polychaetes made in Central America. Material collected by Chamberlin (1919a, b) and Treadwell (1917, 1928a, b, 1941, 1943) may be found at the American Museum of Natural History (New York; contact: Schuh@AMNH.org), the National Museum of Natural History (Washington, DC; contact: bright.cheryl@NMNH.SI.edu), and the Museum of Comparative Zoology (Cambridge, MA; contact: abaldingen@oeb.harvard.edu). The specimens collected by Crossland and examined by Monro (1933a, b) are deposited at the British Museum of Natural History (London; contact: aim@nhm.ac.uk), while most of the material collected by the Allan Hancock Foundation cruises is presently at the Natural History Museum of Los Angeles County (Los Angeles; contact: kfitzhug@nhm.org). The materials collected by Dean (1996a, b, 1998a, b, 2001a, b) may be found at the Museum of Comparative Zoology, the National Museum of Natural History, or the Museo de Zoología, Escuela de Biología, Universidad de Costa Rica (San Pedro, San José; contact: mzucr@biologia.ucr.ac.cr). The Syllidea collected by Russell (1989a, b, 1991, 1995) from Belize have been deposited at the National Museum of Natural History while the Syllidae of Panama collected by Capa *et al.* (2001a, b, c) may be found at the Museo Nacional de Ciencias Naturales de Madrid (Spain; contact: santiagom@mncn.csic.es).

Polychaete researchers usually specialize in a particular taxonomic group rather than working on the fauna of a particular geographic region. For taxonomic questions regarding particular polychaetes from Central America it is best to contact researchers involved with that taxon at Polychaete Researchers Online (PRO) at the Annelid web site: http://biodiversity.uno.edu/~worms/annelid.html.

Numerous additional species have been described mainly from Panama, since January, 2003, most pertinent to Costa Rica is Salazar-Viello (2003) who described a new species of pilargidae common in Pacific Costa Rica, and Dean & Blake (2007) who described eight new species of cirratulidae from Pacific Costa Rica.

Acknowledgments My collections of Costa Rican polychaetes would not have been possible without the continued support of the members of the CIMAR, Universidad de Costa Rica, especially J.A. Vargas and J. Acuña, and their ongoing studies funded by CRUSA. My continuing access to the polychaete literature is due to the long and continuing help of the Department of Invertebrate Zoology, Museum of Comparative Zoology, and the curators H. Levi, D. McHugh, and G. Girabet and curatorial associate A. Johnston. Finally I would like to recognize the guidance and patience of the editors of this volume, Jorge Cortés and Ingo Wehrtmann who pushed me to complete this part and dealt with the numerous drafts of this list.

References

Amaral ACZ, Nonato EF, Petti MA (1994) Contribution of the polychaetous annelids to the diet of some Brazilian fishes. Mém Mus Natl Hist Nat 162:331–337

1. Augener H (1912) Beitrag zur Kenntnis verschiedener Anneliden und Bemerkungen über die nordischen *Nephthys*-Arten und deren epitoke Formen. Arch Naturgesch, Berlin Abt A 78:162–212
2. Augener H (1922) Über litorale Polychaeten von Westindien. Sitz Ges Naturf Freunde Berlin (3–5):38–53
3. Berkeley E, Berkeley C (1939) On a collection of Polychaeta, chiefly from the west coast of Mexico. Ann Mag Nat Hist 3:321–346
4. Berkeley E, Berkeley C (1960) Notes on some Polychaeta from the west coast of Mexico, Panama and California. Can J Zool 38:357–362
5. Böggemann M, Fiege D (2001) Description of seven new species of the genus *Glycera* Savigny, 1818 (Annelida, Polychaeta: Glyceridae). Ophelia 54:29–49
6. Böggemann M (2002) Revision of the Glyceridae Grube 1850 (Annelida: Polychaeta). Abh Senckenberg Naturforsch Ges 555:1–249
7. Capa M, San Martín G, López E (2001a) Description of a new species of *Paraspaerosyllis* (Polychaeta: Syllidae: Syllinae). Proc Biol Soc Wash 114:280–284
8. Capa M, San Martín G, López E (2001b) *Syllinae* (Syllidae: Polychaeta) del Parque Nacional de Coiba, Panamá. Rev Biol Trop 49:103–115
9. Capa M, San Martín G, López E (2001c) Autolytinae, Eusyllinae y Exogoninae (Syllidae: Polychaeta) del Parque Nacional de Coiba, Panamá. Rev Biol Trop 49:621–628

Carey DA (1983) Particle resuspension in the benthic boundary layer induced by flow around polychaete tubes. Can J Fish Aquat Sci 40 (Suppl 1):301–308

10. Caullery M, Mesnil F (1897) Etudes sur la morphologie compare et la phylogenie des espéces chez les *Spirorbis*. Bull Sci France Belg 30:185–233
11. Chamberlin RV (1919a) Pacific coast Polychaeta collected by Alexander Agassiz. Mus Comp Zool Harvard Bull 63:251–276
12. Chamberlin RV (1919b) The Annelida Polychaeta. Mus Comp Zool Harvard, Mem 48:1–514
13. Colbath GK (1989a) Revision of the family Lysaretidae, and recognition of the family Oenonidae Kinberg, 1865 (Enicida: Polychaeta). Proc Biol Soc Wash 102:116–123

14. Colbath GK (1989b) A revision of *Arabella mutans* (Chamberlin, 1919) and related species (Polychaeta: Arabellidae). Proc Biol Soc Wash 102:283–299

Costello MJ, Emblow CS, White R (eds) (2001) European Register of Marine Species. A checklist of the Marine Species in Europe and a Bibliography of Guides to their Identification. Patromoines Naturels 50, 463 p

15. Dauvin J-C, Thiébaut E (1994) Is *Owenia fusiformis* Delle Chiaje a cosmopolitan species? pp. 383–404. *In*: Dauvin J-C, Laubier L, Reish DJ (eds) Actes de la 4ème Conferèrence internationale des Polychètes. Mèm Mus Natl Hist Nat, Paris, p 162

16. Dean HK (1996a) Subtidal benthic polychaetes (Annelida) of the Gulf of Nicoya, Costa Rica. Rev Biol Trop 44 (Suppl 3):69–80

17. Dean HK (1996b) Polychaete worms (Annelida) collected in Golfo Dulce during the Victor Hensen Costa Rica Expedition (1993/1994). Rev Biol Trop 44 (Suppl 3):81–86

18. Dean HK (1998a) A new species of Hesionidae, *Glyphohesione nicoyensis* (Annelida: Polychaeta), from the Gulf of Nicoya, Costa Rica. Proc Biol Soc Wash 111:257–262

19. Dean HK (1998b) The Pilargidae (Annelida: Polychaeta) of the Pacific coast of Costa Rica. Rev Biol Trop 46 (Suppl 6):47–62

20. Dean HK (2001a) Some Nereididae (Annelida: Polychaeta) from the Pacific coast of Costa Rica. Rev Biol Trop 49 (Suppl 2):37–67

21. Dean HK (2001b) Capitellidae (Annelida: Polychaeta) from the Pacific coast of Costa Rica. Rev Biol Trop 49 (Suppl 2):69–84

22. Dean HK (2001c) Marine biodiversity of Costa Rica: The phyla Sipuncula and Echiura. Rev Biol Trop 49 (Suppl 2):85–90

Dean HK, Blake J (2007) *Chaetozone* and *Caulleriella* (Polychaeta cirratulidae) from the Pacific Coast of Costa Rica, with description eight new species. Zootaxa 1451:41–68

23. Dexter DM (1974) Sandy-beach fauna of the Pacific and Atlantic coasts of Costa Rica and Colombia. Rev Biol Trop 22:51–66

24. Dittmann S, Vargas JA (2001) Tropical tidal flat benthos compared between Australia and Central America, pp 275–293. *In*: Reise K (ed) Ecological Comparisons of Sedimentary Shores. Springer-Verlag, Berlin

25. Eibye-Jacobs D (1992) Phyllodocids (Annelida: Polychaeta) of Belize, with the description of three new species. Proc Biol Soc Wash 105:589–613

26. Fauchald K (1972) Benthic polychaetous annelids from deep water off western Mexico and adjacent waters in the eastern Pacific Ocean. Allan Hancock Found., Monogr Mar Biol 7:1–575

27. Fauchald K (1973) Polychaetes from Central American sandy beaches. Bull South Calif Acad Sci 72:19–31

28. Fauchald K (1975) Clave de poliquetos panameños con la inclusión de una clave para todas las familias del mundo. Bol Inst Oceanogr Univ Oriente 14:71–94

29. Fauchald K (1977) Polychaetes from intertidal areas in Panama, with a review of previous shallow-water records. Smithson Contr Zool 221:1–81

30. Fauchald K (1980) Onuphidae (Polychaeta) from Belize, Central America, with notes on related taxa. Proc Biol Soc Wash 93:797–829

31. Fauchald K (1992) A review of the genus *Eunice* (Polychaeta: Eunicidae) based on type material. Smithson. Contr Zool 523:1–422

32. Fitzhugh K (1983) A new species of *Fabricola* and *Fabricia* (Polychaeta: Sabellidae) from Belize. Proc Biol Soc Wash 96:276–290

33. Fitzhugh K (1996). New fanworm species (Polychaeta: Sabellidae: Fabriciinae) in the genus *Pseudofabriciola* Fitzhugh. J Nat Hist 30:1267–1286

34. Friedrich H (1956) Mitteilungen über neue und wenig bekannte Polychaeten aus Mittel- und Südamerika. Senckenberg Biol 37:57–68

35. Gaston GR, Hall J (2000) Lunar periodicity and bioluminescence of swarming *Odontosyllis luminosa* (Polychaeta: Syllidae) in Belize. Gulf & Caribb. Res 12:47–51

36. Glasby CJ (1999) The Namanereidinae (Polychaeta: Nereididae). Part 1, taxonomy and phylogeny. Rec Austr Mus (Suppl 25):1–146

Grassle JF (1973) Variety in coral reef communities, pp 247–270. *In*: Jones OA & Endean R (eds) Biology and Geology of Coral Reefs, Vol. 2: Biology 1. Academic Press, New York
Grassle JP, Grassle JF (1976) Sibling species in the marine pollution indicator *Capitella* (Polychaeta). Science 192:567–569
37. Green KD (1987) Revision of the genus *Sonatsa* (Polychaeta: Maldanidae). Biol Soc Wash Bull 7:89–96
38. Grube AE (1856) Annulata Oerstediana. Enumeratio annulatorum, quae in itinere per Indiam occidentalum et Americam centralem annis 1845–1848 suscepto legit. D.A.S. Oersted, adjectis speciebus nonullis a d. H. Kroeyer in itinere ad Americam meridionalem collectis. Pt. 1. Vidensk Medd Dansk naturh Foren 1856:44–62
39. Grube AE (1857) Annulata Oerstediana. Enumeratio annulatorum, quae in itinere per Indiam occidentalum et Americam centralem annis 1845–1848 suscepto legit. D.A.S. Oersted, adjectis speciebus nonullis a d. H. Kroeyer in itinere ad Americam meridionalem collectis. Pt. 2. Vidensk Medd Dansk naturh Foren 1857:158–166
40. Grube AE (1858) Annulata Oerstediana. Enumeratio annulatorum, quae in itinere per Indiam occidentalum et Americam centralem annis 1845–1848 suscepto legit. D.A.S. Oersted, adjectis speciebus nonullis a d. H. Kroeyer in itinere ad Americam meridionalem collectis. Pt. 3. Vidensk Medd Dansk naturh Foren 1858:105–120
41. Hartman O (1939) Polychaetous annelids. Pt. I. Aphroditidae to Pisionidae. Allan Hancock Pac Exped 7:1–156
42. Hartman O (1940) Polychaetous annelids. Part II. Chrysopetalidae to Goniadidae. Allan Hancock Pac Exped 7:173–287
43. Hartman O (1944a) Polychaetous annelids. Allan Hancock Atlan. Exped Rep 3:3–33
44. Hartman O 1944b Polychaetous annelids. Pt. V. Eunicea. Allan Hancock Pac Exped 10:1–238
45. Hartman O (1947) Polychaetous annelids Part VIII. Pilargiiidae. Allan Hancock Pac Exped 10:483–524
46. Hartman O (1950) Goniadidae, Glyceridae and Nephtyidae. Allan Hancock Pac Exped 15:1–181
47. Hartman O (1957) Orbiniidae, Apistobranchidae, Paraonidae and Longosomidae. Allan Hancock Pac Exped 15:211–393
Hartman O (1959) Catalogue of the polychaetous annelids of the world. Allan Hancock Found Occ Paper 23:1–628
48. Hartmann –Schröder G (1956) Neue Armandia-Arten (Opheliidae, Polychaeta) aus Brasilien und El Salvador. Beitr Neotrop Fauna 1:63–68
49. Hartmann-Schröder G (1959) Zur Ökologie der Polychaeten des Mangroven-Estero-Gebietes von El Salvador. Beitr Neotrop Fauna 1:69–183
50. Hove HA ten (1973) Serpulinae (Polychaeta) from the Caribbean: II. The genus *Sclerostyla*. StudFauna Cur 43:1–21
51. Hove HA ten, Jansen-Jacobs M.J (1984) A revision of the genus *Crucigera* (Polychaeta: Serpulidae); a proposed methodical approach to serpulids, with special reference to variation in *Serpula* and *Hydroides*, pp 143–180. *In*: Hutchings PA (ed) Proceedings of the First Inter Polychaete Conference. The Linnean Society of New South Wales
52. Hutchings P, Glasby C (1986) A revision of the genus *Euthelepus* (Terebellidae: Thelepinae). Rec Austr Mus 38:105–117
53. Imajima M, ten Hove H.A. (1984) Serpulinae (Annelida, Polychaeta) from the Turk Islands, Ponape and Majuro Atoll, with some other new Indo-Pacific records. Proc Jap Soc syst Zool 27:35–66
54. Kinberg JGH (1855) Nya slägter och arter af Annelder. Öfv. Svenska Vetebsk Akad Förh 12:381–388
55. Kinberg JGH (1857) Nya slägter och arter af Annelder. Öfv Svenska Vetebsk Akad Förh 14:11–14
56. Kinberg JGH (1865) Annulata nova. Öfv Svenska Vetebsk Akad Förh 21:559–574
57. Kinberg JGH (1867) Om Amphinomernas systematic. Öfv. Svenska Vetensk. Akad Förh 24:83–91

58. Knight-Jones P, Knight-Jones EW, Dales RP (1979) Spirorbidae (Polychaeta Sedentaria) from Alaska to Panama. J Zool Lond 189:419–458
59. Knight-Jones P (1983) Contributions to the taxonomy of Sabellidae (Polychaeta). Zool J Linn Soc 79:245–295
60. León-Morales R, Vargas JA (1998). Macrofauna of a tropical fjord-like embayment: Golfo Dulce, Costa Rica. Rev Biol Trop 46 (Suppl 6):81–90
61. Macintyre IG, Rützler K, Norris JN, Fauchald K (1978) A submarine cave near Columbus Cay, Belize: a bizarre cryptic habitat. Smiths Contr Mar Sci 12:127–141
62. Maurer D, Vargas JA (1984) Diversity of soft-bottom benthos in a tropical estuary: Gulf of Nicoya, Costa Rica Mar Biol 81:97–106
63. Maurer D, Epifanio C, Dean H, Howe S, Vargas J, Dittel A, Murillo M (1984) Benthic invertebrates of a tropical estuary: Gulf of Nicoya, Costa Rica. J Nat Hist 18:47–61
64. Maurer D, Dean H, Vargas J (1987) Soft-bottom invertebrate communities from the Gulf of Nicoya, Costa Rica. Mem V Simp Biol Mar Univ Autón Baja California Sur 135–141
65. Maurer D, Vargas JA, Dean H (1988). Polychaetous annelids from the Gulf of Nicoya, Costa Rica. Inter Rev ges Hydrobiologie 73:43–59

McHugh D (1997) Molecular evidence that echiurans and pogonophorans are derived annelids. Proc Nat Acad Sci USA 94:8006–8009

Meadows PS, Tait J (1989) Modification of sediment permeability and shear strength by two burrowing invertebrates. Mar Biol 101:75–82

66. Molina Lara OA, Vargas Zamora JA (1995) Poliquetos (Annelida: Polychaeta) del estero de Jaltepeque, El Salvador: una comparación 1959–1991. Rev Biol Trop 43:195–205
67. Monro CGA (1928a) Polychaeta of the families Polynoidae and Acoetidae from the vicinity of the Panama Canal, collected by Dr. C. Crossland and Dr. Th. Mortensen. J Linn Soc Lond 36:553–576
68. Monro CGA (1928b) Papers from Dr. Th. Mortensen's Pacific Expedition, 1914–16. On the Polychaeta collected by Dr. Th. Mortensen off the coast of Panama. Vid Medd Dansk Naturh Foren 63:234–318
69. Monro CCA (1933a) The polychaeta Errantia collected by Dr. C. Crossland at Colón, in the Panama Region, and the Galapagos Islands during the expedition of the S. Y. "St. George". Proc Zool Soc London 1933:1–96
70. Monro CA (1933b) The Polychaeta Sedentaria collected by Dr. C. Crossland at Colón, in the Panama Region, and the Galapagos Islands during the Expedition of the S.Y. 'St. George.' Proc Zool Soc 1933:1039–1092
71. Monro CCA (1939) Polychaeta of the "Rosaura" expedition. Novit Zool 41:345–354
72. Mörch OAL (1863) Revisi critica Serpulidarum. Et Bidrag til Rörormenes Naturhistorie. Naturhist. Tidssk. Copenhagen 1:347–470
73. Mountford NK (1991) Redescription of *Phyllodoce erythrophylla* (Schmarda 1861) and *P. madeirensis* Langerhans, 1880 (Polychaeta: Phyllodocidae), with comments on some related taxa. Ophelia (Suppl 5):157–168

Muniz P, Pires AMS (1999) Trophic strcture of polychaetes in the São Sabastião Channel (southeastern Brazil). Mar Biol 134:517–528

74. Nichols-Driscoll J (1976) Benthic invertebrate communities in Golfo Dulce, Costa Rica, an anoxic basin. Rev Biol Trop 24:281–297

Nonato EF, Petti MAV, Paiva PC (1990) Contribuição dos anelides poliquetas na dieta de crustáceos decápodos braquiúros da região de Ubatuba. Simpósio de Ecossistemas da Costa Sul e Sudeste Brasileira 2, Aguas de Lindóia, Brasil. Publicação ACIESP 71, Vol. 1, São Paulo, pp 224–234

Paiva PC (1993) Trophic structure of a shelf polychaete taxocoenosis in southern Brazil. Cah Biol Mar 35:39–55

75. Paxton H (1998) The *Diopatra chiliensis* confusion – redescription of *D. chiliensis* (Polychaeta, Onuphidae) and implicated species. Zool Scr 27:31–48
76. Perkins TH (1980) Review of species previously referred to *Ceratonereis mirabilis* and new species of *Ceratonereis, Nephtys* and *Goniada*. Proc Biol Soc Wash 93:1–49

77. Perkins TH (1984) Revision of *Demonax* Kinberg, *Hypsicomus* Grube, and *Notaulax* Tauber, with a review of *Megalomma* Johansson from Florida (Polychaeta: Sabellidae). Proc Biol Soc Wash 97:285–368
78. Pettibone MH (1954) Marine polychaete worms from Point Barrow, Alaska, with additional records from the North Atlantic and North Pacific. Proc US Nat Mus 103:203–356
79. Pettibone MH (1969) Revision of the aphroditoid polychaetes of the family Eulepethidae Chamberlin (=Eulepethidae Darboux; = Pareulepidae Hartman). Smith Contr Zool 41:1–44
80. Pettibone MH (1971) Partial revision of the genus *Sthenelais* Kinberg (Polychaeta: Sigalionidae) with diagnoses of two new genera. Smith Contr Zool 109:1–40
81. Pettibone MH (1986) Review of the Iphioninae (Polychaeta: Polynoidae) and revision of *Iphione cimex* Quatrefages, *Gattyana deludens* Fauvel, and *Harmothoe iphionelloides* Johnson (Harmothoinae). Smith Contr Zool 428:1–43
82. Pettibone MH (1989) Revision of the Aphroditoid polychaetes of the family Acoetidae Kinberg (= Polyodontidae Augener) and reestablishment of *Acoetes* Audouin and Milne-Edwards, 1832, and *Euarche* Ehlers, 1887. Smith Contr Zool 464:1–138
83. Pettibone MH (1992) Contribution to the polychaete family Pholoidae Kinberg. Smith Contr Zool 532:1–22
84. Pettibone MH (1997) Revision of the scaleworm genus *Eulagisca* McIntosh (Polychaeta: Polynoidae) with the erection of the subfamily Eulagiscinae and the new genus *Pareulagisca*. Proc Biol Soc Wash 110:537–551
85. Pillai TG, ten Hove HA (1994) On recent species of *Spiraserpula* Regenhardt, 1961, a serpulid polychaete genus hitherto known only from Cretaceous and Tertiary fossils. Bull Nat Hist Mus Lond 60:39–104

Rouse GW, Fauchald K (1997) Cladistics and polychaetes. Zool Scr 26:139–204

Rouse GW, Pleijel F (2001) Polychaetes. Oxford University Press, London. 362 p

86. Rostad T, Hansen KL (2001/2002) The effects of trawling on the benthic fauna of the Gulf of Nicoya, Costa Rica. Rev Biol Trop 49 (Suppl 2):91–95
87. Rouse GW (1994) New species of *Oriopsis* Caullery and Mesnil from Florida, Belize, and Aldabra Atoll (Seychelles), and a new species of *Amphiglena* Claparède from Seychelles (Polychaeta: Sabellidae: Sabellinae). Bull Mar Sci 54:180–202
88. Rouse GW, Gambi MC (1997) Cladistic relationships within *Amphiglena* Claparède (Polychaeta: Sabellidae) with a new species and a redescription of *A. mediterranea* (Leydig). J Nat Hist 31:999–1018
89. Russell DE (1989a) Three new species of *Sphaerosyllis* (Polychaeta: Syllidae) from mangrove habitats in Belize. Zool Scr 18:375–380
90. Russell DE (1989b) A new species of *Odontosyllis* (Polychaeta: Syllidae) from Twin Cays, Belize. Proc Biol Soc Wash 102:768–771
91. Russell DE (1991) Exogoninae (Polychaeta: Syllidae) from the Belizean barrier reef with a key to the species of *Sphaerosyllis*. J Nat Hist Lond 25:49–74
92. Russell DE (1995) Description of a new viviparous species of *Dentatisyllis* (Polychaeta: Syllidae) from Belize with an assessment of growth and variation, and emendation of the genus. Proc Biol Soc Wash 108:568–576
93. Salazar-Vallejo SI (1991) Revisión de algunas eucliméninos (Polychaeta: Maldanidae) del Golfo de California, Florida, Panamá y Estrecho de Magallanes. Rev Biol Trop 39:269–278
94. Salazar-Vallejo SI (1996a) Filodócidos (Polychaeta: Phyllodocidae) del Caribe Mexicano con claves para identificar las especies del Gran Caribe. Rev Biol Trop 44:107–122

Salazar-Vallejo SI (1996b) Lista de especies y bibliografía de poliquetos (Polychaeta) del Gran Caribe. Anales Inst Biol Univ Autón México, Ser Zool 67:11–50

Salazar-Vallejo, SI (2003) Revision of *Synelmis* Chamberlin, 1919 (Annelida, Polychaeta, Pilaridae). Zoosyst 25:17–42

Salazar-Vallejo SI, de León-González JA, Salaices-Polanco H (1988) Poliquetos (Annelida: Polychaeta) de Mexico. Libros Universitarios, Universidad Autónoma de Baja California Sur, 212 p

95. San Martin G, Lopez E, Camacho AJ (1998) First record of a freshwater Pisionidae (Polychaeta): description of a new species from Panama with a key to the species of *Pisione*. J Nat Hist 32:1115–1127
96. Schuster-Dieterichs O (1956) Die Makrofauna am sandigen Brandungsstrand von El Salvador. Senck Biolog 37:1–56
97. Siewing R (1954) Zur Verbreitung von *Pisionidens indica* Aiyar and Alikunhi. Kieler Meeres 10:81–83
Simenstad CA, Cailliet GM (eds) (1986) Papers from the fourth workshop on fish food habits (Gutshop 1984). Environ Biol Fish 16:1–223
98. Solis-Weiss V, Fauchald K (1989) Orbiniidae (Annelida: Polychaeta) from mangrove root-mats in Belize, with a revision of Protoariciin genera. Proc Biol Soc Wash 102:772–792
99. Solis-Weiss V, Fauchald K, Blankensteyn A (1991) Trichobranchidae (Polychaeta) from shallow warm water areas in the western Atlantic Ocean. Proc Biol Soc Wash 104:147–158
100. Treadwell AL (1917) A new species of polychaetous annelid from Panama, with notes on a Hawaiian form. Proc U S Natl Mus 52:427–430
101. Treadwell AL (1928a) A new polychaetous annelid of the genus *Phyllodoce* from the west coast of Costa Rica. Proc U S Natl Mus 74:1–3
102. Treadwell AL (1928b) Polychaetous annelids from the Arcturus Oceanographic Expedition. Zoologica 8:449–485
103. Treadwell AL (1941) Eastern Pacific expeditions of the New York Zoological Society. XXIII. Polychaetous annelids from the west coast of Mexico and Central America. Zoologica 26:17–24
104. Treadwell AL (1943) Biological results of the last cruise of the CARNEGIE. Polychaetous annelids. Sci Res Cruise VII CARNEGIE 1928–1929:30–59
105. Vargas JA (1987) The benthic community of an intertidal mud flat in the Gulf of Nicoya, Costa Rica. Description of the community. Rev Biol Trop 35:299–316
106. Vargas JA (1988) Community structure of macrobenthos and the results of macropredator exclusion on a tropical mud flat. Rev Biol Trop 36:287–308
107. Vargas JA (1989) A three year survey of the macrofauna of an intertidal mud flat in the Gulf of Nicoya, Costa Rica, pp 1905–1919. *In*: Magoon O,Converse M, Miner D, Tobin LT & Clark D (eds) Proceedings of the 6th Symposium on Coastal and Ocean Management. Vol. 2 Am Soc Civil Eng, New York
108. Vargas JA (1996) Ecological dynamics of a tropical intertidal mudflat community, pp 355–371. *In*: Nordstrom KF, Roman CT (eds) Estuarine Shores: Evolution, Environments and Human Alterations. Wiley.
109. Vargas JA, Dean HK,Maurer D & Orellana P (1985) Lista preliminar de invertebrados asociados a los sedimentos del Golfo de Nicoya, Costa Rica. Brenesia 24:327–341
110. Vreeland HV & Lasker HR (1989) Selective feeding of the polychaete *Hermodice carunculata* Pallas on Caribbean gorgonians. J Exp Mar Biol Ecol 129:265–277
111. Watson Russell C (1986) *Paleaequor*, a new genus of polychaete worm (Chrysopetalidae). Rec Aust Mus 38:153–174
112. Williams SJ (1984) The status of *Terebellides stroemi* (Polychaeta; Trichobranchidae) as a cosmopolitan species, based on a worldwide morphological survey, including description of new species, pp 118–142. *In*: Hutchings PA (ed) Proc. 1st Inter. Polychaete Conf, Sydney. The Linnaean Society of New South Wales
113. Wilson RS (1988) A review of *Eteone* Savigny, 1820, *Mysta* Malmgen, 1865 and *Hypereteone* Bergström, 1914 (Polychaeta: Phyllodocidae). Mem Mus Victoria 49:385–431
114. Young DK & Young MW (1982) Macrobenthic invertebrates in bare sand and seagrass (*Thalassia testudinum*) at Carrie Bow Cay, Belize. Smith Contr Mar Sci 12:115–126
115. Young MW & Kritzler H (1987) *Paraeupolymnia*, a new genus of terebellid (Polychaeta: Terebellidae) from Belize. Proc Biol Soc Wash 100:678–690
116. Zibrowius HW (1969) *Hydroides gairacensis* Augener, 1934, a little known serpulid polychaete from Central and South America. Bull Mar Sci 19:366–376

Part 14
Stomatopods

Rita Vargas

Cloridopsis dubia, reported from Pacific Costa Rica (Photo: Arthur Anker)

Abstract A total of 35 species of stomatopod crustaceans are reported for Costa Rica, including the information regarding specimens deposited at the Museo de Zoología, Universidad de Costa Rica, habitat, previous reports for Costa Rica, and distribution. Of these, six occur along the Caribbean coast and 29 along the Pacific coast including six species from Isla del Coco, two of which are endemic.

R. Vargas
Museo de Zoología, Escuela de Biología, Universidad de Costa Rica,
11501-2060 San José, Costa Rica.
e-mail: ritav@biologia.ucr.ac.cr

I.S. Wehrtmann, J. Cortés (eds.) *Marine Biodiversity of Costa Rica, Central America*,

Introduction

Stomatopods, known as “mantis shrimps,” are members of the Class Malacostraca with a body formed by 14 segments and a telson. The body is divided into three parts: cephalon, thorax, and abdomen. The shield-like carapace is small, and is bounded to the first five thoracic segments. Stomatopods differ from the other Eumalacostraca by having three antennular flagella, a mobile and articulated rostrum, gills on the exopodites as primary structures (Hendrickx & Salgado Barragán 1992), five pairs of maxillipeds and three pairs of walking legs, and have a pair of second maxillipeds enlarged for raptorial claw (Schram 1986). Stomatopods have pedunculated eyes formed by omatidia, representing the most developed eyes among the crustaceans (Ruppert & Barnes 1994; Ahyong 1997).

Stomatopods are among the most aggressive and behaviorally complex crustaceans; all are active predators and mark one of the few radiations of obligate carnivores within the Crustacea (Reaka & Manning 1980; Schram 1986; Ahyong 1997). The two methods of prey capture distinguish two broad functional groups: the “smashers” and the “spearers” (Caldwell & Dingle 1976; Schram 1986). These two groups comprise the order Stomatopoda and belong to Subclass Hoplocarida, Class Malacostraca (Martin & Davis 2001).

The study of the stomatopods of the Americas started with collections made by the steamer US Fish Commission Albatross during the winter of 1887–1888. As a result of this expedition, Bigelow (1894) published a report, which included 34 species from both coasts of the Americas. Between 1933 and 1938, the Allan Hancock Foundation carried out expeditions along the Pacific coast, compiling the largest stomatopod collection of the region. The results were published as a monograph authored by Schmitt (1940), including species collected in Costa Rican waters. Another important contribution to our knowledge about the stomatopods from the Pacific coast of Costa Rica was published by Manning (1972a), based on material collected during expeditions (1936–1938) of the Zoological Society of New York. As a result of the cruises carried out by RV “Searcher” in 1972, Manning & Reaka (1979) described three new species from the Pacific coast of Costa Rica, and Reaka & Manning (1980) published a review on distribution, ecology, and zoogeography of the stomatopods of Pacific Costa Rica.

Moreover, Camp & Kuck (1990) added three more stomatopod species to what was known from Isla del Coco, and one additional species for the Pacific coast of Costa Rica. The most recent mayor study of the Eastern Pacific stomatopods is that of Hendrickx & Salgado-Barragan (1992). The Western Atlantic stomatopod fauna was extensively studied by the late Raymond B. Manning, whose monograph of the Western Atlantic fauna remains the standard work for that region (Manning 1969). The most recent classification of the Stomatopoda recognizes 17 families in seven superfamilies (Ahyong 2001) who raised *Hemisquilla ensigera californiensis* Stephenson 1967, to full species status.

Based on material obtained by the cruises of RV Victor Hensen (1993–1994) along the Pacific coast of Costa Rica, Vargas *et al.* (1996) published a list of crustaceans from the study area, including the first report of *Eurysquilla veleronis* (Schmitt 1940) for Costa Rica. Vargas & Cortés (1997) reported a total of 35 stomatopod species for both coasts of the country, including the following reports: *Neogonodactylus oerstedii* (Hansen, 1895), *Pseudosquillisma ciliata* (Fabricius, 1787), *P. oculata* (Brullé, 1837), *Lysiosquillina glabriuscula* (Lamarck, 1818), and *Squilla empusa* Say, 1818 for the Caribbean coast of Costa Rica. *N. curacaoensis* Schmitt, 1924 is a new record informed here.

Four species have been described from materials collected in different localities in Costa Rica: *N. albicinctus* (Manning & Reaka, 1979) (Bahía Herradura); *N. costaricensis* (Manning & Reaka, 1979) (Bahía Herradura); *Nannosquilla canica* (Manning & Reaka, 1979) (Isla del Caño), and *Neocoronida cocosiana* (Manning 1972b), all endemic species collected in different localities on the Pacific coast of Costa Rica.

As a result of the present review, a total of 35 species in 10 families are documented for Costa Rican waters. In the Caribbean coast only four of these are present (Table 14.1). The family with more species is Squillidae (12 species), followed by Gonodactylidae with eight (Species Lists 14.1 and 14.2, included on CD-Rom).

The objective is to summarize the existing information concerning the biodiversity of stomatopods in Costa Rica. Moreover, the present compilation may stimulate future investigations about this important group of crustaceans, especially along the Caribbean coast of Costa Rica, where the reported number of species is considerably lower when compared to Caribbean islands and Colombia. The families were sorted according to Martin and Davis (2001).

Table 14.1 Number of species per family and coast according to Martin and Davis (2001)

Family	Caribbean	Pacific	Total
Tetrasquillidae	n.a.	1	1
Coronididae	n.a.	2	2
Nannosquillidae	n.a.	3	3
Lysiosquillidae	1	1	2
Gonodactylidae	2	6	8
Hemisquillidae	n.a.	1	1
Pseudosquillidae	n.a.	1	1
Eurysquillidae	n.a.	1	1
Squillidae	1	11	12
Parasquillidae	2	2	4
Total	6	29	35

n.a. - information not available.

Specialists

José Salgado Barragán: Instituto de Ciencias del Mar y Limnología, Universidad Nacional Autónoma de México, México - Tropical Eastern Pacific. salgado@icmyl.unam.mx
Marjorie L. Reaka: Department of Biology, University of Maryland, College Park, MD 20742, USA – Tropical Eastern Pacific. mr9@umail.umd.edu
Marylin Schotte: Smithsonian Institution, Washington, DC, USA – Caribbean. schotte.marilyn@nmnh.si.edu
Bernd Werding: Institut für Allgemeine und Spezielle Zoologie der Justus-Liebig-Universitat Stephanostrasse 24, D-6300 Giesen, Germany – Caribbean. bernd.werding@allzool.bio.uni-giessen.de

Collections

Museo de Zoología, Escuela de Biología, Universidad de Costa Rica, mzucr@biologia.ucr.ac.cr
National Museum of Natural History, Smithsonian Institution
Los Angeles County Museum (depositary of Allan Hancock Collection)

Acknowledgments I am very grateful to the editors Jorge Cortés and Ingo Wehrtmann for their help with the manuscript and to improve the language. Thanks go to Majorie Reaka for her suggestions and valuable help with the manuscript. I am grateful to the reviewers who contributed with their commentaries.

References

Ahyong ST (1997) Phylogenetic analysis of the Stomatopoda (Malacostraca). J Crust Biol 17:695x–715
Ahyong ST (2001) Revision of Australian stomatopod Crustacea. Rec Austr Mus Suppl 26:1–326
1. Bigelow RP (1894) Report upon the Crustacea of the Order Stomatopoda collected by the Steamer Albatross between 1885 and 1891, and on other specimens in the US National Museum. Proc US Nat Museum 27:489–550
2. Boone L (1930) Crustacea: Stomatopoda and Brachyura. Scientific results of the Cruises of the Yachts "Eagle" and "Ara" 1921–1928, William K. Vanderbilt, Commanding. Bull Vanderbilt Mar Mus 2:1–228
3. Boyko CB (2000) The rise and fall of *Lysiosquilla desaussurei* and description of *L. manningi* n. sp.: the tale of the type. J Crust Biol 20(special number 2):48–55
Caldwell RL, Dingle H (1976) Stomatopods. Sci Am 234:80–89
4. Camp DK (1973) Stomatopod Crustacea. Mem. Hourglass Cruises. Marine Research Laboratory, Fl Dep Nat Res vol. III, part II. 100 p
5. Camp DK, Kuck HG (1990) Additional records of stomatopod crustacean from Isla del Coco and Golfo de Papagayo, east Pacific Ocean. Proc Biol Soc Wash 103:847–853

6. Hendrickx ME (1995) Estomatópodos. *In*: Fischer W, Krupp F, Schneider W, Sommer C, Carpenter KE, Niem VH (eds) Guía FAO para la identificación de especies para los fines de la pesca. Pacífico Centro-oriental. Vol 1: Plantas e invertebrados. FAO, Roma, 646 p, pp 103–105
7. Hendrickx ME, Salgado-Barragán J (1992) Los Estomatópodos (Crustacea: Hoplocarida) del Pacífico Mexicano. Universidad Nacional Autónoma de México Publ Esp 10, 200 p
8. Holthuis LB (1959) Stomatopod Crustacea of Suriname. Stud Fauna Suriname 3:173–191
9. Manning RB (1961a) Stomatopod Crustacea from the Atlantic of northern South America. University of Southern California, Los Angeles, California, 46 p
10. Manning RB (1961b) A new *Lysiosquilla* (Crustacea: Stomatopoda) from the Gulf of California, with a redescription of *L. decemspinosa* Rathbun. Proc Biol Soc Wash 74:29–35
11. Manning RB (1969) Stomatopod Crustacea of the Western Atlantic. University of Miami, Coral Gables, Florida, 380 p
12. Manning RB (1972a) Eastern Pacific Expeditions of the New York Zoological Society. Stomatopod Crustacea. Zoologica 56:95–113
13. Manning RB (1972b) Three new stomatopod crustaceans of the Family Lysiosquillidae from the eastern Pacific Region. Proc Biol Soc Wash 85:271–278
14. Manning RB (1974) Stomatopods collected by Th. Mortensen in the eastern Pacific region (Crustacea, Stomatopoda). Steenstrupia 3:101–109
15. Manning RB (1977) A monograph of the West African stomatopod Crustacea. Atlantide Report 12:25–181
16. Manning RB (1995) Stomatopod Crustacea of Vietnam: the legacy of Raoul Serène. Crust Res Spec Nr 4:1–339
17. Manning RB, Chace FA (1990) Decapod and stomatopod Crustacea from Ascension Island, South Atlantic Ocean. Smithson Contr Zool 503:1–91
18. Manning RB, Camp DK (1993) Erythrosquilloidea, a new superfamily, and Tetrasquillidae, a new family of stomatopod crustaceans. Proc Biol Soc Wash 106:85–91
19. Manning RB, Reaka ML (1979) Three new stomatopod crustaceans from the Pacific coast of Costa Rica. Proc Biol Soc Wash 92:634–639

Martin JW, Davis GE (2001) An updated classification of the recent Crustacea. Sci Ser Los Angeles County Nat Hist Mus 39:1–124

20. Maurer D, Epifanio CE, Dean HK, Howe S, Vargas JA, Dittel AI, Murillo MM (1984) Benthic invertebrates of a tropical estuary: Gulf of Nicoya, Costa Rica. J Nat Hist 18:47–61
21. Reaka ML, Manning RB (1980) The distributional ecology and zoogeographical relationships of stomatopod Crustacea from Pacific Costa Rica. Smithson Contr Mar Sci 7:1–29
22. Robaina G (1993) Stomatopods. *In*: Cervigon F, Cipriani R, Fischer W, Garibaldi L, Hendrickx M, Lemus AJ, Márquez R, Poutiers JM, Robaina G, Rodríguez B (eds) FAO species identification sheets for fishery purposes. Field guide to the commercial marine and brackish-waters resources of the northern coast of South America. FAO, Roma, pp 103–105

Ruppert EE, Barnes RD (1994) Invertebrate Zoology, 6th ed. Saunder College, Orlando, FL, 1056 p

23. Schmitt WL (1940) The stomatopods of the west coast of America based on collection made by the Allan Hancock Expeditions, 1933–38. Allan Hancock Pac Exped 5:129–225
24. Schotte M, Manning RB (1993) Stomatopod Crustacea from Tobago, West Indies. Proc Biol Soc Wash 106:566–581

Schram FR (1986) Crustacea. Oxford University, Oxford/England, 606 p

25. Vargas R, Cortés J (1997) Biodiversidad marina de Costa Rica: Orden Stomatopoda (Crustacea: Hoplocarida). Rev Biol Trop 45:1531–1539
26. Vargas R, Jesse S, Castro M (1996) List of crustaceans (Decapoda and Stomatopoda) from the Pacific coast of Costa Rica, collected during the RV Victor Hensen Expedition (1993–1994). Rev Biol Trop 44(Suppl 3):97–102
27. Werding B, Müller HG (1990) Estomatópodos (Crustacea: Stomatopoda) de la costa norte de Colombia. Carib J Sci 26:104–121

Part 15
Euphausiids

Iván Castellanos, Eduardo Suárez-Morales, and Álvaro Morales-Ramírez

Stylocheiron maximum, a species reported from both coasts of Costa Rica. (Photo Humberto Bahena)

Abstract Knowledge of the basic aspects of the euphausiid ecology and taxonomy is quite limited on both the Pacific and the Atlantic coasts of Costa Rica. Overall, this group of pelagic crustaceans is poorly known in this tropical region. Currently, up to 20 species of euphausiids have been reported from the Pacific waters of Costa Rica, whereas 14 have been reported from the Costa Rican Caribbean Sea. According to our data, these figures represent 69% and 48% of the total number of

I. Castellanos, E. Suárez-Morales
El Colegio de la Frontera Sur (ECOSUR) Unidad Chetumal, Av. Centenario km 5.5,
A.P. 424, 77000, Chetumal, Quintana Roo, México. e-mail: ivancast@ecosur.mx;
e-mail: esuarez@ecosur.mx

Á. Morales-Ramírez
CIMAR and Escuela de Biología, Universidad de Costa Rica. 11501-2060 San José. Costa Rica.
e-mail: amorales@cariari.ucr.ac.cr

I.S. Wehrtmann, J. Cortés (eds.) *Marine Biodiversity of Costa Rica, Central America*,

species of euphausiids reported from the Eastern Tropical Pacific and the Caribbean Basin, respectively. Most of the zooplankton samples currently available in Costa Rica have been collected in coastal and lagoon systems; this, together with a lack of local specialists are the main reasons why there have been so few reports of these primarily oceanic forms. Most of the reports accounted for herein were generated from the analysis of a few sampling sites in Costa Rican waters. Most of the species reported from Costa Rica are epipelagic forms (18), some of them reach the mesopelagic layers, and only two species are bathypelagic (*Thysanopoda cornuta* and *T. egregia*). The well-known virtue of species of Euphausiacea as oceanographic indicators makes the taxonomic and ecologic study of this group a particularly interesting issue that would allow better knowledge of the zooplankton dynamics in oceanic waters of Costa Rica.

Introduction

Euphausiids are exclusively marine holoplanktonic crustaceans; most of them are epipelagic forms, dwelling the 0–300 m layers in oceanic waters; there is only one species known to have a distributional tendency towards neritic-coastal systems (*Nyctiphanes simplex*). Euphausiids are widely distributed at all latitudes; they tend to be more speciose in the tropical-subtropical regions. After the copepods, the euphausiids are regarded as the most abundant group of crustaceans in the oceanic zooplankton communities. Their numerical abundance in the open tropical oceans often ranges from 2% to 10% of the zooplankton. They occur consistently within this level of abundance in all oceanic areas with the exception of the eastern Canadian Arctic and the Arctic Ocean, where fewer species and smaller numbers occur. However, euphausiids can be overwhelmingly abundant in cold and temperate areas, where they form dense swarms (Mauchline 1980; Antezana & Brinton 1981). Because of this swarming behavior, the density of several species seems to be underestimated; Longhurst (1985) stated that euphausiids represent between 20–40% of the available biomass in the world oceans. At higher latitudes they represent an intermediate link between the primary producers and the upper trophic levels; they are efficient consumers of phytoplankton and suspended particles of organic matter (Brinton 1975), and are part of the diet of fish and whales (Boden *et al.* 1955; Mauchline & Fisher 1969; Antezana 1970; Brinton 1996). Some species are indicators of water masses or of oceanographic conditions.

Account of Surveys on the Euphausiid Fauna of Costa Rica and Adjacent Areas

Euphausiids are among the least known marine zooplankton crustaceans in Costa Rica. The basic aspects of their composition, abundance, distribution, and ecology remain practically unknown. The first reports of this group in Costa Rican waters were generated less than 20 years ago, in the 1980s. A pioneer work was developed in the oceanic waters of the Costa Rican Dome area by Sánchez-Maravilla (1986); this report is exclusively from the analysis of euphausiid juvenile stages (calyptopis and furciliae). On the Caribbean Sea coast the knowledge of the euphausiid fauna is even more limited. In order to have a better understanding about the current status of knowledge of this taxon in Costa Rica and adjacent areas, we divided our account of the literature into two geographic areas, the Pacific coast (Eastern Tropical Pacific), and the Caribbean Sea. For the Pacific coast we included those surveys made from the California Current influence area to the south, including the central coast of the Mexican Tropical Pacific.

Eastern Tropical Pacific. Brinton (1962) provided information about the distribution of 59 species in the Pacific Ocean. Later on, Brinton (1979) studied the influence of the California Current on euphausiid fauna in the Eastern Tropical Pacific. Brinton & Townsend (1980) described the distribution and abundance of nine euphausiid species and their developmental stages in the Gulf of California. The composition of euphausiids and their relation with the oceanic circulation, zones of low oxygen and nutrient concentration, was studied by Reid *et al.* (1978). Gendron (1992) analyzed the spring swarming of *N. simplex* between 1984 and 1992 in the coast of Baja California; it was found that these swarms are related to the reproductive activity. Gómez-Gutiérrez (1996) described the variations of young furciliae in the southwestern coast of Baja California; these stages are more abundant near the coast, in areas with high zooplankton biomass and high concentrations of phytoplankton. In the middle Baja California, Gómez-Gutiérrez *et al.* (1996) and Gómez-Gutiérrez (1995) studied the distribution, abundance, and reproductive cycles of *N. simplex* and other species. Gómez-Gutiérrez & Robinson (1997) observed that *N. simplex, Nematoscelis difficilis, Euphausia eximia, E. pacifica*, and *Thysanoessa spinifera* are the main constituents of diel changes of the macrozooplankton biomass in the western coast of Baja California. Further south, off the coasts of Sinaloa, Sánchez-Osuna & Hendrickx (1984) found *E. lamelligera* and *E. distinguenda* as the most abundant species in the area. Jiménez & Suárez-Morales (1998) analyzed the euphausiids of the surface layers off the central part of the Mexican Pacific coast. Only two species were recorded: *Stylocheiron carinatum* and *E. distinguenda*. The authors suggested that the occurrence of the latter during winter could be associated with the influence of the Equatorial Countercurrent; its occurrence in summer time was considered as an effect of the local influence of the Coastal Current of Costa Rica. Southwards along the same Mexican Pacific coast, López-Cortés (1981, 1990) found that the influence of the Equatorial Countercurrent and the Coastal Current of Costa Rica is revealed in the Gulf of Tehuantepec by the presence of *S. affine* and *E. distinguenda*. Later on, Färber-Lorda *et al.* (1994) observed that the distribution of euphausiids in the Gulf of Tehuantepec is strongly influenced by the hydrographic conditions of this coastal system, and that the strong current associated with the anticyclonic gyre is probably related to the occurrence of *E. distinguenda* in coastal zones. Sameoto *et al.* (1987) studied the distribution of euphausiids at two stations of the Eastern Tropical Pacific and reported 17 species at one site off the coast of Costa Rica, determining *E. eximia* and *E. distinguenda* as being the most abundant species during daytime, plus *E. lamelligera* during the night. This is the most relevant reference for the Pacific euphausiid community of Costa Rica (see Species List 15.2 which is included on the CD-Rom).

Caribbean Sea. One of the earliest surveys on the euphausiid fauna of the western Caribbean Sea is that by Legaré (1961), who reported 14 species for Venezuelan waters including *N. simplex* as a new report for the Atlantic Ocean. It was suggested that the occurrence of this Pacific species in the area is a result of the faunistic transit across the Panama Canal. Later on, Lewis & Fish (1969) found ten species in the insular Caribbean (Barbados), reporting *E. tenera* as the most abundant and widely distributed species in this area.

James (1970) studied samples collected between 1964 and 1969 in the Gulf of Mexico. He provided diagnoses and a key for the identification of 30 species of euphausiids. He also published distribution maps of these species, covering both the southern Gulf of Mexico and part of the western Caribbean Sea (James 1971). These works are a useful reference for the taxonomic study of the euphausiids of the tropical northwestern Atlantic. The euphausiid fauna of the oceanic area of the Mexican Caribbean Sea has been studied recently; Castellanos (1998a) reported *E. tenera, E. americana*, and *S. carinatum* as a group of species representing over 95% of the local euphausiid community. Castellanos (1998b) provided data on the distribution, morphology, and importance of these crustaceans in the Caribbean Sea. Castellanos & Gasca (2002) reported four new faunistic records for the western Caribbean Sea: *Nematobrachion flexipes, S. suhmi, S. affine*, and *S. elongatum*; they provided comments on the absence of euphausiids in the area influenced by local upwelling, one of the most productive zones of the Yucatan shelf. Michel & Foyo (1976) reported 24 species in the entire Caribbean Basin; in their report they included results from a station of the cruise "P6811," made in 1968 in the western Caribbean; this area corresponds to Costa Rican waters. Seven species of euphausiids were reported from this site at depths of 0, 45, and 500 m (see Species List 15.1 which is included on the CD-Rom).

Number of Species Described

The Euphausiacea represents a well-defined group within the Subclass Eucarida. Euphausiids are divided into two families: Bentheuphausiidae, which is monotypic (*Bentheuphausia amblyops*), and Euphausiidae, which contains all the other species known, currently distributed among ten genera (Mauchline 1980). Mauchline (1984) recognized 85 species in the order Euphausiacea. After Mauchline's account, the number increased to 87 because of the reinstating of the taxonomically questioned *S. armatum* on the basis of specimens collected in the Arabian Sea (Mathew 1980), and the description of the deep-living *T. minyops* (Brinton 1987). However, in a more recent taxonomic revision, Baker *et al.* (1990) suggested that *E. similis* var. *armata* is not a species level taxon; hence, the number of euphausiid species considered in our analysis is 86.

The taxonomy of this group has developed into a subspecific level; Brinton (1962, 1975) identified five "forms" of *S. affine* that were separated using their differential geographic distribution and the proportional width of the upper and lower eye lobes. Furthermore, these variations were representative enough to associate each of these forms with distinct patterns related to particular oceanic regions (i.e., California Current, Eastern Equatorial, Western Equatorial, Central Equatorial, and Indo-Australian). A similar intraspecific differentiation was made for three "forms" of *S. longicorne* on the basis of proportional length of the sixth abdominal somite.

Overall, in the waters of Costa Rica we found reports of 29 species of euphausiids. This figure represents almost 34% of the number of species known worldwide. Out of these 29 species, 20 (ca. 69%) have been found in the Pacific waters of Costa Rica and 14 (48%) have been effectively reported from the Costa Rican Caribbean Sea. These numbers are low when compared to the figures reported for the Mexican Pacific (41) and the Mexican Atlantic (34) (see Castellanos & Suárez-Morales 2002). According to the regional reports of Legaré (1961), Lewis & Fish (1969), James (1971), Owre & Foyo (1972), Michel & Foyo (1976), Brinton (1979), and Castellanos (1998a), 30 species have been known to occur in the Caribbean Basin. Therefore, the number of species reported in the Caribbean area of Costa Rica (14) represents almost half the potential reports in this area of the tropical northwestern Atlantic. It is probable that the number of reported species will increase when collections from oceanic areas are analyzed, particularly considering the fact that the major contribution to the number of species in the Costa Rican Pacific (Species List 15.2 which is included on the CD-Rom) resulted from a single sample collected in one oceanic station (Sameoto *et al.* 1987).

Up to 18 species of the euphausiids reported from Costa Rica are considered epipelagic, dwelling mainly in the 0–200 m layer; however, as a result of migratory behavior, several epipelagic forms reach the mesopelagic layers, and only a few are strictly meso or bathypelagic. Two of the recorded species are bathypelagic (*T. cornuta* and *T. egregia*), and only *N. simplex* is known as a coastal species.

E. gibboides, E. tenera, S. carinatum, S. longicorne, and *S. maximum* are species occurring on both coasts of Costa Rica. *N. simplex*, regarded as a Pacific form that crossed over to the Atlantic through the Panama Canal, might be reported from the Caribbean coast of Costa Rica.

Specialists

In general, the marine zooplankton research activities in Costa Rican waters have been developed mainly with a strong emphasis on coastal-neritic areas, which is why the knowledge on the composition, distribution, and abundance of several oceanic taxa such as the euphausiids is still quite limited. Another relevant factor is the lack of local or regional specialists working currently on zooplankton samples from Costa Rica. Therefore, the following list includes colleagues who have analyzed samples from Costa Rica or from the Eastern Tropical Pacific or the Northwestern Tropical Atlantic. Also included are specialists with a worldwide coverage.

Edward Brinton and Annie Townsend: Scripps Institution of Oceanography, University of California, San Diego, USA. Retired (EB) – Worldwide.

Bernadette Casanova: Université de Provence, Marseille, France.

Iván A. Castellanos Osorio: El Colegio de la Frontera Sur (ECOSUR), Chetumal, Quintana Roo, Mexico – Gulf of Mexico and Mexican Caribbean. ivancast@ecosur.mx

Mark J. Gibbons: University of the Western Cape, South Africa – Benguela Current, South Africa.

Jaime Gómez-Gutiérrez: CICIMAR-IPN, La Paz, B.C.S., Mexico – Gulf of California and California Current. jagomezg@ipn.mx

J. Ph. Labat: Observatoire Océanologique, Villefranche-mer Cedex, France – Mediterranean Sea. labat@ccrv.obs-vlf.fr

Bertha A. Lavaniegos-Espejo: CICESE. Ensenada, Baja California, Mexico – Gulf of California and California Current. berlav@cicese.mx

Stephen Nicol: Australian Antarctic Division, Tasmania, Australia – Antarctic Ocean. krill@antidiv.gov.au

Collections

To date there are no collections of Costa Rican euphausiids in Costa Rica. Dr. Manuel Murillo (mmurillo@cariari.ucr.ac.cr), currently at the Universidad de Costa Rica (UCR), put together a collection of euphausiids that has been deposited in the Museo de Zoología at the Universidad de Costa Rica; however, it contains only specimens from the Gulf of California and the coasts of Baja California, Mexico.

A small collection of euphausiids exists in the Universidad Autónoma de Sinaloa, Campus Mazatlan; it includes six species from the area of Sinaloa, northwestern Mexico. The Laboratorio de Invertebrados of the Facultad de Ciencias, UNAM, in Mexico City holds a number of zooplankton samples collected during several cruises carried out in the Costa Rican Dome; the euphausiids contained in these samples remain unstudied. It is expected that the taxonomic analysis of the euphausiids in this particular set of oceanic samples will yield a number of new reports for the zooplankton fauna of the Costa Rican Pacific. The CIMAR, at the Universidad de Costa Rica, has a large collection of zooplankton samples from relatively shallow coastal systems of Costa Rica, mainly from the Pacific coast (Golfo de Nicoya, Golfo Dulce, and Bahía Culebra); a preliminary analysis of these samples by the first author (IC) revealed that these samples contain no adult euphausiids.

Some of the several web sites containing information about these crustaceans are: www.geocities.com/jgomezgu/index.html, built by J. Gómez-Gutiérrez (CICIMAR-IPN, La Paz, Mexico; jagomezg@ipn.mx). This site covers most of the publications related to euphausiids of Mexican waters, and a list of the species recorded in Mexico. Nicol de la Mare (krill@antidiv.gov.au) listed the following sites: www.eco-action.org/dt/timeto.html and www.aad.gov.au/science/AntarcticResearch/AMLR/krill/krill.asp. They are devoted to emphasize the relevance of "krill" in marine trophic webs, especially in the cold seas. Another interesting euphausiid site is www.ecoscope.com/krill4u.htm, by U. Kils (uwekils@aol.com); it deals mainly with the ecology of Antarctic krill and the role of *E. superba*. In www.pac.dfo-mpo.gc.ca/ops/fm/shellfish/Euphausiid/default.htm, different topics related to the biology and ecology of euphausiids are presented, mainly with reference to *E. pacifica*.

Recommendations

Despite the relatively high number of species reported in Costa Rican waters, the current knowledge on the euphausiid fauna can still be described as poor. Most of the species distributions and abundance are merely inferred from surveys in adjacent areas, and nearly 50% of the reports for Costa Rica were based on the analysis of a single oceanic sample. Therefore, surveys covering the far neritic environments and the oceanic realm are urgently needed. The analysis of these samples will yield new reports and will give insight on the distribution and abundance of these crustaceans in Costa Rican waters. The relatively low number of Euphausiacea species and their sharp distributional patterns make them useful in terms of their potential as indicators of oceanographic conditions (Lavaniegos *et al.* 1989; Lavaniegos 1994; Brinton 1996; Biggs *et al.* 1997). For instance, Brinton & Townsend (1980) found that *E. eximia, E. distinguenda, E. tenera, E. lamelligera, E. diomedae, N. gracilis, S. affine*, and *S. carinatum* indicate the influence of waters from the Eastern Tropical Pacific. *E. distinguenda* and *S. affine* are characteristic of the Equatorial Countercurrent (López-Cortés 1990). Recent studies (Gómez-Gutiérrez, *et al.* 1995; Biggs *et al.* 1997; Castellanos & Gasca 1999; Gasca *et al.* 2001) describe the relation of the distribution/abundance of these crustaceans with mesoscale oceanographic processes such as oceanic cyclonic and anticyclonic gyres. In Costa Rica, the knowledge of these aspects of the euphausiid community structure will yield valuable information about different hydrographic processes off Costa Rican coasts.

References

Antezana T (1970) Eufáusidos de la costa de Chile, su rol en la economía del mar. Rev Biol Mar 14:19–27

Antezana T, Brinton E (1981) Euphausiacea. *In*: Boltovskoy D (ed) Atlas del zooplancton sudoccidental y métodos de trabajo con el zooplancton marino. Pub. Esp. INIDEP, Mar del Plata, Argentina, pp 681–698

Baker A de C, Boden BP, Brinton E (1990) A practical guide to the euphausiids of the world. British Museum, Natutral History Publications, London, 96 p

Biggs DC, Zimmerman RA, Gasca R, Suárez-Morales E, Castellanos I, Leben RR (1997) Note on plankton and cold-core rings in the Gulf of Mexico. Fish Bull 95:369–375

Boden BP, Johnson MW, Brinton E (1955) The Euphausiacea (Crustacea) of the North Pacific. Bull Scripps Inst Oceanogr Univ Calif 6:287–400

1. Brinton E (1962) The distribution of Pacific euphausiids. Bull Scripps Inst Oceanogr 8:51–270
Brinton E (1975) Euphausiids of Southeast Asian water. Naga Rep Scripps Inst Oceanogr Univ Calif 4:1–287
2. Brinton E (1979) Parameters relating to the distribution of planktonic organisms, especially euphausiids in the eastern tropical Pacific. Prog Oceanogr 8:125–189
Brinton E (1987) A new abyssal euphausiid, *Thysanopoda minyops*, with comparisons of the eye size, photophores, and associated structures among deep-living species. J Crust Biol 7:636–366
Brinton E (1996) Euphausiacea. Cap. IX. *In*: Gasca R, Suárez E (eds) Introducción al estudio del zooplancton marino. ECOSUR/CONACyT, México, pp 297–342
Brinton E, Townsend AW (1980) Euphausiids in the Gulf of California, the 1957 cruises. CalCOFI Rep. XXI: 211–235
Castellanos I (1998a) Distribución y abundancia de los eufáusidos del estrato superficial del Mar Caribe mexicano. Carib Mar Stud 6:1–11
Castellanos I (1998b) Eufáusidos. Tomo III. *In*: Xacur-Maiza JA (ed) Enciclopedia de Quintana Roo. Chetumal, Q. Roo, Juan A. Xacur-Maiza, México, pp 239–252
Castellanos I, Gasca R (1999) Epipelagic euphausiids (Euphausiacea) and spring mesoscale features in the Gulf of Mexico. Crustaceana 79:391–404
Castellanos I, Gasca R (2002) Eufáusidos (Crustacea: Malacostraca) del centro y sur de Mar Caribe mexicano. Rev Biol Trop 50:77–85
Castellanos I, Suárez-Morales E (2002) Euphausiacea. Cap. 19. *In*: Lorente-Bousquets J, Morrone JJ (eds) Biodiversidad, taxonomía y biogeografía de artrópodos de México: hacia una síntesis de su conocimiento. Vol. III. CONABIO/ECOSUR/Bayer/UNAM, México, pp 401–414
Färber-Lorda J, Lavin MF, Zapatero MA, Robles JM (1994) Distribution and abundance of euphausiids in the Gulf of Tehuantepec during wind forcing. Deep-Sea Res I 41:359–367
Gasca R, Castellanos I, Biggs D (2001) Euphausiids (Crustacea, Euphausiacea) and summer mesoscale features in the Gulf of Mexico. Bull Mar Sci 68:397–408
Gendron D (1992) Population structure of daytime surface swarms of *Nyctiphanes simplex* (Crustacea: Euphausiacea) in the Gulf of California, Mexico. Mar Ecol Prog Ser 87:1–6
Gómez-Gutiérrez J (1995) Distribution patterns, abundance and population dynamics of the euphausiids *Nyctiphanes simplex* and *Euphausia eximia* off the west coast of Baja California, Mexico. Mar Ecol Prog Ser 119:63–76
Gómez-Gutiérrez J (1996) Ecology of early larval development of *Nyctiphanes simplex* Hansen (Euphausiacea) off the southwest coast of Baja California, Mexico. Bull Mar Sci 58:131–146
Gómez-Gutiérrez J, Robinson CJ (1997) Circadian biomass and abundance changes of five euphausiid species along the west coast of Baja California Mexico, December 1993. Sci Mar 61:27–35
Gómez-Gutiérrez J, Palomares-García R, Gendron D (1995) Community structure of the euphausiid populations along the west coast of Baja California, Mexico, during the weak ENSO 1986–1987. Mar Ecol Prog Ser 120:41–51
James BM (1970) Euphausiacean Crustacea. *In*: Pequegnat WE, Chace FA Jr. (eds) Contributions on the Biology of the Gulf of Mexico. Texas A&M Univ. Oceanogr. Studies 1, pp 205–229
James BM (1971) Gulf of Mexico deep sea fauna: Decapoda and Euphausiacea. Folio 20. *In*: Pequegna, WE, Pequegnat LH, Firth RW, James BM, Terrell R (eds) Serial Atlas of the Marine Environment. American Geographic Society, New York, 24 p
Jiménez AM. y E. Suárez-Morales (1998) Eufáusidos (Crustacea) del estrato superficial frente a Michoacán, México, durante dos periodos estacionales. Univ Cienc 14:7–16
Lavaniegos BE (1994) Dispersion and development patterns in larvae of *Nyctiphanes simplex* (Euphausiacea) in the upwelling region off Baja California. Mar Ecol Prog Ser 106:207–225
Lavaniegos-Espejo BE, Lara-Lara JR, Brinton E (1989) Effects of the 1982–83 El Niño event on the euphausiid populations of the Gulf of California. CalCOFI Rep 30:73–87
Legaré JEH (1961) Algunos eufausiáceos del golfo de Paria, golfo de Cariaco y delta del Orinoco, al oriente de Venezuela. Bol Inst Ocean Univ Oriente 1:131–148

Lewis JB, Fish AG (1969) Seasonal variation of the zooplankton fauna on surface waters entering the Caribbean Sea at Barbados. Carib J Sci 9:1–23

Longhurst AL (1985) Relationship between diversity and the vertical structure of the upper ocean. Deep-Sea Res 32:1535–1570

López-Cortés DJ (1981) Algunos aspectos de la distribución y taxonomía de los eufáusidos del Golfo de Tehuantepec, México. *In*: VII Simp. Latinoamer. Oceanog. Biol. Acapulco, Guerrero, México, Noviembre, pp 277–297

López-Cortés DJ (1990) Distribución de la Familia Euphausidae (Euphausiacea: Crustacea) en el Golfo de Tehuantepec, México. Rev Biol Trop 38:21–28

Mathew KJ (1980) Taxonomic status of *Stylocheiron armatum* Colosi, 1917 (Crustacea: Euphausiacea) – a critical evaluation. J Mar Biol Assoc India 22:50–56

3. Mauchline J (1980) The biology of euphausiids. Adv Mar Biol 18:73–623

Mauchline J (1984) Euphausiid, stomatopod and leptostracan crustaceans. No. 30. In: Kermack DM, Barnes RSK (eds) Synopses of the British Fauna (New Series). London, pp 4–53

Mauchline J, Fisher LR (1969) The biology of euphausiids. Adv Mar Biol 7:1–454

4. Michel HB, Foyo M (1976) Caribbean zooplankton. Part 1. Siphonophora, Heteropoda, Copepoda, Euphausiacea, Chaetognatha and Salpidae. Office of Naval Research, Department of Navy, USA, 549 p

Owre HB, Foyo M (1972) Studies on Caribbean zooplankton. Description of the program and results of the first cruise. Bull Mar Sci 22:483–521

5. Reid JL, Brinton E, Fleminger A, Venrick E, McGowan J (1978) Ocean circulation and marine life. *In*: Charnock H, Deacon Sir G (eds) Advances in Oceanography. Plenum, New York, pp 65–130

6. Sameoto D, Guglielmo L, Lewis MK (1987) Day/night vertical distribution of euphausiids in the western tropical Pacific. Mar Biol 96:235–245

Sánchez-Maravilla EE (1986) Distribución y abundancia de las larvas de crustáceos en la región del Domo de Costa Rica. Tesis de Licenciatura, Fac. Ciencias, UNAM, 99 p

Sánchez-Osuna L, Hendrickx ME (1984) Resultados de las campañas *SIPCO* (Sur de Sinaloa, México) a bordo del B/O "El Puma". Abundancia y distribución de los Euphausiacea (Crustacea: Eucarida). An Inst Cienc Mar Limnol UNAM 11:99–106

Part 16
Decapod Crustaceans

Rita Vargas and Ingo S. Wehrtmann

Solenocera agassizi, locally called "camarón fidel," from Pacific Costa Rica (Photo: Ingo S. Wehrtmann)

R. Vargas
Museo de Zoología, Escuela de Biología
Universidad de Costa Rica, 11501-2060 San José, Costa Rica
e-mail: ritav@biologia.ucr.ac.cr

I.S. Wehrtmann
Unidad Investigación Pesquera y Acuicultura (UNIP) of the Centro de Investigación en Ciencias del Mar y Limnología (CIMAR), and Museo de Zoología, Escuela de Biologia,
Universidad de Costa Rica, 11501-2060 San José, Costa Rica.
e-mail: ingowehrtmann@gmx.de

I.S. Wehrtmann, J. Cortés (eds.) *Marine Biodiversity of Costa Rica, Central America*,

Abstract We report the presence of a total of 549 decapod species for Costa Rican marine and coastal waters, including eight species not mentioned previously for Costa Rica. Species number was substantially higher for the Pacific (438 spp.) compared to the Caribbean coast (119 spp). Brachyura (253 spp.), Anomura (116 spp.), and Caridea (114 spp.) comprised the highest number of species. Pacific Costa Rica supports 55% and 52% of Decapoda and Brachyura, respectively, known to occur in the Panamic Province. Species diversity in the Caribbean is more difficult to compare due to the lack of adequate and updated information.

Introduction

The name "Decapoda" refers to the principal diagnostic feature of the representatives of this order: the presence of five pairs of legs (pereiopods). Not all of these five pairs are necessarily used for walking; e.g., the first pair of pereiopods may be modified as large claws (as in alpheid shrimp [Fig. 16.1] and some lobsters), or the last pair is paddle-shaped and used for swimming as in some portunid crabs. Moreover, decapod crustaceans are characterized by the development of a carapace, which encloses the branchial chambers, and by the modification of the first three thoracopods as maxillipeds (McLaughlin 1980).

Decapod crustaceans comprise roughly 10,000 species (Gruner 1993), which occupy a great diversity of marine, freshwater, and semiterrestrial habitats. They are distributed from the tropics to the Arctic and Antarctic. A vast majority (roughly 90%) inhabit marine environments and adjacent brackish waters (Gruner 1993).

Many decapod species are commercially exploited by coastal and offshore fisheries (Fig. 16.2), and these natural resources may play an important role for the economics of the countries involved in such fisheries or aquaculture efforts (Caddy

Fig. 16.1 The snapping shrimp *Alpheus nuttingi*, collected in Cahuita, Caribbean coast of Costa Rica (Photo: Arthur Anker)

Fig. 16.2 A typical deepwater shrimp catch along the Pacific coast of Costa Rica, dominated by the pandalid shrimp *Heterocarpus vicarius* (Photo: Ingo S. Wehrtmann)

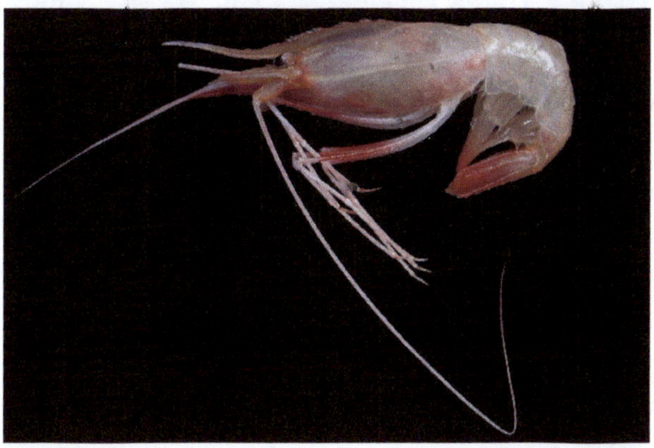

Fig. 16.3 *Heterocarpus vicarius* ("camarón camello"), a commercially exploited deepwater pandalid shrimp from Pacific Costa Rica (Photo: Ingo S. Wehrtmann)

1989). Commercially exploited species include many shrimps and prawns (Penaeidae, Palaemonidae, Pandalidae Fig. 16.3, and Crangonidae), representatives of king crabs (Lithodidae), different types of lobsters (Nephropidae, Palinuridae, and Scyllaridae), and large brachyuran crabs (e.g., *Cancer*, Fig. 16.4, *Callinectes*, and *Chionectes*) (Holthuis 1980a, 1991).

Most decapods are benthic species with a complex life cycle, where larvae hatch from the eggs typically carried by the female during the incubation period under its abdomen (Fig. 16.5). An exception are members of the suborder Dendrobranchiata (Penaeoidea and Sergestoidea), where females release the newly produced eggs directly into the water column. The pelagic larvae differ entirely in both morphology and habits from juvenile and adult conspecifics (for review see Anger 2001). Although most decapods pass through a number of larval stages (nauplius in Dendrobranchiata, zoea, and decapodid), abbreviated development (reduction of the pelagic larval phase) has been observed in many taxa of these crustaceans (Rabalais & Gore 1985).

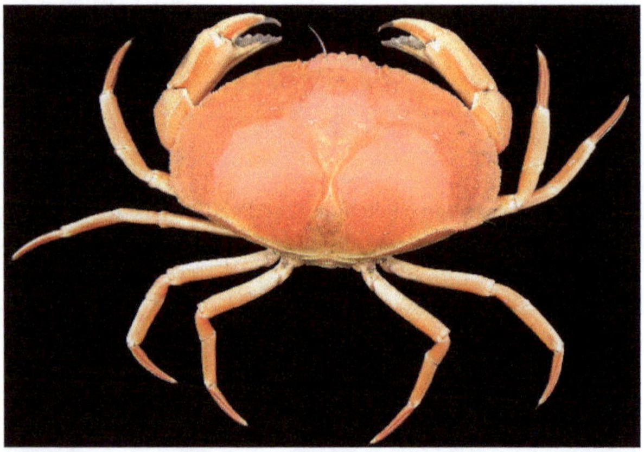

Fig. 16.4 *Cancer johngarthi*, a common bycatch-species of the deepwater shrimp fishery, Pacific Costa Rica (Photo: Ingo S. Wehrtmann)

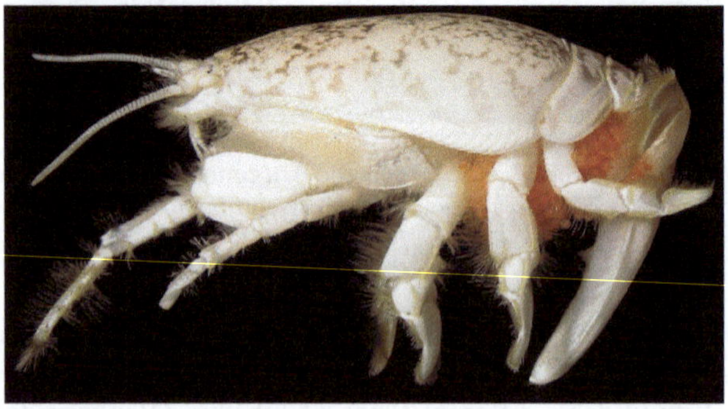

Fig. 16.5 Ovigerous female of *Hippa cubensis*, collected in Parque Nacional Cahuita, Caribbean coast of Costa Rica (Photo: Andrea Bernecker)

Expeditions Visiting Costa Rica

Since the end of the 19th century, several important expeditions (e.g., US Fish Commission Expedition 1891; California Academic Science expedition of 1921 and 1924; New York Zoological Society expeditions in 1937 and 1938; Allan Hancock Foundation's expeditions between 1931 and 1954) visited Costa Rica and collected, among others, marine decapods. Moreover, a considerable number of scientific expeditions stopped, on their way to Galapagos Island, at Isla del Coco, located approximately 600 km off the Pacific coast of Costa Rica (see Hertlein 1963; Hogue & Miller 1981). In 1972, the research vessel "Searcher" of the Janss Foundation operated in Pacific Costa Rica, and the results of this effort contributed considerably to our understanding of the marine fauna of this geographic area (e.g., Reaka & Manning 1980). During 1979–1982, the RV "Skimmer" of the University of Delaware, USA, operated in Pacific Costa Rica, and some results concerning decapods have been presented by De Vries *et al.* (1983), Epifanio & Dittel (1984), and Maurer *et al.* (1984). More recently, the German RV "Victor Hensen" visited Costa Rica in the frame of a joint effort between the "Zentrum für Marine Tropenökologie" (ZMT, Bremen) and the "Centro de Investigación en Ciencias del Mar y Limnología" (CIMAR, Universidad de Costa Rica, San José). This research program attempted to establish the scientific basis for conservation and sustainable management of Pacific coastal areas of Costa Rica, with emphasis in the Golfo Dulce region (Vargas & Wolff 1996). Information on the decapod diversity encountered during the "Victor Hensen"-cruise was published by Vargas *et al.* (1996) and Jesse (1996). As far as we know, the only recent scientific cruise visiting the Caribbean waters of Costa Rica was the RV John Elliott Pillsbury from the University of Miami in January–February 1971 (Voss 1971). They spent 2 days in Costa Rican waters (26–27 January 1971), collecting during this period a total of 11 samples. With the exception of porcellanid crabs (Gore 1974), to our knowledge the results regarding other decapod crustaceans have not been published (Voss [1971] is a narrative of the cruise). However, we included the name of the collected species in our species lists, because Dr. L.B. Holthuis, an international authority in decapod taxonomy, was responsible for the identification of the species during the above-mentioned cruise.

Materials and Methods

The present species lists of the decapod crustaceans so far reported from Costa Rica are based upon a review of the primary literature up to January 2006. Unpublished thesis and project reports are not included, except the cruise report of RV John Elliott Pillsbury (Voss 1971). In eight cases (see section "Results"), we present new and yet unpublished records of these species for Costa Rica; specimens of these new records have been deposited in the Museo de Zoología, Universidad de Costa Rica. The squad lobster *Munidopsis albatrossae* (Anomura, Galatheidae) was included in our list despite the fact that the coordinates indicated in the publication

(Pequegnat & Pequegnat 1973) refer to waters close to the border between Nicaragua and Costa Rica. Due to the proximity, we assume, however, that this species is likely to occur also in Costa Rican waters. We hope that these species lists are reliable indicators of the current status of our knowledge concerning the taxonomic composition of the decapod fauna along both Caribbean and Pacific coasts of Costa Rica. The classification used for the preparation of the tables follows that of the recent update presented by Martin and Davis (2001). To facilitate a rapid orientation of the reader, the species lists were split in three major groups: (1) Shrimps and lobsters: Suborder Dendrobrachiata (Penaeoidea, Sergestoidea); Suborder Pleocyemata with the infraorders Stenopodidea, Caridea, Thalassinidea, and Palinura; (2) Infraorder Anomura; and (3) Infraoder Brachyura.

Results

The following species represent new records for Costa Rica: *Munida flinti* Benedict, 1902, *Gecarcinus planatus* Stimpson, 1898; *Percnon gibbesi* Milne Edwards, 1853; *Haplocarcinus marsupialis* Stimpson, 1859; *Persephona punctata punctata* Linnaeus, 1758, *Mithrax pilosus* Rathbun, 1892; *Xanthus denticulatus* White, 1848; *Uca mordax* (Smith, 1870).

Including these new records, a total of 549 marine decapod species are reported from Costa Rica (Species Lists 16.1–16.6 are included on the CD-Rom). Table 16.1 provides an overview concerning the number of species per systematic unit and region (Pacific and Caribbean). The largest group is Brachyura with a total of 253 species, followed by Anomura (116 spp.) and Caridea (114 spp.). The other systematic units are comprised of 1–34 species. The most important decapod families in Costa Rica in terms of species number (≥40 spp.) are Majidae (58 spp.), Porcellanidae (54 spp.), Xanthidae (50 spp.), and Alpheidae (42 spp.). Species number from the Pacific coast is almost four times higher compared to that from the Atlantic side of Costa Rica (437 vs 119 spp.). With the exception of Stenopodidea, species number is generally substantially higher at the Pacific. A total of 22 families have been reported exclusively from the Pacific coast, while representatives of four families occur at the Caribbean but not at the Pacific coasts. These four families are represented by Dendrobranchiata (Luciferidae), Stenopodidea (Stenopodidae), and Caridea (Rhynchocinetidae, Ogyrididae). A total of ten species are reported to occur at the Pacific as well as at the Caribbean coast of Costa Rica: two species belong to Dendrobranchiata, three species to Caridea, two species to Anomura, and three species to Brachyura. All superfamilies with species of Dendrobranchiata, Stenopodidea, and Palinura inhabiting marine waters are represented by at least one species from Costa Rica. In contrast, so far no representatives of the superfamilies Procaridoidea, Galatheacaridoidea, Bresilioidea, Psalidopodoidea, Stylodactyloidea, Campylonotoidea, Physetocaridoidea (all Caridea), Thalassinoidea (Thalassinidea), Lomisoidea (Anomura), and the brachyuran superfamilies Homalodromioidea, Homoloidea, Cyclodorippoidea,

Table 16.1 Number of decapod species of different systematic units reported from the Pacific and Caribbean of Costa Rica, and species which are represented along both coasts

	Caribbean	Pacific	Pacific + Caribbean	Total species
Dendrobranchiata	10	29	2	37
Stenopodidea	2	0	0	2
Caridea	31	85	2	114
Astacidea	0	1	0	1
Thalassinidea	3	14	0	17
Palinura	4	5	0	9
Anomura	19	99	2	116
Brachyura	50	205	2	253
Total	**119**	**437**	**8**	**548**

Hymenosomatoidea, Retroplumoidea, Bythograeoidea, and Bellioidea have been recorded from Costa Rican waters.

The most important families in terms of species number (≥5 species per family) inhabiting waters of the Caribbean of Costa Rica are Majidae (15 spp.), Porcellanidae (10 spp.), Portunidae (10 spp.), Palaemonidae (7 spp.), Alpheidae (8 spp.), Hippolytidae (8 spp.), Penaeidae (6 spp.), Calappidae (5 spp.), and Xanthidae (5 spp.). Regarding the Pacific, the following families comprise at least 15 species reported from Costa Rican waters: Xanthidae (45 spp.), Porcellanidae (44 spp.), Majidae (43 spp.), Alpheidae (34 spp.), Ocypodidae (28 spp.), Palaemonidae (23 spp.), and Paguridae (17 spp.).

No brachyuran crab is endemic to Costa Rican waters, whereas at least a total of six species of Thalassinidea (Upogebiidae: *Pomatogebia cocosia*) and Anomura (Diogenidae: *Allodardanus rugosus, Cancellus tanneri, Paguristes fecundus*; Paguridae: *Enallopaguropsis janetae*; Porcellanidae: *Pachycheles cocoensis*) are endemic to the Pacific of Costa Rica. Three additional species (Palaemonidae: *Periclimenes murcielagensis*; Upogebiidae: *Upogebia vargasae, U. cortesi*) have been described recently (Williams 1997; Vargas 2000; Williams & Vargas 2000), and so far, they are known only from Costa Rican waters. Since we assumed that the actual distribution of these species might not be restricted to its type localities, we did not treat them as endemic species.

Discussion

Eighty percent of the total number of decapod species reported from Costa Rica (548 spp.) inhabit the Pacific (Table 16.1). This result seems to indicate that the Pacific side of Costa Rica is substantially richer in species compared to the Caribbean. Such a conclusion, however, is misleading and premature, because the sampling effort and the regional coverage of such activities have been much more intense and broader along the Pacific coast of Costa Rica. Not surprisingly, even small and locally restricted surveys at the Caribbean coast of Costa Rica can result in the discovery of a considerable number of decapod species new for Costa Rica (Wehrtmann & Vargas 2003). Moreover, there is an almost complete lack of offshore and deepwater samples, which means that the reported decapod fauna represents almost exclusively the shallow water area of the Caribbean of Costa Rica. Therefore, more collections along and below the continental slope would certainly improve our knowledge on the decapod fauna of the region. We are convinced that a greater sampling effort in the Caribbean will increase the number of decapod species, thus diminishing the substantial difference between species numbers of the Pacific and the Caribbean.

Several authors published information on the decapod fauna of the eastern tropical Pacific (Table 16.2), which serves us to compare our numbers with those from the region. Boschi (2000) estimated the total number of decapods of the entire Panamic Province to be 793 species. Here we report the presence of 438 species from Costa Rica, which represents 55% of the above-mentioned value. A similar situation occurs regarding the brachyuran crabs: 390 species in the Panamic Province (Boschi 2000) versus 205 species or 52% in Costa Rica (Table 16.2). Considering that the coast of Costa Rica covers just a small part of the Panamic Province (from Mexico, Bahía Magdalena, to northern Peru, Bahía Sechura; Boschi 2000), the decapod diversity along the Pacific coast of Costa Rica is surprisingly high and may be attributed to two factors: (1) the Pacific coast of Costa Rica is physically and ecologically quite diverse (see Cortés & Wehrtmann, this volume), and (2) upwelling zones in northern Costa Rica as well as the oceanic Isla del Coco allow the inclusion of other than purely tropical coastal faunal elements, thus enriching the decapod diversity.

The checklist of decapods from Pacific Colombia (Lemaitre & Alvarez León 1992) permits a comparison of species numbers (per suborder and infraorders) with those from Costa Rica (Table 16.2). Interestingly, numbers are similar, in some infraorders even identical. Overall, there are 59 species more in Costa Rica than in Colombia, and the discrepancy is mainly due to the difference in anomuran species: 61 and 99 species in Colombia and Costa Rica, respectively. However, according to Lemaitre & Alvarez León (1992), the decapod fauna of the Pacific coast of Colombia is one of the least known in the tropical eastern Pacific.

Published information concerning the decapod species diversity of mayor systematic units of the Caribbean is sparse. Chace (1972) provided a summary of shrimps collected during the Smithsonian-Bredin Caribbean expeditions. Overall, he mentioned 218 species of Penaeoidea, Sergestoidea, Stenopodidea, and Caridea for the Caribbean. In Costa Rica, 42 species have been reported, representing only 19% of the above-mentioned total of species. Brachyuran species number of Caribbean Costa Rica may be compared with those from the Gulf of Mexico: Powers (1977) reported the presence of 352 species belonging to 22 families, while 50 species or 14% have been documented for the Caribbean coast of Costa Rica (Table 16.1). Both lists (Chace 1972; Powers 1977) are approximately 30 years old, and thus neither includes descriptions of new species nor recent reports on the

Table 16.2 Number of decapod species reported from different regions of the eastern Pacific

Author	Region	Decapoda	Dendro-branchiata	Stenopodidea	Caridea	Astacidea	Thalassinidea	Palinura	Anomura	Brachyura
Wicksten & Hendrickx (2003)	ETP	–	40	5	203	–	–	–	–	–
Hendrickx (1995)	ETP	–	–	–	–	1	32	10	–	–
Lemaitre & Ramos (1992)	Pacific Colombia	–	–	–	–	–	47	–	–	–
Hendrickx & Harvey (1999)	ETP	–	–	–	–	–	–	–	207	–
Boschi (2000)	Panamic Province	793	–	–	–	–	–	–	–	390
Lemaitre & Alavarez León (1992)	Colombia	378	18	0	79	–	13	3	61	201
Present study	Costa Rica	438	29	0	85	1	14	5	99	205

ETP = eastern tropical Pacific

presence of decapod species in the region. As a consequence, the above-mentioned percentages for shrimps and crabs from Costa Rican waters will further diminish when including the recent literature (see also Vargas & Cortés 1999a, b).

Recommendations

One major limitation for the study of the decapod fauna is the fact that just a few people in the region are working in the taxonomy of these interesting animals (see section "Specialists"). More efforts should be directed towards the training of Central American students and other professionals in decapod taxonomy. A regional network would be helpful to organize workshops and courses, for example, to introduce people from different sectors (e.g., students and teachers from colleges and universities, and also employees from institutions related to the management of marine coastal areas) to the identification of the principal decapod groups. Moreover, regional research projects certainly will be helpful to obtain a more complete and integrated picture of the actual diversity of decapods in the area. International funding agencies might be willing to support such a regional approach.

As a result of the obvious difficulties associated with the purchase and the maintenance of a research vessel, almost all Central American countries (exception: Smithsonian Tropical Research Institute in Panama) lack such an important platform necessary for the study of the marine realm. We suggest encouraging the collaboration with international institutions, which are planning to send a research vessel into the Pacific and/or Caribbean region. Even a small number of samples might contribute considerably to our knowledge on the decapod fauna of the region. Furthermore, special attention should be paid to assure that such international research cruises will deposit duplicates in the official national collections (in the case of Costa Rica: Museo Nacional and Museo de Zoología, Universidad de Costa Rica).

Specialists

Here we list experts working in the taxonomy of Costa Rican and/or Central American decapods.

ARTHUR ANKER: Smithsonian Tropical Research Institute, Panama – Alpheidae; worldwide. AnkerA@si.edu

DARRYL FELDER: University of Louisiana at Lafayette, USA – Thalassinidea; worldwide. dlf4517@louisiana.edu

MICHEL HENDRICKX: Instituto de Ciencias del Mar y Limnología, Universidad Nacional Autónoma de México, Mazatlán, Sinaloa; Mexico – Caridea and Brachyura; worldwide. michel@ola.icmyl.unam.mx

RAFAEL LEMAITRE: Smithsonian Institution, Washington, DC; USA – Anomura; worldwide. Lemaitre.Rafael@NMNH.SI.EDU

Joel Martin: Invertebrate Zoology Section (Crustacea), Research & Collections Branch, Natural History Museum of Los Angeles County, Los Angeles, California, USA – Caridea and Brachyura; worldwide. jmartin@nhm.org

Gabriel Ramos: Facultad de Ciencias, Biología Marina, Universidad del Valle, Cali; Colombia – Caridea; America. geramos@hypatia.univalle.edu.co

Rubén Ríos: School of Marine Science, Virginia Institute of Marine Science, The College of William and Mary, Gloucester Point, Virginia, USA – Caridea (especially Caribbean *Synalpheus*); Latin America. rubnrios@vims.edu

Bernd Werding: Institut für Allgemeine und Spezielle Zoologie, Justus-Liebig-Univeristät Giessen, Germany – Porcellanidae; worldwide. bernd.werding@allzool.bio.uni-giessen.de

Mary Wicksten: Department of Biology, Texas A&M University, College Station, Texas, USA – Caridea; worldwide. wicksten@mail.bio.tamu.edu

Collections

The Museo de Zoología, Universidad de Costa Rica, maintains the most complete decapod collection of Costa Rica, comprising around 190 genera with individuals representing a total of 610 species. The following institutions house important collections of decapod species collected in Costa Rican waters:

Los Angeles County Museum: Los Angeles, California, USA – Eastern Pacific (including former Allan Hancock collection); http://crustacea.nhm.org/collection/

National Museum of Natural History: Smithsonian Institution, Washington, DC, USA – Worldwide (type material of Costa Rican decapods); http://www.mnh.si.edu/rc/

Marine Invertebrate Museum: Rosenstiel School of Marine and Atmospheric Science, Miami, Florida, USA – Tropical Atlantic (Caribbean); http://www.rsmas.miami.edu/divs/mbf/invert-museum.html

References

1. Abele LG (1992) A review of the grapsid crab genus *Sesarma* (Crustacea: Decapoda: Grapsidae) in America, with description of a new genus. Smithson Contr Zool 527:1–59

1b. Abele LG, Kim W (1984) Notes on the freshwater shrimps of Isla del Coco with the description of *Macrobrachium cocoense*, new species. Proc Biol Soc Wash 97:951–960

2. Abele LG, Kim W (1986) An illustrated guide to the marine decapod crustaceans of Florida. Dept Environ Reg Tech Series, Florida 8:1–326

3. Abele LG, Kim W (1989) The decapod crustaceans of the Panama Canal. Smithson Contr Zool 482:1–50

Anger K (2001) The Biology of Decapod Crustacean Larvae. *In*: Vonk R (ed) Crustacean Issues 14. A.A. Balkema, Lisse, The Netherlands, 419 p

3b. Baba K 2005. Deep-sea chirostylid and galatheid crustaceans (Decapoda: Anomura) from the Indo-Pacific, with a list of species. Galathea Rep. 20: 1–317.

4. Ball EE (1972) Observation on the biology of the hermit crab, *Coenobita compressus* H. Milne Edwards (Decapoda, Anomura) on the west coast of the Americas. Rev Biol Trop 20:265–273
5. Ball EE, Haig J (1974) Hermit crabs from the tropical eastern Pacific. I. Distribution, color and natural history of some common shallow-water species. Bull South Calif Acad Sci 73:95–104
6. Benedict JE (1902) Description of a new genus and forty-six new species of crustaceans of the family Galatheidae, with a list of the known marine species. Proc US Nat Mus 26:243–334
7. Biffar TA (1972) A study of the eastern Pacific representatives of the genus *Callianassa* (Crustacea, Decapoda, Callianassidae). Ph.D. dissertation, University of Miami, Florida, 281 p
8. Boone L (1926) A new family of Crustacea. Preliminary technical description. Zool Soc Bull 29:73
9. Boone L (1930) Scientific results of the cruises of the yachts "Eagle" and "Ara", 1921–1928, William K. Vanderbilt, commanding. Crustacea: Anomura, Macrura, Schizopoda, Isopoda, Amphipoda, Mysidacea, Cirripedia and Copepoda. Bull Vanderbilt Mar Mus 3:1–221
10. Boone L (1932) The littoral crustacean fauna of the Galapagos Islands. Part II. Anomura. Zoologica 14:1–62

Boschi EE (2000) Biodiversity of marine decapod brachyurans of the Americas. J Crust Biol 20:337–342

11. Boyko CB (2002) A worldwide revision of recent and fossil sand crabs of the Albuneidae Stimpson and Blepharipodidae, new family (Crustacea: Decapoda: Hippoidea). Bull Am Mus Nat Hist 272:1–396
12. Bright DB (1966) Land crabs of Costa Rica. Rev Biol Trop 14:183–203
13. Bright DB, Hogue CL (1972) A synopsis of the burrowing land crabs of the world and list of their arthropod symbionts and burrow associates. Contrib Sci 220:1–58
14. Cabrera PJ, Y Solano L (1996) Tamaños y frecuencias de *Pontonia margarita* (Crustacea: Palaemonidae) asociadas a *Pinctada mazatlanica* (Bivalvia: Pteridae), Costa Rica. Rev Biol Trop 44:915–917

Caddy JF (1989) Marine Invertebrate Fisheries, Their Assessment & Management. Wiley, New York, 752 p

16. Campos E (1993) Systematics and taxonomic remarks on *Pinnotheres muliniarum* Rathbun, 1918 (Crustacea: Brachyura: Pinnotheridae). Proc Biol Soc Wash 106:92–101
17. Campos E, Griffith H (1990) *Clypeasterophilus*, a new genus to receive the small-palped species of the *Dissodactylus* complex (Brachyura: Pinnotheridae). J Crust Biol 10:550–553
18. Campos N, Lemaitre R (1994) A new *Calcinus* (Decapoda: Anomura: Diogenidae) from the tropical western Atlantic and a comparison with other species of the genus from the region. Proc Biol Soc Wash 107:137–150
19. Campos E, Wicksten MK (1997) A new genus for the Central American crab *Pinnixa costaricana* Wicksten, 1982 (Crustacea: Brachyura: Pinnotheridae). Proc Biol Soc Wash 110:69–73
20. Campos E, Díaz V, Gamboa-Contreras JA (1998) Notes on distribution and taxonomy of five poorly known species of pinnotherid crabs from the eastern Pacific (Crustacea: Brachyura: Pinnotheridae). Proc Biol Soc Wash 111:372–381
21. Castaing A, Jiménez JM, Rojas J (1982) Asociación simbiótica entre el cangrejo *Pinnotheres* sp. indt. (Crustacea, Pinnotheridae) y el molusco *Geloina inflata* (Phillipi) (Pelecypoda, Corbiculidae), en Costa Rica. Brenesia 19/20:553–562
22. Castro P (1996) Eastern Pacific species of *Trapezia* (Crustacea, Brachyura: Trapeziidae), sibling species symbiotic with reef corals. Bull Mar Sci 58:531–554

22b. Cervigon F, Cipriani R, Fisher W, Garibaldi L, Hendrickx M, Lemus AJ, Márquez R, Poutiers JM, Robaina G, Rodríguez B (1993) FAO Species Identification Sheets for Fishery Purposes. Field Guide to the Commercial Marine and Brackish Waters Resources of the Northern Coast of South America. FAO, Rome, 513 p

23. Chace FA (1942) Reports on the scientific results of the Atlantis Expedition to the West Indies, under the joint auspices of the University of Havana and Harvard University. Torreia 11:1–105
24. Chace FA (1962) The non-brachyuran decapod crustaceans of Clipperton Island. Proc US Nat Mus 113:605–635
25. Chace FA (1972) The shrimps of the Smithsonian-Bredin Caribbean Expeditions with a summary of the West Indian shallow-water species (Crustacea: Decapoda: Natantia). Smithson Contr Zool 98:1–179
26. Chace FA (1984) The caridean shrimps (Crustacea: Decapoda) of the *Albatros* Philippine Expedition, 1907–1910, Part. 2: Families Glyphocrangonidae and Crangonidae. Smithson Contr Zool 397:1–63

26b. Chace FA Jr, HH Hobbs Jr (1969) The freshwater and terrestrial decapod crustaceans of the West Indies with special reference to Dominica. US Natl Mus Bull 292:1–258

27. Christoffersen ML (1979) Decapod Crustacea: Alpheoidea. Résultats Scientifiques des Campagnes de la "Calypso" Fascicule II. Annales Ser 55:297–377
28. Cortés J (1985) Preliminary observations of *Alpheus simus* Guérin-Méneville, 1856 (Crustacea: Alpheidae): a little-known Caribbean bioeroder. Proc 5th Int Coral Reef Congr, Tahiti 5:351–353
29. Cortés J (1991) Los arrecifes coralinos del Golfo Dulce, Costa Rica: aspectos geológicos. Rev Geol Amér Central 3:15–24
30. Crane J (1947) Eastern Pacific Expeditions of the New York Zoological Society XXXVIII. Intertidal brachygnathous crabs from the west coast of tropical America with special reference to ecology. Zoologica 32:69–95
31. Crane J (1975) Fiddler Crabs of the World. Ocypodidae: genus *Uca*. Princeton University Press, Princeton, NJ, 738 p

De Vries MC, Epifanio CE, Dittel AI (1983) Lunar rhythms in egg hatching of the subtidal crustacean: *Callinectes arcuatus* Ordway (Decapoda: Brachyura). Estuar Coast Shelf Sci 17:717–724

32. Dexter DM (1974) Sandy-beach fauna of the Pacific and Atlantic coast of Costa Rica and Colombia. Rev Biol Trop 22:51–66
33. Effort IE (1971) The species of sand crabs in the genus *Lepidopa* (Decapoda: Albuneidae). Zool Anz Leipzig 186:59–102
34. Efford IE (1973) The distribution of the sand crabs, *Hippa strigillata* (Stimpson) and *Hippa pacifica* (Dana) in the eastern Pacific Ocean (Decapoda: Anomura). Crustaceana 23:119–122
35. Efford IE (1976) Distribution of the sand crabs in the genus *Emerita* (Decapoda: Hippidae). Crustaceana 30:169–183

Epifanio CE, Dittel AI (1984) Seasonal abundance of brachyuran crab larvae in a tropical estuary: Gulf of Nicoya, Costa Rica, Central America. Estuaries 7(4B):501–505

36. Faxon W (1893) Preliminary descriptions of new species of Crustacea. Bull Mus Comp Zool 24:149–220
37. Faxon W (1895) Reports on an exploration off the west coast of Mexico, Central and South America, and off the Galapagos Islands by the U.S. Fish Commission Steamer "Albatros", during 1891. XV. The stalk-eyed Crustacea. Mem Mus Comp Zool 18:1–292
38. Fischer R (1981) Bioerosion of basalt of the Pacific coast of Costa Rica. Senckenberg Marit 13:1–41
39. Fonseca E, AC, J Cortés (1998) Coral borers of the eastern Pacific: the sipunculan *Aspidosiphon (A.) elegans* and the crustacean *Pomatogebia rugosa*. Pac Sci 52:170–175
40. Fransen CHJM (2002) Taxonomy, phylogeny, historical biogeography, and historical ecology of the genus *Pontonia* Latreille (Crustacea: Decapoda: Caridea: Palaemonidae). Zool Verhandl Leiden 336:1–257
41. Fujino T (1972) A new pontoniinid shrimp, *Pontonia spighti* sp. new, associated with a newly described ascidian from the Pacific coast of Costa Rica (Decapoda, Natantia, Pontoniinae). Publ Seto Mar Biol Lab 19:293–301

42. García-Gómez J (1983) Revision of *Iridopagurus* (Crustacea, Decapoda, Paguridae) with the description of new species from American waters. Bull Mar Sci 33:10–54
43. Garth JS (1958) Brachyura of the Pacific coast of America. Oxyrhyncha. Allan Hancock Pac Exped 854 p
44. Garth JS (1959) Eastern Pacific Expedition of the New York Zoological Society. XLIV. Non-intertidal brachygnathous crabs from the west coast of tropical America Part 1: Brachygnatha Oxyrhyncha. Zoologica 44:105–124
45. Garth JS (1961) Eastern Pacific Expeditions of the New York Zoological Society. XLV. Non-intertidal brachygnathous crabs from the west coast of tropical America. Part 2: Brachygnatha Brachyrhyncha. Zoologica 46:133–159
46. Garth JS (1966) Eastern Pacific Expeditions of the New York Zoological Society. XLVI. Oxystomatus and allied crabs from the west coast of tropical America. Zoologica 51:1–14
47. Garth JS (1968) *Globopilumnus xantusii* (Stimpson), n. comb., a stridulating crab from the west coast of tropical America, with remarks on discontinuous distribution of some west American and west African genera of brachyrhynchous crabs. Crustaceana 15:312–318
48. Garth JS (1973) New taxa of brachyuran crabs from deep water off western Peru and Costa Rica. Bull South Calif Acad Sci 72:1–12
49. Garth JS (1974) On the occurrence in the Eastern Tropical Pacific of Indo-West Pacific decapod crustaceans commensal with reef-building corals. Proc 2nd Int Coral Reef Symp, Brisbane 1:397–404
50. Garth JS (1991) Taxonomy, distribution, and ecology of Galápagos Brachyura. *In*: James MJ (ed.) Galápagos Marine Invertebrates. Plenum, New York, pp 123–145
51. Garth JS, Stephenson W (1966) Brachyura of the coast of America. Brachyrhyncha; Portunidae. Allan Hancock Monogr Mar Biol 1:1–154
52. Gordon I (1938) A comparison of the two genera *Albunea* and *Lepidopa* (Crustacea, Anomura), with description of a new species from Singapore. Bull Raffles Mus 14:186–197
53. Gore RH (1970) *Pachycheles cristobalensis*, sp. nov. with notes on the porcellanid crabs of the southern Caribbean. Bull Mar Sci 20:957–970
54. Gore RH (1974) On a small collection of porcellanid crabs from the Caribbean Sea (Crustacea, Decapoda, Anomura). Bull Mar Sci 24:700–721
55. Gore RH (1982) Porcellanid crabs from the coasts of México and Central America (Crustacea, Decapoda, Anomura). Smithson Contr Zool 363:1–34
56. Gore RH, Abele LG (1973) Three new species of porcellanid crabs (Crustacea, Decapoda, Porcellanidae) from the Bay of Panama and adjacent Caribbean waters. Bull Mar Sci 23:569–573
57. Gore RH, Abele LG (1976) Shallow water porcelain crabs from the Pacific coast of Panama and adjacent Caribbean waters (Crustacea, Anomura, Porcellanidae). Smithson Contr Zool 237:1–30
58. Green TM (1985) *Pinnotheres jamesi* synonymized with *P. reticulates* (Decapoda: Brachyura). Proc Biol Soc Wash 98:611–614
59. Griffith H (1987) Taxonomy of the genus *Dissodactylus* (Crustacea: Brachyura: Pinnotheridae) with description of three new species. Bull Mar Sci 40:397–422
Gruner H-E (1993) Wirbellose Tiere. 4. Teil. Arthropoda (ohne Insecta). Gustav Fischer Verlag, Jena, 1 279 p
60. Guinot D (1967) Recheches préliminaires sur les groupements naturels chez les Crustacés Décapodes Brachyoures. II. Les anciens genres *Micropanope* Stimpson et *Medaeus* Dana. Bull Mus Nat Hist Nat 39:345–374
61. Guinot D (1969) Recherches préliminaires sur les groupements naturels chez les Crustacés Décapodes Brachyoures. VII. Les Goneplacidae. Bull Mus Nat Hist Nat 41:241–265
62. Guinot D (1971) Rechercher préliminaires sur les groupementes naturels chez les Crustacés Décapodes Brachyoures. VIII. Synthese et bibliographie. Bull Mus Nat Hist Nat 42:1063–1090
63. Haig J (1956) The Galatheidae (Crustacea: Anomura) of the Allan Hancock Atlantic Expedition with review of the Porcellanidae of the Western North Atlantic. Rep Allan Hancock Atlant Exped 8:1–148

64. Haig J (1957) Four new porcellain crabs from the eastern Pacific. Bull South Calif Acad Sci 56:31–41
65. Haig J (1960) The Porcellanidae (Crustacea: Anomura) of the eastern Pacific. Allan Hancock Pac Exped 24:1–440
66. Haig J (1962) Papers from Dr. Th. Mortensen's Pacific expedition 1914–1916. LXXIX. Porcellanid crabs from eastern and western America. Vidensk Meddel Dansk Naturhist Foren Kjobenhavn 124:171–192
67. Haig J (1968) Eastern Pacific expeditions of the New York Zoological Society. Porcellanid crabs (Crustacea: Anomura) from the west coast of tropical America. Zoologia 53:57–74
68. Haig J (1974) Observations on the lithodid crabs of Peru, with description of two new species. Bull South Calif Acad Sci 73:152–164
69. Haig J (1980) Arthropoda: Crustacea, Superfamily Hippoidea: Families Hippidae and Albuneidae (mole and sand crabs). *In*: Brusca RC (ed.) Common Intertidal Invertebrates of the Gulf of California. University of Arizona Press, Tucson, pp 286–291
70. Haig J, Harvey AW (1991) Three new species of the *Pagurus lepidus* complex (Decapoda, Anomura, Paguridae) from the eastern Pacific. Contr Sci, Nat Hist Mus Los Angeles County 430:1–11
71. Haig J, Provenzano AJ (1965) A new genus and two new species of diogenid hermit crabs (Decapoda, Anomura). Crustaceana 9:199–207
72. Haig J, Hopkins TS, Scandand TB (1970) The shallow water anomuran crab fauna of southwestern Baja California, Mexico. Trans San Diego Soc Nat Hist 16:14–31
73. Harvey AW (1998) Rediscovery and range extension of the Galápagos "endemic" *Pachycheles velerae* (Decapoda: Anomura: Porcellanidae). J Crust Biol 18:746–752
74. Harvey AW, McLaughlin P (1991) Two new hermit crabs of the genus *Pagurus (provenzanoi* group) (Crustacea, Anomura, Paguridae) from the eastern Pacific, with notes on their ecology. Contr Sci, Nat Hist Mus Los Angeles County 425:13–21
75. Hendrickx ME (1984) The species of *Sicyonia* H. Milne Edwards (Crustacea: Penaeoidea) of the Gulf of California, Mexico, with a key for their identification and a note on their zoogeography. Rev Biol Trop 32:279–298
76. Hendrickx ME (1989) Notes on the genus *Ethusa* Roux, 1828, and description of *Ethusa steyaerti* n. sp. (Crustacea, Decapoda, Dorippidae), from the continental shelf of the Gulf of California, Mexico. Bull Mus Nat Hist Nat, Paris 11 (sec A, #2):407–423
77. Hendrickx ME (1995a) Camarones. *In*: Fischer W, Krupp F, Schneider W, Sommer C, Carpenter KE, Niem UH (eds.) Guía FAO para la identificación de especies para los fines de la pesca. Pacífico Centro-oriental. Vol. 1. Plantas e Invertebrados. FAO, Rome, pp 417–537
78. Hendrickx ME (1995b) Checklist of lobster-like decapod crustaceans (Crustacea: Decapoda: Thalassinidea, Astacidea and Palinura) from the eastern tropical Pacific. Anales Inst Biol Univ Nac Autón México, Ser Zool 66:151–163
79. Hendrickx ME (1995c) Anomuros. *In*: Fischer W, Krupp F, Schneider W, Sommer C, Carpenter KE, Niem UH (eds.) Guía FAO para la identificación de especies para los fines de la pesca. Pacífico Centro-oriental. Vol. 1. Plantas e Invertebrados. FAO, Rome, pp 539–564
79b. Hendrickx ME (1995d) Checklist of brachyuran crabs (Crustacea: Decapoda) from the eastern tropical Pacific. Bull Inst Roy Sci Nat Belgique 65:125–150
79c. Hendrickx ME (1996) Los camarones Penaeoidea bentónicos (Crustacea: Decapoda: Dendrobrachiata) del Pacífico mexicano. Comisión Nacional para el Conocimiento y Uso de la Biodiversidad e Inst. Cienc. Mar Limnol., UNAM, México, 147 p
80. Hendrickx M (1997) Los cangrejos braquiuros del Pacífico mexicano (Crustacea: Brachyura: Dromiidae hasta Leucosiidae). Comisión Nacional para el Conocimiento y Uso de la Biodiversidad e Inst. Cienc. Mar Limnol., UNAM, México, 178p
81. Hendrickx ME (1999) Los cangrejos braquiuros del Pacifico mexicano (Crustacea: Brachyura: Majoidea y Parthenopoidea). Comisión Nacional para el Conocimiento y Uso de la Biodiversidad e Inst. Cienc. Mar Limnol., UNAM, México, 274 p

82. Hendrickx ME (2003) The temperate species of genus *Munida* Leach (Crustacea, Decapoda, Galatheidae) in the east Pacific, with the description of a new species and additional records for tropical-subtropical species. Bull Inst Roy Nat Belgique 73:117–138
83. Hendrickx ME, Estrada Navarrete FD (1996) Los camarones pelágicos del Pacífico mexicano (Crustacea: Dendrobranchiata y Caridea). Comisión Nacional para el Conocimiento y Uso de la Biodiversidad e Inst. Cienc. Mar Limnol., UNAM, México, 157 p
84. Hendrickx ME, Harvey AW (1999) Checklist of anomuran crabs (Crustacea: Decapoda) from the eastern tropical Pacific. Belg J Zool 129:363–389
85. Hendrickx ME, Wicksten MK (1989) Los Pandalidae (Crustacea: Caridea) del Pacífico mexicano, con una clave para su identificación. Caldasia (Colombia) 16:71–86
86. Hertlein LG (1963) Contribution to the biogeography of Cocos Island, including a bibliography. Proc Calif Acad Sci 32:219–289

Hogue CL, Miller SE (1981) Entomofauna of Cocos Island, Costa Rica. Atoll Res Bull 250:1–29

87. Holthuis LB (1951) A general revision of the Palaemonidae (Crustacea: Decapoda: Natantia) of the Americas. I. The subfamilies Euryrhynchinae and Pontoniinae. Occ Pap Allan Hancock Found 11:1–332
88. Holthuis LB (1952) A general revision of the Palaemonidae (Crustacea: Decapoda: Natantia) of the Americas, Part. II. The subfamily Palaemoninae. Occ Pap Allan Hancock Found 12:1–396

Holthuis LB (1980a) FAO species catalogue. Vol. 1. Shrimps and prawns of the world. An annotated catalogue of species of interest to fisheries. FAO Fish Synop 125(1):1–261

89. Holthuis LB (1980b) *Alpheus saxidomus* new species, a rock boring snapping shrimp from the Pacific coast of Costa Rica, with notes on *Alpheus simus* Guérin-Méneville, 1856. Zool Medel 55:47–58

Holthuis LB (1991) FAO species catalogue. Vol. 13. Marine lobsters of the world. An annotated and illustrated catalogue of species of interest to fisheries known to date. FAO Fish Synop 125(13):1–292

Jesse S (1996) Demersal crustacean assemblages along the Pacific coast of Costa Rica. A quantitative and multivariate assessment based on the Victor Hensen Costa Rica expedition (1993/1994). Rev Biol Trop 44(Suppl 3):115–134

90. Kim W, Abele LG (1988) The snapping shrimp genus *Alpheus* from the Eastern Pacific (Decapoda: Caridea: Alpheidae). Smithson Contr Zool 454:1–119
91. Kitani H (1997) Postlarval morphology of the western white shrimp *Penaeus occidentales* (Crustacea: Penaeidae). Rev Biol Trop 44:627–630
92. Kropp RK (1989) A revision of the Pacific species of gall crabs, genus *Opecarcinus* (Crustacea; Cryptochiridae). Bull Mar Sci 45:98–129
93. Kropp RK (1990) Revision of the genera of gall crabs (Crustacea: Cryptochiridae) occurring in the Pacific Ocean. Pac Sci 44:417–448
94. Lemaitre R (1981) Shallow-water crabs (Decapoda, Brachyura) collected in the southern Caribbean near Cartagena, Colombia. Bull Mar Sci 31:234–266
95. Lemaitre R (1989) Revision of the genus *Parapagurus* (Anomura: Paguroidea: Parapaguridae), including redescription of the western Atlantic species. Zool Verhandl 253:1–106
96. Lemaitre R (1995) A review of the hermit crabs of the genus *Xylopagurus* A. Milne Edwards, 1880 (Crustacea: Decapoda: Paguridae), including description of two new species. Smithson Contr Zool 570:1–27
97. Lemaitre R, Alvarez-León R (1992) Crustáceos decápodos del Pacífico Colombiano: lista de especies y consideraciones zoogeográficas. An Inst Mar Punta Betín 21:33–76
98. Lemaitre R, McLaughlin PA (1996) Revision of *Pylopagurus* and *Tomopagurus* (Crustacea: Anomura: Paguridae), with the description of new genera and species. Part V. *Anisopagurus* McLaughlin, *Manucomplanus* McLaughlin and *Protoniopagurus* new genus. Bull Mar Sci 59:89–141
99. Lemaitre R, Ramos GE (1992) A collection of Thalassinidea (Crustacea: Decapoda) from the Pacific coast of Colombia, with description of a new species and a checklist of eastern Pacific species. Proc Biol Soc Wash 105:343–358

100. MacPherson E (1988) Revision of the family Lithodidae Samouelle, 1819 (Crustacea, Decapoda, Anomura) in the Atlantic Ocean. Monogr Zool Mar (Inst Cienc Mar, Barcelona) 2:9–153

101. Manning RB, Felder DL (1991) Revision of the American Callianassidae (Crustacea: Decapoda: Thalassinidea). Proc Biol Soc Wash 104:764–792

101b. Markham JC, Donath-Hernández F, Villalobos Hiriart JL, Díaz Barriga AC (1990) Notes on the shallow water marine Crustacea of the Caribbean coast of Quintana-Roo, México. An Inst Biol Univ Nac Autón México, Ser Zool 61:405–446

Martin JW, Davis GE (2001) An updated classification of the recent Crustacea. Sci Ser, Nat Hist Mus Los Angeles County 39:1–124

102. Maurer D, Epifanio CE, Dean HK, Howe S, A Vargas J, Dittel AI, Murillo MM (1984) Benthic invertebrates of a tropical estuary: Gulf of Nicoya, Costa Rica. J Nat Hist 18:47–61

McLaughlin PA (1980) Comparative Morphology of Recent Crustacea. W.H. Freeman, San Francisco, 177 p

103. McLaughlin PA (1981a) Revision of *Pylopagurus* and *Tomopagurus* (Crustacea: Decapoda: Paguridae), with the descriptions of new genera and species: Part I. Ten new genera of the Parapaguridae and a redescription of *Tomopagurus* A. Milne Edwards and Bouvier. Bull Mar Sci 31:1–30

104. McLaughlin PA (1981b) Revision of *Pylopagurus* and *Tomopagurus* (Crustacea: Decapoda: Paguridae), with the descriptions of new genera and species: Part II. *Rhodochirus* McLaughlin and *Phimochirus* McLaughlin. Bull Mar Sci 31:329–365

105. McLaughlin PA (1982) Revision of *Pylopagurus* and *Tomopagurus* (Crustacea: Decapoda: Paguridae), with the descriptions of new genera and species: Part III. *Agaricochirus* McLaughlin, and *Enallopaguropsis* McLaughlin. Bull Mar Sci 32:823–856

106. Méndez M (1981) Claves de identificación y distribución de los langostinos y camarones (Crustacea: Decapoda) del mar y los ríos de la costa del Perú. Bol Inst Mar Perú-Callao 5:1–170

107. Moran DA (1984) Additions to the known anomuran fauna of El Salvador, Central America (Crustacea: Decapoda). J Crust Biol 4:72–84

108. Moran DA, Dittel AI (1993) Anomuran and brachyuran crabs of Costa Rica: annotated list of species. Rev Biol Trop 41:599–617

109. Omori M (1975) The systematics, biogeography, and fishery of epipelagic shrimps of the genus *Acetes* (Crustacea, Decapoda, Sergestidae). Bull Ocean Res Inst Univ Tokyo 7:1–91

110. Pequegnat WE, Pequegnat LH (1973) *Munidopsis albatrossae*, a new species of deep-sea (Galatheidae, Decapoda, Anomura) from the eastern Pacific Ocean. Crustaceana 24:163–168

111. Pérez-Farfante I (1969) Western Atlantic shrimps of the genus *Penaeus*. Fish Bull 67:461–591

111b. Pérez-Farfante, I (1977) American solenocerid shrimps of genera *Hymenopenaeus, Haliporoides, Pleoticus, Hadropenaeus* new genus, and *Mesopenaeus* new genus. Fish Bull 75:261–346

112. Pérez-Farfante I (1971) Western Atlantic shrimps of the genus *Metapenaeopsis* (Crustacea, Decapoda, Penaeidae) with descriptions of three new species. Smithson Contr Zool 79:1–37

113. Pérez-Farfante I, Kensley B (1997) Penaeoid and sergestoid shrimps and prawns of the world. Mém Mus Natn Hist Nat, Paris 175:1–233

114. Pesta O (1931) Ergebnisse der Österreichischen Biologischen Costa Rica Expedition 1930. I. Teil. Crustacea Decapoda aus Costa Rica. Ann Naturhist Mus Wien 45:173–181

115. Powers LW (1977) A catalogue and bibliography to the crabs (Brachyura) of the Gulf of Mexico. Contr Mar Sci (Suppl) 20:1–190

Rabalais NN, Gore RH (1985) Abbreviated development in decapods. *In*: Wenner A, Schram FR (eds.) Larval Growth. Crustacean Issues 2. A.A. Balkema, Lisse, The Netherlands, pp 67–126

116. Rathbun MJ (1918) The grapsoid crabs of America. Bull US Nat Mus 97:1–461
117. Rathbun MJ (1925) The spider crabs of America. Bull US Nat Mus 129:1–613
118. Rathbun MJ (1930) The Cancroid crabs of America of the families Euryalidae, Portunidae, Atelecyclidae, Cancridae and Xanthidae. Bull US Nat Mus 152:1–593
119. Rathbun MJ (1931) A new species of pinnotherid crab from Costa Rica. J Wash Acad Sci 21:262–263
120. Rathbun MJ (1937) The Oxystomatous and allied crabs of America. Bull US Nat Mus 166:1–272
Reaka ML, Manning RB (1980) The distributional ecology and zoogeographical relationships of stomatopod Crustacea from Pacific Costa Rica. Smithson Contr Mar Sci 7:1–29
121. Ríos R (1986) Caridean shrimps of the Gulf of California. V. New records of species belonging to the subfamily Pontomiinae (Crustacea: Decapoda: Palaemonidae). Proc Biol Soc Wash 99:429–434
122. Rodríguez B (1993) Shrimps. *In*: Cervigon F, Cipriani R, Fischer W, Caribaldi L, Hendrickx M, Lemus AJ, Márquez R, Poutiers JM, Robaina G, Rodríguez B (eds.) FAO Species Identification Sheets for Fishery Purposes. Field Guide to the Commercial Marine and Brackish-Waters Resources of the Northern Coast of South America. FAO, Rome, pp 107–131
123. Snyder-Conn E (1980) Arthropoda: Crustacea, Paguroidea and Coenobitoidea. *In*: Brusca R (ed.). Common Intertidal Invertebrates of the Gulf of California. University of Arizona Press, Tucson, pp 275–285
124. Tabash FA (1995) An assessment of pink shrimp (*Penaeus brasiliensis*) populations, in three areas of the Caribbean coast of Costa Rica. Rev Biol Trop 43:239–250
124b. Thoma B, Heard R, Vargas R (2005) A new species of *Parapinnixa* (Decapoda: Brachyura: Pinnotheridae) from Isla del Coco, Costa Rica. Proc Biol Soc Wash 118:543–550
125. Türkay M (1970) Die Gecarcinidae Amerikas. Mit einem Anhang über *Ucides* Rathbun (Crustacea; Decapoda). Senckenberg Biol 51:333–354
126. Umaña R, Chacón D (1994) Asentamiento en estadíos postlarvales de la langosta *Panulirus argus* (Decapoda: Palinuridae), en Limón, Costa Rica. Rev Biol Trop 42:585–594
Vargas JA, Wolff M (1996) Pacific coastal ecosystems of Costa Rica with emphasis on the Golfo Dulce and adjacent areas: a synoptic view based on the RV Victor Hensen expedition 1993/1994 and previous studies. Rev Biol Trop 44(Suppl 3):1–238
127. Vargas JA, Dean HK, Maurer D, Orellana P (1985) Lista preliminar de invertebrados asociados a los sedimentos del Golfo de Nicoya, Costa Rica. Brenesia 24:327–342
128. Vargas R (2000) *Periclimenes murcielagensis*, a new species of shrimp (Crustacea: Decapoda: Palaemonidae) living on black coral from the Pacific coast of Costa Rica. Proc Biol Soc Wash 113:17–23
129. Vargas R, Cortés J (1999a) Biodiversidad marina de Costa Rica: Crustacea: Decapoda (Penaeoidea, Sergestoidea, Stenopodidea, Caridea, Thalassinidea, Palinura) del Caribe. Rev Biol Trop 47:877–885
130. Vargas R, Cortés J (1999b) Biodiversidad marina de Costa Rica: Crustacea: Decapoda (Penaeoidea, Sergestoidea, Caridea, Astacidea, Thalassinidea, Palinura) del Pacífico. Rev Biol Trop 47:887–911
131. Vargas R, Cortés J (2006) Biodiversidad Marina de Costa Rica: Crustacea: Infraorden Anomura. Rev Biol Trop 54:461–488
132. Vargas R, Jesse S, Castro M (1996) Checklist of crustaceans (Decapoda and Stomatopoda), collected during the Victor Hensen Costa Rica expedition (1993/1994). Rev Biol Trop 44(Suppl 3):97–102
132b. Voss GL (1971) Narrative of R/V John Elliott Pillsbury cruise P-7101 to Central America, 20 January–February 1971. School of Marine and Atmospheric Science, University of Miami, Florida, 31 p
133. Wehrtmann IS (1990) Distribution and reproduction of *Ambidexter panamense* and *Palaemonetes schmitti* in Pacific Costa Rica (Crustacea, Decapoda). Rev Biol Trop 38:327–329

134. Wehrtmann IS, Vargas R (2003) New records and range extensions of shrimps (Decapoda: Penaeoidea, Caridea) from the Pacific and Caribbean coasts of Costa Rica, Central America. Rev Biol Trop 51:268–274
135. Werding B (1977) Los porcelánidos (Crustacea: Anomura: Porcellanidae) de la región de Santa Marta, Colombia. An Inst Inv Mar Punta Betín 9:173–214
136. Werding B (1984) Porcelánidos (Crustacea: Anomura: Porcellanidae) de la isla de Providencia, Colombia. An Inst Inv Mar Punta Betín 14:3–16
137. Werding B (1990) Beobachtungen an *Alpheus schmitti* Chace und *Thunor simus* (Guérin-Méneville), zwei gesteinsbohrende Pistolenkrebsen aus dem Karibischen Meer (Crustacea: Alpheidae). Zool Anz 244:87–98
138. Werding B, Haig J (1982) The porcellanids crabs of Isla Gorgona, Pacific coast of Colombia, with a description of *Clastotoechus gorgonensis* sp. nov. (Crustacea: Anomura). An Inst Inv Mar Punta Betín 9:173–214
139. Wicksten MK (1979) New records of the species of *Glyphocrangon* in the northeastern Pacific Ocean (Caridea: Glyphocrangonidae). Proc Biol Soc Wash 92:217–224
140. Wicksten MK (1981) New records of *Steromastis sculpta pacifica* (Faxon) (Decapoda: Polychelidae) in the eastern Pacific Ocean. Proc Biol Soc Wash 93:914–919
141. Wicksten MK (1982a) *Pinnixa costaricana*, a new species of crab from Central America (Brachyura: Pinnotheridae). Proc Biol Soc Wash 95:579–582
142. Wicksten MK (1982b) New records of pinnotherid crabs from the Gulf of California (Brachyura: Pinnotheridae). Proc Biol Soc Wash 95:354–357
143. Wicksten MK (1983) Shallow water caridean shrimps of the Gulf of California, México. Allan Hancock Monogr Mar Biol 13:1–59
144.Wicksten MK (1987) A new species of hippolytid shrimp from the west coast of Mexico. Bull Southern Calif Acad Sci 86:27–33
144b. Wicksten MK (1987) *Nicoya tuberculata*, a new genus and species of spider crab from Pacific Costa Rica (Majidae: Pisinae). Proc Biol Soc Wash 100:691–693
145. Wicksten MK (1989) Ranges of offshore decapod crustaceans in the eastern Pacific Ocean. Trans San Diego Soc Nat Hist 21:291–316
145b. Wicksten MK (1989) A key to the palaeomonid shrimps of the eastern Pacific Region. Bull Southern Calif Acad Sci 88:11–20
146. Wicksten MK (1990) Key to the hippolytid shrimp of the Eastern Pacific Ocean. Fish Bull 88:587–598
147. Wicksten MK (2000) The species of *Lysmata* (Caridea: Hippolytidae) from the Eastern Pacific Ocean. Amphipacifica 11:3–22
148. Wicksten MK, Hendrickx ME (1992) Checklist of penaeoid and caridean shrimps (Decapoda: Penaeoidea, Caridea) from the eastern tropical Pacific. Proc San Diego Soc Nat Hist 9:1–11
148b. Wicksten MK, Hendrickx ME (2003) An updated checklist of benthic marine and brackish water shrimps (Decapoda: Penaeoidea, Stenopodidea, Caridea) from the Eastern Tropical Pacific. *In*: Hendrickx ME (ed.) Contributions to the Study of East Pacific Crustaceans, vol 2. Instituo de Ciencias del Mar y Liminolgia, UNAM, pp 49–76
149. Wicksten MK, Vargas R (2001) A new species of *Thor* (Crustacea: Decapoda: Caridea) from the tropical Eastern Pacific. Proc Biol Soc Wash 114:139–144
150. Williams A (1974) The swimming crabs of the genus *Callinectes* (Decapoda: Portunidae). Fish Bull 72:685–798
151. Williams A (1984) Shrimps, Lobsters and Crabs of the Atlantic Coast of the Eastern United States, Maine to Florida. Smithsonian Institution Press, Washington, DC, 550 p
152. Williams A (1986) Mud shrimps, *Upogebia*, from the eastern Pacific (Thalassinoidea: Upogebiidae). San Diego Soc Nat Hist Mem. 14:1–60
153. Williams A (1993) Mud shrimps, Upogebiidae from the western Atlantic (Crustacea: Decapoda: Thalassinoidea). Smithson Contr Zool 544:1–77

154. Williams A (1997) Two new species and a range extension of mud shrimps, *Upogebia*, from Pacific Costa Rica and Mexico (Decapoda: Thalassinidea: Upogebiidae). Proc Biol Soc Wash 110:617–623
155. Williams A, Vargas R (2000) A new species of mud shrimp, *Upogebia cortesi*, from the Pacific Costa Rica (Decapoda: Thalassinidea: Upogebiidae). Proc Biol Soc Wash 113:13–16
156. Wolff T (1961) Animal life from a single abyssal trawling. Galathea Rep 5:129–162
157. Zimmerman TL, Martin JW (1999) Brachyuran crabs of Cocos Island (Isla del Coco), Costa Rica: Leucosiidae, Calappidae, and Parthenopidae, with descriptions of two new species. J Crust Biol 19:643–668

Part 17
Shallow Water Mysids

W. Wayne Price, Richard W. Heard, and Rita Vargas

Bowmaniella dissimilis, a species reported from the Caribbean coast of Costa Rica. (Photo: Ernst Peebles)

W.W. Price
University of Tampa, Dept. of Biology, Tampa, FL 33606, USA.
e-mail: wprice@ut.edu

R.W. Heard
University of Southern Mississippi, Institute of Marine Sciences, Dept. of Coastal Sciences, 703 East Beach Blvd., P.O. Box 7000, Ocean Springs, MS 39566-7000, USA.
e-mail: richard.heard@usm.edu

R. Vargas
Museo de Zoología, Escuela de Biología, Universidad de Costa Rica, 11501-2060 San José, Costa Rica.
e-mail: ritav@biologia.ucr.ac.cr

Abstract The mysid fauna of Costa Rica is understudied and poorly known. Only five species, four from the Pacific and one from the Caribbean, are currently known. Based on our preliminary observations, species richness for mysids along the Costa Rican Caribbean coast may be similar to the 22 species reported from the Panamanian Caribbean coast. The diversity of mysids from the Pacific coast of Costa Rica should greatly exceed the species reported to date and many of the species from this area may be new to science. An intensive survey of the Costa Rican Mysida is needed to properly determine the diversity, endemicity, and zoogeography of this group in Costa Rican waters.

Introduction

Members of the order Mysida are shrimp-like crustaceans, often called "opossum shrimps," because ovigerous females carry developing larvae within a subthoracic marsupium or brood pouch formed by pairs of overlapping oostegites. The family Mysidae is the dominant one of this order in shallow water brackish and marine habitats. This family is characterized by the presence of a statocyst on each uropodal endopod. The only shallow water mysids that lack statocysts are the cave and subterranean species belonging to the families Lepidomysidae and Stygiomysidae. Euphausids or krill are superficially similar to mysids, but lack oostegites and uropodal statocysts, and generally have photophores and conspicuous thoracic gills, two characteristics absent in Mysida.

Mysid shrimp are found in most aquatic environments; more than 90% are marine, occurring in oceanic, coastal, and estuarine habitats, while the rest inhabit brackish, fresh, and subterranean waters. More than 75% of the species live on or near the sea floor, and the remainder are pelagic. The great majority of mysids are strictly free-living; however, some live in association with a variety of invertebrates (Mauchline 1980; Roast *et al.* 1998; Wittmann 1999).

Mysids represent an abundant hyperbenthic component of the fauna of coastal ecosystems where they may serve as major producers and consumers (Mauchline 1980). They have the potential to structure phytoplankton, zooplankton, and meiofaunal communities (Fulton 1982; Mees & Jones 1997), and serve as an important food source for higher trophic levels (Mees *et al.* 1994). Mysids are commonly used in laboratory toxicity testing (Nimmo & Hamaker 1982). Available evidence indicates that they are highly sensitive to toxic substances (Roast *et al.* 1998), and that this sensitivity occurs at levels that are likely to exist in the environment (Odenkirchen & Eisler 1988).

The order Mysida includes about 1,000 known species and about 160 genera (Price *et al.* 2002a); however, Wittmann (1999) estimated that more than 3,000 species remain to be described. More than twice as many described species are from the northern hemisphere as compared to the southern hemisphere. Since the best known mysid faunas occur in the waters of the Caribbean, North Atlantic, Mediterranean, Black Sea, and Japan, all regions in the northern hemisphere, the diversity imbalance may be due to differences in sampling effort (Wittmann 1999). Using data from continental shelf areas, Wittmann showed that species numbers and diversity are highest in the tropics and decrease with increasing latitude.

Mysida Previously Known from Costa Rica

The knowledge of the mysid fauna in the waters of Costa Rica is quite limited; only one record is known from the Caribbean coast (Species List 17.1 is included on the CD-Rom), and four species are reported from the Pacific coast (Species List 17.2 is included on the CD-Rom). From the Caribbean, Heard & Price (2006) recorded *Bowmaniella dissimilis* (Coifmann, 1937), a senior synonym of *B. brasiliensis* and *B. floridana*, from the surf zones of sandy beaches in Manzanillo and Puerto Vargas. This species is common in the shallow coastal waters of the northwestern Atlantic and, in the tropics, has been previously reported from Panama, Tobago and Brazil (Heard & Price 2006).

Antromysis anophelinae was described by Tattersall (1951) from specimens collected in burrows of the land crab *Cardisoma crassum* Smith, 1870 from Río Aranjuez, Puntarenas on the Pacific coast. He speculated that the burrows, which were supratidal, contained near-fresh water and that *A. anophelinae* is a brackish water species with a wide range of salinity tolerance. Tattersall's description is the only report for this species.

Heteromysis panamaensis was described by Tattersall (1967) from Bahía Salinas, Costa Rica, and Taboguilla, Taboga, and Pontadora (Perlas Islands), Panama. Most collections were made on sandy/shelly bottoms in depths of 5–18 m. However, at Taboga, several specimens were taken from the "innermost spiral" of gastropod shells harboring hermit crabs (*Dardanus* sp.). Vannini *et al.* (1993) reported several species of *Heteromysis* Smith, 1873, all from the Pacific Ocean, living in association with hermit crabs of the genus *Dardanus* Paulson, 1875. *H. panamaensis* has not been reported since its original description.

Dexter (1974) reported *Bowmaniella* spp. A, B, and C as well as *Metamysidopsis* sp. in her description of sandy-beach fauna of the Pacific and Atlantic coasts of Costa Rica and Colombia. From the western coast of Costa Rica, *Bowmaniella* spp. were taken from Boca de Barranca, Jacó, Playa Cocal and Playa Espadilla, Quepos. Examination of Dexter's material from the former three collection sites revealed that all specimens are *B. banneri* Băcescu, 1968. This species has been reported previously from sandy beaches at La Jolla and Solana Beach, California and San Felipe, Baja California, Mexico (Holmquist, 1975). *Metamysidopsis* sp. was collected from Playa Tamarindo and Playa Cocal. Two species of *Metamysidopsis* Tattersall, 1951 have been reported from the Pacific coast of the Americas: *Metamysidopsis elongata elongata* (Holmes, 1900) from California and *M. pacifica* (Zimmer, 1918) from Chile and Panama (Dexter 1972).

Since only five species of mysids have been reported from both coasts of Costa Rica, an examination of literature from the Caribbean Sea and the tropical eastern Pacific may provide a general idea of the composition of the mysid fauna of Costa Rica. The shallow water mysid fauna of the Caribbean region is better known than that of the eastern Pacific, due mainly to a series of investigations by Brattegard (1969, 1970a, b, 1973, 1974a, b, 1975, 1980). Over 90 species are reported from the Caribbean. Most of the intensively surveyed areas are insular and have a species richness of about 20 species: Bahamas, 26 species (Brattegard, 1969, 1970a;

Bowman *et al.* 1984; Bowman 1985; Modlin 1987a; Băcescu 1991; Bamber 2000; Pesce & Iliffe 2002); Turks and Caicos, 23 species (Bowman *et al.* 1984; Price & Heard 2000; Price & Heard 2004); Florida Keys, 21 species (Brattegard 1969, 1970a, b, 1973; LaRoe 1971; Modlin 1987b); Cuba, 22 species (Băcescu 1968a, b, 1970; Băcescu & Orghidan 1971, 1977; Băcescu & Ortíz 1984; Ortíz 1988; Ortíz & Lalana 1988, 1993; Ortíz *et al.* 1997), and the Cayman Islands, 20 species (Price *et al.* 2002b).

Although much of the mysid fauna along the Central American Caribbean coast still remains largely unknown, studies have been performed in Belize and Panama. Modlin (1987c) reported 16 species from coral reef and mangrove communities in waters surrounding Carrie Bow Cay, Belize. Sampling a small area of the coastline of Panama, Brattergard (1974b) recorded 22 species. From adjacent waters of Colombia, Brattegard (1973, 1974a) found 33 species in the most intensive survey of mysids along the continental coastline within the Caribbean.

Mysids of the shallow waters of the eastern Pacific coast from Mexico to Panama are virtually unknown. No surveys comparable to the ones cited for the Caribbean have been conducted. Besides the four species recorded from Costa Rica (Species List 17.1 is included on the CD-Rom), only seven more are reported: *Coifmanniella mexicana* (W.M. Tattersall, 1951), *M. pacifica* (Zimmer, 1918) (Dexter 1972), and *Siriella panamensis* W.M. Tattersall, 1951 from Panama; *Mysidopsis californica* W.M. Tattersall, 1932, *Mysidium rickettsi* Harrison & Bowman, 1987 and *S. aequiremis* Hansen, 1910 from Mexico (Tattersall 1951; Tattersall 1969); and *S. gracilis* Dana, 1852 from the coasts of Panama and Mexico (Tattersall 1951). Except for *Antromysis* Creaser, 1936, the other genera reported from Pacific waters of Central America are represented along the Caribbean coast. Although no species are found on both coasts, one Pacific species is reported from more distant localities in the Atlantic. *C. mexicana* was originally described from the northeastern Gulf of Mexico as *Gastrosaccus mexicanus* (Tattersall, 1951), and has been identified from samples taken in continental shelf waters off the southeastern US Atlantic coast (Heard & Price 2006).

The authors have made a few shallow water collections on both coasts of Costa Rica. Mysids were collected at three areas on the Caribbean coast (Playa Manzanillo, Puerto Viejo, Puerto Vargas) and two areas on the Pacific coast (Islas Murciélago and Puerto Caldera). Various collecting methods and gear types were employed including fine mesh dredge nets (kicknets), yabby pumps (hand-held suction devices), epibenthic sleds, dredges, and rock and algae washing techniques.

Preliminary examination of the samples collected indicates that the species richness of mysids on the Caribbean coast of Costa Rica may be similar to that found by Brattegard (1974b) in Panama (22 species). The diversity of mysids from the Pacific coast should greatly exceed the four species reported to date, and many of the species from this area may be new to science. An intensive survey of the Mysida is needed to provide a more complete picture of the diversity, endemicity, and zoogeography of this group in Costa Rican waters.

Specialists

The authors are currently studying the mysid fauna of Costa Rica. Other researchers have worked on the taxonomy of this group from areas with influences on Costa Rican waters. Names, affiliations, and e-mail addresses of these workers as well as their geographic regions of expertise are listed below.

Elva Escobar-Briones: Universidad Nacional Autonoma de Mexico, Mexico City, Mexico – Southern Gulf of Mexico, Caribbean Sea. escobri@mar.icmyl.unam.mx

Richard W. Heard: The University of Southern Mississippi, Ocean Springs, Mississippi – Northwestern Atlantic, Caribbean Sea. richard.heard@usm.edu,

Manuel Ortiz: Universidad de la Habana, La Habana, Cuba – Caribbean Sea. normano@comuh.uh.cu

W. Wayne Price: University of Tampa, Tampa, Florida, USA – Northwestern Atlantic, Caribbean Sea. wprice@ut.edu

Collections

Although there are no formal collections of mysids in Costa Rica, samples collected by the authors reside in the Museo de Zoología at the Universidad de Costa Rica, San José, Costa Rica. More than 100 lots of unidentified mysids taken from Atlantic and Pacific coasts of Panama are housed in the Smithsonian Institution, National Museum of Natural History. The zooplankton collection of El Colegio de la Frontera Sur (ECOSUR) in Chetumal, Mexico, contains a set of sorted, unidentified mysids from the western Caribbean Sea, including samples from reef and coastal systems.

Acknowledgments We wish to express our appreciation to Jorge Cortés and José Vargas of the Universidad de Costa Rica, who arranged for the collection trips to the Caribbean and Pacific coasts of Costa Rica between 1998 and 2001. Odalisca Breedy, Jorge Cortés, Leonora Rodríguez, and José Vargas kindly assisted us in the collection of specimens from both coasts. We also wish to thank R. Lemaitre and J. C. Walker, Smithsonian Institution, National Museum of Natural History for the loan of specimens of *B. banneri*.

References

Băcescu M (1968a) Heteromysini nouveaux des eaux Cubaines: trois espèces nouvelles de *Heteromysis* et *Heteromysoides spongicola* n.g.n.sp. Rev Roum Biol Zool 13:221–237

Băcescu M (1968b) Étude des quelques Leptomysini (Crustacea Mysidacea) des eaux du Brésil et de Cuba; description d'un genre et de cinq autres taxons nouveaux. Ann Mus Stor Nat Genova 77:232–249

1. Băcescu M (1968c) Contributions to the knowledge of the Gastrosaccinae psammobionte of the tropical America, with the description of a new genus (*Bowmaniella*, n.g.) and three new species of its frame. Trav Mus Hist Nat "Gr Antipa" 8:355–373
Băcescu M (1970) New spongicolous *Heteromysis* of the Caribbean Sea (*H. gomezi* n.sp. and *H. mariani* n. sp.). Rev Roum Biol 15:11–16
Băcescu M (1991) *Mysidium iliffei* n.sp. et *Amathimysis sarbui* n.sp., Mysidacés nouveaux dans le secteur des Antilles. Rev Roum Biol Biol Anim 36:3–8
Băcescu M, Orghidan T (1971) *Antromysis cubanica* n.sp. et *Spelaeomysis nuniezi* n.sp., Mysis cavernicoles nouvelles de Cuba. Rev Roum Biol Zool 16:225–231
Băcescu M, Orghidan TN (1977) New contribution to the knowledge of troglobian mysids of Cuba: *Antromysis juberthiei* n.sp. *In*: Orghidan T Nunez A, Jiménez V Decou, Negra S & Vina Bayes N (eds) Résultats des expéditions biospéléologiques cubano-roumaines à Cuba, Vol. 2. Ed. Acad. Rep. Socialiste Romania, Bucuresti, pp 263–265
Băcescu M, Ortíz M (1984) Contribution to the knowledge of the Mysidacea (Crustacea of the Cuban insular shelf waters. Trav Mus Hist Nat "Gr Antipa" 24:15–23
Bamber R (2000) A new species of *Heteromysis* (Crustacea: Mysidacea) from a "Blue Hole" in the Bahamas. Spec Div 5:129–134
Bowman TE (1985) *Heteromysoides dennisi*, a new mysidacean crustacean from Cemetery Cave, Grand Bahama Island. Proc Biol Soc Wash 98:945–948
Bowman TE, Iliffe TM, Yager J (1984) New records of the troglobitic mysid genus *Stygiomysis*: *S. clarkei*, new species, from the Caicos Islands, and *S. holthuisi* (Gordon) from Grand Bahama Island (Crustacea: Mysidacea). Proc Biol Soc Wash 97:637–644
Brattegard T (1969) Marine biological investigations in the Bahamas. 10. Mysidacea from shallow water in the Bahamas and southern Florida. Part 1. Sarsia 39:7–106
Brattegard T (1970a) Marine biological investigations in the Bahamas. 11. Mysidacea from shallow water in the Bahamas and southern Florida. Part 2. Sarsia 41:1–35
Brattegard T (1970b) Mysidacea from shallow water in the Caribbean Sea. Sarsia 43:111–154
Brattegard T (1973) Mysidacea from shallow water on the Caribbean coast of Colombia. Sarsia 54:1–65
Brattegard T (1974a) Additional Mysidacea from shallow water on the Caribbean coast of Colombia. Sarsia 57:47–86
2. Brattegard T (1974b) Mysidacea from shallow water on the Caribbean coast of Panama. Sarsia 57:87–108
Brattegard T (1975) Shallow water Mysidacea from the Lesser Antilles and other Caribbean regions. Stud Fauna Curaçao 47:102–115
Brattegard T (1980) *Platymysis facilis* gen. et sp. nov. (Crustacea: Mysidacea: Heteromysini) from the Saba Bank, Caribbean Sea. Sarsia 65:49–52
Coifmann I, (1937) Misidacei raccolti dalla R. Corvetta Vettor Pisani negli anni 1882–85. Ann Mus Zool Univ Napoli (n.s.) 7:1–14
Creaser EP (1936) XI. Crustaceans from Yucatan. Carnegie Inst Wash Publ 457:117–132
Dana JD (1852) United States Exploring Expedition during the years 1838–1842, under the command of Charles Wilkes, U.S.N. 13. Crustacea, pt. 1, Philadelphia, Pennsylvania, USA, 685 p
Dexter DM (1972) Comparison of the community structures in a Pacific and an Atlantic Panamanian sandy beach. Bull Mar Sci 22:449–462
3. Dexter DM (1974) Sandy-beach fauna of the Pacific and Atlantic coasts of Costa Rica and Colombia. Rev Biol Trop 22:51–66
Fulton RS (1982) Predatory feeding of two marine mysids. Mar Biol 72:183–191
Hansen HJ (1910) The Schizopoda of the Siboga Expedition. Siboga Exped 37:1–123
Harrison EB, Bowman TE (1987) *Mysidium rickettsi*, a new species of mysid from the Gulf of California (Crustacea: Mysidacea: Mysidae). Proc Biol Soc Wash 100:674–679
4. Heard RW, Price WW (2006) Revision of *Bowmaniella* sensu Băcescu, 1968 (Crustacea: Mysida: Mysidae: Gastrosaccinae): a taxonomic conundrum. Zootaxa 1269:1–29
Holmes SJ (1900) Synopsis of California stalk-eyed Crustacea. Occ Pap Calif Acad Sci 7:1–262

5. Holmquist C (1975) A revision of the species *Archaeomysis grebnitzkii* Czerniavsky and *A. maculata* (Holmes) (Crustacea, Mysidacea). Zool Jb Syst 102:51–71

LaRoe ET (1971) The culture and maintenance of the loliginid squids *Sepioteuthis sepioidea* and *Doryteuthis plei*. Mar Biol 9:9–25

Mauchline J (1980) The biology of mysids. Adv Mar Biol 18:1–369

Mees J, Jones MB (1997) The hyperbenthos. Oceanogr Mar Biol Ann Rev 35:221–255

Mees J, Abdulkerim AZ, Hamerlynk O (1994) Life history, growth and production of *Neomysis integer* in the Westerschelde estuary (S.W. Netherlands). Mar Ecol Prog Ser. 109:43–57

Modlin RF (1987a) Heteromysini from Grand Bahama Island: description of *Heteromysis agelas*, new species, first description of male *H. floridensis*, and notes on *H. guitarti* (Crustacea: Mysidacea). Proc Biol Soc Wash 100:296–301

Modlin RF (1987b) *Heteromysis kensleyi* and *H. coralina*, new species from the shallow waters off Looe Key, Florida (Mysidacea: Heteromysini). Proc Biol Soc Wash 100:653–658

Modlin RF (1987c) Mysidacea from shallow waters in the vicinity of Carrie Bow Cay, Belize. Central America, with descriptions of two new species. J Crust Biol 7:106–121

Nimmo DR Hamaker TL (1982) Mysids in toxicity testing: a review. Hydrobiol 93:171–178

Odenkirchen EW, Eisler R (1988) Chlorpyrifos hazards to fish, wildlife and invertebrates: a synoptic review. US Fish Wildlife Serv Biol Rep 85(1.13): 34

Ortíz M (1988) Una nueva especie de misidáceo del género *Bowmaniella* (Crustacea: Mysidacea), de aguas cubanas. Rev Inv Mar 9:3–7

Ortíz M, Lalana R (1988) Lista de especies y bibliografía de los crustáceos de Cuba. II. Cirripedia, Phyllocarida, Pancarida, Mysidacea, Tanaidacea y Cumacea. Rev Inv Mar 9:11–19

Ortíz M, Lalana R (1993) *Mysidopsis cojimarensis*, una nueva especie de misidáceo marino (Crustacea, Mysidacea) de Cuba. Carib J Sci 29:50–53

Ortíz M, García-Debrás A, Pérez A (1997) Nuevo género y nueva especie de misidáceo cavernícola (Crustacea: Peracarida, Mysidacea) de Cuba, Avicennia 6/7:21–28

Paulson O (1875) Studies on the Crustacea of the Red Sea with notes regarding other seas. Part 1. Podophthalmata and Edriophthalmata (Cumacea). S.V. Kul'zhenko, 83 Malo-Zhitomirskaya St. Kiev. Translation published for NSF, Washington, D.C., and Smithson. Inst., USA, by The Israel Program for Scientific Translations, 1961, 164 p

Pesce G, Iliffe T (2002) New records of cave-dwelling mysids from the Bahamas and Mexico with description of *Palaumysis bahamensis* n. sp. (Crustacea: Mysidacea). J Nat Hist 36:265–278

Price WW, Heard RW (2000) Studies on the Crustacea of the Turks and Caicos Islands, British West Indies IV. *Heteromysis (Heteromysis) spottei*, a new species (Peracarida: Mysidacea: Mysidae) from Pine Cay. Proc Biol Soc Wash 113:88–94

Price WW, Heard RW (2004) Studies on the Crustacea of the Turks and Caicos Islands, British West Indies V. Records of mysids from Pine Cay, Fort George Cay, Water Cay, and adjacent waters. Gulf Carib Res 16:147–159

Price WW, Anderson G, Heard RH (2002a) Mysida taxa and identification key. http://tidepool.st.usm.edu/mysids/mysidataxa.html

Price WW, Heard RW, Harris JT, McCoy CMR (2002b) Crustacea of the Cayman Islands, British West Indies I. Records of mysids from shallow water non-reef habitats. Gulf Carib Res 14:35–52

Roast SD, Thompson RS, Widdows J, Jones MB (1998) Mysids and environmental monitoring: a case for their use in estuaries. Mar Freshwat Res 49:827–832

Smith SI (1870) Notes on American Crustacea, No. 1, Ocypodoidea. Trans Conn Acad Art Sci 2:113–176

Smith SI (1873) Report upon the invertebrate animals of Vineyard Sound and the adjacent waters, with an account of the physical characters of the region. Report Comm. Fish., 1871 and 1872, pp 295–747

6. Tattersall OS (1967) A survey of the genus *Heteromysis* (Crustacea: Mysidacea) with descriptions of five new species from tropical coastal waters of the Pacific and Indian Ocean with a key for the identification of the known species of the genus. Trans Zool Soc Lond 31:57–193

Tattersall OS (1969) A synopsis of the genus *Mysidopsis* (Mysidacea, Crustacea) with key for the identification of its known species and descriptions of two new species from South African waters. J Zool Lond 158:63–79

Tattersall WM (1932) Contributions to the knowledge of the Mysidacea of California I. On a collection of Mysidae from La Jolla, California. Univ Calif Publ Zool 37:301–314

7. Tattersall WM (1951) A review of the Mysidacea of the United States National Museum. Bull US Nat Mus 201:1–292

Vannini M, Innocenti G, Ruwa RK (1993) Family group structure in mysids, commensals of hermit crabs (Crustacea). Trop Zool 6:189–205

Wittmann KJ (1999) Global Biodiversity in Mysidacea, with Notes on the Effects of Human Impact. *In*: Schram FR, von Vaupel Klein JC (eds) Crustaceans and the Biodiversity Crisis. Proc. 4th Int. Crust. Congr. Brill, Leiden, pp 511–525

Zimmer C (1918) Neue und wenig bekannte Mysidaceen des Berliner Zoologischen Museums. Mittl Zool Mus Berlin 9:15–26

Part 18
Cumaceans

Iorgu Petrescu, Richard W. Heard, Rita Vargas, and Odalisca Breedy

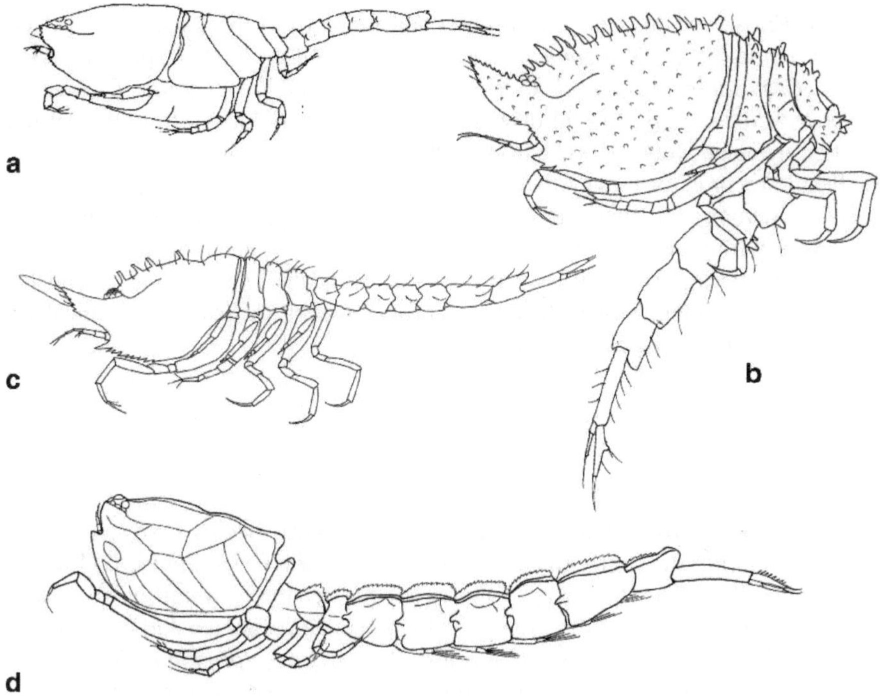

Cumaceans collected in Costa Rica. A. *Cyclapsis* sp. C., adult female; Pacific. B. *Cumella spinifera*, immature male; Caribbean. C. *Cyclapsis pustulata*, adult male; present on both coasts. D. *Cumella spinifera*, immature female; Caribbean. (Illustration: Iorgu Petrescu)

I. Petrescu
"Grigore Antipa" National Museum of Natural History, Department of Aquatic fauna, Kiseleff 1, Bucharest, 011341, Romania
e-mail: iorgup@antipa.ro

R.W. Heard
Department of Coastal Sciences, University of Southern Mississippi, Ocean Springs, MS 39566-7000, USA
e-mail: richard.heard@usm.edu

R. Vargas and O. Breedy
Museo de Zoología, Escuela de Biología, Universidad de Costa Rica, 11501-2060 San José, Costa Rica. e-mails: ritav@biologia.ucr.ac.cr; odalisca@racsa.co.cr

I.S. Wehrtmann, J. Cortés (eds.) *Marine Biodiversity of Costa Rica, Central America*,

Abstract Eight cumacean species, five from shallow water (*Cumella spinifera, Coricuma nicoyensis, Cyclaspis breedyae, C. dolera*, and *C. vargasae*), and three from deepwater (*?Diastylis tenebricosa, Makrokylindrus menziesi* and *Vemakylindrus costaricanus*), have been previously reported from the coastal and marine waters off Costa Rica. A preliminary study of the estuarine and marine Cumacea of Costa Rica conducted between 1998 and 2003 resulted in the collection of 14 species from Caribbean and 13 species from Pacific. Three families (Bodotriidae, Nannastacidae and Diastylidae) and ten genera were represented in these collections. The species from the Caribbean coasts are represented by the genera *Cyclaspis, Vaunthompsonia, Cumella, Elassocumella,* and *Schizotrema* and those from the Pacific by the genera *Coricuma, Cyclaspis, Leptocuma, Cumella* and *Diastylis*. Unexpectedly, eight species (*C. pustulata, C. jonesi, C. garrityi, C. gomoiui, C. medeeae, C. ocellata, C. ruetzleri*, and *C. somersi*), all of which were previously known from the western Atlantic, are tentatively reported from both coasts of Costa Rica. These apparent new Pacific records for these Atlantic species, especially those for *C. pustulata* and *C. ocellata*, require additional study (e.g., DNA fingerprinting) and careful morphological comparison with type material before any definitive taxonomic conclusions can be made. The 25 species (8 records from previous studies and the 17 new records established in the present study) now known from Costa Rican waters are listed systematically in tabular form with ecological information.

Introduction

Brief overview of the Order Cumacea: About 1,400 species representing 120 genera in eight families currently comprise the peracarid order Cumacea (Băcescu & Petrescu 1999). The members of this order are relatively small malacostracan crustaceans, usually 1–10 mm in length, which occur in brackish and marine habitats throughout the world. One species, *Almyracuma proximoculi* Jones & Burbanck, 1959, has been reported from freshwater conditions (Jones & Burbanck 1959). Cumaceans are known to occur from the intertidal zone to depths greater than 8,000 m (Băcescu & Petrescu 1999). Cumaceans are primarily benthic and/or epibenthic organisms, but they can also become part of the hypoplankton during mating and nocturnal migrations. The members of most cumacean families occur and feed in the first few centimeters of surface sediment; however, many members of the highly derived family Nannastacidae have evolved as cryptic creeping forms associated with algae, sponges, corals, and other epibenthic organisms. Some nannastacid species have highly derived, piercing mouth parts suggesting that they may be micropredators. This is especially true for members of the genera *Campylaspis, Procampylaspis*, and some other related genera that occur in shallow water tropical environments. Like the orders Tanaidacea and Isopoda, Cumacea release their young from the brood pouch (marsupium) as “mancas,” early stages that are characterized by having their last pair of thoracic appendages (legs) undeveloped and absent.

The distinctive morphology of cumaceans allows them to be easily separated from other malacostracan Crustacea. Some of these distinguishing features include: (a) a bulbous carapace composed of the fused dorsal parts of the cephalon (head) and first three thoracic somites (segments), (b) the first three pairs of thoracic legs modified for feeding (maxillipeds), (c) a long thin abdomen, and (d) a single pair of elongate uropods (often giving a "forked tail" impression). Members of the order display considerable sexual dimorphism. Females have a large brood chamber, a character that allies the Cumacea with other members of the Superorder Peracarida. With one known exception, female cumaceans lack pleopods, while in males the number of pleopods can vary from 0 to 5 pairs. Other important specific characters include the development of the antenna, the degree of development for the exopods, setation of the uropods, and the ornamentation of the carapace, which can be quite variable in some families. Some of the characters used to separate the genera and families are the morphology of the mouth parts, the number of male and female exopods, the presence or absence of a distinct telson, and the number or absence of male pleopods. Some important references on cumaceans include Calman (1912), Stebbing (1912), Fage (1951), Jones (1969, 1976), Băcescu (1988, 1992a), and Băcescu & Petrescu (1999). Two websites are also devoted to this crustacean group: http://www.coms.usm.edu/cumacean/ and http://nature.umesci.maine.edu/cumacea.html.

Historical background: Published information on the Cumacea of the Pacific coast of Costa Rica is limited to seven identified species: four shallow water bodotriids, *Coricuma nicoyensis* Watling & Breedy, 1988, *Cyclaspis breedyae* Petrescu & Heard, 2004, *C. dolera* Zimmer, 1943, and *C. vargasae* Petrescu & Heard, 2004; and three diastylids, *Vemakylindrus costaricanus* (Băcescu, 1961), *Makrokylindrus menziesi* Băcescu, 1962, and *Diastylis tenebricosa* Jones, 1969. Dexter (1974) recorded "Bodotriidae" from the Pacific beaches at Playita Blanca, Tamarindo, Sámara, Jacó, and Playa Cocal, and an "unidentified cumacean" at La Punta (Puntarenas). Until the recent description of *Cumella spinifera* Petrescu & Heard, 2004, the listing of "Bodotriidae" by Dexter (1974) from a beach at Cahuita was the only published record for a cumacean from the Caribbean coast of Costa Rica.

Cyclapsis dolera was the first cumacean described and reported from Costa Rican waters. Its description is based on two specimens, an adult and juvenile female, collected from the inshore waters of the northern Pacific (Zimmer 1944). Later *Coricuma nicoyensis* was described from the upper reaches of the Golfo de Nicoya (Watling & Breedy 1988). Vargas (1989) presented information on the reproduction, seasonal occurrence, and abundance of *C. nicoyensis* in a study conducted at the type locality, the lower intertidal zone on a mud flat at Punta Morales. Three deepwater diastylids, *V. costaricanus, M. menziesi*, and ?*D. tenebricosa* collected from depths ranging from 3,400 to 3,718 m off the west coast of Costa Rica were reported by Băcescu (1961, 1962) and Jones (1969). Recently, Petrescu & Heard (2004) described two shallow water species *Cyclaspis breedyae* and *C. vargasae*, from Puerto Caldera (Golfo de Nicoya) and Isla San José (Islas Murciélago), respectively.

Besides the works of Zimmer (1944), Băcescu (1961, 1962), Jones (1969), Watling & Breedy (1988), and Petrescu & Heard (2004) taxonomic works on the Cumacea from the tropical and subtropical American Pacific coast (Mexico to southern Peru) are restricted to those of Zimmer (1943a), Donath-Hernández (1987a, b), and Watling & McCann (1997).

Although there has been only one previous published report (Petrescu & Heard 2004) of cumaceans identified to the generic or species level (i.e., *C. serrata*) from the Caribbean coast of Costa Rica, there are several relevant reports dealing with their taxonomy and distribution from the Caribbean region. These include Donath-Hernández (1988) for the coast of Mexico, Petrescu & Iliffe (1992), and Petrescu (1996) for the Bahamas, Petrescu *et al.* (1993, 1994) for Jamaica, and Petrescu (2002) for the coast of Belize. Notwithstanding our overall knowledge of the cumaceans of the Caribbean, along the Central American coasts, it is still incomplete.

Preliminary survey of Costa Rican Cumacea: Between 1998 and 2003 a preliminary survey of the malacostracan Crustacea of Costa Rica was conducted. Collections were limited to three areas on the Caribbean coast (Playa Manzanillo, Puerto Viejo, and Puerto Vargas) and four areas on the Pacific coast (Islas Murciélago, Punta Morales, Puerto Caldera, and Boca Coronado). With the exception of three of the stations at Islas Murciélago, most of the collections were made at depths of less than 5 m. Most of the cumaceans collected during this survey are new to Costa Rica. The results of this preliminary study and a review of the previously known species of Cumacea from Costa Rican waters are the subject of this report.

Specimens were collected by Odalisca Breedy, Jorge Cortés, Richard Heard, Leonora Rodríguez, José Vargas, and Rita Vargas during several trips to the Caribbean and Pacific coasts in 1998–2003 (Table 18.1). Various collection methods and gear types were employed including fine-mesh dredge nets (kicknets), yabby pumps (hand-held suction devices), epibenthic sleds, dredges, and rock and algae washing techniques.

Cumacea occurred in samples from two areas on the Caribbean coast (Puerto Viejo and Puerto Vargas) and from three areas on the Pacific coast (Islas Murciélago, Punta Morales, and Caldera). On the Caribbean coast, specimens were collected in depths of less than 2 m, whereas at or near Islas Murciélago station depths ranged from 1 to 35 m, but most specimens came from less than 4 m. At Caldera and Punta Morales, cumaceans were collected in depths of less than 2 and 5 m, respectively (see Table 18.1).

A total of 21 species representing eight genera and three families were identified in the collections that we made from these areas between 1998 and 2003. They included at least four undescribed species, three belonging to the bodotriid genus *Cyclaspis* and one to the nannastacid genus *Cumella* (Petrescu & Heard 2004, F.E. Donath-Hernández, personal communication). Two other species may also be new, but further study is needed to determine whether or not they are ecophenotypic variants of known taxa.

Based on our collections and previous records (Zimmer 1944; Băcescu 1961, 1962; Watling & Breedy 1988), at least 25 species of Cumacea, representing three

Table 18.1 Costa Rican coastal sites where marine or estuarine Cumacea were collected between 1998 and 2003

Coast	Collection site	Depth (m)	Date	Methods
Pacific	Islas Catalina, Islas Murciélago (Sta.7), off southeast shore.	5–6	07 May 1999	Epibenthic sled
Pacific	Isla San José, Islas Murciélago (Sta. 1), South Beach	1–2	05 May 1999	Rock washings, dredge net
Pacific	Approx. 1 km east of Isla San José, Islas Murciélago	20–25	04 May 1999	Diving
Pacific	Approximately 2–3 km east of Isla San José (Sta. 8) Islas Murciélago	25–35	08 May 1999	Diving
Pacific	Punta Morales, Golfo de Nicoya, front Playa Blanca	1–5	18 April 1998	Epibenthic sled
Pacific	Puerto Caldera, Golfo de Nicoya, at beach in front	1–1.5	09 May 1999	Dredge net
Caribbean	Puerto Vargas, back reef area with coral rubble, some submerged vegetation, and sand, Sta. 4	1–1.5	25-–26 May 1998 18-–19 November 1999 09 August 2001	Rock washings, dredge net
Caribbean	Puerto Viejo, backreef area with coral rubble, sand, some submerged vegetation, algae	0.5–1.5	18 November 1999 09 August 2001	Rock washings, dredge net

families and ten genera, are now known from Costa Rican waters (Species Lists 18.1 and 18.2 are included on the CD-Rom). Species belonging to the two widely distributed genera *Cyclaspis* (six species) and *Cumella* (nine species) constitute over 70% of the species reported from both coasts of Costa Rica. Of the 14 species now known from the Costa Rican Caribbean coast, seven are previously known from the Caribbean coasts of Mexico and Belize (Donath-Hernández 1988; Petrescu 2002).

Eight species, *Cyclaspis jonesi, C. pustulata, Cumella garrityi, C. gomoiui, C. medeeae, C. ocellata, C. ruetzleri*, and *C. somersi*, which were previously known from the western Atlantic, are tentatively reported from the eastern Pacific for the first time. Although morphologically similar to their western Atlantic counterparts, there is a possibility that some of the Pacific materials may represent sibling species. The Pacific specimens, especially those of *Cumella ocellata* and *Cyclaspis pustulata*, will require careful additional study, possibly employing molecular techniques, before any definitive conclusions concerning their taxonomic status can be made.

The genus *Elassocumella* is only known from the western Atlantic, and *Coricuma* is currently considered endemic to the Costa Rican Pacific coast (Golfo de Nicoya). *Cyclaspis breedyae* and *C. vargasae* are currently known only from Costa Rican waters; however, this is probably due to the limited sampling of Cumacea along the Pacific coast of Central America. The remaining genera known for Costa Rican waters have relatively broad geographical distributions (Species Lists 18.1 and 18.2 are included on the CD-Rom).

An unexpectedly large number of species, all of which were previously known only from the western Atlantic, also appear to occur on Costa Rican Pacific coast. Also, there are several new taxa represented in our Costa Rican collections. Of these, two species of *Cyclaspis* from the Pacific coast and one species of *Cumella* from the Caribbean, were recently described by Petrescu & Heard (2004), and another species (*Cyclaspis* sp. C) is currently being described (F. Donath-Hernández, 2003, personal communication).

Our preliminary collections and observations, which were limited to and based on only a few shallow water sites (0–35 m), indicate that a large part of the Costa Rican cumacean fauna remains unknown. We believe that a more comprehensive sampling program utilizing fine-mesh epibenthic sleds and benthic grabs, especially in depths of 5–200 m, will result in the discovery of many additional cumacean species from both the Caribbean and Pacific coasts in Costa Rica.

Specialists

The following specialists have worked on Cumacea or are currently studying Cumacea from the Caribbean and/or tropical eastern Pacific regions are:

F. Eduardo Donath-Hernández: Miami, USA – edonath@caribnaz.org

Richard W. Heard: University of Southern Mississippi, Department of Coastal Sciences, 703 East Beach Blvd, P.O. Box 7000, Ocean Springs, MS 39566-7000, USA – NW Atlantic, eastern Pacific, Antarctic, and sub-Antarctic regions. richard.heard@usm.edu

Iorgu Petrescu: "Grigore Antipa" National Museum of Natural History, Department of Aquatic Fauna, Kiseleff 1, Bucharest, 011341, Romania – Worldwide iorgup@antipa.ro

Daniel Roccatagliata: Departamento de Biodiversidad y Biología Experimental, Facultad de Cs. Exactas y Naturales, Universidad de Buenos Aires, Ciudad Universitaria-Núñez, C1428EHA, Buenos Aires, Argentina – Western Atlantic, eastern Pacific, Antarctic, and sub-Antarctic regions. rocca@bg.fcen.uba.ar

Les Watling: Darling Marine Center, University of Maine, Walpole, Maine, 04573, USA – Worldwide. watling@maine.edu

Collections

Museums housing cumacean collections from the Caribbean and tropical eastern Pacific include:

- Museo de Zoología, Universidad de Costa Rica, San José (Costa Rica)
- Los Angeles County Museum, Los Angeles, California (USA)
- National Museum of Natural History, Smithsonian Institution, Washington, DC (USA)
- Gulf Coast Research Laboratory Museum, University of Southern Mississippi, Ocean Springs, Mississippi (USA)
- "Grigore Antipa" National Museum of Natural History, Bucharest (Romania)

Acknowledgments We wish to thank Jorge Cortés, José Vargas, and Ingo Wehrtmann of the Centro de Investigación en Ciencias del Mar y Limnología (CIMAR) and Escuela de Biología, Universidad de Costa Rica, for their support and help during the collection of material and the preparation of the manuscript. Daniel Roccatagliata kindly read over earlier drafts of the manuscript and made constructive comments and corrections. We also express our appreciation to the anonymous external reviewers for their constructive comments on the manuscript. This material is based in part upon work supported by the National Science Foundation under grant No. DEB-0529749.

References

1. Băcescu M (1961) Deux espèces nouvelles de *Makrokylindrus* sous-genre *Vemakylindrus* n.sg. (Crustacés Cumacés) des eaux tropicales du Pacifique (côte américaine). Rev Roum Biol Ser Zool 6:325–333
2. Băcescu M (1962) Contribution à la connaissance du genre *Makrokylindrus* Stebbing (Crustacea, Cumacea). Espèces nouvelles recueillies au cours des campagnes du Lamont Geological Observatory de New York. Abyssal Crustacea. 1. Vema Research Series, Columbia University, pp 209–222
3. Băcescu M (1971) New Cumacea from the littoral waters of Florida (Caribbean Sea). Trav Mus Hist Nat "Grigore Antipa" 11:5–24

Băcescu M (1988) Cumacea. Part 1. *In*: Gruner HE, Holthuis LB (eds) Crustaceorum Catalogus. SPB Academic Publishing, Bucharest, Romania, pp 1–173

Băcescu M (1992a) Cumacea. Part II. *In*: Gruner HE, Holthuis LB (eds) Crustaceorum Catalogus. SPB Academic Publishing, Bucharest, Romania, pp 175–468

4. Băcescu M (1992b) Deux espéces nouvelles de *Cumella* (Crustacea, Cumacea) des grottes sous- marines de Bermuda. Trav Mus Hist Nat "Grigore Antipa" 32:257–262
5. Băcescu M, Muradian Z (1977) Species of the genus *Cumella* (Cumacea, Nannastacidae) from the Western Tropical Atlantic. Trav Mus Hist Nat "Grigore Antipa" 18:89–102

Băcescu M, Petrescu I (1999) Ordre de Cumacés (Cumacea Krøyer, 1846). *In*: Forest J (ed) Traité de Zoologie, 7 (3a) Crustacés Péracarides. Mém. Inst. Océanogr. Monaco 19:391–428

Calman WT (1912) The Crustacea of the Order Cumacea in the collection of the United States National Museum. Proc US Nat Mus 41:603–676

Dexter DM (1974) Sandy-beach fauna of the Pacific and Atlantic coasts of Costa Rica and Colombia. Rev Biol Trop 22:51–66

6. Donath-Hernández FE (1987a) Distribution of the Cumacea (Crustacea, Peracarida) of the Bahía de Todos Santos, Baja California, Mexico. Cien Mar 13:35–52

Donath-Hernández FE (1987b) First record and enlargement of geographic distribution of four species of Cumacea for the Gulf of California (Crustacea, Cumacea). Cien Mar 13:70–74

7. Donath-Hernández FE (1988) Cumacea from the Gulf of Mexico and the Caribbean Sea (Crustacea, Peracarida). I: Descriptions of known species, new records and range extensions. Carib J Sci 24:44–51

Fage L (1951) Cumacés. *In*: P. Lechevalier (ed) Faune de France 54. Paris, France, pp 1–136

8. Jones NS (1969) The systematics and distributions of Cumacea from depths exceeding 200 meters. Galathea Rep 10:99–180

Jones NS (1976) British Cumaceans. Synopses of the British Fauna (new series). Linn Soc Lond 7:1–62

Jones NS, Burbanck WD (1959) *Almyracuma proximoculi* gen. et sp. nov. (Crustacea, Cumacea) from brackish water of Cape Cod, Massachusetts. Biol Bull 116:115–124

Petrescu I (1996) Cumaceans (Crustacea: Peracarida) from Abaco Island (Bahamas). Trav Mus Nat Hist nat "Grigore Antipa" 36:157–183

9. Petrescu I (2002) Cumacea (Crustacea: Peracarida) from Belize. Trav Mus Nat Hist Nat "Grigore Antipa" 44:141–203

Petrescu I, Iliffe TM (1992) Contributions to the knowledge of the Cumacea species from the British Blue Holes, Andros Island (Bahamas). Trav Mus Hist Nat "Grigore Antipa" 32:283–301

10. Petrescu I, Sterrer W (2001) Cumacea (Crustacea) from shallow waters of Bermuda. Ann Nat Mus Wien 103:89–128

Petrescu I, Iliffe TM, Sarbu S (1993) Contributions to the knowledge of Cumacea (Crustacea) from the littoral waters of Jamaica Island, including the description of three new species (I). Trav Mus Hist Nat "Grigore Antipa" 33:373–395

11. Petrescu I, Iliffe TM, Sarbu S (1994) Contributions to the knowledge of the cumaceans (Crustacea) from Jamaica II. Five new species of the genus *Cumella*. Trav Mus Hist Nat "Grigore Antipa" 34:347–367

12. Petrescu I, Heard RW (2004) Three new Cumacea (Crustacea: Peracarida) from Costa Rica. Zootaxa 721:1–12

13. Roccatagliata DC (1985) Three new species of the genus *Cyclaspis* (Cumacea) from the south-west Atlantic with a redescription of *Cyclaspis platymerus* Zimmer, 1944. Crustaceana 49:177–192

14. Roccatagliata DC, Moreira PS (1986) Four *Cyclaspis* species (Cumacea) from the South American Atlantic coast. Crustaceana 52:61–77

Stebbing TRR (1912) Cumacea (Sympoda). Das Tierreich 39:1–218

Vargas J (1989) Seasonal abundance of *Coricuma nicoyensis* Watling and Breedy (Crustacea, Cumacea) on an intertidal mud flat in the Gulf of Nicoya, Costa Rica. Rev Biol Trop 37:207–211

15. Watling L (1991a) Rediagnosis and revision of some Nannastacidae (Crustacea: Cumacea). Proc Biol Soc Wash 104:751–757

16. Watling L (1991b) Revision of the cumacean family Leuconidae. J Crust Biol 11:569–582

17. Watling L, Breedy O (1988) A new cumacean (Crustacea) genus from beaches of Golfo de Nicoya, Costa Rica. Rev Biol Trop 36:527–533

Watling L, McCann LD (1997) Cumacea. *In*: Blake JA, Scott PH (eds) Taxonomic Atlas to the Benthic Fauna of the Santa Maria Basin and Santa Barbara Channel. Vol. 11, Santa Barbara Museum of Natural History, Santa Barbara, California, pp 121–180

18. Zimmer C (1936) California Crustacea of the Order Cumacea. Proc US Nat Mus 83:423–439

19. Zimmer C (1943a) Cumaceen des Stillen Ozeans. Arch Nat 12:130–174

20. Zimmer C (1943b) Über neue und weniger bekannten Cumaceen. Zool Anz 141:148–167

21. Zimmer C (1944) Cumaceen des tropischen Westatlantiks. Zool Anz 144:121–137

22. Zimmer C (1980) Cumaceans of the American Atlantic boreal shores (Crustacea: Peracarida). Smithson Contr Zool 302: 1–29

Part 19
Tanaidaceans

Richard W. Heard, Odalisca Breedy, and Rita Vargas

Discapseudes colombiensis, a species collected from rotten wood in the mangroves of Térraba, southern Pacific coast of Costa Rica (Photo: Ingo S. Wehrtmann)

Abstract Little information is available of the crustacean order Tanaidacea in Costa Rican coastal waters. There are published reports for four species: *Parapseudes pedispinis*, *Neotanais armiger*, *N. pfaffi*, and "*Anatanais*? sp." from Pacific coast, and an unpublished report for the genera *Leptochelia* and *Apseudes* from the Caribbean. Preliminary data obtained from a limited shallow water survey (1998–2001), restricted to just four areas on each coast, indicate the presence of a relatively diverse tanaidacean fauna, especially on the Caribbean coast. This survey was restricted largely to shoreline sites usually in depths of less than 2 m; for this reason, the tanaidaceans collected probably only represent a fraction of those

R.W. Heard
Department of Coastal Sciences, University of Southern Mississippi, Ocean Springs, 703 East Beach Drive, Mississippi 39564–7000, USA.
e-mail: richard.heard@usm.edu

O. Breedy and R. Vargas
Museo de Zoología, Escuela de Biología, Universidad de Costa Rica, 11501-2060 San José, Costa Rica.

I.S. Wehrtmann, J. Cortés (eds.) *Marine Biodiversity of Costa Rica, Central America*,

occurring in Costa Rican waters. Preliminary examination of this material revealed the presence of several undescribed tanaidacean taxa. This would indicate that more extensive benthic sampling, especially at depths greater than 2 m, should yield many additional unknown species. The current status of the Tanaidacea previously reported from coastal and oceanic environments of Costa Rica is discussed, and a systematic list and synoptic table for the currently known species is presented.

Introduction

Crustaceans belonging to the Order Tanaidacea (Crustacea: Malacostraca: Peracarida) are poorly known and understudied in most oceanic regions of the world, which includes the Caribbean and Pacific coasts of Costa Rica. Published information on the tanaidaceans from Costa Rican waters is limited to five reports for the Pacific (Menzies 1953; Wolf 1956; Gardiner 1975a; Nunomura 1978) and to a single unpublished report by Breedy (1986) for the Caribbean.

Menzies (1953) in his major work on the apseudomorphan Tanaidacea from California to Peru, reported *Parapseudes pedispinis* (Boone 1923) from Parker Bay (= Bahía Santa Elena) and Playa Blanca, Costa Rica. Most of his Costa Rican specimens were found associated with coral (Menzies 1953). In an extensive monograph on the deepwater family Neotanaidae, Wolf (1956) reported two species, *Neotanais armiger* Wolf, 1956 and *N. pfaffi* Wolf, 1956 both from the same station at a depth of 3,590 m off the Costa Rican Pacific coast (Wolf 1956). An unidentified species, "*Antanais*? sp." belonging to the family Tanaidae was reported by Nunomura (1978) from Isla del Coco.

Two genera from the Caribbean were mentioned by Breedy (1986) in an unpublished study on peracarids of Cahuita coral reef, *Apseudes* Leach, 1814 and *Leptochelia* Dana, 1849, that constitutes the only information available for the east coast of Costa Rica.

Brief Overview of the Crustacean Order Tanaidacea (Modified from Heard 2002)

Tanaidaceans are usually considered to be a minor order of the crustacean class Malacostraca. Most are small (2–5 mm long) but adults range in size from 0.5–120 mm. Gamô (1984) described the largest known species, *Gigantapseudes maximus*, which was collected from deep water off the Philippines.

The order presently contains one extinct (Anthracocaridomorpha) and three extant suborders (Apseudomorpha, Neotanaidomorpha, and Tanaidomorpha) (Sieg 1980b; Guţu & Sieg 1999). Of the extant suborders, the widely distributed Tanaidomorpha, whose species are nearly all tube dwellers, contains the largest number and most highly derived species. Its members are widely distributed and occur in freshwater to hadal environments. The Apseudomorpha, which contains some of the larger and more plesiomorphic members of the order, is represented worldwide at nearly all depths (intertidal to hadal). The Neotanaidomorpha, represented by only a single family, is confined to deep water (continental slope or hadal) throughout the world's oceans. This small suborder has characters intermediate between the other two extant suborders.

Prior to 1980 the extant species were contained within two suborders, the Monokonophora and Dikonophora, the former presently represented by the Apseudomorpha and the latter now represented by the Neotanaidomorpha and Tanaidomorpha. At present there are 24 recognized families, 12 within the Apseudomorpha, one within the Neotanaidomorpha, and 11 within the Tanaidomorpha (Larsen & Wilson 2002, Larsen 2005, Guţu 2006, Anderson 2008, Anderson *et al.* 2008).

Several taxonomic changes at the family level have recently occurred. The families Anourpodidae and Tanapseudidae were recently synonymized with the families Parapseudidae and Kalliapseudidae, respectively (see Guţu 2001; Hansknecht *et al.* 2002), and a new apseudomorphan family, Numbakullidae, was recently described from Australian waters. In their revision of the tanaidomorphan superfamily Paratanaoidea, Larsen & Wilson (2002) diagnosed two new families, resurrected the families Leptognathiidae and Agathotanaidae, and synonymized the family Typhlotanaidae with the Nototanaidae.

Morphology and Ecology

General morphological characteristics for the Tanaidacea include a carapace formed via fusion of the first two thoracic somites, chelate second thoracopods, six free thoracic somites bearing peraeopods, five abdominal somites, sometimes fused or partially fused with or without pleopods and a pleotelson with a pair of terminal or subterminal uropods. As in the other members of the superorder Peracarida, tanaidaceans nearly always brood their young, usually within the brood pouch (marsupium) of the female. Subsequently, mancas (first and second juvenile stages), which lack a sixth or last pair of peraeopods, emerge from the marsupium to begin their epibenthic existence.

Tanaidaceans are known from the intertidal zone to hadal depths exceeding 9,000 m (Guţu & Sieg 1999). Although some euryhaline species such as *Sinelobus stanfordi* (Richardson, 1901), and *Teleotanais gerlachi* Lang, 1956 are capable of living in freshwater conditions (Gardiner 1975b; Sieg 1976; Sieg & Heard 1983); most species occur in marine environments. In fact, the majority of the approximately 800 recognized species (Guţu & Sieg 1999) have been reported living at depths below 300 m (Sieg 1983a, 1986a). In some deepwater habitats, tanaidaceans are among the most diverse and abundant assemblages present (Sieg 1986b; Wilson 1987). Although relatively few species occur in estuaries, the tanaidaceans present may constitute one of the more bionomically significant groups present and often occur in high densities, in some instances up to146,000 individuals/m^2 (Delille *et al.* 1985).

Preliminary Surveys on Costa Rican Tanaidacea

During 1998–2001, preliminary surveys were carried out in shallow Costa Rican waters. Tanaidacea were collected from four of the eight areas sampled along the coasts, two on the Caribbean and two on the Pacific. The Caribbean sites were at Puerto Vargas and Puerto Viejo, where specimens representing the families Apseudidae, Metapseudidae, Pagurapseudidae, Parapseudidae, Sphyrapidae, Tanaidae, and Leptocheliidae were collected in water depths of 0.5–1.5 m employing rock washing techniques or a dredge net adjacent to fringing corals reefs and in submerged seagrass beds. On the Pacific coast tanaidaceans were collected in two areas: an oligohaline, intertidal mangrove forest near the mouth of Río Terraba (SW coast), and the shores and waters surrounding Islas Murciélago, a small archipelago in the north near the border with Nicaragua.

At the brackish mangrove habitat in the Río Térraba delta, sampling was done by hand on muddy substratum with associated rotting wood; only a single brackish water tanaidacean, *Discapseudes colombiensis* (see Fig. 19.0), was collected from this sampling site. In contrast, from the waters along shores or in the immediate vicinity of Islas Murciélago, specimens representing several families (Tanaidae, Leptocheliidae, Parapseudidae) were collected. In shallow water (less than 2 m) rock and algae washing techniques and a fine mesh dredge net were employed, and in deeper water (3–35 m) an epibenthic sled and benthic dredge were employed.

We have not been able to study much of the tanaidacean collected during this limited preliminary survey; however, we have sorted a variety of tanaidacean taxa, including additional members of the families Leptocheliidae, Apseudidae, Parapseudidae, Tanaidae, and new records for the families Pagurapseudidae and Sphyrapidae. Some of these have been identified to the species level and are mentioned here.

Specimens of an apparently undescribed species of *Apseudes* was collected at depth of about 1.5 m in soft sediment associated submerged vegetation at Puerto Vargas. This species is fragile and elongate, a characteristic of deepwater members of the family. Only two specimens were collected and more material, especially adult males, is needed before it can be properly examined and described.

Specimens of the parapseudid, *D. colombiensis*, were collected in the muddy sediments and associated rotten wood from a brackish mangrove forest located in the Río Térraba estuarine system on the southern Pacific coast. The original description of *D. colombiensis* was based on two female specimens collected from Punta Soldado in Bahía de Buenaventura, Colombia. This occurrence represents the second record for this tube-dwelling apseudomorph and the first report from Costa Rica.

A member of the sphyrapid genus *Sphyrapoides* occurred on the Caribbean coast of Costa Rica. *Sphyrapoides* is the only genus of the family Sphyrapidae known to occur in shallow water (<20 m). The type species *S. bicornis* Guţu & Iliffe, 1998 was described from Bermuda (Guţu & Iliffe 1998) and recently Guţu & Heard (2003) described *S. tuberfrons* from Grand Cayman Island. During our survey a single adult specimen, similar and possibly conspecific with the species from Grand Cayman, was collected with a fine mesh dredge net at a depth of approximately 1 m behind the fringing reef at Puerto Vargas. Also, an unidentified species was recently reported from the Caribbean coast of Mexico (Suárez-Morales *et al.* 2004; García-Madrigal *et al.* 2005).

The pagurapseudid, *Pagurotanais bouryi* was relatively common at Puerto Vargas during November 1999. This highly derived species, which was previously only known form Cuban waters (Bouvier 1918; Guţu 1996), was found occupying small gastropod shells among coral rubble between the shore and the fringing reef. Morphological adaptations for living in the shells of dead gastropods by this and other members of its subfamily (Pagurapseudinae) include an asymmetrical body and chelae, and anterior peraeopods bearing specialized, sucker-like, hold-fast structures.

The neuter and females of two as yet unidentified species of the leptocheliid genus *Pseudoleptochelia*, one from the Caribbean at Puerto Vargas and the other from Islas Murciélago on the Pacific, were collected during our 1998–2001 preliminary survey. The specimens from Puerto Vargas resemble those of females of *P. mortenseni* Lang, 1973, a species described from the southeastern Caribbean Sea.

Specimens of an unidentified female collected on the Pacific coast in shallow waters at Islas Murciélago, which are somewhat similar to the Caribbean material. These specimens are definitely not referable to *Pseudoleptochelia filum* (Simpson 1853), a species that has previously been reported from the coasts of El Salvador and Perú (Sieg 1983a). Instead they appear too more closely allied to the Caribbean specimens. The females of these species, like those of *P. mortenseni*, are characterized by having their antennules short with first article broad dorsally. As in most leptocheliids, the males for the species of *Pseudoleptochelia* collected from the Costa Rican Caribbean and Pacific coasts are required before the taxonomic status of these two species can be determined with certainty. An apparently undescribed species of *Pseudoleptochelia*, possibly conspecific with the species from the Costa Rican Caribbean, has been reported recently from Florida (Heard *et al.* 2004) and Mexican Caribbean (Suárez-Morales *et al.* 2004; García-Madrigal *et al.* 2005).

Review of the Status of Tanaidacea Previously Reported from Costa Rica

Pacific Coast: During May 1999, we collected over 50 specimens of *P. pedispinis* from rock washings in shallow water at Isla San José, Islas Murciélago. Based on material from Laguna Beach, California, Boone (1923) described and designated this species as the type of *Dalapseudes*, however, Menzies (1953) synonymized this genus with *Parapseudes* in his redescription of the species. The species is known from California, USA (Boone 1923; Menzies 1953) and from the Pacific coasts of Mexico, Costa Rica, Colombia, and Ecuador (Menzies 1953). Other than our record of *P. pedispinous* from Islas Murciélago, there is no additional information on the distribution or the taxonomic status of the species previously reported and described from the Pacific coast of Costa Rica. This species is presently known from California southwest to Ecuador (Boone 1923; Menzies 1953).

The deepwater neotanaidomorphans, *N. armiger* and *N. pfaffi*, which originally described from off Central America (Wolf 1956), were redescribed by Gardiner (1975a), who also reported *N. armiger* from both Pacific and North West Atlantic (Gulf of Mexico). In the eastern Pacific, this species is presently known from the west coast of the United States (45° N) southward off Costa Rica, Panama, Columbia, Perú, to Chile (40° S) in depths ranging from 2,000 to 6,100 m (Gardiner 1975a).

N. pfaffi was reported originally by Wolf (1956) from off the coasts of Costa Rica (type locality), El Salvador, Honduras, and Perú in depths of 2,599 to 3,950 m (Wolf 1956). It was redescribed and illustrated by Gardiner (1975a).

Since the specimens of "*Anatanais*? sp." examined by Nunomura (1979) were damaged, additional collections are needed from Isla del Coco to clarify the identification of this species. During May 1999, we collected tanaids from the waters of the Islas Murciélago and tentatively placed them in the *Parazeuxo–Anatanais* complex, but they have not studied in detailed. A careful comparison of Nunomura's specimens with those from Islas Murciélago is needed to determine if the two taxa are conspecific.

Caribbean Coast: The material collected from artificial substrata studies conducted by Breedy (1986) was reexamined and four tanaidaceans, two apseudomorphans: *Apseudes cf. intermedius* and *Apseudomorpha* nr. *glebosus* (Menzies 1953), and two species of *Leptochelia*: *Leptochelia dubia*, and *L. forresti*, were identified.

Apseudes intermedius sensu *lato* is a relatively small (2–4 mm), variable, and apparently widely distributed species (Sieg 1983b) or species complex. The type specimens (two female syntypes) were originally described from "St. Vincent Island" in the Cape Verde Islands off Northwest Africa (Hansen 1895). Specimens attributable to this species are known from depths ranging of 1 to over 30 m and are widely distributed on live bottoms habitats in Brazil, the Caribbean, Gulf of Mexico, and Mediterranean (Morocco), and off Mexico (Sieg 1983a; R. Heard, personal observations). A closely related and possibly conspecific species, *A. bermudeus* Băcescu, 1980, has been reported from Bermuda (Băcescu 1980), and tentatively reported from Florida waters by Heard *et al.* (2004). *A. intermedius* also appears to have affinities with *A. tropicalis* Miller, 1940 known from the Hawaiian Islands, and especially with *A. garthi* Menzies, 1953 known from the eastern Pacific in the Gulf of California, Mexico (Menzies 1953). Some authors (Sieg 1983a) considered *A. garthi* as a junior synonym of *A. intermedius*, and Băcescu (1961) designated it as a subspecies, *A. intermedius garthi*.

There is a great deal of variability in size and the development of specimens attributed to populations of *A. intermedius* sensu *lato*. Studies on this species, or species complex, from the Caribbean, the Gulf of Mexico, and South Florida indicate that the males are protandric hermaphrodites (Heard *et al.* 2004; Suárez-Morales *et al.* 2004; García-Madrigal *et al.* 2005). One of us (RWH) has examined over 1,000 specimens from the Caribbean, the Gulf of Mexico, and South Florida and no terminal males have been recognized indicating that the males are protandric hermaphrodites.

Specimens closely resembling the metapseudid, *A. glebosus* Menzies, 1953 occurred commonly on the shallow live bottoms and artificial microhabitat studies adjacent to the fringing reef at Puerto Vargas on the Caribbean coast. Although there are some minor morphological differences between *A. glebosus* and the Caribbean specimens, it has not been determined whether or not the Caribbean populations represent an undescribed geminate species. Menzies (1953) originally placed *A. glebosus* in the new genus *Imitapseudes*, which is now considered a junior synonym of *Apseudomorpha* (see Gardiner 1973). Since it is known from California to as far south as Isla de Plata, Ecuador, there is a good possibility that *A. glebosus* sensu Menzies (1953) may occur in Costa Rican Pacific waters. Recently, a closely related but distinctly different species was found in algal washing of the southeast coast of Florida (T. Hansknecht and R. Heard, personal observation)

L. dubia, which is also known under the name *L. savignyi* Krøyer 1842 by some authors, has a nearly worldwide distribution (Lang 1973; Sieg 1983a; Dojiri & Sieg 1997), but some previous records may represent closely related sibling species (Ishimaru 1985; Larsen 2001). It was reported from the coast of El Salvador (Sieg 1976) as *L. savignyi*, but later Sieg (1983a) listed it under the name *L. dubia*. This nomenclatural problem exists because it was originally described as two species in the same publication (Krøyer 1842). The description of *L. dubia* is based on a specimen from Brazil and that of *L. savignyi* on a specimen from the Mediterranean. In Krøyer's original publication, the name *L. savignyi* appeared first in the text. Sieg (1983a), however, chose to recognize the name *L. dubia*, apparently because it came first alphabetically; however, his reasoning has not been followed by other authors (e.g., Ishimaru 1985). Dojiri & Sieg (1997) rediagnosed and illustrated the species based on material from California (USA). *L. dubia* was common in both the collections of Breedy (1986) and those made by us during the preliminary survey. This species appears to be one of the most common and abundant tanaidaceans occurring in the warm waters of the northwestern Atlantic (Heard *et al.* 2004; Suárez-Morales *et al.* 2004; García-Madrigal *et al.* 2005; R. Heard, personal obsservations). Several size morphs attributable to *L. dubia* have been observed, but since so many male and female stages are known for some leptocheliid species, further careful study is needed to determine whether a single polymorphic species or one or more closely related cryptic species are present in the Costa Rican waters.

Specimens of the highly sexually dimorphic species, *L. foresti*, were collected in rock and algae washings made at Puerto Vargas and Puerto Viejo. *L. foresti* was described from an adult male specimen collected at Antigua in the Caribbean. Stebbing originally placed this species in the monotypic genus *Dolichochelia* Stebbing, 1896, but he later synonymized it with the genus *Leptochelia* (for synonymies see Sieg 1983a). Like *L. dubia*, *L. foresti* appears to be common in reef, back reef, and grass bed habitats along the southeastern Caribbean coast of Costa Rica. It occurred in both collections of Breedy (1986) and our survey. This species is quite common in shallow water habitats throughout the warm waters of the Gulf of Mexico and Caribbean Sea (Heard *et al.* 2004; Suárez-Morales *et al.* 2004; García-Madrigal *et al.* 2005; R. Heard, personal observations). Like *L. dubia*, considerable polymorphism and several size morphs attributable to this species were observed during our preliminary study of Costa Rican material, and there remains the possibility that *L. foresti* sensu *lato* may represent a complex of closely related cryptic species in the Caribbean region. Specimens collected from Costa Rica, the Mexican Caribbean, and south Florida have dark pigment spots, a rare condition for members of the family Leptocheliidae (Heard *et al.* 2004; Suárez-Morales *et al.* 2004; García-Madrigal *et al.* 2005).

Synoptic Species Lists 19.1 and 19.2 (both included on CD-Rom), present taxonomic listings of the Tanaidacea currently known from Costa Rican waters, including the new records and data established during this study is presented. These tables also include distribution data and relevant references to the taxa treated.

Important Publications, Specialists, and Collections

Some important sources of information on the taxonomy, systematics, distribution, evolution, or biology of Tanaidacea are: Anderson (2008), Anderson *et al.* (2008) Dojiri & Sieg (1997), Gardiner (1975a, b), Gutu (1996), Gutu (2005), Gutu & Sieg (1999), Hansen (1913), Heard *et al.* (2004), Holdich & Jones (1983), Larsen 2001; Larsen (2006), Larsen & Wilson (2002), Shiino [1978] (1979), Sieg (1980a, b, 1983a, b, 1986a, b, c), and Sieg & Winn (1981).

Specialists

Currently working in the region:

ODALISCA BREEDY: Museo de Zoología, Universidad de Costa Rica, San José, Costa Rica – Central America, primarily on shallow waters species. odalisca@racsa.co.cr

MODEST GUŢU: "Grigore Antipa" National Museum of Natural History, Bucharest, Romania – Worldwide, works primarily on Apseudomorpha. mgutu@antipa.ro

RICHARD HEARD: University of Southern Mississippi, Department of Coastal Sciences, Ocean Springs, Mississippi, USA – Worldwide, works primarily on Apseudomorpha and shallow water Tanaidomorpha. richard.heard@usm.edu

KIM LARSEN: Texas A & M University, Galveston, Texas, USA – Worldwide, works primarily on Tanaidomorpha and Neotanaidomorpha. tanaids@hotmail.com

RITA VARGAS: Museo de Zoología, Universidad de Costa Rica, San Jose, Costa Rica – Central America, works primarily on shallow water species. ritav@biologia.ucr.ac.cr

KÁTIA CHRISTOL DOS SANTOS: Universidade de São Paulo, Instituto Oceanográfico, São Paulo, Brazil – Northeastern Brazil, works primarily on Apseudomorpha and Neotanaidomorpha. kchristol@ceres.io.usp.br

Thomas Hansknecht, Barry Vittor & Associates: Mobile, Alabama, USA. – Warm temperate an tropical western Atlantic, works primarily on Apseudomorpha and Neotanaidomorpha

Collections

The main collections containing specimens of Costa Rican Tanaidacea are housed in the following museums:

- Gulf Coast Research Laboratory Museum, University of Southern Mississippi, Ocean Springs, MS (USA.)
- Los Angeles County Museum, Los Angeles, CA (USA).
- National Museum of Natural History, Smithsonian Institution, Washington, DC (USA).
- Museo de Zoología, Universidad de Costa Rica, San Jose (Costa Rica).

Conclusions

If one extrapolates based on the currently known species in deep waters of the Northeast Atlantic, a large part of which still remains unsurveyed, and with the less well-studied regions of the world's oceans, we conservatively estimate that well over 2,000 species remain to be discovered. Recent molecular research by Larsen (2001) suggests that even this estimate might be significantly underestimated. Following this reasoning, and the fact the tanaidacean fauna of the Costa Rican Caribbean and Pacific has been so poorly studied, we conservatively estimate that the number of species listed in this report represents less than 20% of the total species present in the country's estuarine and continental shelf habitats. Based on previous studies in deepwater habitats, the tanaidacean fauna in depths greater than 200 m off the Caribbean and Pacific coasts of Costa Rica may be represented by well over 100 species.

Recommendations

Making collections with large box corers and benthic grabs are labor and equipment-intensive, and consequently prohibitively expensive. To economically sample and collect tanaidaceans and other small benthic and epibenthic crustaceans (e.g., Cumacea, Mysida, Isopoda, Amphipoda) occurring in Costa Rican subtidal near shore waters (2–20 m), small anchor dredges and epibenthic sleds should be employed. These can be made relatively inexpensively (US$50–250) and they are capable of being employed from both small and large vessels. Because this gear is relatively light, a wench is not required for its retrieval, which can be done by hand when the boat of the vessel has stopped or is going very slowly.

Aknowledgments This material is based in part upon work supported by the National Science Foundation under grant No. DEB-0529749.

References

Anderson, G (2008) Tanaidacea taxa and literature, 28 May, 2008. http://peracarida.usm.edu/fmi/iwp/cgi?-db=Tanaidacea&-loadframes [accessed 1 August 2008]

Anderson, G, Heard RW, Larsen K, Drumm DT (October 2007) Tanaidacea Homepage, http://www.usm.edu/gcrl/tanaids/

Băcescu M (1961) Contribution a la connaissance des tanaidaces de la Méditerranée orientale, 1: Les Apseudidae et Kalliapseudidae des côtes d'Israel. Bull Res Coun Is, Zoology 10B:137–170

Băcescu M (1980) *Apseudes bermudeus*, n. sp. from caves around Bermuda Islands. Acta Adriaica 21:401–407

1. Boone PL (1923) New marine tanaid and isopod Crustacea from California. Proc Biol Soc Wash 36:147–148
2. Bouvier EL (1918) Sur un petitecollectionde crustacés de Cuba offerte au Muséum par M. deBoury. Bull Mus Nat, Paris 24:6–15
3. Breedy O (1986) Contribución al estudio de los microcrustaceos bentónicos (Isopoda y Tanaidacea) en el arrecife coralino del Parque Nacional Cahuita, Limón, Costa Rica. Tesis Licenciatura Biología, University of Costa Rica, 107 p

Delille D, Guidi LD, Soyer J (1985) Nutrition of *Allotanais hirsutus* (Crustacea, Tanaidacea) at Kerguelen Islands, pp 378–380. *In*: Siegfried WR, Condy PR, Laws RM (eds) Antarctic Nutrient Cycles and Food Webs, Springer Verlag, Berlin

Dojiri M, Sieg J (1997) The Tanaidacea. Chapter 3, pp 181–268. *In*: Blake JA, Scott PH (eds) Taxonomic Atlas of the Benthic Fauna of the Santa Maria Basin and Western Santa Barbara Channel, Vol. 11. The Crustacea Part 2. The Isopoda, Cumacea and Tanaidacea. Santa Barbara Museum of Natural History, Santa Barbara, CA.

Gamô S (1984) A new remarkably giant tanaid, *Gigantapseudes maximus* sp. nov. (Crustacea) from the abyssal depths far off southeastern of Mindanao, the Philippines. Scient Rep Yokohama Natl Univ Sec II 31:1–12

García-Madrigal S, Heard RW, Suárez-Morales E (2005) Records of and observations on tanaidaceans (Peracarida) from shallow waters of the Caribbean coast of Mexico. Crustaceana 77:1153–1177

Gardiner LF (1973) New species of the genera *Synapseudes* and *Cyclopoapseudes* with notes on morphological variation, post marsupial development, and phylogenetic relationships within the family Metapseudidae (Crustacea: Tanaidacea). Zool J Linn Soc 53:25–58

4. Gardiner LF (1975a) The systematics, postmarsupial development and ecology of the deep-sea family Neotanaidae (Crustacea: Tanaidacea). Smithson Contr Zool 170:1–265

Gardiner LF (1975b) A fresh- and brackish-water tanaidacean, *Tanais stanfordi* Richardson, 1901, from a hypersaline lake in the Galapagos Archipelago, with a report on West Indian specimens. Crustaceana 29:127–140

5. Guţu M (1996) The synoptic table and key to superspecific taxa of recent Apseudomorpha (Crustacea, Tanaidacea). Trav Mus Hist Nat "Grigore Antipa" 35:135–146

Guţu, M (2006) New Apseudomorph Taxa of the World Ocean: Crustacea, Tanaidacea. Curtea Veche, Bucharest, Romania, 318p

Guţu M, Heard RW (2003) A new genus and species of a new family of apseudomorph tanaidaceans (Crustacea: Percarida) form Australian waters. Trav Mus Hist Nat "Grigore Antipa" (2002) 44:93–103

6. Guţu M, Iliffe TM (1998) *Sphyrapoides bicornis*, new genus and new species from a submarine scarpment in the Bahamas and some remarks on Sphyrapinae (Crustacea, Tanaidacea). Trav Mus Hist Nat "Grigore Antipa" 35:29–48
7. Guţu M, Ramos GE (1995) Tanaidaceans (Crustacea, Peracarida) from the waters of Colombian Pacific with the description of two new species. Trav Mus Hist Nat "Grigore Antipa" 35:29–48

Guţu M, Sieg J (1999) Ordre tanaidacés (Tanaidacea Hansen, 1895). Mém Inst Océanogr, Monaco 19:353–389

Hansknecht T, Heard RW, Bamber R (2002) *Tanapseudes gutui*, a new species of apseudomorphan Tanaidacea (Crustacea: Peracarida) from the Caribbean Sea and the taxonomic status of the family Tanapseudidae Băcescu, 1978. Gulf Carib Res 14:67–74

Hansen H (1895) Isopoden, Cumaceen und Stomatopoden der Plankton Expedition. II. Ordnung: Tanaidacea. Ergebnisse der Plankton-Expedition 2:(G.C.): 49–50

Hansen HJ (1913) Crustacea Malacostraca. II. The Order Tanaidacea. The Danish Ingolf Expedition 3, IV Copenhagen, 145 p

Heard RW (2002) An annotated checklist and bibliography for the order Tanaidacea (Crustacea: Malacostraca: Percarida) reported form the Pacific coasts of the Americas (Alaska to Chile) and associated off shore islands, pp 369–383. In: Hendrickx ME (ed) Contributions to the study of East Pacific Crustaceans. Inst Cienc Mar Limnol, UNAM, 383 p

Heard RW, Hansknecht T, Larsen K (2004) Guide to the Tanaidacea (Crustacea: Peracarida) from Florida waters to depths of 200 m. Final Suppl Rep Florida Dept Environ Protect, Tallahassee, Florida, 157 p

Holdich DM, Jones JA (1983) Tanaids: Keys and Notes for the Identification of the Species. Synopses of British Fauna 27. Cambridge University Press, London, 98 p

Ishimaru S-I (1985) A new species of *Leptochelia* (Crustacea, Tanaidacea) from Japan, with redescription of *L. savignyi* (Krøyer, 1842). Publ Seto Mar Biol Lab 30:241–267

Krøyer H (1842) Nye Arteraf Slagten *Tanais*. Kroyer Naturhist Tids 6:167–188

Lang K (1956a) Neotanaidae nov. fam., with some remarks on the phylogeny of the Tanaidacea. Ark Zool 2:469–475

Lang K (1956b) Tanaidacea aus Brasilien, gesammelt von Prof. Dr. A. Remane und Dr. S. Gerlach. Kieler Meer 12:249–260

Lang K (1973) Taxonomische und phylogenetische Untersuchungen über die Tanaidaceen (Crustacea). 8. Die Gattungen *Leptochelia* Dana, *Paratanais* Dana, *Heterotanais* G.O. Sars und *Nototanais* Richardson. Dazu einige Bemerkungen über die Monokophora und ein Nachtrag. Zool Scripta 2:197–229

Larsen K (2001) Morphological and molecular investigation of polymorphism and cryptic species in tanaid crustaceans: Implications for tanaid systematics and biodiversity estimates. Zool J Linn Soc 131:353–379

Larsen K (2005) Deep-Sea Tanaidacea (Peracarida) from the Gulf of Mexico. Brill, Leiden, 381 p.

Larsen K, Wilson GDF (2002) Tanaidacean phylogeny, the first step: the superfamily Paratanaidoidea. J Zool Syst Evol Res 40:205–222

8. Menzies R (1953) The apseudid Chelifera of the eastern tropical and north temperate Pacific Ocean. Bull Mus Comp Zool 107:442–496
9. Nunomura N (1978) Tanaidaceans and anthuridean isopods collected on the Presidential Cruise of 1938. Proc Biol Soc Wash 91:936–952

Shiino SM (1979) (1978) Tanaidacea collected by French scientists on board the survey ship "Marion-Dufresne" in the regions around the Kerguelen Islands and other subantarctic islands in 1972, '74, '75, '76. Sci Rep Shima Marineland 5:1–122

Sieg J (1976) Crustacea, gesammelt von Professor Dr. W. Noodt an den Küsten El Salvadors und Perus. Stud Neotr Fauna Env 11:65–85

Sieg J (1980a) Taxonomische Monographie der Tanaidae Dana 1849 (Crustacea: Tanaidacea). Abh Senck Natur Ges 537:1–267

Sieg J (1980b) Sind die Dikonophora eine polyphyletische Gruppe? Are the Dikonophora a polyphyletic group?. Zool Anz 205:401–416

Sieg J (1983a) Evolution of Tanaidacea. Crust Iss 1:229–256

Sieg J (1983b) Tanaidacea. *In*: Gruner HE and Holthuis LB (eds) Crustaceorum Catalogus, Part 6. W Junk, The Hague, 552 p

Sieg J (1986a) Biology of the Antarctic Seas 18. Crustacea Tanaidacea of the Antarctic and Subantarctic. 1. On material collected at Tierra del Fuego, Isla de los Estados, and the west coast of the Antarctic Peninsula. Ant Res Ser 45:1–180

Sieg J (1986b) Tanaidacea (Crustacea) von der Antarktis und Subantarktis. II. Tanaidacea gesammelt von Dr. J. W. Waegele während der Deutschen Antarktis Expedition 1983. Mittl Zool Mus Kiel II (4): 1–80

Sieg J (1986c) Distribution of the Tanaidacea: synopsis of the known data and suggestions on possible distribution patterns. Crust Iss 4:165–194

Sieg J, Heard RW (1983) Tanaidacea (Crustacea: Peracarida) of the Gulf of Mexico 3. On the occurrence of *Teleotanais gerlachi* Lang, 1956 (Nototanaidae) in the Eastern Gulf. Gulf Res Rep 7:267–271

Sieg J, Winn RN (1981) The Tanaidae (Crustacea: Tanaidacea) of California, with a key to the world genera. Proc Biol Soc Wash 94:315–343

Suárez-Morales E, Heard RW, García-Madrigal MS, Oliva JJ, Escobar E (2004) Catálogo de los tanaidáceos (Crustacea: Peracarida) del Caribe Mexicano. CONACYT/SEMARNA/ECOSUR. México, 121 p

Wilson GDF (1987) Crustacean communities of the Manganese Nodule Province. Report for NOAA, Of. Ocean & Coast. Res Manag, 35 p

10. Wolf T (1956) Six new abyssal species of *Neotanais* (Crust. Tanaidacea). Vidensk Medd Dansk Natur Forensing København 118:41–52

Part 20
Isopods

Richard C. Brusca and Ingo S. Wehrtmann

Ancinus brasiliensis, a species reported from the Caribbean coast of Costa Rica (Photo: Leslie Harris)

Abstract The order Isopoda is diverse (~10,300 described species worldwide) and common in almost all environments – marine, terrestrial, and freshwater. A total of 78 marine species have been reported from Costa Rica (46 species from the Caribbean coast and 34 from the Pacific), comprising 6 suborders and 24 families. Two species occur on both coasts: *Excirolana braziliensis* and *Cirolana parva*. Numerous additional species are expected to occur in Costa Rican waters, probably three times this number, but due to the paucity of research on this taxon in this region only these 78 can be included in our species lists. Only two broad surveys have been accomplished for Costa Rican isopods: Brusca & Iverson's (1985) survey of the Pacific coast, and an unpublished survey of the isopods of Cahuita (Brusca & White, in preparation). Representatives of at least eight families (Anthuridae, Ancinidae, Sphaeromatidae, Cirolanidae, Corallanidae, Aegidae, Gnathiidae, Holognathidae) occur on both the Pacific and Caribbean coasts. The most specious families of isopods on Costa Rican shores are: Asellota (16 species), Anthuridae (14 species), Sphaeromatidae (13 species), and Cirolanidae (seven species).

Introduction

The order Isopoda is one of the nine orders in the crustacean superorder Peracarida. Peracarids are the "marsupial crustaceans," distinguished from the three other eumalacostracan superorders (Hoplocarida, Syncarida, Eucarida) by the presence of a ventral marsupium constructed by specialized thoracic coxal endites called oöstegites, in which the developing embryos are housed (except in Thermosbaenacea, which carry broods beneath the carapace). All peracarids undergo direct development with brooding, hence true larval forms do not occur in this superorder. Isopods can be distinguished from other peracarids (and crustaceans in general) by their dorsoventrally flattened body (except in Anthuridea and Phreatoicidea); compact head with unstalked compound eyes; seven free thoracomeres (five in Gnathiidea); one pair of maxillipeds; appendages never chelate; six pairs of biramous pleonal appendages, including five pairs of platelike respiratory/natatory pleopods and a single pair of fanlike or sticklike uropods; and biphasic molting (i.e., posterior half of body molts before anterior half).

R.C. Brusca
Arizona-Sonora Desert Museum, 2021 N. Kinney Rd., Tucson, AZ 85743, USA.
e-mail: rbrusca@desertmuseum.org

I.S. Wehrtmann
Unidad Investigación Pesquera y Acuicultura (UNIP) of the Centro de Investigación en Ciencias del Mar y Limnología (CIMAR) and Museo de Zoología, Escuela de Biología, Universidad de Costa Rica, 11501-2060 San José, Costa Rica.
e-mail: ingowehrtmann@gmx.de

All isopods possess one of the two fundamental morphologies, being "short-tailed" or "long-tailed" (Brusca & Wilson 1991). In the more primitive, short-tailed isopods the telsonic region is small, positioning the anus and uropods terminally or subterminally on the pleotelson (Phreatoicidea, Asellota, Microcerberidea, Oniscidea, Calabozoidea). The more derived, long-tailed isopods have the telsonic region greatly elongated, thus shifting the anus and uropods to a subterminal position on the pleotelson (Flabellifera, Anthuridea, Gnathiidea, Epicaridea, Valvifera).

In the sea, isopods compare in ecological importance to the related Amphipoda and Tanaidacea, notably as abundant intermediate links in food chains. They typically predominate, along with tanaids, bivalves, and polychaetes, in soft bottom sediment samples from continental shelves. On many coasts, isopods may constitute the majority of prey items consumed by nearshore fishes (Wallerstein & Brusca 1982). In the Arctic region, they are one of the primary food items of gray whales (R.C. Brusca, personal observation). Intertidal isopods are predominantly benthic and cryptic, living under rocks, in crevices, empty shells and worm tubes, and among sessile and sedentary organisms, such as algae, sponges, hydroids, ectoprocts, mussels, urchins, barnacles, and ascidians. Some burrow in natural substrates including mud, sand, soft rocks, and driftwood, and some burrowers, such as the *Limnoria* (the gribbles) and *Sphaeroma*, can do extensive damage to pilings and wooden boats. In the tropics, some species of *Sphaeroma* burrow into mangroves, weakening the prop roots and causing them to break more easily, which typically stimulates the growth of multiple new rootlets, leading to the classic stairstep structure of red mangrove prop roots (Perry & Brusca 1989). Several species are important scavengers on shore wrack or dead animals (e.g., *Ligia*, *Tylos*). Cirolanids, corallanids, and tridentellids are voracious carnivores, functioning both as predators and scavengers. Epicarideans are all parasites on other crustaceans, cymothoids are all parasites on fishes, and aegids and gnathids are "temporary parasites" on fishes. Some invertebrate parasites, notably acanthocephalans, use isopods as intermediate hosts.

Of 10,300 described isopod species, about 4,500 are terrestrial/freshwater and 5,800 marine. Isopods occur in nearly every environment on Earth – from the littoral zone to the greatest depths of the sea (10,000 m or more), in lakes, streams, rivers, springs, cave systems, and water entrapments in bromeliads (see www.nmnh.si.edu/iz/isopod/; Brusca & Brusca 2003). Terrestrial species are found nearly everywhere, even in the driest deserts. Many aquatic species are parasites. The body of isopods (Fig. 20.1) is usually dorsoventrally flattened and composed of three regions (see Brusca & Iverson 1985; Brusca & Brusca 2003): (1) the cephalon or cephalothorax (head), bearing compound eyes (some species have greatly reduced eyes), two pairs of antennae, mandibles, two pairs of maxillae, and the maxillipeds; (2) the pereon, composed of those thoracic segments not fused with the head, its segments (pereonites) bearing paired uniramous legs, or pereopods (with few exceptions, isopods have seven pairs of pereopods); and (3) pleon (abdomen) with its free or variously fused segments, or pleonites. All isopods have the sixth pleonite fused with the telson, thus forming a pleotelson. The pleonal appendages are ventral swimming and gas-exchange structures called pleopods.

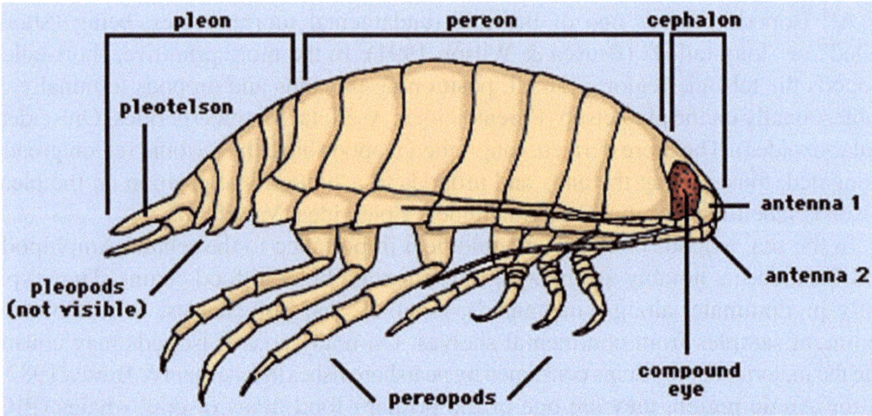

Fig. 20.1 Schematic isopod with morphological characteristics (from http://tolweb.org/tree)

Most isopods are dioecious, and generally males and females can be distinguished by the form of the second (or the first and second) pleopods (which bear gonopods, or appendices masculinae, in males), presence of a pair of penes on the ventrum of the last pereonite in males, and presence of oostegites in mature females.

Limited information on the marine isopods of Costa Rica is available. Brusca & Iverson (1985) produced a guide to the isopods of the Pacific coast of Costa Rica, including a total of 37 species of 14 families; these numbers include actual recorded species as well as those which are expected to occur along the Pacific coast of Costa Rica. Guzmán *et al.* (1988) studied the diel and seasonal occurrence, density, size, and sex ratio of *Excorallana tricornis occidentalis* from Isla del Caño. Several studies have addressed the effects of root-boring isopods of the genus *Sphaeroma* on mangroves along the Pacific coast of Costa Rica (Villalobos *et al.* 1985; Perry 1988; Perry & Brusca 1989). Dexter (1974) examined the macroscopic infauna of sandy beaches along both the Pacific and the Atlantic coasts of Costa Rica; the isopods *Excirolana braziliensis* (as *Cirolana salvadorensis*) and *Exosphaeroma* sp. (as *E. diminutum*) were the most abundant species, occurring on both coasts. Jiménez & Vargas (1990) reported on the parasitic bopyrid isopod *Probopyrus pandalicola* infesting the caridean shrimp *Palaemonetes schmitti* (reported as *P. hiltonii*) on the Pacific coast (also see Campos & Campos 1989). Markham (1992) summarized known distributions for the bopyrid isopods of the tropical eastern Pacific. Regarding the Caribbean coast of Costa Rica, Breedy & Murillo (1995) collected seven isopods on artificial habitats, but the validity of their identifications has not been established. More recently, Wetzer & Bruce (1999) described a new genus and species of sphaeromatid isopods from Caribbean Costa Rica. Brusca & White (in preparation) conducted an 18-month study of isopod diversity on the Cahuita Reef, and data from that work are included in this part (incorporated into Species List 20.1 which is included on the CD-Rom).

Isopod Diversity in Costa Rica

A total of 78 isopod species have been reported from both coasts of Costa Rica (Species Lists 20.1 and 20.2 are included on the CD-Rom); two species (*E. braziliensis* and *C. parva*) have been collected from both coasts. At least 16 additional species are expected to occur in shallow waters along the Pacific coast of Costa Rica, but have not yet been collected there (Brusca & Iverson 1985); they are not included in this part. The isopod fauna of Costa Rica's offshore benthic habitats has yet to be studied in any detail. Overall, it is reasonable to assume that species diversity of the Isopoda of both Pacific and Caribbean coasts of Costa Rica is considerably higher than reflected by the reports so far published, probably three times the reported diversity to date.

Costa Rican isopods are represented by 6 suborders and 24 families. Representatives of eight families (Anthuridae, Ancinidae, Sphaeromatidae, Cirolanidae, Corallanidae, Aegidae, Gnathiidae, Holognathidae) occur on both coasts. The most specious families of isopods in Costa Rican waters are: Asellota (16 species), Anthuridae (14 species), Sphaeromatidae (13 species), and Cirolanidae (seven species).

The only place on the Caribbean coast of Costa Rica that has been examined for its isopod fauna is Cahuita National Park, which is well-known for its nearshore fringing reef. Unlike offshore barrier reefs, which are characterized by extensive coral growth, complex zonation, and large deep inner lagoons, fringing reefs are small in size, shallow, and under the influence of terrestrial/coastal runoff and sedimentation. In the case of Cahuita, coastal runoff from nearshore and upland disturbance (deforestation and soil erosion, agriculture, etc.) has powerfully impacted the reef for many years. Most of the Cahuita reef suffers from high sediment load and eutrophication. Sedimentation decreases water clarity, which disrupts endosymbiotic photosynthesis reducing coral growth. Coral recruitment is reduced because of the coverage by sediments and increased growth of non-coral organisms, which further reduces the settlement of coral planula larvae (Cortés & Risk 1985; Cortés 1994). Agricultural pollution adds excessive nutrients to the system and further drives eutrophication. It is well-known that moderate levels of nutrients lead to changes in community structure where scleractinian corals are outcompeted for space by algae and benthic filter feeders such as sponges, bryozoans, and tunicates (Pastorok & Bilyard 1985; Tomascik & Sander 1987; Wittenberg & Hunte 1992; Naim 1993; Chabanet *et al.* 1995). This change in community structure may result in increased bioerosion of the reef framework reducing habitat complexity which, in turn, affects other organisms (Sano *et al.* 1987). At higher nutrient inputs, coral growth may be reduced due to inorganic phosphate, which prevents calcification (Kinsey & Davies 1979).

Cahuita National Park (9°18′ N, 82°7′ W) was established in 1970 and includes a small 240 ha barrier reef and a 1,100 ha coastal lowland wet forest (Cortés & Risk 1985). Cahuita has coral growth relatively close to the shore and scattered patch reefs within the lagoon. On the reef crest, spur and groove formations reach 10 m in depth (Risk *et al.* 1980). Shoreward from the reef crest, the lagoon varies in width from 400 to 1,100 m and has a depth of less than 10 m. The lagoon has scattered boulders and rocky outcrops, coral heads, scattered small patch reefs, large rubble areas, sand/mud flats, and sea grass beds. The reef crest itself is approximately 4 km long with well-developed buttresses (spur and groove formations), but live coral coverage and diversity are low within the lagoon and outer reef zones.

Despite the anthropogenic impacts and destruction of corals, the isopod fauna of the Cahuita reef area is quite diverse (45 species), more diverse than some studied "pristine" reefs such as Carrie Bow Cay, Belize (Brusca, Kensley & White, unpublished data, 2007). However, species composition differs markedly between Cahuita and "pristine" offshore reefs, in that the Cahuita community is strongly dominated by herbivores and detritivores as opposed to a predominance of carnivores on offshore reefs. At Cahuita, species richness is positively associated with habitat type, and most species occur in algal-covered coral rubble habitats. There is no change in species richness at Cahuita between the wet and the dry season. Of the 45 isopod species known from the Cahuita reef, one, *Paracerceis caudata* (an herbivore), is the dominant species in all samples taken, except dry-season algal-covered rubble samples.

Specialists

Richard C. Brusca: Arizona-Sonora Desert Museum, 2021 N. Kinney Rd., Tucson, AZ 85743, USA – Eastern tropical and temperate Pacific, Caribbean. rbrusca@desertmuseum.org

Wendy Moore: University of Arizona, Tucson, AZ 85721, USA – Eastern tropical and temperate Pacific, central Pacific. wmoore@ag.Arizona.edu

Marilyn Schotte: National Museum of Natural History, Smithsonian Institution, Washington, DC, USA – Caribbean, northwestern Atlantic. Schotte.Marilyn@nmnh.si.edu

Regina Wetzer: Los Angeles County Natural History Museum, 900 Exposition Blvd., Los Angeles, CA 90007, USA – Eastern tropical and temperate Pacific, Caribbean.rwetzer@nhm.org

Collections

The Museo de Zoología of the Universidad de Costa Rica houses a small collection of Costa Rican isopods. The largest collections are those of the Los Angeles County Museum of Natural History and the US National Museum of Natural History (Smithsonian Institution).

Recommendations

The isopod fauna of Costa Rica has been studied mainly by non-Costa Rican investigators. There is currently no isopod specialist working in Costa Rica. As a consequence, material deposited in public collections (in Costa Rica) is scarce. Joint projects as well as regional workshops with specialists on this group of crustaceans would help increase the interest of students and scientists in Costa Rica.

Acknowledgments The second author (Ingo S. Wehrtmann) thanks Jorge Cortés (CIMAR, Universidad de Costa Rica) for his support with literature. Both authors thank two anonymous reviewers for helpful suggestions.

References

Breedy O, Murillo MM (1995) Isópodos (Crustacea: Peracarida) de un arrecife del Caribe de Costa Rica. Rev Biol Trop 43:219–229

Brusca RC, Brusca GJ (2003) Invertebrates, 2nd ed. Sinauer, Assoc Sunderland, MA, 936 p

1. Brusca RC, Iverson EW (1985) A guide to the marine isopod Crustacea of Pacific Costa Rica. Rev Biol Trop 33(Suppl 1):1–77

Brusca RC, Wilson GDF (1991) A phylogenetic analysis of the Isopoda with some classificatory recommendations. Mem Queensland Mus 31:143–204

Campos E, de Campos AR (1989) Epicarideos de Baja California: distribución y notas ecológicas de *Probopyrus pandalicola* (Packard, 1879) en el Pacífico oriental. Rev Biol Trop 37:29–36

Chabanet P, Dufour V, Galzin R. (1995) Disturbance impact on reef fish communities in Reunion Island (Indian Ocean). J Exp Mar Biol Ecol 188:29–48

Cortés J (1994) A reef under siltation stress: a decade of degradation. *In*: Ginsburg RN (compiler) Proceedings of the Colloquium on Global Aspects of Coral Reefs: Health, Hazards and History, 1993. Rosenstiel School of Marine and Atmospheric Science, University of Miami, Miami, Florida, pp 240–246

Cortés J, Risk MJ (1985) A reef under siltation stress: Cahuita, Costa Rica. Bull Mar Sci 36:339–356

Dexter DM (1974) Sandy-beach fauna of the Pacific and Atlantic coasts of Costa Rica and Colombia. Rev Biol Trop 22:51–66

Guzmán HM, Obando VL, Brusca RC, Delaney PM (1988) Aspects of the population biology of the marine isopod *Excorallana tricornis occidentalis* Richardson, 1905 (Crustacea: Isopoda: Corallanidae) at Caño Island, Pacific Costa Rica. Bull Mar Sci 43:77–87

2. Jiménez MP, Vargas VM (1990) *Probopyrus pandalicola* (Isopoda: Bopyridae) infesting *Palaemonetes hiltonii* (Crustacea: Caridae) along the Pacific coast of Costa. Rica Rev Biol Trop 38:457–462

Kinsey DW, Davies PJ (1979) Effects of elevated nitrogen and phosphorous on coral reef growth. Limnol Oceanogr 24:935–940

Markham JC (1992) The Isopoda Bopyridae of the Eastern Pacific – missing or just hiding? Proc San Diego Soc Nat Hist 17:1–4

3. Markham JC (2004) New species and records of Bopyridae (Crustacea: Isopoda) infesting species of the genus *Upogebia* (Crustacea: Decapoda: Upogebiidae): the genera *Orthione* Markham, 1988, and *Gyge* Cornalia & Panceri, 1861. Proc Biol Soc Wash 117:186–198

4. Menzies RJ, Frankenberg D (1967) Systematics and distribution of the bathyl-abyssal genus *Mesosignum* (Crustacea: Isopoda). Antarctic Res Ser Biol Antarctic Seas III 11 (1579):1–261

Naim O (1993) Seasonal responses of a fringing reef community to eutrophication (Reunion Island, Western Indian Ocean). Mar Ecol Prog Ser 99:137–151

5. *Nierstrasz HF, Brender a Brandis GA (1931) Papers from Dr. TH. Mortensen's Pacific Expedition 1914*–16. 57. Epicaridea 2. Videns. Meddel. Dansk Naturhist. Forensing København 91:147–225

Pastorok RA, Bilyard GR (1985) Effects of sewage pollution on coral-reef communities. Mar Ecol Prog Ser 21:175–189

Perry DM (1988) Effects of associated fauna on growth and productivity in the red mangrove. Ecology 69:1064–1075

Perry DM, Brusca RC (1989) Effects of the root-boring isopod *Sphaeroma peruvianum* on red mangrove forest. Mar Ecol Prog Ser 57:287–292

6. Richardson H (1912) Descriptions of two new isopods, an *Apseudes* and a *Munnopsis*, both from the Galapagos islands. Proc US Nat Mus 43:159–162

Risk MJ, Murillo MM, Cortés J (1980) Observaciones biológicas preliminares sobre el arrecife coralino en el Parque Nacional de Cahuita, Costa Rica. Rev Biol Trop 28:361–382

Sano M, Shimizu M, Nose Y (1987) Long-term effects of destruction of hermatypic corals by *Acanthaster planci* infestation on reef fish communities at Iriomote Island, Japan. Mar Ecol Prog Ser 37:191–199

7. Sch*i*oedte JC, Meinert FWS (1881) Symbolae ad Monographiam Cymothoarum Crustaceorum Isopodum Familiae 2. Anilocridae. Naturhist Tidsskr (3) 13:1–166

Tomascik T, Sander F (1987) Effects of eutrophication on reef-building corals. II. Structure of scleractinian coral communities on fringing reefs, Barbados, West Indies. Mar Biol 94:53–75

Villalobos CR, Cruz GA, Cruz RA (1985) Notas sobre la biología de *Sphaeroma terebrans* Bate, 1866 (Sphaeromatidae: Isopoda) en el manglar de Pochote, Provincia de Puntarenas, Costa Rica. Brenesia 24:287–296

Wallerstein BR, Brusca RC (1982) Fish predation: a preliminary study of its role in the zoogeography and evolution of shallow water idoteid isopods (Crustacea: Isopoda: Idoteidae). J Biogeogr 9:135–150

8. Wetzer R, Bruce NL (1999) A new genus and species of sphaeromatid isopod (Crustacea) from Atlantic Costa Rica. Proc Biol Soc Wash 112:368–380

Wittenberg M, Hunte W (1992) Effects of eutrophication and sedimentation on juvenile corals. I. Abundance, mortality and community structure. Mar Biol 112:131–138

9. Wolff T (1962) The systematics and biology of bathyal and abyssal Isopoda Asellota. Galathea Rep 6:1–320

Part 21
Gammaridean Amphipods

John M. Foster, Sara E. LeCroy, Richard W. Heard, and Rita Vargas

Ceradocus sheardi collected in Puerto Vargas, Caribbean coast of Costa Rica (Photo: Sara E. LeCroy)

Abstract Although gammaridean amphipods number over 7,900 species worldwide, only 43 marine and estuarine species have previously been reported from Costa Rican waters. The majority of the records is from the Pacific, resulting from J.L. Barnard's work on material collected during the Allan Hancock Pacific Expeditions; a scant handful of records is from the Caribbean coast. In addition, one terrestrial and one freshwater amphipod species have been reported from Costa Rica. Five gammaridean species from Costa Rican waters have not been found elsewhere to date and may prove to be endemic; a sixth possibly endemic species from Isla del Coco has also been reported once (questionably) from the Caribbean

J.M. Foster, S.E. LeCroy, R.W. Heard
University of Southern Mississippi, Dept. of Coastal Sciences, 703 East Beach Drive, Ocean Springs, MS 39564, USA; Fax: 228/872-4204.
e-mail: (JMF) beachbugs@aol.com; (SEL) sara.lecroy@usm.edu; (RWH) richard.heard@usm.edu

R. Vargas
Universidad de Costa Rica, Escuela de Biología, Museo de Zoología, 11501-2060 San José, Costa Rica.
e-mail ritav@biologia.ucr.ac.cr

I.S. Wehrtmann, J. Cortés (eds.) *Marine Biodiversity of Costa Rica, Central America*,

coast of Colombia. Recent collecting efforts on both coasts of Costa Rica have resulted in records of at least 20 gammaridean amphipod taxa previously unreported from that country. These records are based upon a preliminary examination of the samples and many more records are anticipated as the remainder of the material is identified. A list of species and literature references, as well as information on distribution, depth range, and habitat type, are presented in tabular form for all gammaridean amphipods known from Costa Rica to date.

Introduction

Amphipods are peracarid crustaceans that occupy marine, freshwater, terrestrial, and subterranean environments. They are found in a wide variety of habitats ranging from leaf litter in mountain forests to the deep sea, and play a number of roles in the communities in which they live. Many species are grazers or detritivores, others are filter-feeders, still others are predators or scavengers. Some species are infaunal burrowers in mud or sand, some are epibenthic species, occurring on the surface of the sediment or other substrates, some are fouling species that build tubes or other domiciles on hard structures such as pilings, rocks, or ships, and some are planktonic species, spending all of their lives in the water column. Although many species are free-living, there are a number that are commensals or parasites on other invertebrates (e.g., colomastigid and leucothoid amphipods in marine sponges) and even some vertebrates (e.g., cyamid amphipods on porpoises and whales) (Bousfield 1973; Nelson 1980; Thomas 1993; Bellan-Santini 1998; Bellan-Santini & Ruffo 1998). Amphipods also form a critical component of the diet of many species of fish (Nelson 1979) and even some whales (Oliver & Slattery 1985).

Worldwide, amphipods number in excess of 7,900 species (Vader 2003). There are three major suborders, the Caprellidea, the Hyperiidea, and the Gammaridea, as well as a fourth, much smaller suborder, the Ingolfiellidea. A recent revision by Myers & Lowry (2003) proposes an alternative higher-level classification within the Order Amphipoda. According to their new classification, there are still four suborders of amphipods, but the composition of two of those suborders has changed. The suborders Hyperiidea and Ingolfiellidea, which were not included in their analysis, remain unchanged; however, a number of families are removed from the suborder Gammaridea and, along with the former members of the suborder Caprellidea, placed within the resurrected suborder Corophiidea. Thus, if their classification is accepted, the order Amphipoda would include the four suborders Hyperiidea, Ingolfiellidea, Gammaridea, and Corophiidea.

The Caprellidea (339 species [Vader 2003]) include the skeleton shrimps and the whale lice. The Hyperidea (248 species [Vader 2003]) are comprised of typically large-eyed, planktonic amphipods, some of which live symbiotically with pelagic salps and jellyfish. The Gammaridea (7,275 species [Vader 2003]) contain the common land-hoppers, beach fleas, and scuds. This large suborder, comprised of approximately 125 families (Martin & Davis 2001) and over 1,000 genera

(Bousfield 1973; Barnard & Karaman 1991; Bellan-Santini 1999), contains virtually all of the terrestrial, freshwater, and subterranean taxa, as well as the majority of the marine species. The Ingolfiellidea, containing 2 families, 5 genera, and 39 species (Vonk & Schram 2003), are tiny, elongate, interstitial inhabitants. The amphipods reported in this paper are all members of the suborder Gammaridea, although unidentified representatives of two of the other three suborders (Caprellidea and Hyperiidea) have also been recorded in recent samples collected from Costa Rican waters (RWH, personal observation, 2001).

Amphipods have no carapace, and gammarideans, in particular, are typically laterally compressed, although subcylindrical forms are also common. They have a separate head with two pairs of antennae, the first often with a short accessory flagellum in addition to the main flagellum, and a ventral complex or bundle of paired mouthparts. This mouthpart bundle is comprised of a pair of maxillipeds, two pairs of maxillae, a lower lip or labium, a pair of mandibles, and an upper lip or labrum. The head also often bears an anteriorly or anteroventrally projecting rostrum. The legs or pereopods are attached ventrally to the seven pereonal segments, one pair per segment. The first two pairs of pereopods are often modified as grasping appendages or gnathopods and are frequently enlarged, especially in males. In female amphipods, as in those of most other peracarid crustaceans, there is a marsupium or brood pouch, located on the pereon between the bases of the legs, in which the eggs and developing young are carried. Posterior to the pereon is a strongly muscular body region called the pleon, comprised of three segments called pleonites. Attached ventrally to the pleonites are the pleopods, which may be used either for swimming or for propelling feeding currents toward the mouthparts. The next three segments comprise the urosome, with its attached uropods. The three pairs of uropods each have a basal peduncle and usually bear two rami, although these rami may be variously reduced or even absent in some taxa. The rami often possess spines and setae and may be tubular or flattened and expanded in shape. The terminal appendage is the telson, which is attached to the posterior margin of the third urosomal segment. Usually laminar or fleshy in shape, the telson may be cleft, pointed, rounded, or truncate apically. The pereon, pleon, and urosome may also possess dorsal ornamentation in the form of spines or teeth (adapted from Bousfield 1973; Barnard & Karaman 1991).

Previous Surveys in Costa Rica

A review of the literature indicates that only 43 species of marine and estuarine gammaridean amphipods have been previously reported from Costa Rica. This published information is limited primarily to species occurring on the Pacific coast, and most of these records were the result of work done by J.L. Barnard on samples collected during the Allan Hancock Pacific Expeditions (*Velero* III, IV). These cruises were conducted from 1932–1941 (*Velero* III) and 1949–1952 (*Velero* IV), and encompassed a wide variety of localities and habitats between Oregon, on the northwest coast of the United States, and Chile (Barnard 1954a, b, 1960, 1969, 1971, 1979, 1980; Barnard & Barnard 1982). Costa Rican localities sampled during these cruises include Puerto Culebra, Playa Blanca, and Bahía Salinas. Other published reports on marine and brackish water gammaridean Amphipoda from Pacific Costa Rican waters are those of Stebbing (1903, 1906, 1908), Myers (1968a, b), and Dexter (1974).

There are far fewer records in the literature for gammaridean amphipods from the Caribbean coast of Costa Rica. Dexter (1974) reported *Haustorius* sp., *Pseudorchestoidea ?biolleyi* (Stebbing, 1908) (as *Orchestoidea ?biolleyi*) and an unidentified phoxocephalid species from Caribbean beaches; however, *P. biolleyi*, although originally described from the Pacific coast, remains one of the few definitively identified species previously recorded from the Costa Rican Caribbean coast. It was reported from Limón by Dexter (1974) and also by Bousfield (1982) in his review of the genus *Pseudorchestoidea*; however, he apparently was unaware that this location is on the Caribbean coast and represented the distribution of this species as being Pacific only. Only one other gammaridean species, *Ampelisca agassizi* (Judd, 1896) (Barnard 1954a, b, as *A. vera*), has been reported from the Caribbean coast of Costa Rica. An additional Caribbean coast species, a new species of *Cerapus* Say, 1817, is currently being described by Ortiz & Thomas (J.D. Thomas, personal communication, 2002).

Two non-marine gammarideans have been described from inland Costa Rican habitats. Stebbing (1903) described *Hyalella faxoni* Stebbing, 1903, a freshwater species collected at an altitude of 2,400m on Volcan Reventado, and Lindeman (1990, 1991) described a new genus and species of talitrid, *Cerrorchestia hyloraina* Lindeman, 1990, found in forest leaf litter at an altitude of 2,000m. Although the former species has since been synonymized with *H. azteca* (Saussure, 1858) (Barnard & Barnard 1983), Gonzalez & Watling (2002) suggest that it may be a valid species.

In addition, the Pacific coast of Costa Rica is the type locality for three semiterrestrial estuarine or marine talitrid species. Two beach species, *Talorchestia fritzi* Stebbing, 1903 and *P. biolleyi* (Stebbing, 1908), were described from Isla del Coco and Puntarenas, respectively (Stebbing 1903, 1908), and the estuarine species *Chelorchestia costaricana* (Stebbing, 1906) was described from a mangrove habitat at Boca Jesus María (Stebbing, 1906).

Several species of gammaridean amphipods that have been described from Costa Rican waters have not been found elsewhere to date. These include *Gammaropsis dubia* (Shoemaker, 1942) (Aoridae), *Pseudomegamphopus barnardi* Myers, 1968 (Neomegamphopidae), *Metharpinia oripacifica* Barnard, 1980 (Phoxocephalidae), *Microphoxus minimus* Barnard, 1960 (Phoxocephalidae), and *C. costaricana* (Talitridae). However, these taxa are either marine sand-dwelling forms (the former four species) or mangrove mud-dwelling forms (the latter species), and it seems likely that their apparently restricted distribution is the result of limited sampling rather than true endemism. An additional species, the semiterrestrial *T. fritzi* from Isla del Coco, may prove to be endemic, although it has also been questionably reported from the Atlantic coast of Colombia by Dexter (1974) (as *Orchestia ?fritzi*). Other species that occur in Costa Rica are more widespread and none have been listed as endemics (Species Lists 21.1 and 21.2, both included in the CD-Rom).

Between 1998 and 2001, a preliminary survey of the malacostracan Crustacea of Costa Rica was conducted by Odalisca Breedy, Leonora Rodríguez, José A. Vargas, Rita Vargas, and Jorge Cortés (Universidad de Costa Rica), and Richard W. Heard (University of Southern Mississippi). This survey was limited to four areas on the Caribbean coast (Puerto Limón, Puerto Vargas, Puerto Viejo, and Manzanillo) and four areas on the Pacific coast (Islas Murciélago, Punta Morales (Golfo de Nicoya), Caldera (Golfo de Nicoya), and Boca Coronado) (Table 21.1). With the exception of three of the stations at Islas Murciélago, these collections were all made at depths of less than 2 m. Crustaceans were collected with dredge nets, yabby pumps (hand-held suction devices), epibenthic sleds, dredges, fine-mesh dredge nets (kicknets), and rock and algae washing techniques. Although most of the amphipod fractions from these collections remain unsorted and unidentified, preliminary observations indicate that a rich and diverse marine gammaridean fauna occurs there. These collections contain over 10,000 amphipod specimens representing approximately 100 species belonging to at least 22 families (Amphilochidae, Ampithoidae, Anamixidae, Aoridae, Colomastigidae, Corophiidae, Cyproideidae, Eusiridae, Haustoriidae,

Table 21.1 Costa Rican coastal sites where marine and estuarine gammaridean Amphipoda were collected between 1998 and 2001

Site	Depth (m)	Date	Substratum
Pacific Coast			
Islas Murciélago			
Catalina Island	5–6	May 1999	Sand, rubble
Cocinera Island	1–3	May 1999	Coral rock, sand, gravel
San José Island			
South beach	1–3	May 1999	Coral rock, sand, algae
1 km east of island	20–25	May 1999	Soft sand, mud
2–3 km east of island	35	May 1999	Soft sand, mud
Golfo Nicoya			
Caldera			
Just north of Port of Caldera	1–1.5	April 1998; May 1999	Open sand beach
Punta Morales	3–5	November 1999	Sand, silt
Boca Coronado, Osa, Puntarenas			
Near Golfo Dulce at mouth of Terraba River		July 2000; August 2000	Mangrove, mud, rotten wood
Caribbean Coast			
Manzanillo Beach	1–1.5	November 1999	Sand
Puerto Limón	3–5	April 1998	Sand, silt
Puerto Viejo	0.5–1.5	August 2001	Coral rock, grass, sand
Puerto Vargas	1–1.5	November 1999	Coral rock, sand

Hyalidae, Isaeidae, Ischyroceridae, Leucothoidae, Melitidae, Oedicertodidae, Phliantidae, Platyschnopidae, Phoxocephalidae, Podoceridae, Stenothoidae, Synopidae, and Talitridae). A limited number of identifications based on a small part of this material has resulted in the addition of at least 20 taxa to the list of species known from Costa Rica and these are included in Species Lists 21.1 and 21.2 (see CD-Rom). Undoubtedly, continued taxonomic work on these and additional collections from other marine habitats, especially in deeper water, will increase the confirmed species and family list and may result in the discovery of additional undescribed taxa.

Specialists

People with expertise on and collections of gammaridean Amphipoda from the Caribbean, tropical eastern Pacific and neotropical freshwater regions include:

Adam Baldinger: Museum of Comparative Zoology, 26 Oxford Street, Harvard University, Cambridge, Massachusetts 02138, USA – Caribbean. abaldinger@oeb.harvard.edu

John M. Foster: University of Southern Mississippi, Department of Coastal Sciences, 703 East Beach Dr., P.O. Box 7000, Ocean Springs, MS 39566–7000, USA – Caribbean. Beachbugs@aol.com

Michael Gable: Department of Biology, Eastern Connecticut State University, Willimantic, Connecticut 16226-2295, USA – Caribbean. gable@ecsuc.ctstateu.edu

Exequiel R. González: Universidad Católica del Norte, Coquimbo, Chile – Neotropical freshwater. egonzale@ucn.cl

Richard W. Heard: University of Southern Mississippi, Department of Coastal Sciences, 703 East Beach Dr., P.O. Box 7000, Ocean Springs, MS 39566-7000, USA – Caribbean; tropical eastern Pacific. richard.heard@usm.edu

Traudl Krapp: Museum Alexander Koenig, Adenauerallee 160, D-53113 Bonn, Germany – Caribbean. Traudl.Krapp@uni-bonn.de

Eric A. Laso-Wasem: Division of Invertebrate Zoology, Peabody Museum of Natural History, Yale University, 170 Whitney Avenue, P.O. Box 6666, New Haven, Connecticut 16511-8161, USA – Caribbean. eric.lazo-wasem@yale.edu

Sara E. LeCroy: University of Southern Mississippi, Department of Coastal Sciences, 703 East Beach Dr., P.O. Box 7000, Ocean Springs, Mississippi 39566-7000, USA – Caribbean. sara.lecroy@usm.edu

Manuel Ortiz: Center for Marine Studies, Faculty of Biology, University of Havana, Cuba – Caribbean. ortiztouzet@yahoo.com

Sandro Ruffo: Museo Civico di Storia Naturale, Lungadige Porta Vittoria, 9, 37129 Verona, Italy – Caribbean. sandro.ruffo@comune.verona.it

James D. Thomas: National Coral Reef Institute, Nova Southeastern University Oceanographic Center, 8000 North Ocean Drive, Dania Beach, Florida 33004, USA – Caribbean, tropical eastern Pacific. thomasjd@nova.edu

Ronald Vonk: Institute for Biodiversity and Ecosystem Dynamics, University of Amsterdam, Mauritskade 57, 1092 AD Amsterdam, The Netherlands – Caribbean. vonk@science.uva.nl

Collections

Museums, in addition to the Museo de Zoología (http://museo.biologia.ucr.ac.cr) of the Universidad de Costa Rica, with amphipod collections from the Caribbean and tropical eastern Pacific include (but are not limited to):

Western Atlantic, Caribbean & Gulf of Mexico collections:

Center for Marine Research, University of Havana, Calle 16 #114 entre 1ra y 3ra, Playa, Ciudad de la Habana, Cuba; http://www.uh.cu/centros/cim

Worldwide collections:

Gulf Coast Research Laboratory Museum, University of Southern Mississippi, P.O. Box 7000, Ocean Springs, Mississippi 39566, USA; www.usm.edu/gcrlmuseum

Museo Civico di Storia Naturale di Verona, Lungadige Porta Vittoria, 9, Verona, Italy; www.museostorianaturaleverona.it

National Museum of Natural History, Smithsonian Institution, 10th Street and Constitution Ave., NW, Washington, D.C. 20560, USA; www.mnh.si.edu

Natural History Museum of Los Angeles County, 900 Exposition Blvd., Los Angeles, California 90007, USA; www.nhm.org

Peabody Museum of Natural History, Division of Invertebrate Zoology, Yale University, 170 Whitney Avenue, P.O. Box 6666, New Haven, Connecticut 16511-8161, USA; www.peabody.yale.edu

Zoological Museum Amsterdam, University of Amsterdam, Department of Crustaceans; P.O. Box 94766; 1090 GT Amsterdam; The Netherlands; www.science.uva.nl/zma

Zoological Museum, University of Copenhagen, 13 Universitetsparken DK2100, Copenhagen, Denmark; www.zmuc.dk

Acknowledgments We wish to thank Jorge Cortés, José Vargas, and Ingo Wehrtmann of the Centro de Investigación en Ciencias del Mar y Limnología (CIMAR) and Escuela de Biología, Universidad de Costa Rica, for their support and help during the collection of the material and the preparation of the manuscript. Odalisca Breedy and Leonora Rodríguez (Universidad de Costa Rica) kindly assisted in the collection of specimens, and Wim Vader (Tromsø Museum, Norway) graciously provided information on numbers of amphipod species from his poster presentation at the XIth International Colloquium on Amphipoda in Tunis. We are also grateful to the two reviewers for their many valuable comments on the initial manuscript.

References

1. Barnard JL (1954a) Amphipoda of the family Ampeliscidae collected in the eastern Pacific Ocean by the *Velero III* and *Velero IV*. Allan Hancock Pac Exped 18:1–137
2. Barnard JL (1954b) Amphipoda of the family *Ampeliscidae* collected by the *Velero III* in the Caribbean Sea. Allan Hancock Atl Exped Rep 7:1–13
3. Barnard JL (1959) Part II. Estuarine Amphipoda, pp 13–69. *In*: J.L. Barnard, D.J. Reish (eds) Ecology of Amphipoda and Polychaeta of Newport Bay, California. Allan Hancock Found Publ Occ Pap 21:1–106
4. Barnard JL (1960) The amphipod family Phoxocephalidae in the eastern Pacific Ocean, with analyses of other species and notes for a revision of the family. Allan Hancock Pac Exped 18:1–375

5. Barnard JL (1964) Los anfípodos bentónicos marinos de la costa occidental de Baja California. Rev Soc Mex Hist Nat 24:205–274
6. Barnard JL (1967) New species and records of Pacific Ampeliscidae (Crustacea: Amphipoda). Proc US Natl Mus 121:1–20
7. Barnard JL (1969) A biological survey of Bahia de Los Angeles Gulf of California, Mexico, IV. Benthic Amphipoda. Trans San Diego Soc Nat Hist 15:175–228
Barnard JL (1971) Gammaridean Amphipoda from a deep-sea transect off Oregon. Smithson Contr Zool 61:1–86
8. Barnard JL (1979) Littoral gammaridean Amphipoda from the Gulf of California and the Galapagos Islands. Smithson Contrib Zool 271:1–149
9. Barnard JL (1980) Revision of *Metharpinia* and *Microphoxus* (marine phoxocephalid Amphipoda from the Americas). Proc Biol Soc Wash 93:104–135
10. Barnard JL, Barnard CM (1982) Revision of *Foxiphalus* and *Eobrolgus* (Crustacea: Amphipoda: Phoxocephalidae) from American oceans. Smithson Contrib Zool 372:1–35
Barnard JL, Barnard CM (1983) Freshwater Amphipoda of the World, Parts I and II. Hayfield Assoc, Mt. Vernon, Virginia, 830 p
Barnard JL, Karaman GS. (1991) The families and genera of marine gammaridean Amphipoda (except marine gammaroids). Rec Australian Mus (Suppl 13): 1–866
11. Barnard JL, Thomas JD. (1987) New species of *Neomegamphopus* from tropical America (Crustacea: marine Amphipoda). Proc Biol Soc Wash 100: 147–163
Bellan-Santini D (1998) Ecology. *In*: Ruffo S (ed) The Amphipoda of the Mediterranean, Part 4. Localities and Map, Addenda to Parts 1–3, Key to Families, Ecology, Faunistics and Zoogeography, Bibliography, Index. Mém Inst Océanogr, Monaco, Vol 13, pp 869–893
Bellan-Santini D, Ruffo S (1998) Faunistics and Zoogeography. *In*: Ruffo S (ed) The Amphipoda of the Mediterranean, Part 4. Localities and Map, Addenda to Parts 1–3, Key to Families, Ecology, Faunistics and Zoogeography, Bibliography, Index. Mém Inst Océanogr., Monaco, Vol 13, pp 895–911
Bellan-Santini D (1999) Ordre des amphipods (Amphipoda Latreille, 1816). *In*: Grassé PP, Forest J (eds) Traité de Zoologie, Tome VII, Fas. IIIA. Crustacés Péracarides. Mem Inst Oceanogr. Monaco, Vol 19, pp 93–176
12. Bousfield EL (1973) Shallow-Water Gammaridean Amphipoda of New England. Cornell University Press, Ithaca, New York. 312 p
13. Bousfield EL (1982) The amphipod superfamily Talitroidea in the northeastern Pacific region. I. Family Talitridae: systematics and distributional ecology. Natl Mus Canada Publ Biol Oceanog 11:1–73
14. Bousfield EL, Chevrier A (1996) The amphipod family Oedicerotidae on the Pacific coast of North America. Part 1. The *Monoculodes* and *Synchelidium* generic complexes: systematics and distributional ecology. Amphipacifica 2:75–148
15. Bousfield EL, Hoover PM (1997) The amphipod superfamily Corophioidea on the Pacific coast of North America. Part V. Family Corophiidae. Corophiinae, new subfamily. Systematics and distributional ecology. Amphipacifica 2:67–139
16. Dexter DM (1974) Sandy-beach fauna of the Pacific and Atlantic coasts of Costa Rica and Colombia. Rev Biol Trop 22:51–66
17. Dexter DM (1979) Community structure and seasonal variation in intertidal Panamanian sandy beaches. Estuar Coast Mar Sci 9:543–558
18. Dickinson JJ (1982) The systematics and distributional ecology of the family Ampeliscidae (Amphipoda: Gammaridea) in the northeastern Pacific region. I. The genus *Ampelisca*. Natl Mus Canada Publ Biol Oceanogr 10:1–39
Gonzalez ER, Watling L (2002) Redescription of *Hyalella azteca* from its type locality, Vera Cruz, Mexico (Amphipoda: Hyalellidae). J Crust Biol 22:173–183
19. Jarrett NE, Bousfield EL (1994) The amphipod superfamily Phoxocephaloidea on the Pacific coast of North America. Family Phoxocephalidae. Part II. Subfamilies Pontharpiniinae, Parharpiniinae, Brolginae, Phoxocephalinae, and Harpiinae. Systematics and distributional ecology. Amphipacifica 1:71–150

20. Just J (1977) *Amphyllodomus incurvaria* gen. et sp. n. (Crustacea, Amphipoda), a remarkable leaf-cutting amphithoid from the marine shallows of Barbados. Zool Scrip 6:229–232
21. LeCroy SE (1995) Amphipod Crustacea III. Family Colomastigidae. Mem Hourglass Cruises 9:1–139
22. LeCroy SE (2000) An illustrated identification guide to the nearshore marine and estuarine gammaridean Amphipoda of Florida, Vol 1. Families Gammaridae, Hadziidae, Isaeidae, Melitidae and Oedicerotidae. Annual Report, October 2000 (FDEP Contract No. WM724). Florida Department of Environmental Protection, Tallahassee, FL, 195 p
23. LeCroy SE (2002) An illustrated identification guide to the nearshore marine and estuarine gammaridean Amphipoda of Florida, Vol 2. Families Ampeliscidae, Amphilochidae, Ampithoidae, Aoridae, Argissidae and Haustoriidae. Ann Rep, March 2002 (FDEP Contract No. WM724). Florida Department of Environmental Protectection, Tallahassee, FL, 213 p

Lindeman D (1990) New terrestrial amphipods (Crustacea: Amphipoda: Talitridae) from Mexico and Central America. Can J Zool 68:2323–2337

Lindeman D (1991) Natural history of the terrestrial amphipod *Cerrorchestia hyloraina* Lindeman (Crustacea: Amphipoda: Taltridae) in a Costa Rican cloud forest. J Nat Hist 25:623–638

Martin JW, Davis GE (2001) An updated classification of the recent Crustacea. Nat Hist Mus Los Angeles County Sci Ser 39:1–124

24. McKinney LD (1980) Four new and unusual amphipods from the Gulf of Mexico and Caribbean Sea. Proc Biol Soc Wash 93:83–103
25. Mills EL (1962) Amphipod crustaceans of the Pacific coast of Canada. Nat Hist Pap Natl Mus Canada 15:1–21
26. Myers AA (1968a) Some Aoridae (Amphipoda: Gammaridea) collected by the Hancock Expeditions to the eastern Pacific, 1931–1941. Pac Sci 22:497–506
27. Myers AA (1968b) A new genus and two new species of gammaridean Amphipoda from Central America. J Linn Soc London Zool 47:527–531

Myers AA, Lowry JK (2003) A phylogeny and a new classification of the Corophiidea Leach, 1814 (Amphipoda). J Crust Biol 23:443–485

Nelson WG (1979) Experimental studies of selective predation on amphipods: consequences for amphipod distribution and abundance. J Exp Mar Biol Ecol 38:225–245

Nelson WG (1980) Amphipods: little-known crustaceans. Sea Frontiers 26:138–144

Oliver JS, Slattery PS (1985) Destruction and opportunity on the sea floor: effects of gray whale feeding. Ecology 66:1965–1975

28. Ruffo S (1954) Nuovi anfipodi raccolti nel Venezuela dal Prof. G. Marcuzzi. Mem Mus Civ Stor Nat Verona 4:117–125
29. Shoemaker CR (1933) Two new genera and six new species of Amphipoda from Tortugas. Pap Tortugas Lab 28:247–256
30. Shoemaker CR (1934) Two new species of *Corophium* from the west coast of America. J Wash Acad Sci 24:356–360
31. Shoemaker CR (1942) Amphipod crustaceans collected on the Presidential Cruise of 1938. Smithson Misc Coll 101:1–52
32. Shoemaker CR (1948) The Amphipoda of the Smithsonian-Roebling Expedition to Cuba in 1937. Smithson Mis Coll 110:1–15
33. Shoemaker CR (1949) The amphipod genus *Corophium* on the west coast of North America. J Wash Acad Sci 39:66–82
34. Shoemaker CR (1956) A new genus and two new species of amphipods from Dry Tortugas, Florida. J Wash Acad Sci 46:61–64
35. Stebbing TRR (1903) Amphipoda from Costa Rica. Proc US Nat Mus 26:925–931
36. Stebbing TRR (1906) A new Costa Rican amphipod. Proc US Nat Mus 31:501–504
37. Stebbing TRR (1908) A new amphipod crustacean, *Orchestoidea biolleyi*, from Costa Rica. Proc US Nat Mus 34:241–244
38. Thomas JD (1993) Identification manual for marine Amphipoda (Gammaridea): I. Common coral reef and rocky bottom amphipods of South Florida. Final Report, September 1993 (FDEP Contract No. SP290). Florida Department of Environmental Protection, Tallahassee, FL, 83 p

39. Thomas JD, Barnard JL (1983) The Platyischnopidae of America (Crustacea: Amphipoda). Smithson Contr Zool 375:1–33
40. Thomas JD, Barnard JL (1985) *Perioculodes cerasinus*, n. sp., the first record of the genus from the Caribbean Sea (Amphipoda: Oedicerotidae). Proc Biol Soc Wash 98(1):98–106
41. Thomas JD, Barnard JL (1986) New genera and species of the *Megaluropus* group (Amphipoda: Megaluropidae) from American Seas. Bull Mar Sci 38:442–476

Vader W (2003) How many species of Amphipods? Poster, XIth Internat. Colloq. Amphipoda, Tunis, 26–27 March 2003

Vonk R, Schram FR (2003) Ingolfiellidea (Crustacea, Malacostraca, Amphipoda): a phylogenetic and biogeographic analysis. Contrib Zool 72:39–72

42. Present study

Part 22
Hyperiid Amphipods

Rebeca Gasca

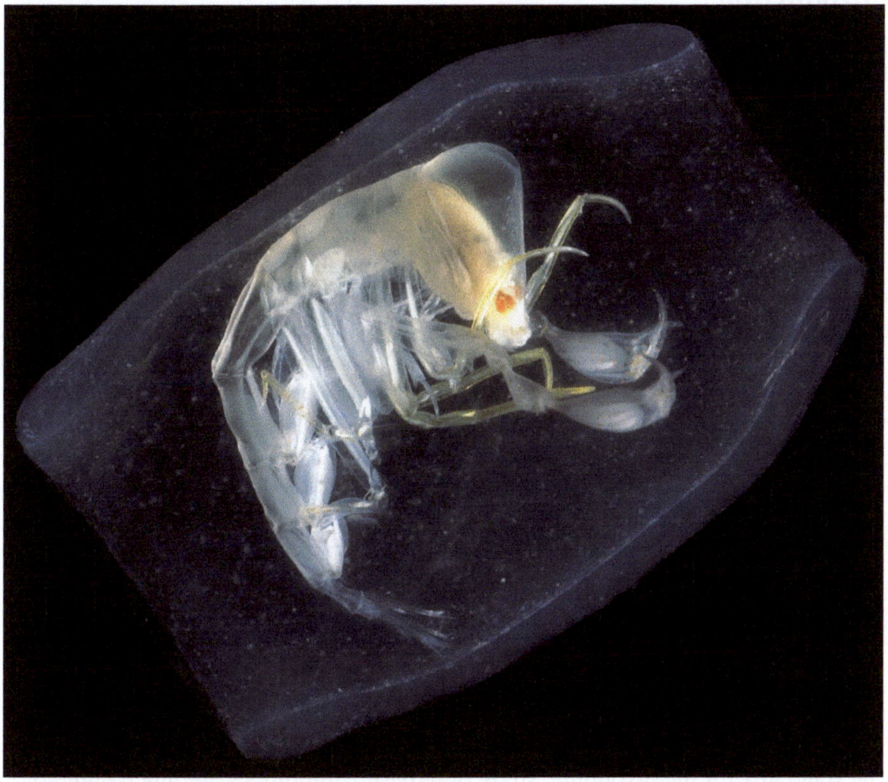

A salp with *Phronima sedentaria*, a species from Pacific Costa Rica (Photo: Humberto Bahena Basave)

R. Gasca
El Colegio de la Frontera Sur (ECOSUR), Apdo. Postal 424, Chetumal, Quintana Roo, México
e-mail: rgasca@ecosur.mx

I.S. Wehrtmann, J. Cortés (eds.) *Marine Biodiversity of Costa Rica, Central America*,

Abstract The knowledge of the hyperiid amphipods in Costa Rican waters is quite limited. The estimated number of species currently recognized is near 277; there are 222 known from the Pacific Ocean and 175 from the Atlantic. The number of species reported herein for Costa Rican waters is 34, all obtained from samples collected in waters of the Pacific side, and most of them generated quite recently. Previous records are revised and corrected. There are no records for the Caribbean coast of Costa Rica. Based on the regional (eastern tropical Pacific, western Caribbean Sea) faunistic records, the potential number of hyperiid species that could be found in Costa Rican waters is over 124.

Introduction

The Suborder Hyperiidea comprises exclusively pelagic amphipods. Most species are oceanic, frequently found as parasites or commensals of gelatinous zooplankton (Harbison *et al.* 1977; Laval 1980). They are more easily collected at nighttime since they occupy lower layers of the water column during the maximum daylight intensity hours (Répelin 1978). However, many species dwell within the 0–100 m layer without performing extensive vertical migrations.

Taxonomically, they are divided into two infraorders: the Physosomata with seven families, and the Physocephalata with 16 families (Vinogradov *et al.* 1996). Hyperiids are distributed from the surface to abyssal depths, although the bulk of the more primitive Physosomata lives mainly in deep waters. Most Physocephalata, which can perform vertical migrations to deeper layers, are surface dwellers; because they are captured by routine plankton samplings, the Physocephalata are relatively well-known. They are a group that includes forms susceptible to changes in hydrographic conditions and are considered to be oceanographic indicators (Répelin 1978). Most species have a wide distribution or are cosmopolitan, although true cosmopolitanism has been questioned by Shih (1986). Therefore, more detailed taxonomic works are required in little known areas, particularly the tropical seas.

Taxonomic Status/Problems

The standards for the taxonomic study of hyperiid amphipods are much higher now than they were a few decades ago. Microscopic dissection or manipulation is required to identify members of this group; the structure of the appendages including the mouthparts is used in this process (Vinogradov 1999). Some nominal species have been considered invalid by recent authors; most of the confusion is derived from the description of immature specimens. Furthermore, several hyperiid nominal genera or species (*v. gr. Scina*, *Primno*, *Brachyscelus*, *Lycaea pulex*) contain taxonomically complex groups that represent unresolved problems; as a consequence of this constraint, the literature sometimes contains erroneous or doubtful records (Zeidler 1998). Therefore, it is probable that there are more (or fewer) species than those currently known.

In order to make a complete faunal analysis of these amphipods, Vinogradov (1999) recommended the use of plankton nets with mouth opening over 1 m^2 and meshes less than 0.5 mm. This suggests that a considerable part of the hyperiid fauna might be underestimated in areas where plankton trawls are not made with this kind of gear.

Species Richness

The overall number of species recognized by Shih & Chen (1995), Vinogradov *et al.* (1996), Zeidler (1998), Vinogradov (1999), and Shih & Hendrycks (2003) is near 277. There are 222 species known to dwell in the Pacific Ocean and 175 in the Atlantic. Of course, these figures represent estimates mainly because of the taxonomic uncertainty of several species, as mentioned above. Many species are true cosmopolitan forms, distributed in the tropical-temperate belt (Zeidler 1998); only a few species seem to be confined to a particular ocean. However, according to Vinogradov (1999), hyperiids can be divided roughly into warm-water and cold-water species; therefore, their distribution seems to be limited latitudinally or by depth. Most warm-water hyperiids are circumtropical (including tropical/subtropical belt). Therefore, it is not expected that the species richness will be significantly different when comparing the Pacific and the Atlantic waters of Costa Rica. There are few works dealing with this group in areas adjacent to Costa Rica, and the information is analyzed separately for the Pacific and the Atlantic waters.

Knowledge of the hyperiid amphipod fauna of the tropical northwestern Atlantic Ocean and of the eastern tropical Pacific is quite limited; the regional species composition is merely guessed from surveys in adjacent areas. Although some extrapolation can be performed in order to predict which species could occur in these areas, it is important to examine local and regional records/collections to verify their composition and abundance and to determine the main distributional aspects of the group in terms of their relation with the hydrological conditions both horizontally and vertically. Direct records of hyperiid amphipods in Costa Rican waters are scarce. However, an analysis of the literature records from the adjacent areas of both the tropical northwestern Atlantic and the eastern tropical Pacific can provide a general information base of the hyperiid community composition of Costa Rica, as follows.

Caribbean Sea and Gulf of Mexico: Overall, the Atlantic hyperiid fauna is less well-known than that of the Indo-Pacific region (Vinogradov 1999). In fact one of the most complete studies on the western Atlantic Hyperiidea has been published recently by Vinogradov (1999); it covers mainly the southwestern Atlantic region and contains keys and diagnoses of more than 188 epipelagic and deep-living species, including 37 Gammaridea and 151 Hyperiidea. Actually, nearly 30% of these hyperiid species have not been recorded in waters of the tropical Atlantic. Previous works in the northwestern tropical Atlantic include those by Vosseler (1901), Pearse (1913), Shoemaker (1945, 1948), Springer & Bullis (1956), Fage (1960), Yang (1960), Evans (1961), Grice & Hart (1962), Bullis & Thompson (1965), Hopkins (1966), Lewis & Fish (1969), Shih (1969), Gillespie (1971), Moore & Sander (1977, 1979), and Stuck *et al.* (1980). Most of them contain records of no more than ten species. Gasca & Shih (2001, 2003) recorded over 50 species of the Hyperiidea in the westernmost Caribbean Sea. Currently, the number of species known to exist in the Caribbean Sea is 56. There are no previous records of Hyperiidea in the Atlantic waters of Costa Rica, but this (over 56 species) is the potential figure of species that could be distributed in this area.

Eastern tropical Pacific: In this region, most of the knowledge on this taxon is limited to the subtropical-temperate California region; the number of species in this area is over 124 (Brusca 1967; Bowman 1978; Siegel-Causey 1982; Brinton *et al.* 1986; Lavaniegos & Ohman 1999; Shih & Hendrycks 2003). Relatively little is known about the hyperiid fauna of the tropical Pacific subregion. The only surveys dealing with this group in Costa Rican waters of the Pacific are those of Salmán-Palacios (1985) and Shih & Hendrycks (2003). The former author recorded 18 species in the oceanic waters of the Costa Rican Dome; the same records were cited by Vicencio & Fernández (1996). More than half (ten) of these records were not considered valid for inclusion in the species account (see Species List 22.1 which is included on the CD-Rom) because the identification of these taxa seem to be erroneous and/or the descriptions and illustrations provided are insufficient to determine the species. The species records in this situation are: *Vibilia peronii* Milne-Edwards, 1830, it is not possible to assign the name to a valid species because it was described from an immature specimen. *Themistella steenstrupi* Bovallius, 1887, the valid name for this species is *T. fusca* (Dana, 1852), but the pereopod II illustrated by Salmán-Palacios (1985) suggests that this is *Lestrigonus bengalensis* instead. *Hemithyphis tenuimanus* Claus, 1879 is not identifiable with the figures presented, and the drawings do not correspond to this species. *Brachyscelus crusculum* Bate, 1861 is probably a different species of the same genus, based on the illustration of the uropods. The correct name of *Oxycephalus piscatoris* is *O. piscator* Milne-Edwards, 1830, but the shape of the telson, the urosomite, and of the uropods do not agree with this species. *Primno macropa* Guérin-Méneville, 1836 is probably *P. brevidens* Bowman, 1978 based on the denticles of pereopod V, and it is included as this species in the Species List 22.1 (see CD-Rom) is included on the CD-Rom. *Hyperia medusarum* (O.F. Müller, 1776) is bipolar in distribution and is restricted to north of 30° N latitude in the Pacific Ocean; this is a dubious record, which must be confirmed. *Hyperietta luzoni* (Stebbing, 1888) is probably a different species based on the shape of the second segment of pereopods I and II presented in the corresponding illustration. *Leptocotis spinifera* Streets, 1877 is a junior synonym of *L. tenuirostris* (Claus, 1871), which is included under this name in the list. *Sympronoe parva* is currently *Pararonoe parva* Claus, 1879, although the structure of the second segment of pereopod VI is not well illustrated; this identification could be correct, as this species is known to be distributed in the area. *Rhabdosoma armatus* is currently *R. armatum* (Milne-Edwards, 1840), the

Fig. 22.1 *Cranocephalus scleroticus*, a species from Pacific Costa Rica (Photo: Rebeca Gasca)

genus is the same, but illustrations suggest a different species. The specimens on which these unique records for Costa Rica were based on should be reexamined in order to define the taxon to which they really belong. Furthermore, Salmán-Palacios (1985) recorded *Bougisia ornata*, which is known only from the Mediterranean (Laval 1966) and the Tasman Sea (Zeidler 1998). It has been included in this species account mainly because this is a monotypic genus with easily distinguishable characters; however, the record should be confirmed.

The examination of nearly 35 surface zooplankton samples collected in Bahía Culebra, Costa Rican Pacific, yielded up to 20 reports of hyperiids that are included in the Species List 22.1 (see CD-Rom). These specimens are deposited, yet uncatalogued, in the Collection of Zooplankton of El Colegio de la Frontera Sur, Chetumal, Mexico. Therefore, the 32 reports for the Pacific waters of Costa Rica (Fig. 22.1) represent at least 25% of the number of species known in the adjacent subtropical regions of the Pacific, which are likely to be found in the oceanic waters of Costa Rica as well. It is expected that this figure will grow after the analysis of more samples and also when deeper, oceanic areas off Costa Rica are surveyed.

Collections/Databases

There are no voucher collections of hyperiid amphipods in Costa Rica. However, there are a large number of samples collected in different parts of the country. Several of the records included in this work were generated upon examination of these samples. The zooplankton collection of El Colegio de la Frontera Sur (ECOSUR) contains faunistic and geographic data of 56 species of hyperiids collected in 75 localities from over 360 neritic and oceanic samples of the western Caribbean Sea, which are reported in Gasca & Shih (2001, 2003). A database linked to this reference collection is available at ECOSUR, in Chetumal, Mexico.

Specialists

There are no experts on this group currently working in Costa Rica or in Central America. There are, however, several researchers who have studied the hyperiid fauna of adjacent areas and are familiar with the regional fauna; other researchers have a worldwide overview. Names are arranged alphabetically, followed by their affiliation and the area related to Costa Rican waters in which they have worked; when available, an e-mail address is added for each name.

REBECA GASCA: El Colegio de la Frontera Sur (ECOSUR), Unidad Chetumal, Mexico – Tropical Atlantic and Pacific. rgasca@ecosur-qroo.mx

ED A. HENDRYCKS: Canadian Museum of Nature, Ontario, Canada – Pacific. ehendrycks@mus-nature.ca

BERTHA A. LAVANIEGOS-ESPEJO: CICESE, Ensenada, BC, Mexico – California Current. berlav@cicese.mx

CHANG-TAI SHIH: Canadian Museum of Nature, Ontario, Canada – Worldwide. ctshih@mail.tfrin.gov.tw

GEORGYI VINOGRADOV: Russian Academy of Science, Moscow, Russia – Worldwide

Mikhail E. VINOGRADOV: Russian Academy of Science, Moscow, Russia – Worldwide

WOLFGANG ZEIDLER: South Australian Museum, Adelaide, Australia – Worldwide. zeidler.wolfgang@saugov.sa.gov.au

Acknowledgments My gratitude to Álvaro Morales from CIMAR (Universidad de Costa Rica) for granting the zooplankton samples analyzed. Karina Rodríguez and Iván Castellanos sorted the samples. Three anonymous reviewers made important improvements and suggested relevant additions to this manuscript.

References

1. Bowman TE (1978) Revision of the pelagic amphipod genus *Primno* (Hyperiidea: Phrosinidae). Smithson Contrib Zool 275:1–23

Bullis HR, Thompson JR (1965) Collections by the exploratory fishing vessels Oregon, Silver Bay, Combat, and Pelican made during 1956 to 1960 in the southwestern North Atlantic. US Dep Int Spec Sci Rep Fish 510:1–131

Brinton E, Fleminger A, Siegel-Causey D. (1986) The temperate and tropical biotas of the Gulf of California. CalCOFI Rep 27:228–266

Brusca GJ (1967) The ecology of pelagic Amphipoda. I. Species accounts, vertical zonation and migration of Amphipoda from the waters off Southern California. Pac Sci 21:382–393

Evans F (1961) The planktonic Crustacea of the Petula transatlantic expedition. Proc Linn Soc Lond (Zool) 172:189–202

Fage L (1960) Oxycephalidae. Amphipodes pélagiques. Dana Rep 52:1–145

Gasca R, Shih C-t (2001) Hyperiid amphipods of surface waters of the western Caribbean Sea (1991). Crustaceana 74:489–499

Gasca R, Shih C-t (2003) Hyperiid amphipods of Banco Chinchorro. Bull Mar Sci 73:37–43

2. Gasca R (2001) Personal observations: Bahía Culebra, Costa Rica

Grice GD, Hart AD (1962) The abundance, seasonal occurrence and distribution of the epizooplankton between New York and Bermuda. Ecol Monogr 32:287–309

Gillespie MC (1971) Analysis and treatment of zooplankton of estuarine waters of Louisiana. *In*: Louisiana Wildlife and Fisheries Commission (eds) Cooperative Gulf of Mexico Estuarine Inventory and Study USA (Louisiana), pp 107–175

Harbison GR, Biggs DC, Madin LP (1977) The associations of Amphipoda Hyperiidea with gelatinous zooplankton – II. Associations with Cnidaria, Ctenophora and Radiolaria. Deep-Sea Res 24:465–468

Hopkins TL (1966) The plankton of the St. Andrew Bay system, Florida. Publ Inst Mar Sci Univ Tex 11:12–64

Laval P (1966) *Bougisia ornata*, genre et espéce nouveaux de la famille des Hyperiidae (Amphipoda, Hyperiidea). Crustaceana 10:210–218

Laval P (1980) Hyperiid amphipods as crustacean parasitoids associated with gelatinous zooplankton. Oceanogr Mar Biol Ann Rev 18:11–56

Lavaniegos BE, Ohman MD (1999) Hyperiid amphipods as indicators of climate change in the California Current. *In*: Schram FR, von Vaupel Klein JC (eds) Crustaceans and the Biodiversity Crisis. Proceedings of the 4th International Crustacean Congress, Amsterdam, The Netherlands, pp 489–509

Lewis JB, Fish AG (1969) Seasonal variation of the zooplankton fauna of surface waters entering the Caribbean Sea at Barbados. Carib J Sci 9:1–23

Moore E, Sander F. (1977) A study of the offshore zooplankton of the tropical western Atlantic near Barbados. Ophelia 16:77–96

Moore E, Sander F (1979) A comparative study of zooplankton from oceanic, shelf, and harbor waters of Jamaica. Biotropica 11:196–206

Pearse AS (1913) Notes on certain amphipods from the Gulf of Mexico, with description of new genera and new species. Proc US Nat Mus 43:369–379

3. Répelin R (1978) Les Amphipodes pelagiques du Pacifique occidental et central. Trav Docum, ORSTOM 86:1–381

4. Salmán-Palacios C (1985) Sistemática y distribución de los hipéridos pelágicos (Crustacea: Amphipoda) del Domo de Costa Rica. Tesis Licenciatura, Fac. Ciencias, UNAM, 132 p

Shih C-T (1969) The systematics and biology of the family Phronimidae (Crustacea: Amphipoda). Dana Rep 74:1–100

Shih C-T (1986) Biogeography of oceanic zooplankton. *In*: Pierrot-Bults AC, van der Spoel S, Zahuranec BJ, Johnson RK (eds) Pelagic Biogeography. Proceedings of the International Conference, The Netherlands. UNESCO Tech Pap Mar Sci 49, Paris, pp 250–253

5. Shih C-T, Chen, Q-C (1995) Zooplankton of China Seas (2). The Hyperiidea (Crustacea: Amphipoda). China Ocean Press, Beijing, 295 p

6. Shih C-T, Hendrycks EA (2003) A new species and new records of the genus *Vibilia* Milne Edwards, 1830 (Amphipoda: Hyperiidea: Vibiliidae) occurring in the eastern Pacific Ocean. J Nat Hist 37:253–296

Shoemaker CR (1945) The Amphipoda of the Bermuda Oceanographic Expeditions, 1929–1931. Zoologica 30:185–266

Shoemaker CR (1948) The Amphipoda of the Smithsonian-Roebling Expedition to Cuba in 1937. Smithson Miscell Coll 110:1–15

Siegel-Causey D (1982) Factors determining the distribution of hyperiid Amphipoda in the Gulf of California. Ph.D. dissertation, University of Arizona, Tucson, Arizona, 535 p

Springer S, Bullis HR Jr (1956) Collections of the OREGON in the Gulf of Mexico. US Dep Inter Spec Sci Rep Fish 196:1–134

Stuck KC, Perry HM, Fish AG (1980) New records of Hyperiidea (Crustacea: Amphipoda) from the North Central Gulf of Mexico. Gulf Res Rep 6:359–370

Yang WT (1960) A study of the subgenus *Parahyperia* from the Florida Current (genus *Hyperia*; Amphipoda: Hyperiidea). Bull Mar Sci Gulf Carib 10:11–39

Vicencio MEA, Fernández MAA (1996) Zooplancton del Domo de Costa Rica: taxonomía y biogeografía. Rev Biol Trop 44:631–642

Vinogradov G (1999) Amphipoda. *In*: Boltovskoy D (ed) South Atlantic Zooplankton. Backhuys, Leiden, The Netherlands, pp 1141–1240

7. Vinogradov ME, Volkov AF, Semenova TN (1996) Hyperiid amphipods (Amphipoda, Hyperiidea) of the World Oceans. Science Publishers, Lebanon, NH, 632 p

Vosseler J (1901) Die Amphipoden der Plankton-Expedition. I. Theil. Hyperiidea 1. Ergebnisse der Plankton-Expedition der Humboldt-Stiftung 2 G.e.: I–VII, 1–129

8. Zeidler W (1992) Hyperiid amphipods (Crustacea: Amphipoda: Hyperiidea) collected recently from eastern Australian waters. Rec Aust Mus 44:85–133

Zeidler W (1998) Pelagic Amphipods (Crustacea: Amphipoda: Hyperiidea) collected from eastern and southeastern Australian waters by the CSIRO research vessel "Warreen" during the years 1938–1941. Rec S Aust Mus Monogr Ser 4:1–143

Part 23
Barnacles

Robert Van Syoc

Barnacles associated to mangroves in Térraba, Pacific Costa Rica (Photo: Ingo S. Wehrtmann)

Abstract Fifty species of Cirripedia are listed as known from Costa Rica through published accounts or museum collections. Of these, nine species inhabit both coasts of Costa Rica, while 11 are found only in the Caribbean and 30 only in the Pacific. The published records of Cirripedia from Costa Rica generally mirror

R.Van Syoc
Department of Invertebrate Zoology and Geology, California Academy of Sciences, Golden Gate Park, San Francisco, California 94118–4599, U.S.A.
e-mail: bvansyoc@calacademy.org

I.S. Wehrtmann, J. Cortés (eds.) *Marine Biodiversity of Costa Rica, Central America*,

those of adjacent regions, most notably Panama. Comparison of the compiled list of Costa Rican cirripeds with that of Laguna (1985) for Panama shows great similarity, with the exception of certain symbiotic forms. Many of the species are circumtropical or cosmopolitan fouling forms, or widespread regional taxa. However, a careful field survey of more cryptic forms should yield many more species, some simply unrecorded by prior workers, some undescribed, and perhaps even some short-range endemics.

Introduction

Cirripedia, commonly known as barnacles, are the only group of crustaceans with permanent external calcareous shells. The external shell is retained throughout the life of the adult barnacle and it offers protection from biotic and physical environmental pressures. The shell, combined with planktonic larval dispersal in most species, provide Cirripedia with adaptive tools to invade and persist in widespread, diverse, and physiologically challenging environments.

Cirripedia are perhaps most familiar to the layman for the ability of the larvae of some species to attach themselves to boat hulls where they metamorphose into adult form and so travel about with vessels from port to port. This "fouling" habit is largely responsible for the cosmopolitan, or circumtropical distributions of several species of Cirripedia, especially those adapted to physiologically rigorous environments such as estuaries and harbors.

However, most species of Cirripedia are either unable to survive the physiological rigors of long-distance travel, or have highly specific substratum requirements for larval settlement. Certain Cirripedia species are obligate symbionts of other animals. Examples of host taxa include Cetacea, Scleractina, Antipatharia, Alcyonacea, Porifera, Crustacea, Hydrophiliidae, and Cheloniidae. The cirriped species symbiotic with highly mobile taxa, such as Cetacea and Cheloniidae, tend to be widespread or nearly cosmopolitan, reflecting the range of their potential host species. On the other hand, Cirripedia that are symbionts of sessile organisms, such as sponges and corals, may be relatively restricted in their geographic ranges and depth, depending on the host requirements.

The list of Cirripedia taxa for Costa Rica includes many species that are wide-ranging symbionts of mobile hosts, fouling species with broad ranges resulting from their habit of riding on the hulls of boats or drifting flotsam (e.g., Zullo 1992), as well as species with limited ranges due to specific host or habitat requirements.

Species-specific habitats and local occurrences of Cirripedia within the countries of Central America have not been broadly documented. This paucity of precise information is reflected in the species list table column for countries of occurrence in Central America. Cosmopolitan or circumtropical Cirripedia species probably occur throughout Central America. Fouling species may, or may not, be present in various harbors and estuaries in Central America. Similarly, species ranging north and south of Costa Rica and other Central American countries have often not been specifically recorded in those countries. These may show up in a thorough biotic inventory of coastal organisms. Species included in Species Lists 23.1 and 23.2 (on the CD-Rom) are usually represented by actual specimen records, taken from published literature or museum specimen databases. However, in some cases I have included species with ranges that span Costa Rica, or "cosmopolitan" species.

Cirripedia morphology and taxonomy: Historically, barnacle taxonomy has been based on various characters of the calcareous external shells. In addition, the flexible cuticular characters present on the thoracic limbs (cirri) and mouth parts may offer good species level taxonomic cues. Although a recent morphometric study of shell and flexible cuticular characters by Gomez Daglio (2003) shows that environmental plasticity may confound species level taxonomy of some species complexes of shore barnacles, these taxonomic characters are generally reliable. Several good introductions to barnacle taxonomy are available. Perhaps the most celebrated, and thorough detailed descriptions, are in the monographic works of Charles Darwin (1852, 1854). Laguna (1988), and Gittings *et al.* (1986) have provided more recent taxonomic guides to Panamanian and Gulf of Mexico barnacles, respectively. The Mexican faunal list of Young and Ross (2000) reviews the basic adult and larval morphology of cirripeds, as well as the natural history of major groups.

Although no single comprehensive modern review of all Cirripedia taxa exists, Newman and Ross (1976) compiled a listing of balanomorph barnacle names, with synonyms and references to taxonomy, ecology, and evolution for each species. Zevina (1978, 1981) monographed the pedunculate cirripedia of Scalpellomorpha and Lepadomorpha (Zevina 1980), respectively. An overview of Cirripedia taxa can be found in Newman (1996). Pitombo (2004) revised the Balanidae, adding two new genera.

Broadly distributed and short-range endemic species, various reproductive strategies and modes of dispersal, and possession of a durable calcareous shell are qualities that have made Cirripedia useful as "model" organisms in a great number of studies on various aspects of evolution and ecology. Cirripedia have commonly been subjects for the study of morphological plasticity (e.g., Lively 1986, 1999; Van Syoc 1992), biogeography (e.g., Newman 1986; Laguna 1990; Van Syoc 1996), and genetic divergence (e.g., Van Syoc 1994, 1995).

Unlisted species, a comparison with Panama records: Laguna (1985) compiled a list of 44 cirriped species for Panama by surveying museum collections, published accounts, and collecting specimens at 60 field stations. His field work led to the discovery of a previously undescribed species from the eastern Pacific. He noted that this species, *Euraphia eastropacensis* (a sibling species of the Caribbean *E. rhizophora*), is also found in museum collections from the Pacific coast mangrove lagoons of Costa Rica. Similarly, cryptic symbiotic species on Laguna's list such as *Heteralepas* sp. and *Octolasmis* sp. (decapod crustacean symbionts), *Megabalanus stultus*, *Hexacreusia durhami*, *Ceratoconcha* sp., and *Megatrema madreporarum* (Scleractinia symbionts) may be found in Costa Rica with a careful field survey. Other taxa in Laguna's list that one would expect to find in Costa Rica include *Lithotrya dorsalis* (a Caribbean species that burrows in coral rubble and soft rock), *E. rhizophorae* (Caribbean mangrove inhabitant), *M. vinaceus* (an eastern Pacific species known to foul boat hulls), *Chthamalus angustitergum* and *Tetraclitella divisa* (Caribbean intertidal taxa).

Ross and Perreault (1999) described *Newmanella kolosvaryi* from the Caribbean coast of Panama. They also suggest that *N. radiata* does not range as far west as Central America. This would conflict with a published record of *N. radiata* in Panama by Laguna (1985). It is possible that *N. radiata* of Laguna (1985) is the *N. kolosvaryi* of Ross & Perrault (1999). Closer examination of Caribbean Costa Rican tetraclitids will be necessary to determine which species, if any, of *Newmanella* (or *T. divisa*, another common Caribbean species) occur there.

The collection of the California Academy of Sciences (CAS) contains one specimen lot from Isla del Coco that has been tentatively identified as *C. anisopoma*. This material is labeled as collected in 1932 by Leo G. Hertlein. This is either an error in collection locality recording, or the specimens represent a range extension from the Gulf of California, Mexico. Alternatively, the specimens may not be *C. anisopoma*, but an undescribed sibling species.

Undescribed species: New species of Cirripedia are commonly found in biotic surveys of tropical marine environments. Recent field work by Gomez Daglio (2003) in Baja California Sur, Mexico led to the discovery of a new genus of tetraclitid (Gomez Daglio & Van Syoc, 2006). As more areas are subjected to biotic surveys, with subsequent sample identifications by appropriate taxonomists, we should expect many new species discoveries along with regional faunal range extensions for known taxa.

There are certainly undescribed species of Cirripedia in Costa Rica. These will most likely be species that bear some superficial similarity to known species and therefore are lumped into those taxa. Close scrutiny of morphological characters may be sufficient to identify these as new species. However, some may require molecular-level analysis to differentiate them from sibling species (e.g., *C. proteus* and other *Chthamalus* spp.; Hedgcock 1979; Dando & Southward 1980; Laguna 1985). Additionally, there will probably be a number of undescribed cryptic species, usually symbionts of other organisms (Van Syoc & Winther 1999).

Specialists

The following is a partial list of workers who have published taxonomic papers on Cirripedia in the eastern tropical Pacific and Caribbean. It is not an exhaustive list, and is not meant to be exclusive. Those seeking to identify Costa Rican cirripeds should consult the current scientific literature as a guide to active researchers in the field.

Liza Gomez Daglio: University of California at Merced, Merced, CA, USA – lgomezdaglio@ucmerced.edu

William A. Newman: Scripps Institution of Oceanography, La Jolla, CA, USA – wnewman@ucsd.edu

Fabio Pitombo: Universidade Federal Rural do Rio de Janeiro (UFRRJ), Brazil – fbpnit@urbi.com.br
Robert J. Van Syoc: California Academy of Sciences, San Francisco, CA, USA – rvansyoc@calacademy.org

Collections

The research collections listed below are known to have Cirripedia material from Central America:

- Museo de Zoología de la Universidad de Costa Rica, 11501-2060 San José, Costa Rica.
- California Academy of Sciences, Department of Invertebrate Zoology and Geology, Golden Gate Park, San Francisco, CA 94118–4599 USA.
- Scripps Institution of Oceanography, Benthic Invertebrates Collection, University of California San Diego, 9500 Gilman Dr., La Jolla, CA 92093 USA.
- United States National Museum, Smithsonian Institution, Crustacea Section, Washington, D.C. 20560 USA.
- Museu Nacional/UFRJ, Departamento de Invertebrados, 20940–040, Rio de Janeiro, RJ, Brazil.
- British Museum of Natural History, London, England.

Acknowledgments Three anonymous reviewers made many good suggestions and the manuscript is much improved for their additions and corrections. However, any errors of fact or interpretation must remain mine alone.

References

1. Abbott DP (1987) Observing Marine Invertebrates. Drawings from the laboratory. Stanford University Press, Stanford, 380 p
2. Dando PR, Southward AJ (1980) A new species of *Chthamalus* (Crustacea: Cirripedia) characterized by enzyme electrophoresis and shell morphology: with a revision of other species of *Chthamalus* from the western shores of the Atlantic Ocean. J Mar Biol Ass UK (1980) 60:787–831

Darwin CR (1852) A monograph on the sub-class Cirripedia, with figures of all the species. The Lepadidae, or, pedunculated cirripedes. Ray Society, London, 400 p

Darwin CR (1854) A monograph on the sub-class Cirripedia, with figures of all species. The Balanidae, (or sessile cirripedes); the Verrucidae, etc. Ray Society, London, 684 p

Gittings SR, Dennis GD, Harry HW (1986) Annotated guide to the barnacles of the northern Gulf of Mexico. Texas A & M Sea Grant College Program 86–402, 36 p

Gomez Daglio LE (2003) Sistemática de los Balanomorphos (Cirripedia, Thoracica) de la región sur de la península de Baja California, México. M.Sc. thesis, Inst Polit Nac Cent Interdisc Cien, March, 90 p

Gomez Daglio L, Van Syoc R (2006) A new genus and species of high intertidal barnacle (Cirripedia, Tetraclitidae) from Baja California Sur, Mexico. Zootaxa 1118:57–68

Hedgcock D (1979) Biochemical genetic variations and evidence of speciation in *Chthamalus* barnacles of the tropical Eastern Pacific Ocean. Mar Biol 54:207–214

3. Henry DP, McLaughlin PA (1967) A revision of the subgenus *Solidobalanus* Hoek (Cirripedia Thoracica) including a description of a new species with complemental males. Crustaceana 12(1):43–58
4. Henry DP, McLaughlin PA (1975) The barnacles of the *Balanus amphitrite* complex (Cirripedia, Thoracica). Zool Verhandel 141:203–212
5. Henry DP, McLaughlin PA (1986) The recent species of *Megabalanus* (Cirripedia: Balanomorpha) with special emphasis on *Balanus tintinnabulum* (Linnaeus) sensu lato. Zool Verhandel 235:1–69

Laguna JE (1985) Systematics, ecology and distribution of barnacles (Cirripedia; Thoracica) of Panama, including an analysis of provincialism in the tropical eastern Pacific. M.Sc. thesis, University of California, San Diego, 234 p

6. Laguna JE (1988) *Euraphia eastropacensis* (Cirripedia, Chthamaloidea), a new species of barnacle from the tropical eastern Pacific: morphological and electrophoretic comparisons with *Euraphia rhizophorae* (de Oliveira) from the tropical western Atlantic and molecular evolutionary implications. Pac Sci 41:132–140

Laguna JE (1990) Shore barnacles (Cirripedia, Thoracica) and a revision of their provincialism and transition zones in the tropical eastern Pacific. Bull Mar Sci 46:406–424

Lively CM (1986) Predator-induced shell dimorphism in the acorn barnacle *Chthamalus anisopoma*. Evolution 40:232–242

Lively CM (1999) Developmental strategies in spatially variable environments: barnacle shell dimorphism and strategic models of selection. *In*: Tollrian R, Harvell CD (eds.). The Ecology and Evolution of Inducible Defenses. Princeton University Press, Princeton, NJ/Chichester, pp 245–258

Newman WA (1986) Origin of the Hawaiian marine fauna: dispersal and vicariance as indicated by barnacles and other organisms. *In*: Heck K, Gore R (eds) Crustacean Issues 4. Balkema, Rotterdam, pp 21–49

Newman WA (1996) Sous-classe des Cirripedes (Cirripedia Burmeister, 1834). Super-ordres des Thoraciques et des Acrothoraciques. (Thoracica Darwin, 1854 - Acrothoracica Gruvel, 1905). *In*: Forest J (ed) Zoological Treatise. Anatomy, Systematics, Biology. Vol. 7: Crustacea. Part 2: General Information (Continuation) and Systematics. Masson Paris, Milan/Barcelona, pp 453–540

7. Newman WA, Ross A (1976) Revision of the balanomorph barnacles; including a catalogue of the species. Mem San Diego Soc Nat Hist 9:1–108
8. Pilsbry HA (1916) The sessile barnacles (Cirripedia) contained in the collections of the U.S. National Museum Bull US Nat Mus 93:1–366

Pitombo FB (2004) Phylogenetic analysis of the Balanidae (Cirripedia, Balanomorpha). Zoologica Scripta 33:261–276

9. Pitombo B, Ross A (2002) A checklist of the intertidal and shallow-water sessile barnacles of the eastern Pacific, Alaska to Chile. *In*: Hendrickx ME (ed) Contributions to the Study of East Pacific Crustaceans, Vol. 1. ICMyL, UNAM, México, pp 97–107
10. Ross AR, Newman WA (2001) The Catophragmidae: members of the basal balanomorph radiation. Sessile Organisms 18:77–91.

Ross AR, Perreault RT (1999) Revision of the Tetraclitellinae and description of a new species of Newmanella Ross from the tropical western-Atlantic Ocean (Cirripedia: Tetraclitoidea). Sessile Organisms 15:1–8

11. Sutherland JP (1987) Recruitment limitation in a tropical intertidal barnacle: *Tetraclita panamensis* (Pilsbry) on the Pacific coast of Costa Rica. J Exp Mar Biol Ecol 113:267–282
12. Young PS (1989) *Ceratoconcha paucicostata*, a new species of coral-inhabiting barnacle (Cirripedia, Pyrgomatidae) from the western Atlantic. Crustaceana 56:193–199

Young PS, Ross A (2000) Crustacea: Cirripedia. *In*: Llorente Bousquets J, González Soriano E, Papavero N (eds) Biodiversidad, Taxonomía y biogeografía de artrópodos de méxico: Hacia una síntesis de su conocimiento. Volumen II. UNAM, Ciudad de México, pp 213–238

Van Syoc RJ (1992) Living and fossil populations of a western Atlantic barnacle, *Balanus subalbidus* Henry, 1974, in the Gulf of California region. Proc San Diego Soc Nat Hist 12:1–7

Van Syoc RJ (1994) Genetic divergence between subpopulations of the eastern Pacific goose barnacle *Pollicipes elegans*: mitochondrial cytochrome c oxidase subunit 1 nucleotide sequences. Molec Mar Biol Biotech 3:338–346

Van Syoc RJ (1995) Barnacle mitochondrial DNA: determining genetic relationships among species of *Pollicipes*. *In*: Schram F, Høeg J (eds) New Frontiers in Barnacle Evolution. Crustacean Issues: 10. A.A. Balkema, Rotterdam, pp 269–296

Van Syoc RJ (1996) Barnacles of Rocas Alijos. *In*: Schmeider R (ed) Rocas Alijos: Scientific Results from the Cordell Expeditions. Monographiae Biologicae 75. Dumont HJ, Werger MJA (series eds) Kluwer, Dordrecht, pp 299–304

Van Syoc RJ, Winther R (1999) Sponge-inhabiting barnacles of the Americas: a new species of *Acasta* (Cirripedia, Archaeobalanidae), first record from the eastern Pacific, including discussion of the evolution of cirral morphology. Crustaceana 72:467–486

13. Villalobos CR (1980) Variations in population structure in the genus *Tetraclita* (Crustacea: Cirripedia) between temperate and tropical populations 3. *T. stalactifera* in Costa Rica. Rev Biol Trop 28:193–201

Zevina GB (1978) A new classification of the family Scalpellidae Pilsbry (Cirripedia, Thoracica). Zool Zh Akad Nauk SSSR 57:1343–1352 (in Russian)

14. Zevina GB (1980) A new classification of Lepadomorpha (Cirripedia). Akad Nauk SSSR Zool Zh 59:689–698 (in Russian)

Zevina GB (1981) Barnacles of the suborder Lepadomorpha (Cirripedia, Thoracica) of the world ocean. Part 1: Family Scalpellidae. Guides to the fauna of the USSR: Zool Inst Acad Sci USSR, Leningrad. 127:406 p

15. Zullo VA (1984) New genera and species of balanoid barnacles from the Oligocene and Miocene of North Carolina. J Paleo 58:1312–1338

Zullo VA (1992) *Balanus trigonus* Darwin (Cirripedia, Balaninae) in the Atlantic Basin: an introduced species? Bull Mar Sci 50:66–74

Part 24
Copepods

Álvaro Morales-Ramírez and Eduardo Suárez-Morales

A calanoid copepod (*Arietellus* sp.) from Pacific Costa Rica (Photo: Iván Castellanos)

Abstract The marine copepod fauna of Costa Rica has been studied from samples obtained in different environments including reef areas, coastal systems, and fully oceanic zones. Up to 7 (Calanoida, Cyclopoida, Harpacticoida, Monstrilloida, Mormonilloida, Poecilostomatoida, Siphonostomatoida) out of the 11 orders of Copepoda that are currently recognized are present in Costa Rica. The total number of species known in Costa Rican waters is 209. Most records are of planktonic forms, and nearly 10% of the copepod species richness known in the country is represented by benthic Harpacticoida. There is, however, a substantial asymmetry in the number of copepod species comparing the Costa Rican Pacific (164 species) with the Caribbean coast (45 species). Excluding the strictly benthic forms, the proportion represented by these numbers compared to the corresponding regional fauna is also unequal: the Costa Rican copepod fauna includes 49.6% of the species known in the eastern tropical Pacific (ETP) and adjacent areas, and only 8% of the species recorded in the Caribbean Basin. Overall, 20 species of benthic Harpacticoida represent the first information of this order in Costa Rica and some of them are endemic forms. Despite the fact that copepods are probably the best known marine zooplankton group in Costa Rica, it is recommended that both ecological and taxonomic studies should continue as it is likely that new surveys particularly from relatively neglected areas such as the oceanic epipelagic and meso-bathypelagic environments will increase substantially the faunistic lists.

Introduction

Copepods are a widely distributed group of crustaceans; they are the most abundant metazoans in the world (Humes 1994; Boxshall & Halsey 2004). In the marine environment, copepods constitute a high proportion (60–80%) of the zooplankton biomass; in some circumstances this figure can increase up to 95% or 100% when they form unispecific aggregations with overwhelming densities (Fleminger 1976; Hamner & Carleton 1979; Ambler *et al.* 1991). Copepods are also a major constituent of the deep scattering layers, and acoustic methods have been used to estimate their abundance and to study their migratory pattern in the water column (Greenlaw 1979; Burd *et al.* 1992). Copepods can be found

Á. Morales-Ramírez
Centro de Investigación en Ciencias del Mar y Limnología (CIMAR) y Escuela de Biología, Universidad de Costa Rica, Ciudad, Universitaria, 11501-2060 San José, Costa Rica.
e-mail: alvaro.morales@ucr.ac.cr

E. Suárez-Morales
El Colegio de la Frontera Sur-Chetumal Ar. Centenario km 5.5 cp. 77900. A.P. 425 Chetumal, Quintana Roo, México.
e-mail: esuarez@ecosur.mx

in the water column from the surface layers to the abyssal depths and even in highly specialized environments such as the hydrothermal vents (Humes 1991). These microcrustaceans have colonized quite successfully the benthic environment, where they can be very abundant and diverse.

Because of their abundance in the pelagic realm, copepods are the main food source for many other zooplankters, including the larvae of commercially important fish species. Oceanographic discontinuities (fronts, mesoscale gyres, upwellings) favor the concentration of copepods, thus enhancing the local productivity. High densities of copepods are normally associated with upwelling areas (Arcos & Fleminger 1986; Suárez-Morales 1997). Many copepod species or assemblages have been regarded as indicators of water masses or oceanographic conditions (Björnberg 1981; He *et al.* 1992). For instance, the occurrence of *Calanoides carinatus* in Brazilian surface waters and of *C. philippensis* in Indonesian coastal waters is associated with the influence of upwelling.

The horizontal distribution of the pelagic copepods is determined, as that of the other zooplankters, by currents and movements of the water masses; however, at a smaller scale, it has been suggested that in many cases these zooplankters may show adaptive behavior to remain in a specific area with favorable conditions, which is related to their vertical migration. In general, marine planktonic copepods can be divided into two large groups in terms of their migratory behavior along the water column: (1) with low to moderate migrations and (2) with extended migrations including the meso or bathypelagic layers. Most epipelagic forms can be included in the first group; they dwell between the surface and 200m. Other species, usually large-sized forms, are able to migrate 1,500m or more along the water column (see Chen 1986), whereas others remain in a more or less defined layer. At the family level, the highest diversity of marine planktonic copepods can be found in the epipelagic layer, where representatives of at least 23 families have been recognized. There are, however, many other families exclusive to meso and bathypelagic depths (Boxshall & Halsey 2004). Benthic copepods are represented mainly by a diverse group at the order level, the Harpacticoida, a taxon, which has not been studied consistently in Costa Rica; we have included only a brief section about the few faunistic records available, most of them representing new taxa, some of them supposedly endemic to Costa Rica.

Systematics: Copepods were known by the earliest scientific observers, including Aristotle, Leeuwenhoek, Hooke, and Linnaeus. Milne-Edwards (1840) proposed the name Copepoda making reference to their oar-shaped legs. Latreille (1802) divided the crustaceans into two subclasses: Entomostraca and Malacostraca; free-living copepods were accommodated within the first group and the parasitic forms were classified separately. Dana (1848) divided the free-living copepods into five families: Cyclopidae, Harpactidae, Calanidae, Corycaeidae, and Miracidae. Claus (1866) added the consideration of the mouth appendages and created the Gnathostomata for the free-living forms and Siphonostomata for the parasitic copepods. Later on, Giesbrecht (1892) developed a large systematic analysis by which he separated the most primitive groups under the suborder Gymnoplea, and the other forms were grouped under the Podoplea; the parasitic and semiparasitic groups were left apart. Based on the antennule morphology, the Gymnoplea were divided into two groups: Amphaskandria and Heterarthandria. The Podoplea were also divided into two categories: Ampharthandria and the Isokerandria. Sars (1901, 1903) proposed a classification in which all copepods (free-living, parasitic, and semiparasitic) were included, each based on a type genus: Calanoida (*Calanus*), Cyclopoida (*Cyclops*), Harpacticoida (*Harpacticus*), Notodelphyoida (*Notodelphys*), Lernaeopodoida (*Lernaea*), Caligoida (*Caligus*), and Monstrilloida (*Monstrilla*). Some of them are recognized currently as orders. Damkaer (2002) presented a detailed historical analysis of the development of the systematics and classification of the Copepoda; this account includes data from the earliest observations to the first stages of the systematic copepodology at the beginning of the 19th century.

Huys & Boxshall (1991) established ten orders that represent the most widely accepted systematic arrangement at this level: Platycopioida, Calanoida, Misophrioida, Harpacticoida, Monstrilloida, Mormonilloida, Gelyelloida, Cyclopoida, Siphonostomatoida, and Poecilostomatoida. Out of these, the Calanoida, Cyclopoida, Harpacticoida, Monstrilloida, Mormonilloida, Poecilostomatoida, and Siphonotomatoida have planktonic representatives. The order Thaumatopsylloida was recently proposed by Ho *et al.* (2003) to become the 11th order of the Copepoda. It is based on the family Thaumatopsyllidae; this group was previously contained in the Cyclopoida and includes, like the Monstrilloida, basically semiparasitic forms with planktonic adults.

After the detailed descriptions and illustrations by several authors, the taxonomy of marine plankton copepods has advanced rapidly during the last few decades. Morphological analyses were expanded with the discovery of cuticular pore patterns, and the use of SEM to separate closely related species (Mauchline 1988; Hulsemann & Fleminger 1990). Methods of numerical taxonomy and evaluation of meristic characters were used to explore the relations among the copepod lineages (Huys & Boxshall 1991). Recently, the molecular and genetic analyses have solved many controversial issues such as the presence of cryptic species complexes sharing a sympatric oceanic environment (Bucklin 1986, 1998). This approach has interesting potential as an aid to taxonomy, but it is not likely that it will substitute comparative morphology as the main tool of taxonomy in Copepoda.

Account of Surveys on the Marine Copepod Fauna of Costa Rica and Adjacent Areas

Copepods are probably the most studied zooplankton group in Costa Rica. The basic aspects of their composition, abundance, distribution, and ecology have been described in several works that include different marine environments. Some of the first records of this group in Costa Rican waters were generated less than 20 years ago. In order to have a better understanding of the current status of the knowledge of this taxon in Costa Rica and adjacent areas, we divided our account of the extant literature into two geographic areas: the Pacific coast (eastern tropical Pacific) and the Caribbean Sea. For the Pacific coast, we included in this analysis those surveys developed from the central coast of the Mexican tropical Pacific to the south down to Costa Rica. For the Caribbean Sea we included some representative works carried out in the southern Gulf of Mexico and in the Caribbean Basin.

Eastern Tropical Pacific (ETP): Although not strictly within the Pacific tropical zone, the copepod fauna of the North Pacific Central Gyre is one of the most species-rich provinces of the world oceans (McGowan & Walker 1979, 1985), where more than 228 planktonic species of Copepoda have been recorded. Other works about the calanoid copepod fauna of the California Current Region (Fleminger 1964, 1967) are also relevant to mention in this contribution for their coverage of the eastern Pacific fauna and the inclusion of many widely distributed species. These surveys indicate 176 species in the California Current Region. This area was considered by Mauchline (1998) as one of the best known for copepods in the world oceans. The copepod fauna of the Baja California area was studied, among others, by Palomares-García *et al.* (1998), and Hernández-Trujillo & Esqueda-Escárcega (2002); the number of species recorded in these works reaches 197. The overall ETP regional species accounts have grown recently by the addition of several new species (Markhaseva 1995, 1997; Markhaseva & Ferrari 1996).

Hence, it is considered that the number of species of pelagic Copepoda known to dwell in the ETP is close to 280; this figure is a gross estimation and was used herein only as a frame for diversity comparisons.

The earliest survey on the pelagic copepod fauna of the Costa Rican Pacific we have knowledge of is that published by Johnson (1964), who described a new species of the coastal-estuarine genus *Pseudodiaptomus* (*P. wrighti*) from the ETP, noting that its range stretches from Baja California, Mexico to Guayaquil, Ecuador; he found this species in the Costa Rican Golfo de Nicoya. After a two-decade gap, Guzmán & Obando (1988) surveyed the zooplankton community dwelling off Isla del Caño and included a general analysis of the copepods, at a group level only. Rao & Sameoto (1988) studied the relations between the copepods and phytoplankton in Costa Rican waters of the ETP (specifically in the Costa Rica Dome); they reported that copepods tend to be more abundant in areas with higher concentration of phytoplankton cells along the water column. The composition, abundance, and distribution of oceanic copepods of the Costa Rica Dome was studied by Suárez-Morales & Gasca (1989a); they found 41 species of Calanoida in this oceanic area related to a highly productive upwelling system in the oceanic ETP. They recorded *Eucalanus giesbrechti* as the most abundant copepod species; later on, similar results were obtained by Vinogradov *et al.* (1991) and Vinogradov & Shishkina (1994) in the same area, although they corrected the identity of the dominant species as *E. inermis* Giesbrecht. It is noteworthy that according to the latter authors, *E. inermis* can constitute up to 90–95% of the zooplankton biomass in the 200–600 m layer in this upwelling system. The work by Sameoto (1986) included the analysis of one sampling station in the Costa Rica Dome (7°19′ N, 83°25′ W); the number of species recorded in this single site is 91, including Calanoida, Cyclopoida, Poecilostomatoida, and Harpacticoida. Only five species remained unidentified (*Temora* sp., *Amallothrix* sp., *Labidocera* sp., *Cephalophanes* sp.,

Lubbockia sp.). Contrasting with the results by Suárez-Morales & Gasca (1989a) and Vinogradov *et al.* (1991), Sameoto (1986) found *Paracalanus parvus* and *Oithona* sp. as the most abundant forms. This is probably related to the different seasons and depths sampled; the upwelling featuring the oceanographic conditions in the Costa Rica Dome also has sharp seasonal variations that are likely to produce important changes in the copepod community structure. Morales-Ramírez & Vargas (1995) recorded 12 species (nine calanoids, two harpacticoids, one poecilostomatoid) from plankton samples collected in the Golfo de Nicoya. Morales-Ramírez (1996) provided a list of the 54 pelagic copepod species collected in the Golfo de Nicoya, Golfo Dulce, and Bahía de Coronado. Morales-Ramírez (2001) discussed different aspects of the diversity of copepods in Costa Rica, both in the Pacific and the Caribbean coasts; he presented an account of 80 species that included 67 pelagic forms, which are included in Species Lists 24.1 and 24.2 (both included on the CD-Rom). The most diverse families of pelagic copepods recorded to date in the Pacific waters of Costa Rica are the Pontellidae, followed by Paracalanidae and Eucalanidae (Species List 24.2; see CD-Rom). Suárez-Morales & Morales-Ramírez (2001) reported recently *Acartia negligens* in Costa Rican waters as a new record for the ETP and Costa Rica. Later on, Suárez-Morales & Morales-Ramírez (2003) described *Cymbasoma concepcionae*, a new species of monstrilloid copepod collected in one of the coastal systems on the Pacific waters of Costa Rica.

The benthic harpacticoids of the Pacific coast of Costa Rica have been surveyed by Mielke (1992, 1994a, b, 1995, 1997); he described several species of different genera. Some of these new species are likely to be endemic to Costa Rica; also, the diversity of some of these taxa in Costa Rica (i.e., *Schizopera*: see Mielke 1995) suggests that the local fauna of benthic Harpactioida is quite speciose. Overall, he published records or descriptions of 17 species of benthic Harpacticoida from the Pacific coast of Costa Rica (see Species List 24.2 on CD-Rom), thus providing the first information on the composition of this complex group of copepods in this country. The harpacticoid fauna of the adjacent Pacific coast of Panama has also been studied by Mielke (1990a, b).

Caribbean Sea and adjacent areas: The copepods of the Caribbean Sea have been surveyed by several authors; the most relevant works are those published by Owre & Foyo (1964a, b); they recorded 142 species from samples collected in the eastern sector of the Caribbean Sea, between 15°00′–14°55′ N and 67°05′–64°50′ W and between 0 and 1,300 m depth. Studies on neritic and coastal systems of the Caribbean are by Legaré (1961, 1964) in the Venezuelan area, Coker & González (1960) and González & Bowman (1965) in embayments of Puerto Rico; the latter authors provided illustrations and identification keys for nearly 20 estuarine-coastal species. In one of the most relevant regional taxonomic accounts of the marine oceanic Copepoda, Park (1970) recorded 178 species, most of them from the Caribbean Sea. Björnberg (1971) analyzed the general distributional patterns of the plankton in the Caribbean and included a section on the planktonic copepods. Michel & Foyo (1976) reviewed the previous records of Copepoda in the Caribbean and estimated that 450 species dwell in this basin; in their report they included partial results from a single station of the cruise “P6811” made in 1968 in the western Caribbean. This particular station is in Costa Rican waters; 20 species of copepods were reported from this single site based on samples collected at depths of 0, 45, and 500 m. However, the authors explicitly state that their analysis was restricted to a selected group of species. They indicated the presence of other species as "other Calanoida" and "other Cyclopoida"; the latter probably includes poecilostomatoid forms as well. De la Cruz (1971) recorded 16 species in the southern Gulf of Mexico. Campos (1980) found 108 species in the southern Gulf of Mexico and part of the Caribbean Sea; she found differences in the copepod composition between the two regions. Later on, Campos (1982) provided a list of 115 species recorded in neritic waters of Cuba. Suárez-Morales & Gasca (1989b) recorded 82 species in the Yucatan Channel area. Campos & Suárez-Morales (1994) revised, among many others published before 1990, taxonomic and distributional works referring to particular copepod taxa such as that of Deevey (1979) on *Bathypontia*, Ferrari (1975) on *Oncaea*, or Fleminger (1979) on *Labidocera*, and prepared a list of 190 species with brief descriptions, identification keys, and illustrations of these species. Reid (1990) published a list of free-living copepods, including 217 pelagic species, known from the northwestern tropical Atlantic. Suárez-Morales & Gasca (2000b) surveyed the oceanic copepod fauna of the Mexican Caribbean Sea and recorded 89 species in surface waters (0–50 m). The reef copepods of the western Caribbean were surveyed by Suarez-Morales & Gasca (2000a); they recorded 45 species. Overall, Suárez-Morales & Gasca (1998) & Suárez-Morales *et al.* (2000b) listed 201 species for the westernmost part of the Caribbean Sea.

In the Caribbean waters of Costa Rica, Morales & Murillo (1996) surveyed the zooplankton community dwelling at Cahuita coral reef and included a general analysis of the pelagic copepods, at a group level only. Recently, Morales-Ramírez (2001) discussed the biodiversity of pelagic copepods in waters around the Cahuita coral reef, distinguishing only 13 species. No more studies are known of this and adjacent regions of the Caribbean coast of Costa Rica

The benthic harpacticoid copepods of the Caribbean coast of Costa Rica were studied by Mielke (1992, 1993, 1994c, 1995), who recorded seven species, two of the genus *Karllangia*, one of *Nitocra*, three of *Phyllopodopsyllus*, and one of *Longipedia*. The Harpacticoida of the Caribbean coast have been less studied than the Pacific coast of Costa Rica (17 species), thus representing the same tendency found in the planktonic copepods. Another work from the adjacent Caribbean coast of Panama includes the description of two species of the cylindropsyliid harpacticoid genus *Psammopsyllus* (Mielke 1983), probably present in Costa Rica as well.

Number of Species Described

Copepods are represented by an estimated number of over 11,500 species (Humes 1994); following this calculations and the fact that nearly two thirds (>4,250) of these species were described in the last 27 years, we have an increase rate of *ca.* 157 species/year. Therefore, considering the time elapsed since this calculation was made (14 years), the figure might have risen, conservatively, to over 13,500 species. However, the description of new taxa in this group and the highly dynamic taxonomy of several taxa makes it impossible to state a conclusive figure on the number of species known. It has been stated recently that the marine realm is one of the most productive habitats in terms of taxonomic novelties during the last few decades (Boxshall & Halsey 2004). Mauchline (1998) presented a complete analysis of the biology of the Calanoida, which is the most diverse group of pelagic copepods. He included all the published information on this group before 1997; moreover, the global distribution, synonymy, vertical distribution, and different issues about their development and ecology were included in this comprehensive work. This author recognized about 1,460 species of marine planktonic Calanoida; according to Boltovskoy *et al.* (1999) the number of species in this order is closer to 2,000.

On the other hand, the most recent accounts of the diversity of marine Harpacticoida published by Bodin (1988, 1997) comprise over 1,300 recognized benthic and pelagic species plus many more taxa with an uncertain status.

Latitudinal Diversity

The latitudinal copepod species richness is variable and does not necessarily increase at lower latitudes or oceanwards; this is suggested by results of different works along the ETP. Up to 176 taxa have been recorded in several studies along the Baja California peninsula (24–27°), including mostly oceanic areas (Palomares-García *et al.* 1998; Hernández-Trujillo 1999). Off the coast of Jalisco, mainly in neritic or near-oceanic waters, Suárez-Morales *et al.* (2000a) found 44 species (20° N); southwards (17–18° N) along the same coast, Suárez-Morales & Zurita (1991) recorded 28 species, whereas Alameda (1980) found 73 species in the Tehuantepec area (14–16° N). Suárez-Morales & Gasca (1989a) and Sameoto (1986) recorded over 100 species in the Costa Rica Dome (7–11° N). An accurate figure for the Dome cannot be provided due to the unidentified species recorded as "sp." in Sameoto's work. These latitudinal differences have been attributed to deeper sampling, to the distance to the coast, and to the local productivity-oceanographic status of the various areas (McGowan & Walker 1985; Suárez-Morales *et al.* 2000a).

Diversity of Copepoda in Costa Rica

Overall, in waters of Costa Rica, we found records of 209 species of marine copepods: 185 were planktonic and 24 benthic. Out of the 185 planktonic species, only 37 have been recorded in Costa Rican Caribbean waters (Species List 24.1 on CD-Rom) and up to 149 from the Pacific side of Costa Rica (Species List 24.2 on CD-Rom). The number of species known in this country is relatively low when compared to the figures reported in adjacent areas such as the Mexican Pacific (251), the Mexican Atlantic (277), the California Current System (over 200 species), or the Caribbean Sea (*ca.* 450 species).

The number of pelagic forms recorded in the Pacific waters of Costa Rica (excluding the benthic Harpacticoida) (185) represents about 55% of the estimated species richness of copepods in the central and eastern tropical Pacific (McGowan & Walker 1979, 1985; Sameoto 1986; Suárez-Morales & Gasca 1989a, 2000; Suárez-Morales *et al.* 2000a).

According to the regional reports by Owre & Foyo (1964a, b), Michel & Foyo (1976), Campos & Suárez-Morales (1994), and Suárez-Morales *et al.* (2000b), the number of species of pelagic Copepoda in the Caribbean seems to be around 400–450. Therefore, the number of pelagic species recorded in the Caribbean area of Costa Rica (45) represents around one tenth of the potential records in this area of the tropical northwestern Atlantic. It is foreseeable that the number of species will increase when sample collections from deeper oceanic areas are analyzed.

Up to 22 species occur along both coasts of Costa Rica. A total of 118 species of planktonic copepods of Costa Rica are epipelagic, dwelling mainly in the 0–200 m layer; several epipelagic forms reach the mesopelagic layers. Only a few (13 species) are strictly meso or bathypelagic forms, dwelling in the 200–1,000 m layer.

Specialists

In general, the research activities in the Costa Rican waters have focused on coastal-neritic areas, with relatively low species richness. This is why knowledge on composition, distribution, and abundance of oceanic copepods is quite limited. Another relevant factor is the scarceness of national specialists working with material from Costa Rica; this is a problem that is shared with other countries in Latin America (Suárez-Morales & Gómez-Aguirre 1996). Currently, most of the information about this group in Costa Rica has been produced by a single researcher, who is the first author of this contribution (A. Morales-Ramírez). The following list includes active colleagues who have worked directly with planktonic copepods from Costa Rica or marginally on material from the ETP or the northwestern tropical Atlantic.

José N. Álvarez-Cadena: Instituto de Ciencias del Mar y Limnología, UNAM, Unidad Académica Puerto Morelos, A.P. 1153, Cancún, Quintana Roo, 77500, Mexico – Tropical Pacific and Atlantic. inac@mar.icmyl.unam.mx

Frank D. Ferrari: National Museum of Natural History, Smithsonian Institution, M.Sc., Washington, DC 20013-7012, USA – Tropical and subtropical Atlantic and Pacific. Ferrari.Frank@msc.si.edu

Samuel Gómez-Noguera: Instituto de Ciencias del Mar y Limnología, UNAM, Unidad Académica Mazatlán, Mazatlán, Sinaloa, Mexico – Tropical Harpacticoida, Pacific. samuelgomez@ola.icmyl.unam.mx

Sergio Hernández-Trujillo: CICIMAR-IPN, La Paz, BCS, Mexico – Tropical and subtropical Pacific. strujil@ipn.mx

Elena L. Markhaseva: Academy of Sciences, Zoological Institute, Lab Marine Biology, Universitetskaya nab 1, St. Petersburg, Russia – Tropical and subtropical Pacific. site: www.zin.ru

Wolfgang Mielke: Institute for Zoology and Anthropology, University of Göttingen, Göttingen, Germany – Harpacticoida of Costa Rica, worldwide

Álvaro Morales-Ramírez: CIMAR and Escuela de Biología, Universidad de Costa Rica, San Pedro, 11501-2060 San José, Costa Rica – Estuarine and oceanic copepods of Costa Rica (Pacific and Caribbean). alvaro.morales@ucr.ac.cr

Ricardo Palomares-García: CICIMAR-IPN, La Paz, BCS, Mexico – Tropical and subtropical Pacific. rpalomar@ipn.mx

Eduardo Suárez-Morales: El Colegio de la Frontera Sur (ECOSUR). Unidad Chetumal, A.P. 424, Chetumal, Quintana Roo 77000, Mexico – Tropical Pacific and Atlantic. esuarez@ecosur-qroo.mx

Taisoo Park: Scripps Institution of Oceanography. Center of Coastal Research, La Jolla, California, USA – Tropical and subtropical Atlantic and Pacific. Retired

Collections

There is one collection of Costa Rican copepods in Costa Rica, established by Álvaro Morales-Ramírez at the Universidad de Costa Rica. The CIMAR, Universidad de Costa Rica, has a large collection of unstudied zooplankton samples from relatively shallow coastal systems of Costa Rica, mainly from the Pacific coast (Golfo de Nicoya, Golfo Dulce, Bahía Culebra). A preliminary analysis by the second author (E. Suárez-Morales) revealed that these samples contain a diverse fauna and several new local records (Suárez-Morales & Morales-Ramírez 2001), and also a new species of Monstrilloida (Suárez-Morales & Morales-Ramírez 2003).

The Laboratorio de Invertebrados of the Facultad de Ciencias, UNAM, in Mexico City holds zooplankton samples collected in the Costa Rica Dome; the copepods in these samples were studied by Suárez-Morales & Gasca (1989a), but there are several original samples that have not been examined yet. It is expected that the analysis of this particular set of samples will yield a number of new records for the copepod fauna of the Costa Rican Pacific.

Some of the sites with information about these crustaceans are: http://www.nmnh.si.edu/iz/copepoda, managed and improved by Chad Walter; this service is currently being provided from the National Museum of Natural History, Smithsonian Institution, Washington, DC. The site includes a link to the most relevant bibliographical source about the Copepoda, the Wilson Copepod Library, with over 40,000 titles. The World Association of Copepodologists (WAC) has recently put a web page on line (http://www.copepoda.uconn.edu) in which many different topics related to the present, past, and future of copepodology can be read and even discussed. The WAC has an official newsletter

(Monoculus) that is distributed both electronically and in printed form among the members of the association. It offers links to related sites of copepodologists in which each of them provide different kind of information about their work and advances in copepodology.

Problems

The current knowledge on the Costa Rican copepod fauna is still limited. Most of the species richness is merely guessed from surveys developed in adjacent areas, and nearly 50% of the species diversity was reported from the analysis of a single oceanic sample. Therefore, it is recommended to start or continue sampling plans covering not only the neritic-coastal environments, but also the oceanic realm. The analysis of these samples will yield a number of new records and will provide insights concerning the distribution and abundance of these crustaceans in Costa Rica.

Recent studies (Huntley *et al.* 2000; López-Salgado *et al.* 2000) in the northwestern tropical Atlantic intended to relate the distribution/abundance of pelagic copepods with mesoscale oceanographic processes such as oceanic cyclonic and anticyclonic gyres. In Costa Rica, the knowledge about copepods is expected to yield valuable information about oceanographic processes at different scales, the oceanic upwelling area of the Costa Rica Dome being one of the most interesting systems.

References

Alameda GM (1980) Sistemática y distribución de los copépodos (Crustacea) del Golfo de Tehuantepec (Mexico). B.Sc. thesis, Fac Sci, UNAM, 121 p

Ambler JW, Ferrari FD, Fornshell JA (1991) Population structure and swarm formation of the cyclopoid copepod *Dioithona oculata* near mangrove cays. J Plankton Res 113:1257–1272

Arcos F, Fleminger A (1986) Distribution of filter-feeding calanoid copepods in the eastern equatorial Pacific. CalCOFI Rep 27:170–186

1. Bednarski M (2001) Macrozooplankton of Culebra Bay, Costa Rica with emphasis on copepods. M.Sc. thesis, ZMT, University of Bremen, Germany, 155 p

Björnberg TKS (1971) Distribution of plankton relative to the general circulation system in the area of the Caribbean Sea and adjacent regions. Symposium on Investigations and Resources of the Caribbean Sea and Adjacent Regions, UNESCO, Paris, pp 343–356

Björnberg TKS (1981) Copepoda. *In*: Boltovskoy D (ed) Atlas del zooplancton del Atlántico sudoccidental y métodos de trabajo con zooplancton marino. INIDEP, Mar del Plata, Argentina, pp 587–679

Bodin P (1988) Catalogue des Nouveaux Copepodes Harpacticoides Marins. Université de Bretagne Occidentale, Brest, 288 p

Bodin P (1997) Catalogue of the new marine harpacticoid copepods (1997 edition). Studiedoc. Koninkl Belgisch Inst Natuurwetensch (Doc Travail, Inst Royal Sci Nat Belgique) 89:1–304

Boltovskoy D, Gibbons MJ, Hutchings L, Binet D (1999) General biological features of the South Atlantic. *In*: Boltovskoy D (ed) Zooplankton of the Southwestern Atlantic. Backhuys, Leiden, pp 1–42

Boxshall GA, Halsey SH (2004) An Introduction to Copepod Diversity, vol 166. Part I. The Ray Society, London, 421 p

Bucklin A (1986) The genetic structure of zooplankton populations. Pelagic biogeography UNESCO Tech Pap Mar Sci 49:33–41

Bucklin A (1998) Molecular analysis of biogeographic patterns and processes: the copepod *Nannocalanus minor* in the N. Atlantic Ocean IOC-UNESCO Workshop Rep 142:78–85

Burd BJ, Thomson RE, Jamieson GS (1992) Composition of a deep scattering layer overlying a mid-ocean ridge hydrothermal plume. Mar Biol 113:517–526.

Campos A (1980) Distribución y abundancia relativa de los copépodos planctónicos en el Golfo de México y el Mar Caribe. Cienc Biol 5:57–74

Campos A (1982) Lista de especies de copépodos planctónicos de aguas cubanas. Poeyana 24:1–27

Campos A, Suárez-Morales E (1994) Copépodos pelágicos del Golfo de México y el Mar Caribe. I. Sistemática y Biología. CONACYT/ CIQRO, México, 368 p

Chen YQ (1986) The vertical distribution of some pelagic copepods in the eastern tropical Pacific. CalCOFI Rep. 27:205–227

Claus, C (1866) Die Copepoden Fauna von Nizza. Ein Beitrag zur Charakteristik der Form und anderen Abänderungen "im Sinne Darwin's," Marburg & Leizig, 34 p

Coker RE, González JG (1960) Limnetic copepod population of Bahía Fosforescente and adjacent waters, Puerto Rico. J Elisha Mitchell Sci Soc 76:8–28

Damkaer DM (2002) The Copepodologist's cabinet: a biographical and bibliographical history. Am Phil Soc, Mem 240:300

Dana JD (1848) Conspectus Crustaceroum quae in orbis terrarium circumnavigatione "Carolo Wilkes" e Classe Republicae Foederatae Duce, lexit et descripsit Jacobus D. Dana. Collectorum, 1. Proc Am Acad Arts Sci 1:149–155

Deevey GB (1979) *Bathypontia* Sars (Copepoda: Calanoida): eight species, two new, from the Caribbean Sea and Gulf of Mexico. Proc Biol Soc Wash 92:724–742

De la Cruz A (1971) Zooplancton de la región sureste del Golfo de México. Ciencias (Cuba), Ser 4(24):1–55

Ferrari FD (1975) Taxonomic notes on the genus *Oncaea* (Copepoda, Cyclopoida) from the Gulf of Mexico and northern Caribbean Sea. Proc Biol Soc Wash 88:217–232

2. Ferrari F, Bowman TEB (1980) Pelagic copepods of the Family Oithonidae (Cyclopoida) from the east coast of Central and South America. Smithson Contrib Zool 312:1–27

Fleminger A (1964) Distributional atlas of calanoid copepods in the California current region, part II. CalCOFI Atlas 7:1–213

Fleminger A (1967) Taxonomy, distribution and polymorphism in the *Labidocera jollae* group, with remark on evolution within the group (Copepoda: Calanoida). Proc US Nat Mus 120:1–61

Fleminger A (1976) Evidence of a species swarm in a Caribbean planktonic copepod species group. Am Zool 16:269

Fleminger A (1979) *Labidocera* (Copepoda, Calanoida): new and poorly known Caribbean species with a key to species in the western Atlantic. Bull Mar Sci 29:170–190

Giesbrecht W (1892) Systematik und Faunistik der pelagischen Copepoden des Golfes von Neapal und der angrezenden Meeres-Abschnitte. Fauna und Flora des Golfes von Neapel und der angrenzenden Meeres-Abschnitte. Zoologische Station zu Neapel. XIX. Monographie. Friedländer & Sohn, Berlin, pp 1–831

González JG, Bowman TE (1965) Planktonic copepods from Bahia Fosforescente, Puerto Rico, and adjacent waters. Proc US Nat Mus 117:241–304

Guzmán HM, Obando VL (1988) Diversidad y abundancia diaria y estacional del zooplancton marino de la Isla del Caño, Costa Rica. Rev Biol Trop 36:139–150

Greenlaw CF (1979) Acoustical estimation of zooplankton populations. Limnol Oceanogr 24:226–241

Hamner WM, Carleton JH (1979) Copepod swarms: attributes and role in coral reef ecosystems. Limnol Oceanogr 24:1–14

He DH, Wang CS, Liu HB, Yang GM (1992) A study on zooplankton distribution patterns and indicator species in Kuroshio upstream area and adjacent East China Sea. Acta Ocean Sin 11:237–254

Hernández-Trujillo S (1999) Variability of community structure of Copepoda related to El Niño 1982–83 and 1987–88 along the west coast of the Baja California Peninsula, Mexico. Fish Oceanogr 8:284–295

Hernández-Trujillo S, Esqueda-Escárcega G (2002) La diversidad de copépodos marinos en México. Oceánides 17:65–69

Ho J-S, Dojiri M, Hendler G, Deets GB (2003) A new species of Copepoda (Thaumatopsyllidae) symbiotic with a brittle star from California, USA, and designation of a new Order Thaumatopsylloida. J Crust Biol 23:582–594

Hulsemann K, Fleminger A (1990) Taxonomic value of minute structures on the genital segment of *Pontellina* females (Copepoda: Calanoida). Mar Biol 105:99–108

Humes AG (1991) Zoogeography of copepods at hydrothermal vents in the eastern Pacific Ocean. Bull Plankt Soc Japan, Spec Vol (1991):383–389

Humes AG (1994) How many copepods? Hydrobiol 292/293:1–7

Huntley ME, González A, Zhu Y, Zhou M, Irigoien X (2000) Zooplankton dynamics in a mesoscale eddy-jet system off California. Mar Ecol Progr Ser 201:165–178

Huys R, Boxshall GA (1991) Copepod Evolution. The Ray Society, London 159:1–468

3. Johnson MW (1964) On a new species of Pseudodiaptomus from the west coast of Mexico, Costa Rica and Ecuador (Copepoda). Crustaceana 7:33–41

Latreille PA (1802) Historie naturalle générale et particulière des Crustacés et des Insectes. París, 468 p

Legaré JEH (1961) Estudios preliminares del zooplancton en la región de Cariaco. Bol Inst Oceanogr Univ Oriente 1:191–218

Legaré JEH (1964) The pelagic Copepoda of eastern Venezuela. 1. The Cariaco Trench. Bol Inst Oceanogr, Univ Oriente 3:15–81

López-Salgado I, Gasca R, Suárez-Morales E (2000) La comunidad de copépodos (Crustacea) en los giros a mesoescala en el occidente del Golfo de México (julio, 1995). Rev Biol Trop 48:169–179

Markhaseva EL (1995) A new species of *Paracomantenna* from the eastern tropical Pacific (Crustacea, Copepoda: Aetideidae). Zoosyst Ross 3:229–233

Markhaseva EL (1997) A new species of *Pseudochirella* from the eastern tropical Pacific (Crustacea: Copepoda: Aetideidae). Zoosyst Ross 6:31–32

Markhaseva EL, Ferrari FD (1996) Three new species of *Ryocalanus* from the eastern tropical Pacific (Crustacea, Copepoda: Ryocalanidae). Zoosyst Ross 4:63–70

Mauchline J (1988) Taxonomic value of pore pattern in the integument of calanoid copepods (Crustacea). J Zool 214:697–749

Mauchline J (1998) The biology of calanoid copepods Adv Mar Biol 33:1–710

McGowan JA, Walker PW (1979) Structure in the copepod community of the North Pacific Central Gyre. Ecol Monogr 49:195–226

McGowan JA, Walker PW (1985) Dominance and diversity maintenance in an oceanic system. Ecol Monogr 55:103–118

4. Michel HB, Foyo M (1976) Caribbean Zooplankton. Part 1. Siphonophora, Heteropoda, Copepoda, Euphausiacea, Chaetognatha and Salpidae. Office of Naval Research, Department of the Navy, USA, 549 p

Mielke W (1983) Two new *Psammopsyllus* species (Copepoda) from the Caribbean coast of Panama. Stud Neotrop Fauna Environ 18:101–109

Mielke W (1990a) *Zausodes septimus* Lang, 1965 und *Enhydrosoma pericoense* nov. spec., zwei bentische Ruderfußkrebse (Crustacea, Copepoda) aus dem Eulitoral von Panama. Microfauna Mar 6:139–156

Mielke W (1990b) A new species of *Psammotopa* (Copepoda) from the Pacific coast of Panama. Crustaceana 59:69–75

5. Mielke W (1992) Six representatives of the Tetragonicipitidae (Copepoda) from Costa Rica. Microfauna Mar 7:101–146

6. Mielke W (1993) Species of the taxa *Orthopsyllus* and *Nitocra* (Copepoda) from Costa Rica. Microfauna Mar 8:247–266

7. Mielke W (1994a) New records of two copepod species (Crustacea) from the Pacific coast of Costa Rica. Microfauna Mar 9:55–60

8. Mielke W (1994b) *Microcanuella bisetosa* gen. n., sp. n., a new taxon of Canuellidae (Crustacea, Copepoda) from the Costa Rican Pacific coast, with remarks on *Galapacanuella beckeri*. Bijdr Dierk 64:55–64
9. Mielke W (1994c) Two co-occurring new *Karllangia* species (Copepoda: Ameiridae) from the Caribbean coast of Costa Rica. Rev Biol Trop 42:141–153
10. Mielke W (1995) Species of the taxon *Schizopera* (Copepoda) from the Pacific coast of Costa Rica. Microfauna Mar 10:89–116
11. Mielke W (1997) New findings of interstitial Copepoda from Punta Morales, Pacific coast of Costa Rica. Microfauna Mar 11:271–280

Milne-Edwards H (1840) Ordre des Copepodes. *In*: Historie naturelle des Crustaces comprenant l'Anatomie, la Physiologie et la Classification de ces Animaux, 3. Paris, pp 411–529

12. Morales-Ramírez A (1996) Checklist of copepods from Gulf of Nicoya, Coronado Bay and Golfo Dulce, Pacific coast of Costa Rica, with comments on their distribution. Rev Biol Trop 44(Suppl 3):103–113
13. Morales-Ramírez A (2001) Biodiversidad marina de Costa Rica, los microcrustáceos: Subclase Copepoda (Crustacea: Maxillopoda). Rev Biol Trop 49(Suppl 2):115–133

Morales-Ramírez A, Murillo MM (1996) Distribution, abundance and composition of coral reef zooplankton, Cahuita National Park, Limon, Costa Rica. Rev Biol Trop 44:619–630

14. Morales-Ramírez A, Vargas-Zamora JA (1995) Especies comunes de copépodos pelágicos (Crustacea: Copepoda) del Golfo de Nicoya, Costa Rica. Rev Biol Trop 43:207–218

Owre HB, Foyo M (1964a) Report on a collection of Copepoda from the Caribbean Sea. Bull Mar Sci Gulf Caribb 14:359–372

Owre HB, Foyo M (1964b) Plankton of the Florida Current. Part IX. Additions of the list of Copepoda, with descriptions of two rare species. Bull Mar Sci Gulf Caribb 14:342–358

Palomares-García R, Suárez-Morales E, Hernández-Trujillo S (1998). Catálogo de los copépodos (Crustacea) pelágicos del Pacífico Mexicano. CICIMAR-ECOSUR, México, 352 p

Park TS (1970) Calanoid Copepods from the Caribbean Sea and Gulf of Mexico. 2. Two new species and new records from plankton samples. Bull Mar Sci 20:472–546

14b. Park TS (1994) Taxonomy and distribution of the marine calanoid copepod family Euchaetidae. Bull Scripps Inst Oceanogr Univ Calif, San Diego 29:1–203

Rao DVS, Sameoto D (1988) Relationship between phytoplankton and copepods in the deep tropical Pacific Ocean off Costa Rica. Bull Mar Sci 42:85–100

Reid JW (1990) Continental and coastal free-living copepoda (Crustacea) of Mexico, Central America and the Caribbean region. *In*: Navarro D, Robinson JG (eds) Diversidad biológica en la Reserva de la Biosfera de Sian Ka'an, Quintana Roo, México. CIQRO/University of Florida, Chetumal, México, pp 175–213

15. Sameoto DD (1986) Influence of the biological and physical environment on the vertical distribution of mesozooplankton and micronekton in the eastern tropical Pacific. Mar Biol 93:263–279

Sars GO (1901) Copepoda, Calanoida. Crustacea of Norway 4:1–28

Sars GO (1903) Copepoda, Calanoida. Crustacea of Norway 4:145–171

Suárez-Morales E (1997) Pelagic copepod assemblages during spring upwelling off the Yucatan Peninsula (1985). Int Oceanogr Comm (SCOR/IOC-UNESCO) Workshop Rep 142:345–352

16. Suárez-Morales E, Gasca R (1989a) Copépodos calanoides epiplanctónicos del Domo de Costa Rica (julio-agosto, 1982). Cienc Mar 15:89–102

Suárez-Morales E, Gasca R (1989b) Copépodos epiplanctónicos (Crustacea) del Canal de Yucatán (mayo-junio, 1984). Carib J Sci 25:196–202

Suárez-Morales E, Gasca R (1998) Updated checklist of the free-living marine Copepoda (Crustacea) of Mexico. An Inst Biol, Univ Nac Autón Méx, Ser Zool 69(1):105–119

Suárez-Morales E, Gasca R (2000a) The planktonic copepod community at Mahahual reef, western Caribbean. Bull Mar Sci 66:255–267

Suárez-Morales E, Gasca R (2000b) Epipelagic copepod assemblages in the Western Caribbean Sea. Crustaceana 73:1247–1257

Suárez-Morales E, Gómez-Aguirre S (1996) Aspectos históricos y perspectivas del estudio de zooplancton marino en México. Ciencia 47:215–223

17. Suárez-Morales E, Morales-Ramírez A (2001) Nuevo registro de *Acartia (Planktacartia) negligens* Dana, 1849 (Copepoda, Calanoida) en el Pacífico Tropical Oriental. Rev Biol Trop 49:1286
18. Suárez-Morales E, Morales-Ramírez A (2003) A new species of *Cymbasoma* (Crustacea: Copepoda: Monstrilloida) from the Pacific coast of Costa Rica, Central America. Proc Biol Soc Wash 116:206–214

Suárez-Morales E, Zurita JC (1991) Copépodos epipelágicos de la zona costera central de Michoacán, México (invierno-verano, 1983). Univ Cienc (UJAT) 8:25–32

Suárez-Morales E, Franco-Gordo C, Saucedo M (2000a) On the pelagic copepod (Crustacea: Copepoda) community of the central Mexican tropical Pacific (autumn, 1990). Crustaceana 73:751–761

Suárez-Morales E, Reid JW, Gasca R (2000b) Copepoda. *In*: Llorente J, González E, Papavero N (eds) Biodiversidad, taxonomía y biogeografía de artrópodos de México: hacia una síntesis de su conocimiento. Vol. II. UNAM/CONABIO/Bayer, México, pp 172–190

19. Vinogradov M, Shushkina A (1994) A study of vertical distribution of the North Pacific zooplankton based on quantitative estimations from the deep manned submersible (DMS) "Mir." Trud Inst Okean 131:41–63

Vinogradov ME, Shushkina EA, Gorbunov AY, Shashkov NL (1991) Vertical distributions of mesozooplankton and macroplankton in the vicinity of the Costa Rica Dome. Oceanology 31:759–769

20. Walter C (1989) Review of the new world species of *Pseudodiaptomus* (Copepoda: Calanoida) with a key to the species. Bull Mar Sci 45:590–628

Part 25
Sea-Spiders

Roger N. Bamber

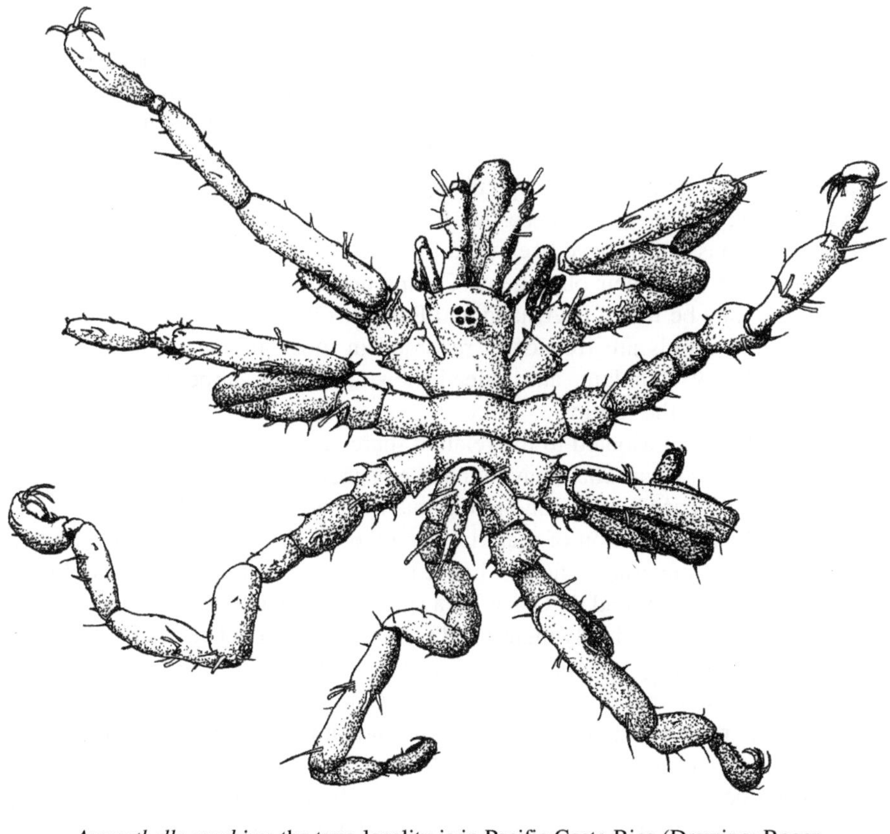

Ammothella symbius: the type-locality is in Pacific Costa Rica (Drawing: Roger N. Bamber)

R.N. Bamber
Department of Zoology, The Natural History Museum, Cromwell Road, London SW7 5BD, United Kingdom.
e-mail: r.bamber@nhm.ac.uk

I.S. Wehrtmann, J. Cortés (eds.) *Marine Biodiversity of Costa Rica, Central America*,

Abstract Records of pycnogonids, or sea-spiders, from the waters of Costa Rica are relatively scarce. Including those known *ex. litt.*, only 12 species, representing seven genera and four families, have been recorded. Ten of these species are from Pacific waters and only two from Caribbean waters (with no species in common); three are from deepwater and nine from shallow water (including the littoral). While knowledge of the Central and South American Pacific pycnogonid fauna is as yet poor, and so predictions of the likely fauna are hardly possible, many more species are to be expected from the Caribbean coasts.

Introduction

The Class Pycnogonida, known as sea-spiders, has little evident affinity to other classes within the Phylum Arthropoda. Pycnogonids are a fully marine group, found from the littoral to the deepsea and from the tropics to the poles. They commonly inhabit sessile epifaunal communities, in some cases marine plants, while a few are infaunal, or parasitic, or bathypelagic. Sea-spiders are typically eight-legged, with three pairs of cephalic appendages, namely palps, chelifora, and ovigerous legs; any or all of these last three may be absent in adults of certain genera, conditions which have been used historically, rightly or wrongly, to determine many aspects of their phylogeny. Comprehensive description of the biology of this group may be found in Arnaud and Bamber (1987).

While pycnogonids are frequently recorded in most marine surveys, they are rarely numerically abundant, and indeed their abundance appears lowest in the shallower waters of the tropics.

This may have contributed to the relative scarcity of records of pycnogonids in the waters of Costa Rica. There having been no specific collecting for pycnogonids in Costa Rica, all records are serendipitous and most are from shallow waters. Child (1979) described a collection of material from the Pacific and Caribbean coasts of Central America, material which is housed in the collections of the Zoological Museum of the Universidad de Costa Rica (marked ZMUCR below; those marked ZMUCR* were not specifically mentioned from Costa Rica in Child's paper). Costa Rica is the type locality for *Ammothella symbius* Child, 1979.

Stock (1986) collated all records from the Caribbean, and references to Costa Rican material prior to that date can be found in that paper; none of the new material he describes (from "Pillsbury" and "Gerda" cruises) are from Costa Rica, although he does record *Pallenopsis kemfi* Stock, 1975 from the Caribbean coast of Nicaragua. His listing of all records gives an indication of what further species may be expected from the Caribbean coast of Costa Rica. The Pacific coasts of Central and South America are relatively poorly studied, and it is not feasible to suggest which further species may be anticipated from the western Costa Rican waters. However, records to date are largely incidental. Specific searching for pycnogonids off the coasts of Costa Rica will undoubtedly reveal further species, like, for example, the as yet unidentified species of *Pentapycnon* photographed in Punta Cocles, Caribbean of Costa Rica (Fig. 25.1) or the other unidentified species of *Anoplodactylus* sp. from Playa Tamarindo, Pacific Costa Rica (Fig. 25.2).

Fig. 25.1 Unidentified species of *Pentapycnon* from Punta Cocles, Caribbean of Costa Rica (Photo: Ingo S. Wehrtmann)

Fig. 25.2 Unidentified species of *Anoplodactylus* from Playa Tamarindo, Pacific Costa Rica (Photo: Ingo S. Wehrtmann)

Of the 12 species recorded *in litt.* or otherwise from Costa Rica (Species Lists 25.1 and 25.2 which are included on the CD-Rom), ten are from Pacific waters and only two from Caribbean waters (with no species in common); three are from deepwater and nine from shallow water (including the littoral). Seven genera are recorded. The families represented are the Ammotheidae Dohrn, 1881; Callipallenidae Hilton, 1942; Phoxichilidiidae Sars, 1891; and Nymphonidae Wilson, 1878. No species from the families Endeididae Norman, 1908; Austrodecidae Stock, 1954; Colossendeidae Hoek, 1881 Rhynchothoracidae Fry, 1978; or Pycnogonidae Wilson, 1878 have yet been identified from Costa Rica (but see Fig. 25.1).

While the shallow species other than *Anoropallene palpida* are typically known from hard substrata (epifaunal and fouling communities), the local records rarely specify habitat.

Specialists

Roger Bamber: Department of Zoology, The Natural History Museum, Cromwell Road, London SW7 5BD, United Kingdom – Worldwide. R.Bamber@nhm.ac.uk

Franz Krapp: Zoologisches Forschungsinstitut und Museum Alexander Koenig, Bonn, Germany – Worldwide.

Helen Turpaeva: P.P. Shirshov Institute of Oceanology, 36 Nakhimovsky Prospect, 117997 Moscow, Russia – Worldwide.

Collections

The Museo de Zoología, Universidad de Costa Rica, contains the pycnogonid material cited above. Other museums with large, international collections of pycnogonids (i.e., The Smithsonian Institute; The Natural History Museum, London; The Zoölogisch Museum, Amsterdam) may have material from Costa Rica and Central America; however, I am unaware of the extent and completeness of these collections regarding the Central American region.

Acknowledgments I am grateful to Ingo Wehrtmann of the Universidad de Costa Rica for information on the collections at the Zoological Museum (ZMUCR).

References

Arnaud F, Bamber RN (1987) The biology of Pycnogonida. Adv Mar Biol 24:1–96

1. Arnaud F, Child CA (1988) The South African museum's Meiring Naude Cruises, Part 17: Pycnogonida. Ann South Afr Mus 98:121–187
2. Child CA (1979) Shallow-water Pycnogonida of the Isthmus of Panama and the coasts of Middle America. Smithson Contrib Zool 293:1–86

3. Child CA (1992) Pycnogonida of the Southeast Pacific biological oceanographic project (SEPBOP). Smithson Contrib Zool No 526:1–43
4. Cole LJ (1904) Pycnogonida of the West Coast of North America. Harriman Alaska Exped 10:249–330
5. Hall HVM (1912) Studies on Pycnogonida. Ann Rep Laguna Mar Lab 1:1–99
6. Hall HVM (1913) Pycnogonida from the coast of California with descriptions of two new species. Univ Calif Publ Zool 11 (6):127–142
7. Helfer H (1938) Einige neue Pantopoden aus der Sammlung des Zoologischen Museums in Berlin. Sitzungsb. Gesell. Naturforsch. Freunde Berlin 1937(2):162–185
8. Stock JH (1955) Pycnogonida from the West Indies, Central America and the Pacific coast of North America. Papers from Dr Th. Mortensen's Pacific Expedition 1914–1916. Vidensk Medd Dansk Naturh Foren 117:209–266
9. Stock JH (1968) Pycnogonida collected by the Galathea and Anton Bruun in the Indian and Pacific Oceans. Vidensk Medd Dansk Naturh Foren 131:7–65
10. Stock JH (1975) Pycnogonida from the continental shelf, slope, and deep sea of the tropical Atlantic and East Pacific. Biological results of the University of Miami deep-sea expeditions, 108. Bull Mar Sci 24:957–1092

Stock JH (1986) Pycnogonida from the Caribbean and the Straits of Florida. Biological results of the University of Miami deep-sea expeditions. Bull Mar Sci 38:399–441

Part 26
Marine Insects

Monika Springer

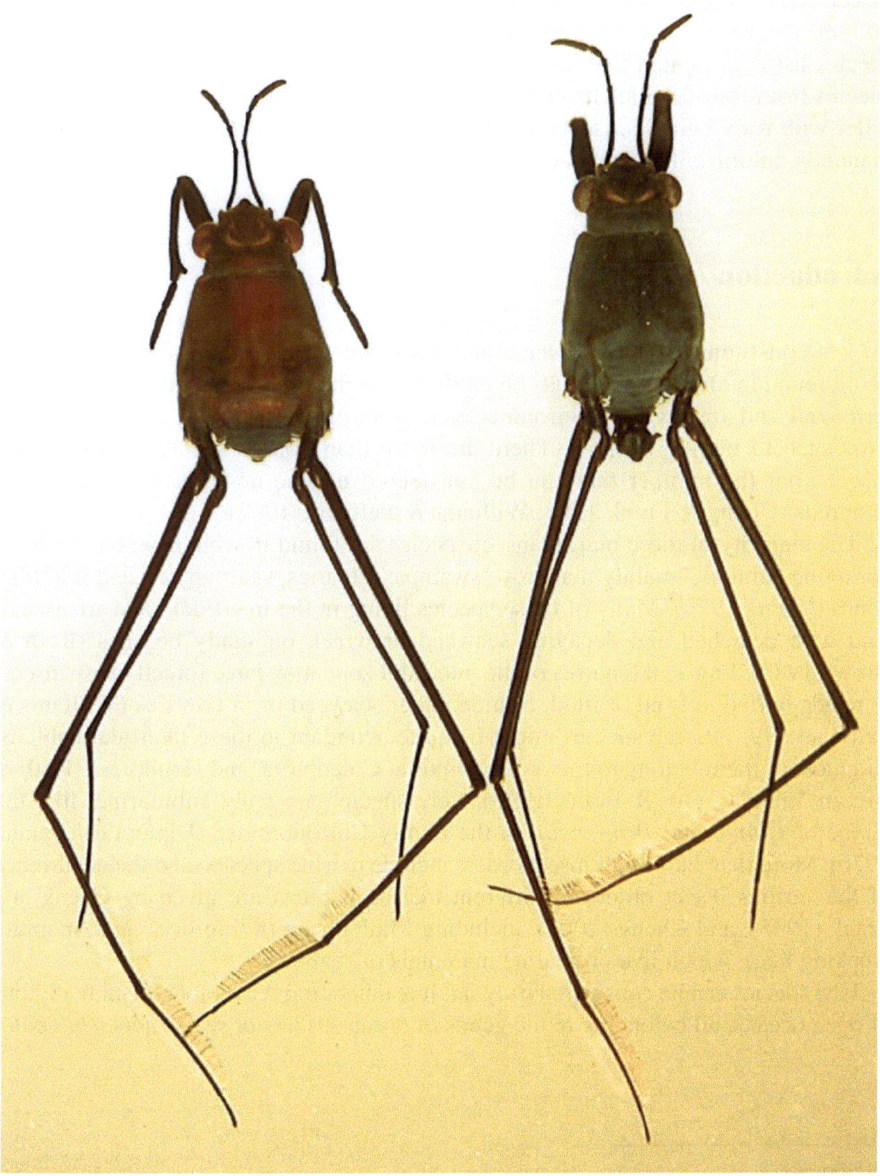

Female (left) and male (right) of the seaskater, *Halobates sobrinus* (Photo: Bernald Pacheco)

Abstract Although insects are one of the most diverse groups of organisms on earth, relatively few species are found associated with marine habitats. The majority of these marine species live in intertidal or coastal habitats and most of them belong to the orders Hemiptera, Diptera, and Coleoptera. Little is known about the marine insects of Central America and no survey of marine insect species has been carried out for Costa Rica. Therefore, this part presents a short discussion on each insect order with marine species that are most likely to occur in the area. Nearly nothing is known for the Caribbean coast of Costa Rica. For the Pacific coast a species list is presented that includes one marine species of Collembola, and eight species from four families from the order Hemiptera. This last one is also the only order with truly pelagic insects, the sea skaters *Halobates* (Gerridae), able to permanently colonize the open ocean.

Introduction

Insects constitute the most diverse and abundant group in the animal kingdom, being found in almost all habitats on earth. Given their unparalleled success in both terrestrial and freshwater environments, it is surprising that not more are found associated to marine habitats. There are more than 30,000 species of freshwater insects, but the number that can be considered marine amounts to only several hundreds (Cheng & Frank 1993; Williams & Feltmate 1992).

The majority of these marine insect species are found in what have been termed "bridging habitats," mainly mangrove swamps, estuaries, saltmarshes, and intertidal zones (Cheng 1976). Many of those species living in the intertidal zone are associated with detached and decaying seaweed or wrack on sandy beaches (Roth & Brown 1980). Physical features of the intertidal zone may have forced many insects to reside buried in sand or mud, or hide under seaweed or in crevices (Williams & Feltmate 1992). Some species might be quite abundant in these intertidal habitats, and most of them belong to the orders Diptera, Coleoptera, and Hemiptera (Roth & Brown 1980; Cheng & Frank 1993). Few species are truly submarine, like the larvae of *Clunio* and *Pontomyia*, in the family Chironomidae, Diptera (Neumann 1976). More than half of all insect orders include marine species, and the occurrence of the various insect orders in different marine habitats are given by Cheng and Frank (1993), and Cheng (2003), including Mallophaga (biting lice) and Anoplura (sucking lice), which live on marine mammals or seabirds.

Few species can be considered truly marine, inhabiting the remote offshore regions of open oceans, all belonging to the genus of ocean striders or sea skaters, *Halobates*

M. Springer
Escuela de Biología, Universidad de Costa Rica, San Pedro de Montes de Oca, 11501-2060 San José Costa Rica.
e-mail: springer@biologia.ucr.ac.cr

(Gerridae, Hemiptera). They were discovered some 180 years ago, and a comprehensive work was published by Herring (1961). Cheng (1985) presents a paper on the biology of the genus, and also provides a list of the distribution of the 44 species known at that date. Most of these species have been found near coasts, and only five live on the surface of the offshore open ocean, in tropical or subtropical waters.

In the 1920s and 1930s, several hypotheses were proposed for the fascinating question of why insects are so poorly represented in the sea (e.g. Buxton 1926; Pruthi 1932). Other authors providing possible explanations for the failure of insects to dominate the seas include Usinger (1957), Cheng (1976), Hinton (1976), and Norris (1991). One theory that has been suggested by some of these authors is the possible competition with the highly successful and well-established crustaceans in fully marine environments. Another aspect that has been discussed is the fact that living at large distances from shore requires development of total independence from land (Barton & Smith 1984) by developing a fully aquatic life cycle, which is a trait not common among freshwater insects (Williams & Feltmate 1992).

Although several papers have been published recently on the life history, distribution, and behavior of some selected groups of marine insects (e.g. Birch *et al.* 1979; Cheng 1989a, b, Cheng & Frank 1993; Soong *et al.* 1999), we still know little about the biology and ecology of the majority of marine insect species. Especially poorly known are tropical seas, where most undescribed species will probably be found, and where surveys will reveal important new information. For Costa Rica and the Central American region no surveys on marine insects have been carried out yet, and reports of marine species are difficult to obtain. Therefore, it is also difficult to state whether a species found only in Costa Rica could be declared as endemic.

In the next section I present those taxa of each order that are most likely to be found in the area, and a species list is presented only for species from the orders Collembola and Hemiptera reported from the Pacific coast. No species list can be presented for the other orders, or for the Caribbean coast, due to the lack of information.

Orders with Marine Species Most Likely to Exist in Costa Rica

Order Collembola: There are several families of springtails that include marine species, living in wrack, salt marshes, and littoral subsoil. A list of marine species of Collembola, with notes on their habitat and distribution is provided by Joosse (1976), as well as keys to the different genera. Christiansen & Bellinger (1988) reviewed the knowledge of the littoral Collembola of the continental coasts north of Panama, described several new species and presented a key to the marine Collembola of North and Central America. They described *Anurida mara* from Costa Rica, where it was collected on damp rocks in the intertidal zone along the central Pacific coast (Species List 26.1 is included on the CD-Rom). This species has not been recorded yet for other Central American countries, and might be endemic to Costa Rica.

Order Hemiptera: Several families of this order include species associated with marine habitats, and those known to occur in the Panamic Province are Gelastocoridae,

Gerridae, Mesoveliidae, Saldidae, and Veliidae. A checklist of marine species of these families from the eastern tropical Pacific is provided by Polhemus and Manzano (1992). Members of each of these families are regularly found in freshwater habitats of Costa Rica, and from the last four, several marine species have been recorded for the Pacific coast of the country (Species List 26.1).

In the family Gelastocoridae only the genus *Nerthra* occurs in marine habitats and the marine species *Nerthra rugosa* has been recorded from Panama (Polhemus & Manzano 1992), but not yet from Costa Rica. Its known distribution (Mauritius, Belize, Brazil, and Florida) might indicate that, with growing sampling effort, this species could be found in many more countries, including Costa Rica. Specimens have been found in mangrove swamps, mud flats, or associated with beach drift, and intertidal algae (Polhemus & Manzano 1992).

Another family that has been collected from many freshwater habitats in Costa Rica is the family Corixidae or water-boatmen. Since there are 12 genera with several species found in saline environments in many parts of the world (Scudder 1976), it is likely that marine species of this family will be reported from Costa Rica in the future.

In the group of water-striders with true marine representatives, species from the families Gerridae, Veliidae, and Mesoveliidae are reported from Costa Rica (Species List 26.1). The seaskaters *Halobates* (Gerridae) have been considered, by many authors, the only truly pelagic insects, able to colonize permanently the open ocean, and to complete their entire life cycle there. The only species collected, and recorded to occur off the coasts of Costa Rica is *Halobates sobrinus* from the Pacific (Species list 26.1). No coastal *Halobates* species are reported from the Caribbean, although the pelagic *H. micans* is common in tropical Atlantic waters (Cheng 1989b; Andersen & Polhemus 1976). This is the only cosmopolitan species in the genus, being found in all three major oceans (Pacific, Indian, and Atlantic). The greatest diversity of the genus occurs in the Indo-Malaysian and Australasian regions, which is thought to be the evolutionary center of the genus (Cheng 1985). Thirty-nine species have coastal distributions, usually in sheltered localities, and endemism appears to be high. The five species that live in the open ocean are all found in tropical or subtropical waters, where distribution and abundance may be influenced by surface water temperatures and surface currents (Cheng 1985, 1989a; Cheng & Schulenberger 1980). Even though they usually skate on open water away from shore, specimens may be found in the intertidal zone or on shore after storms (Roth & Brown 1980). Associated with mangrove swamps, two more species of gerrids are reported from Costa Rica, both in the genus *Rheumatobates* (Species List 26.1). Another genus that might be found is *Telmatometroides*, since it has been recorded for the Pacific coast of Panama and some countries in South America (Polhemus & Manzano 1992).

From the family Veliidae, the neotropical genus *Trochopus*, with five known species, is confined to the Central and South American coasts (Cheng & Roussis 1998; Andersen & Polhemus 1976). The species *Trochopus plumbeus* is the most widely distributed, and probably occurs along the Costa Rican Caribbean coast as well (Cheng personal communication, 2003). Another genus from this family, *Husseyella*, with three marine species, was reported by Anderson and Polhemus (1976) from estuaries and mangroves of the Central American coast. One specimen (nymph) of an unidentified *Husseyella* species has been collected recently in an estuary lagoon from the northern Caribbean coast of Costa Rica. *Microvelia inquilina* was found in crab holes of *Ucides occidentalis* from the Pacific coast of Costa Rica, and might be endemic to the country (Species list 26.1).

Each of the families mentioned also have freshwater-inhabiting species and some of these species are euryhaline, thus occasionally invading the marine habitat, like *Mesovelia mulsanti*, which regularly turns up as single specimens in samples from mangrove estuaries and bays (Polhemus & Manzano 1992). Since little is known of the ecology and habitat preferences of many species, it is still quite a subjective judgment, which species can be considered obligatorily marine associated.

Order Coleoptera: Even though this order contains the largest number of species described in the Class Insecta, relatively few are associated with marine habitats. Doyen (1976) mentioned at least 50 species of beetles representing 11 families as obligatorily marine. The majority was found in wrack on sandy beaches and never or seldom submerged. A few, however, lived on algae-covered rock and were submerged daily. Marine beetles are typically benthic in habit and most are specialized members of largely terrestrial families, primarily Carabidae and Staphylinidae (Doyen 1976). Most typical freshwater families, like Dryopidae, Elmidae, or Psephenidae, have few or no marine species at all. In the Mesoamerican region, two species of Limnichidae, *Martinius tellipontis* and *Throscinus crotchi*, are known from mangrove swamps and intertidal zones.

Among those that can live in intertidal rock pools are the free-swimming members of the families Dytiscidae and Hydrophilidae, which have many species that occasionally invade brackish estuaries, or splash pools in rocks (Doyen 1976). Both families are diverse and abundant in Costa Rican freshwater habitats, and some species will surely be found associated with saline waters in the future. The Family Hydraenidae has also been reported as locally abundant in small saline pools and in estuaries (Doyen 1976), although not as diverse as Dytiscidae and Hydrophilidae. Three genera of this family have been collected from freshwater habitats in Costa Rica, and it is possible that species will be collected from saline habitats also.

In the large Family Staphylinidae (rove beetles), about 1% of the species are known to be confined to seashore habitats, and this is the dominant family of marine Coleoptera (Doyen 1976). A list and keys to the genera with marine members are given by Moore and Legner (1976). On the Pacific coast of North America this family comprises a large percentage of the marine insect fauna, both in number of species and in number of individuals, but still little is known about their biology and ecology (Moore & Legner 1976). The fauna of many parts of the world is relatively unexplored and it is represented by mostly undescribed species and even genera. Roth and Brown (1980) distinguished three groups of marine staphylinids, based upon their habitats: (1) the obligates, associated with rocky shores and large boulders, which are inundated daily or twice a day by the sea; (2) the wrack species (the most common), associated with rotting seaweed and debris; and (3) the salt and mud flat species. Many species can be expected to live in the different marine environments of Costa Rica.

The next most abundant family by number of species is the Carabidae (ground beetles), with about 15 strictly marine members, most of them belonging to genera, which are exclusively marine (Doyen 1976). Ground beetles are predaceous, and they have been found in a variety of marine habitats, including wrack, wet sand, burrows in sandy or clay soil, mud flats, and from rock crevices at the upper level of the barnacle-algae zone (Roth & Brown 1980). Closely related to the ground beetles are the tiger beetles, Cicindelidae, and several species of the genus *Cicindela* have been recorded from beaches of the Gulf of California (Roth & Brown 1980). Although they are not strictly intertidal, they prefer this habitat and may be considered opportunists, moving in and out with the tides. In the Gulf of California, one species has been collected 10 km from land at night where it was landing on the water surface, feeding upon arrow worms (Roth & Brown 1980).

Other families that have been recorded from the intertidal zone include the Tenebrionidae (darkling ground beetles), Histeridae (hister beetles), Ptilidae (feather-winged beetles), Melyridae (soft-winged flower beetles), and Anthicidae (antlike flower beetles). Most of them have been found in association with rotting seaweed or under debris (Roth & Brown 1980), and they all occur in Costa Rica in terrestrial habitats.

Although marine Coleoptera seem to be concentrated in temperate and cold seas (Doyen 1976), many species can be expected to be found in the different marine habitats of Costa Rica and Mesoamerica, with more specific sampling effort. Our knowledge is still fragmentary, and as Doyen (1976) stated: "whether the lack of records of intertidal forms in the tropics is a sampling artifact or reflects an actual absence is uncertain. There seems to be no biological reason for the absence of a rich intertidal Coleoptera fauna in the tropics".

Order Diptera: Adult dipterans are probably the most common and active insects in the intertidal zone and their larvae might be associated with moist sand, mud flats, decomposing seaweed, wrack, and with algae or barnacle-covered rocks (Roth & Brown 1980). Most of these marine dipteran species belong to families that are commonly associated with freshwaters during their development, like Chironomidae, Ceratopogonidae, Culicidae, Tabanidae, Ephydridae, Empididae, and Tipulidae. Many members of these families have been collected as adults along the coasts of Costa Rica. More detailed studies of their biology are necessary in order to find out which of these species are obligatorily associated with marine habitats, since many of them can be found there as visitors.

Of the extremely diverse family of nonbiting midges, the Chironomidae, only the subfamilies Chironominae, Orthocladiinae, and Clunioinae are found in marine habitats. The majority is primarily restricted to the intertidal zone, where the larvae feed on algae, but the larvae of a few species are found on the bottom of the open sea, at depths up to 20 m (Hashimoto 1976). In Costa Rica, the first two subfamilies are common in freshwater habitats and it is likely that marine species will be found in some of the diverse marine habitats of Costa Rica. One could be *Thalassomya pilipes*, a species found from California to the Galapagos, with larvae living in silken tubes among algae-encrusted rocks (Roth & Brown 1980), but many of them will probably be new to science.

The quite bothersome adults of biting midges, family Ceratopogonidae, are common along the beaches of both coasts of Costa Rica. The three genera that include marine species, *Culicoides* (50–60 marine species), *Leptoconops*, and *Dasyhelea*, occur in all major regions of the world (Linley 1976), and their larvae live on algae-covered rocks and in mangrove swamps, where they feed on algae and diatoms (Roth & Brown 1980).

In coastal areas, many species of mosquitoes (Culicidae) are often found, although less than 5% of some 2,500 described species regularly breed in brackish waters (O'Meara 1976). The habitats of those "tideland species" include saltmarshes, mangrove swamps, and coastal rockpools. Nine major genera include those species, and of these *Aedes*, *Anopheles*, *Culex*, and *Culiseta* are common in Costa Rican freshwater habitats. Several species, especially from the genus *Deinocerites*, have been found to breed in crab holes (O'Meara 1976).

Members of the Family Tabanidae (horse flies and deer flies) include many species living in coastal areas with some of them developing in saltmarshes, brackish pools, and tidal overwash areas (Axtell 1976). In Costa Rica and Central America, marine species of the genera *Chrysops* and *Tabanus* could be found, as well as several genera and species from the tribe Diachlorini, which is especially diverse in the neotropical region.

The herbivorous larvae of the beach flies (Canaceidae) live among algae on wave-splashed rocks, while the adults fly in swarms over wave-washed rocks and the adjacent sand and wrack (Roth & Brown 1980). *Canaceoides* species are found along the Pacific coast of North and South America (Wirth 1969). Adults of this family have been collected along the coasts of Costa Rica, as have the seaweed flies (Coelopidae). These are found in many parts of the world, especially in temperate and subarctic regions, with their larvae living in rotting seaweed (Roth & Brown 1980).

In the Family Ephydridae (shoreflies), 36 genera are known to include marine species, and they might be extremely abundant in saline waters (Simpson 1976). *Lipochaeta* is found on both the Pacific and Atlantic coasts, in saline habitats (mud flats and wet sand), *Mosillus tibialis* is known from Canada south to Florida and California, the West Indies and Ecuador, and *Paralimna decipiens* is recorded as neotropical (Roth & Brown 1980).

Two other families occurring in Costa Rica, and likely to have marine species, are Empididae (dance flies) and Dolichopodidae (long-headed flies). Both might be collected from algae-covered rocks at the water's edge, and from seaweed or on mud flats (Roth & Brown 1980). Adults from the families Sphaeroceridae (*Leptocera*), Tethinidae (*Tethina*), and Anthomyiidae (*Fucellia*) have been found swarming over rotting seaweed in California (Roth & Brown 1980). Marine species from these families could be found in Costa Rica since adult specimens have been collected here, although the biology of the larvae is unknown (Zumbado personal communication, 2003).

Order Hymenoptera: The Hymenoptera that may be found in the intertidal zone are minute parasitic wasps whose hosts are the eggs, larvae, and pupae of other insects. In the Gulf of California, adults of the families Braconidae and Pteromalidae are reported associated with marine habitats (Roth & Brown 1980). The first were observed on algae-covered rocks, which are covered during the high spring tides, and the latter were found at low-tide crawling about on green algae mixed with tiny barnacles where dolichopodid fly larvae may be their host. Ants have been collected on sandy beaches, but these species are opportunists, foraging in from above the supralittoral zone, and can probably not be considered marine. Although an extensive survey of the Hymenoptera of Costa Rica has been carried out, no marine species has been found yet (Hanson personal communication, 2003).

Specialists

Few scientists work specifically with marine insects. One of them is Lanna Cheng, who edited a comprehensive work on this group in 1976. She also published a marine insect homepage together with other authors (http//www.unk.edu/marineinsects or http//entomology.unl.edu/marine_insects/marinehome.html), and continues to actively research marine insects at the University of California, San Diego (lcheng@ucsd.edu). The main orders of insects with families that include marine species may have several taxonomists for each family, although not specializing in marine species. Information about the specialists working on a specific taxonomic group in Costa Rica or the region of Central America can be obtained from the author and from Paul Hanson (phanson@biologia.ucr.ac.cr), Universidad de Costa Rica. Also the following specialists can provide information on specific groups:

John T. Polhemus: Colorado, USA – Heteroptera. polhemus@stripe.colorado.edu

Rob Roughly: University of Manitoba, Canada. Coleoptera. rob_roughley@umanitoba.ca

William Shepard: San Francisco, USA – Coleoptera. s12647@saclink.csus.edu

Manuel Zumbado: National Institute for Biodiversity (INBio, http//www.inbio.ac.cr), Costa Rica. Diptera. mzumbado@inbio.ac.cr

Collections

Little collection of marine insects has been done in Costa Rica and Central America, and specimens are practically nonexistent in the different entomology collections of Costa Rica. Various foreign collections might contain specimens that have been collected associated with some marine habitats in Costa Rica and other Central American countries, but this information is widespread and difficult to obtain. A complete collection of the aquatic insects from freshwater habitats from Costa Rica and some other Central American countries has been established since 1992 at the Zoology Museum of the University of Costa Rica (http://www.biologia.ucr.ac.cr). This collection is preserved in alcohol, and includes the different aquatic life stages from more than 300 genera in 11 insect orders. Efforts are planned for the future to include marine species in this collection. Further research with more specific collection, together with cooperation from several taxonomists is necessary in order to obtain a more complete list of the marine insect species found in Costa Rica and other Central American countries.

References

Andersen NM, Polhemus JT (1976) Water-striders (Hemiptera: Gerridae, Veliidae, etc.). *In*: Cheng L (ed) Marine Insects. North Holland, Amsterdam, pp 187–224

Axtell RC (1976) Coastal horse flies and deer flies (Diptera: Tabanidae). *In*: Cheng L (ed) Marine Insects. North Holland, Amsterdam, pp 415–445

Barton DR, Smith SM (1984) Insects of extremely small and extremely large aquatic habitats. *In*: Resh VH, Rosenberg DM (eds) The Ecology of Aquatic Insects. Praeger Scientific, New York, pp 459–483

Birch MC, Cheng L, Treherne JE (1979) Distribution and environmental synchronization of the marine insect *Halobates robustus* in the Galápagos Islands. Proc R Soc Lond, Ser B 206:33–52

Buxton PA (1926) The colonization of the sea by insects: with an account of the habits of *Pontomyia*, the only known submarine insect. Proc Zool Soc Lond 1926:808–814

Cheng L (1976) Insects in marine environments. *In*: Cheng L (ed) Marine Insects. North Holland, Amsterdam, pp 1–4

1. Cheng L (1985) Biology of *Halobates* (Heteroptera: Gerridae). Ann Rev Entomol 30:111–135

Cheng L (1989a) Factors limiting the distribution of *Halobates* species. *In*: Ryland JS, Tyler PA (eds) Reproduction Genetics and Distributions of Marine Organisms. Olsen & Olsen, Fredenborg, Denmark, pp 357–362

Cheng L (1989b) Biogeography and phylogeny of the seaskater *Halobates*. Chinese J Ocean Limnol 7:233–239

Cheng L (2003) Marine insects. *In*: Resh VH, Cardé RT (eds) Encyclopedia of Insects. Academic, St. Louis, MO, pp 679–682

Cheng L, Frank JH (1993) Marine insects and their reproduction. Oceanogr Mar Biol Annu Rev 31:479–506

Cheng L, Schulenberger E (1980) Distribution and abundance of *Halobates* species (Insecta: Heteroptera) in the eastern tropical Pacific. Fish Bull 78:579–591

Cheng L, Roussis V (1998) Sex attractant in the marine insect *Trochopus plumbeus* (Heteroptera: Veliidae): a preliminary report. Mar Ecol Prog Ser 170:283–286.

2. Christiansen K, Bellinger P (1988) Marine littoral Collembola of North and Central America. Bull Mar Sci 42:215–245

Doyen JT (1976) Marine beetles (Coleoptera: excluding Staphylinidae). *In*: Cheng L (ed) Marine Insects. North Holland, Amsterdam, pp 497–520

Hashimoto H (1976) Non-biting midges of marine habitats (Diptera: Chironomidae). *In*: Cheng L(ed) Marine Insects. North Holland, Amsterdam, pp 377–414

Herring JL (1961) The genus *Halobates* (Hemiptera: Gerridae) Pacif Insects 3:223–305

Hinton HE (1976) Enabling mechanisms. Proc 15th Int Congr Entom, Washington, DC, pp 71–83

Joosse ENG (1976) Littoral apterygotes (Collembola and Thysanura). *In*: Cheng L(ed) Marine Insects. North Holland, Amsterdam, pp 151–186

Linley JR (1976) Biting midges of mangrove swamps and saltmarshes (Diptera: Ceratopogonidae). *In*: Cheng L(ed) Marine Insects. North Holland, Amsterdam, pp 335–376

Moore I, Legner EF (1976) Intertidal rove beetles (Coleoptera: Staphylinidae). *In*: Cheng L (ed) Marine insects. North Holland, Amsterdam, pp 521–551

Neumann D (1976) Adaptations of chironomids to intertidal environments. Ann Rev Entomol 21:387–414

Norris KR (1991) General biology. *In*: CSIRO (ed) The Insects of Australia, 2nd ed, Vol I. Melbourne University Press, Canberra, pp 68–108

O'Meara GF (1976) Saltmarsh mosquitoes (Diptera: Culicidae). *In*: Cheng L (ed) Marine Insects. North Holland, Amsterdam, pp 303–333

Pruthi HS (1932) Colonization of the sea by insects. Nature 130:312

3. Polhemus JT (1975) New estuarine and intertidal water-striders from Mexico and Costa Rica (Hemiptera: Gerridae: Mesoveliidae). Pan-Pac Entom 51:243–247
4. Polhemus JT, Cheng L (1976) A new *Rheumatobates* from Costa Rica (Hemiptera: Gerridae). Pan-Pac Entom 52:321–323
5. Polhemus JT, Hogue CL (1972) Two new *Microvelia* from crabholes in Costa Rica. (Hemiptera: Veliidae). Contr Sci 224:1–6
6. Polhemus JT, Manzano MR (1992) Marine Heteroptera of the eastern tropical Pacific (Gelastocoridae, Gerridae, Mesoveliidae, Saldidae, Veliidae). In: Quintero D, Aiello A (eds) Insects of Panama and Mesoamerica. Selected Studies. Oxford University Press, Oxford
7. Roth V, Brown W (1980) Arthropoda: Insecta (Insects). Chapter 22. *In*: Brusca RC (ed) Common Intertidal Invertebrates of the Gulf of California. University of Arizona Press, Tucson, Arizona, pp 326–346

Scudder GGE (1976) Water-boatmen of saline waters (Hemiptera: Corixidae). *In*: Cheng L(ed) Marine insects. North Holland, Amsterdam, pp 263–289

Simpson KW (1976) Shore flies and brine flies (Diptera: Ephydridae). *In*: Cheng L(ed) Marine Insects. North Holland, Amsterdam, pp 465–495

Soong K, Chen GF, Chao JR (1999) Life history studies of the flightless marine midge *Pontomyia* spp. (Diptera: Chironomidae). Zool Stud 38:466–473

Usinger RL (1957) Marine insects. Geol Soc Am Mem 67:1177–1182

Williams DD, Feltmate BW (1992) Aquatic Insects. CAB International, Wallingford, UK, 358 p

Wirth WW (1969) The shore flies of the genus *Canaceoides* Cresson (Diptera: Canaceidae). Proc Calif Acad Sci 4th Ser 36:551–570

Part 27
Chitons

Enrico Schwabe and Ingo S. Wehrtmann

Acanthopleura granulata from the Caribbean coast of Costa Rica (Photo: Ingo Wehrtmann)

E. Schwabe
Zoologische Staatssammlung München, Münchhausenstrasse 21, D-81247 München, Germany.
e-mail: enrico.schwabe@zsm.mwn.de

I.S. Wehrtmann
Unidad Investigación Pesquera y Acuicultura (UNIP) of the Centro de Investigación en Ciencias del Mar y Limnología (CIMAR), and Museo de Zoología, Escuela de Biología, Universidad de Costa Rica, 11501-2060 San José, Costa Rica.
e-mail: ingowehrtmann@gmx.de

Abstract The chiton fauna of Costa Rica, with exception of some sporadic reports, was never revised. The present review is based on literature data as well as on material examined from the collections of the Museo de Zoología, Universidad de Costa Rica (UCR), and the Zoological State Collection Munich (ZSM). The total number of verified chiton species of Costa Rica is 31, of which 23 occur only in the Pacific and 7 only in the Caribbean Sea. One species is reported from both sides of Costa Rica. Disjunctive distributions may indicate that there may be several (up to 12) other chiton species in Costa Rica. The only species endemic to Costa Rica refers to *Ischnochiton victoriae*, so far only reported from Isla del Coco.

Introduction

The bilateral symmetrical, dorsoventral flattened Polyplacophora (chitons) are exclusively marine mollusks with eight, usually overlapping plates. Chitons are distributed worldwide and have their highest abundance in the warm temperate zones. Their bathymetrical distribution ranges from the splash zone (e.g., *Acanthopleura*) down to hadal depths (e.g., *Ferreiraella*). Chitons generally live on all kinds of hard substratum, but sand-dwellers are also known. Since many species of Polyplacophora are photonegative, they are often found cryptically.

The plates are surrounded by a fleshy girdle (perinotum), which may cover them completely. A complex muscle system enables the chitons to roll up ventrally. Chitons vary in length from a few millimeters to about 15 cm. The world's largest species is the North Pacific *Cryptochiton stelleri* (von Middendorff, 1847), which may attain a length of more than 30 cm.

The broad, fleshy foot is situated ventrally. Chitons lack a true head; their cephalic region is eyeless and without tentacles. The anus is situated posteriorly, from where rows of ctenidia extend laterally on both sides of the foot in anterior direction. The perinotum is dorsally beset densely with different kinds of growth (scales, bristles, hairs, needles, corpuscles) and sensory organs (see Leise 1986, 1988).

The upper valve layer (tegmentum) is the thickest one, often colorful and mostly sculptured. The tegmentum layer is divided into the lateral areas, the pleural area, and the jugal area (the latter two together build the central area). The articulamentum extends ventrally under the tegmentum in anterior and lateral direction. Anteriorly it builds two thin processes, the apophyses, which are overlapped by the preceding valve. Laterally, it forms slitted or unslitted insertion plates, which embeds the valves in the girdle.

The moderately large chiton radula is a good diagnostic feature and can reach a length of at least one third of the full body length. Approximately 25–150 rows of teeth are situated on a thin radula membrane, with usually 17 teeth per row. In general, the diet of chitons consists of diatoms, detritus, and fleshy and encrusting algae. However, carnivorous, xylophagorous, and spongivorous species are also known.

The best known polyplacophoran sensory organs are the aesthetes, which have photoreceptoric, secretoric, and probably additional mechano- and chemosensory functions (Boyle 1974; Fischer 1978, 1988; Baxter & Sturrock 1987). The aesthetes may bear large lenses (e.g., *Tonicia*) or be black pigmented (e.g., *Callochiton*).

The most striking muscular elements of adult Polyplacophora are the eight paired sets of dorsoventral muscle units, which insert at the shell plates and each consist of various distinct subunits of mainly obliquely running muscle bundles (Wingstrand 1985). Apomorphic for chitons is the dorsally situated musculus rectus, which runs underneath the shell plates in anterior-posterior direction. Laterally, the body is engulfed by the circular enrolling muscle (Wanninger 2001).

Most polyplacophoran species have separate sexes and are free spawners with external fertilization. Accordingly, brooding is observed, and even ovovivipary and hermaphrodism have been described (Eernisse 1986). The chorion processes of mature eggs often show a characteristic, species-specific ornamentation (Sirenko 1993). The trochophore-like larva is free-swimming and non-feeding (i.e., lecithotrophic). The planktonic stage can last from a few hours to several days (Strathmann & Eernisse 1987).

Compared to representatives of other molluscan classes, metamorphosis appears as a gradual process in chitons and follows initial settlement of the competent larva. Subsequently, the prototroch and all larval neuronal components are lost, the dorsoventral axis of the animal flattens, and the juvenile starts its creeping life stage (Wanninger 2001; Friedrich *et al.* 2002).

Discussion

The chiton fauna of Costa Rica has not been revised previously. The lack of specific studies in Costa Rica and inaccurate literature citations make it difficult to obtain a clear and complete picture of the chiton fauna occurring in Costa Rican waters.

Hertlein (1963) was the first to revise the existing literature and to prepare an annotated checklist of the chitons of the Isla del Coco, mentioning four species for this island. Ferreira (1987) suggested a misidentification of two of these species, and added another four species to the Isla del Coco checklist. Overall, a total of eight species are reported for Isla del Coco (von Martens 1902; Biolley 1907; Boone 1933; Hertlein 1963; Fischer 1981; Montoya 1983; Ferreira 1987), representing roughly 25% of the 32 species mentioned for Costa Rican waters (Species Lists 27.1 and 27.2 are included on the CD-Rom). Species numbers from Isla del Coco are comparable to those from Galapagos Islands (10 species; Smith & Ferreira 1977). Moreover, four of these species have been reported from both Galapagos and Isla del Coco (*?Stenoplax petaloides*, *Chiton goodalli*, *Acanthochitona angelica*, *A. hirudiniformis*).

While the chiton fauna of Panama has been studied by several authors (Smith 1961; Ferreira 1974; Aviles 1985; Skoglund 1989), such information is scarce for continental Costa Rica. Pacific Costa Rica (including Isla del Coco) is represented by 24 species, while eight species have been reported from the Caribbean coast (see Species Lists 27.1 and 27.2 are included on the CD-Rom). Only one species

Fig. 27.1 *Callistoplax retusa* observed at Isla Murciélagos, Pacific Costa Rica (Photo: Ingo Wehrtmann)

Fig. 27.2 *Acanthochitona hirudiniformis*, a species reported from Pacific Costa Rica (Photo: Ingo Wehrtmann)

(*S. boogii*) occurs along both coasts of Costa Rica. *I. victoriae*, so far described exclusively from Isla del Coco, seems to be the only endemic chiton species from Costa Rica.

The 31 species listed in the Species Lists 27.1 and 27.2 (included on the CD-Rom) were compiled from published literature and recent collections made by the second author. These samples comprise 20 species and have been deposited in the Museo de Zoología, UCR, and the ZSM. Of these, seven were from the Caribbean and 13 from the Pacific coast of Costa Rica, including the first record of *Chaetopleura roddai* for Costa Rica.

In addition to the 31 species reported from Costa Rica (Figs. 27.1 and 27.2), 12 species of chitons are so far only known from the neighboring countries (Table 27.1). Special micro-habitats such as coral rubble, algae, and deepwater

Table 27.1 Taxa of chitons that are known from the neighboring countries and thus may also occur in Costa Rica

Species	World distribution[a]	Central America[b]	Oceanic distribution[c]	Depth[d] (m)	Habitat/ Community[e]	Comments & References[f]
Class POLYPLACOPHORA						
Order CHITO NIDA						
Family LEPTOCHITONIDAE						
Leptochiton americanus Kaas & Van Belle, 1985	EP	P	bp,eb,mp	400–1400	n.a.	The species recorded from 44°39 N to 21°23 S, closest record to Costa Rica from the type locality: S of the Gulf of Panama (5°58.8 N 81°38.2 W) [16]
Leptochiton incongruus (Dall, 1908)	EP	P	bp,eb,mp	589–3612	mb	Beside the type locality from the Gulf of Panama (07°09.45 N 80°50 W), it is also recorded from Mexico (15°40 N 95°20 W) [5]
Family FERREIRAELLIDAE						
Ferreiraella scrippsiana (Ferreira, 1980)	EP	P	bp,eb	2507–4000	sunken wood	The species was originally described from Mexico (22°30.8 N 110°03.8 W) and also recorded from the Panama Basin (05°09.8 N 81°41.2 W) [23]
Family ISCHNOCHITONIDAE						
Callistochiton portobelensis Ferreira, 1976	Car	P	eb,it,st	0–5	rb,rub	Beside the type locality (Panama, Portobelo) the species is also recorded from SW Florida [19]
Chaetopleura hanselmani (Ferreira, 1982)	EP		eb,it,st	0–17	n.a.	The species was described as ranging from 23°13 N and 6°57 S [8]
Ischnochiton kaasi Ferreira, 1987	Car	P	eb,it,st	0.5–3	n.a.	It was originally described from Bocas del Toro, Panama [18]
Stenoplax mariposa (Bartsch MS, Dall, 1919)	EP	N,P	eb,it,st	0–20	rb	Distribution unclear as it may confused with the other species from this list [18]
Stenoplax rugulata (Sowerby in Broderip & Sowerby, 1832)	EP	N,P	eb,it,st	0–20	rb	Distribution unclear as it may confused with the other species from this list [18]
Family LEPIDOCHITONIDAE						
Lepidochitona beanii Carpenter, 1857	EP		eb,ep,it,st	0–230	rb	The species is recorded from Peru and from S-California and a lot of localities in between [9]

Family CRYPTOPLACIDAE						
Choneplax lata (Guilding, 1829)	Car	B,H,P	eb,it,st	0–3 m	rb,sb,cr	The species is recorded from Honduras [22] and from Panama [4]
Family ACANTHOCHITONIDAE						
Acanthochitona arragonites (Carpenter, 1857)	EP	ES	eb,it,st	0–20	rb	The species was found in Mexico and Ecuador [26]
Acanthochitona imperatrix Watters, 1981	EP		eb,st	c. 17	n.a.	It is recorded from lower California to Galapagos Islands [26]

n.a. = information not available

[a]Car = Caribbean; EP = eastern Pacific

[b]B = Belize; ES = El Salvador; H = Honduras, N = Nicaragua; P = Panama

[c]bp = bathypelagic; eb = epibenthic; ep = epipelagic; it = intertidal; mp = mesopelagic; st = subtidal

[d]Uppermost and lowermost occurrences (in meters)

[e]cr = coral reefs; mb = muddy bottom; rb = rocky bottom; rub = rubble bottom; sb = sandy bottom

[f]Numbers in parenthesis refer to numbers used in the reference list

zones have been sampled scarcely in Costa Rica. Thus, we expect that the number of chitons will certainly increase with further studies.

Due to the poor knowledge of the chiton diversity in Costa Rican waters, ecological studies often concern relatively large-sized species, which are easy to identify and to collect. Therefore, information, e.g., on habitat preferences, are concentrated on a few large and common species, such as *Chiton stokesii*, *C. tuberculatus*, *C. marmoratus*, and *Acanthopleura granulata* (Houbrick 1969; Glynn 1970; Fischer 1981; Miller 1983; Cruz & Sotela 1984; Jörger *et al.* 2008).

Specialists

Caribbean and Central America

CEDAR I. GARCÍA-RÍOS: Departamento de Biología, Colegio Universitario de Humacao, Humacao, Puerto Rico – cgarcia@cuhwww.upr.clu.edu

Worldwide

ENRICO SCHWABE: Zoologische Staatssammlung München, Münchhausenstrasse 21, D-81247 München, Germany – enrico.schwabe@zsm.mwn.de
BORIS I. SIRENKO: Zoological Institute, Russian Academy of Sciences, Saint Petersburg 199034, Russia – marine@zin.ru
HERMANN L. STRACK: National Natuurhistorisch Museum; Department of Invertebrates - 2300 RA Leiden The Netherlands – h.l.strack@planet.nl
YVES TERRYN: Acaciastraat 44, B-9000 Gent, Belgium – loricata@pandora.be

Collections

The Museo de Zoología, Universidad de Costa Rica, contains a fairly good collection of chitons from Costa Rica. Other museums with large collections of mollusks (e.g., Smithsonian Institute, California Academy of Science) may have material from Costa Rica and Central America; however, we are unaware of the extent and completeness of these collections regarding the Central American region.

References

Aviles EMC (1985) Contribution to the knowledge of the malacological fauna of the island of Rey, Archipelago of Las Perlas, Panama. La Conchiglia 17(192–193):3–5

Baxter JM, Sturrock MG (1987) The ultrastructure of aesthetes in *Tonicella marmorea* (Polyplacophora; Ischnochitonina) and a new functional hypothesis. J Zool Lond 211:589–604.

1. Biolley P (1907) Mollusques de l'Isla del Coco. Résultats d'une expédition faite en janvier 1902 - du 11 au 16 - sous les auspices du Gouvernement de Costa Rica. Mus Nac Costa Rica, 30 p

Boone L (1933) Scientific Results of Cruises of the Yachts "Eagle" and "Ara", 1921–1928, William K. Vanderbilt, Commanding Coelenterata, Echinodermata and Mollusca. Bull Vanderbilt Mar Mus 4:1–217

Boyle PR (1974) The aesthetes of chitons. II. Fine structure in Lepidochitona cinereus (L.). Cell Tiss Res 153:383–398

2. Bullock RC (1985) The *Stenoplax limaciformis* (Sowerby, 1832) species complex in the New World (Mollusca: Polyplacophora: Ischnochitonidae). Veliger 27:291–307
3. Bullock RC (1988) The genus *Chiton* in the New World (Polyplacophora: Chitonidae). Veliger 31:141–191

Cruz RA, Sotela A (1984) Contribución a la biología de *Chiton stockesii* (sic) (Polyplacophora: Chitonidae) de Punta Pochote, Puntarenas, Costa Rica. Rev Biol Trop 32:61–68

4. Cubit J, Williams S (1983) The invertebrates of Galeta reef (Caribbean Panama). A species list and bibliography Atoll Res Bull 269:260–272
5. Dall WH (1908) Reports on the dredging operations off the West Coast of Central America to the Galapagos, to the West Coast of Mexico, and in the Gulf of California, in charge of Alexander Agassiz, carried on by the US Fish Commission Steamer "Albatross" during 1891, Lieut. Commander Z.L. Tanner, U.S.N., Commanding, XXXVIII, and Reports on the Scientific Results of the Expedition to the Eastern Tropical Pacific in charge of Alexander Agassiz, by the US Fish Commission Steamer "Albatross", from October, 1904, to March, 1905, Lieut. Commander L.M. Garrett, U.S.N., Commanding, XIV. Reports on the Mollusca and Brachiopoda. Bull Mus Comp Zool Harv Coll 43:205–487

Eernisse DJ (1986) The genus *Lepidochitona* Gray, 1821 (Mollusca: Polyplacophora) in the northeastern Pacific Ocean (Oregonian and Californian provinces). Zool Verhand Leiden 228:1–52

6. Ferreira AJ (1974) The genus *Lepidozona* in the Panamic Province, with the description of two new species (Mollusca: Polyplacophora). Veliger 17:162–180
7. Ferreira AJ (1979) The genus *Callistochiton* Dall, 1879 (Mollusca: Polyplacophora) in the eastern Pacific, with the description of a new species. Veliger 21:444–466
8. Ferreira AJ (1982a) A new species of *Calloplax* (Mollusca: Polyplacophora) in the eastern Pacific. Veliger 24:321–324
9. Ferreira AJ (1982b) The family Lepidochitonidae Iredale, 1914 (Mollusca: Polyplacophora) in the eastern Pacific. Veliger 25:93–138
10. Ferreira AJ (1983) The genus *Chaetopleura* Shuttleworth, 1853 (Mollusca: Polyplacophora) in the warm-temperate and tropical eastern Pacific, southern California to Peru, with the description of two new species. Veliger 25:203–224
11. Ferreira AJ (1983a) The chiton fauna of the Revillagigedo Archipelago, Mexico. Veliger 25:307–322
12. Ferreira AJ (1986) A revision of the genus *Acanthopleura* Guilding, 1829 (Mollusca: Polyplacophora). Veliger 28:221–279
13. Ferreira AJ (1987) The chiton fauna of Cocos Island, Costa Rica (Mollusca: Polyplacophora) with the description of two new species. Bull South Cal Acad Sci 86:41–53

Fischer FP (1978) Photorezeptor cells in chiton aesthetes. Spixiana 1:209–213

Fischer R (1981) Bioerosion of basalt of the Pacific Coast of Costa Rica. Senckenbg marit 13:1–41

Fischer FP (1988) The ultrastructure of the aesthetes in *Lepidopleurus cajetanus* (Polyplacophora: Lepidopleurina). Am Malac Bull 6:153–159

Friedrich S, Wanninger A, Brückner M & Haszprunar G. (2002) Neurogenesis in the mossy chiton, *Mopalia muscosa* (Gould) (Polyplacophora): evidence against molluscan metamerism. J Morph 253:109–117

Glynn PW (1970) On the ecology of the Caribbean chitons *Acanthopleura granulata* Gmelin and *Chiton tuberculatus* Linné: density, mortality, feeding, reproduction, and growth. Smithson Contrib Zool 66:1–21

14. Hertlein LG (1963) Contribution to the biogeography of Cocos Island, including a bibliography. Proc Cal Acad Sci (ser. 4) 32(8):219–289

15. Houbrick JR (1969) A survey of the littoral marine mollusks of the Caribbean coast of Costa Rica. Veliger 11:4–23

Jörger K, Meyer R, Wehrtmann IS (2008) Species composition and vertical distribution of chitons (Mollusca: Polyplacophora) in a rocky intertidal zone of the Pacific coast of Costa Rica. J Mar Biol Ass UK 88:807–816

16. Kaas P,.Van Belle RA (1985) Monograph of Living Chitons (Mollusca: Polyplacophora) 1, Order Neoloricata: Lepidopleurina. E.J. Brill/W. Backhuys, Leiden. 240 p

17. Kaas P, Van Belle RA (1987) Monograph of Living Chitons (Mollusca: Polyplacophora) 3, Ischnochitonidae: Chaetopleurinae, Ischnochitoninae (pars), additions to Vols. 1 & 2. E.J. Brill/ W. Backhuys, Leiden, 302 p

18. Kaas P, Van Belle RA (1990) Monograph of Living Chitons (Mollusca: Polyplacophora) 4, Suborder Ischnochitonina: Ischnochitonidae: Ischnochitoninae (continued). Additions to Vols. 1, 2 and 3. E.J. Brill/W. Backhuys, Leiden, 298 p

19. Kaas P, Van Belle RA (1994) Monograph of Living Chitons (Mollusca: Polyplacophora) 5, Suborder Ischnochitonina: Ischnochitonidae: Ischnochitoninae (concluded) Callistoplacinae; Mopaliidae; Additions to Vols. 1–4. E.J. Brill/ W. Backhuys, Leiden, 402 p

20. Keen AM (1971). Sea Shells of Tropical West America. Marine Mollusks from Baja California to Peru, 2nd ed, Stanford, CA, 1064 p

Leise EM (1986) Chiton integument: development of sensory organs in juvenile *Mopalia muscosa*. J Morph 189:71–87

Leise EM (1988) Sensory organs in the hairy girdles of some mopaliid chitons. Am Malac Bull 6:141–151

21. Leloup E (1955) *Chiton (Radsia) goodalli* Broderip, 1832 = *Chiton (Radsia) chierchiae* Nierstrasz, 1908. Bull Inst R Sci Nat Belg 31:1–10

22. Lyons WG (1988) A review of Carribbean Acanthochitonidae (Mollusca: Polyplacophora) with descriptions of six new species of *Acanthochitona* Gray, 1821. Am Malac Bull 6:79–114

Miller AC (1983) A comparison of the species richness and trophic roles of gastropods and chitons on rocky shores of temperate and tropical west America. Veliger 26:62–68

Montoya M (1983) Los moluscos marinos de la Isla del Coco, Costa Rica, I. Lista anotada de especies Brenesia 21:325–353

Sirenko BI (1993) Revision of the system of the order Chitonida (Mollusca: Polyplacophora) on the basis of correlation between the type of gills arrangement and the shape of the chorion processes. Ruthenica 3:93–117

23. Sirenko BI (1997) Position in the system and the origin of deep-water chitons of the family Ferreiraellidae (Mollusca: Polyplacophora). Ruthenica 7:77–89

Skoglund C (1989) Additions to the Panamic Province chiton (Polyplacophora) literature - 1971 through 1988. Festivus 21:78–91

Smith AG (1961) Four species of chitons from the Panamic Province (Mollusca: Polyplacophora). Proc Cal Acad Sci 4 (30) (4):81–90

24. Smith AG (1977) Rectification of west coast chiton nomenclature (Mollusca: Polyplacophora). Veliger 19:215–258

25. Smith AG & Ferreira AJ (1977) Chiton fauna of the Galápagos Islands. Veliger 20:82–97.

Strathmann MF & Eernisse DJ (1987) Phylum Mollusca, Class Polyplacophora. *In*: Strathmann MF (ed) Reproduction and Development of Marine Invertebrates of the Northern Pacific Coast. University of Washington Press, Seattle, Washington, pp 205–219

von Martens E (1902) Die Meeres-Conchylien der Cocos-Insel. Sitzungsb Ges naturforsch Freund Berlin 6:137–141

Wanninger A (2001) Micromorphology and gene expression in muscle and shell development of the Mollusca. Diss. Dr., Fak Biol Ludw -Max -Univ. München, Germany, 176 p

26. Watters GT (1990) A review of the recent eastern Pacific Acanthochitoninae (Mollusca: Polyplacophora: Cryptoplacidae) with the description of a new genus, *Americhiton*. Veliger 33:241–271

Wingstrand KG (1985) On the anatomy and relationships of recent Monoplacophora. Galathea Rep 16:7–94

Part 28
Benthic, Shelled Gastropods

Leonora Rodríguez-Sevilla, Rita Vargas, and Jorge Cortés

Cyphoma gibbosum on a seafan in the coral reef of Parque Nacional Cahuita, Caribbean of Costa Rica (Photo: Jorge Cortés)

Abstract Here we present a compilation of 1,198 benthic, shelled gastropod (prosobranch and pulmonate) species records from Costa Rica published in the scientific literature. Of these, 389 species are from the Caribbean and 818 from the Pacific; 9 species are common to both coasts. In terms of families, the Caribbean has 63, while the Pacific has 75. Including a list of 84 unpublished species (identified by specialists), the total of benthic, shelled gastropods for Costa Rica is 1,282, and the number of species from the Pacific coast increases to 902. Compared to a study of Colombia's Caribbean coast, our list includes 60% of the species reported there, but only 31% of the 2,682 species reported for the Panamic province. The difference in species number between the Caribbean and Pacific of Costa Rica may be due to several facts, including the length of the coast (the Caribbean coast is five times shorter), number of expeditions (many more on the Pacific), and difference in coastal morphology (the Caribbean less complex). Compared to other areas, Costa Rica has lower species numbers, which may be due to patchy sampling (mostly in shallow waters with few samples from deeper than 100 m), and to the study of only macromolluscs. All of this indicates that more sampling is needed on both sides of Costa Rica and in the other Central American countries.

Introduction

Molluscs are the second most specious group of invertebrates in the world, and three fourths of them are gastropods (Morton 1960; Ruppert & Barnes 1994). Mollusc collecting has been popular for centuries, and amateurs have contributed significantly to the advancement of mollusc studies. There are inmense collections both in museums and in private hands. Molluscs are also important as food source, for the curio trade and in some traditional practices (use of dyes or the shells themselves, e.g., as musical instruments).

The Caribbean and Pacific coasts of Costa Rica measure approximately 212 and 1,254 km in length, respectively, totaling 1,466 km. The Caribbean coast of Costa Rica is relatively straight, with strong surf, sandy beaches on the north section, and only two small mangrove areas, one in the central section and one in the south; tide range is less than 50 cm. The southern section, from Limón to the border with

L. Rodríguez- Sevilla
Museo de Zoología, Escuela de Biología, Universidad de Costa Rica, 11501-2060 San Pedro, San José, Costa Rica.
e-mail: leonorars@racsa.co.cr

R. Vargas
Apdo. 3144-1000, San José, Costa Rica.
e-mail: ritav@cariari.ucr.ac.cr

J. Cortés
Centro de Investigación en Ciencias del Mar y Limnología (CIMAR), and Escuela de Biología, Universidad de Costa Rica, 11501-2060, San José Costa Rica.
e-mail: jorge.cortes@ucr.ac.cr

Panama, is formed by rocky headlands of coralline origin, alternating with sandy beaches. In that section, coral reefs with lagoons and seagrass beds predominate (Vargas & Cortés 1999a; Coll *et al.* 2001; Cortés & Jiménez 2003a). The Pacific coast is longer and more irregular. It has different types of environments such as estuaries, mangroves, muddy bottoms, rocky areas, sandy beaches, and coral reefs, and the tides reach up to 4 m (Vargas & Cortés 1999b; Cortés & Jiménez 2003b).

Malacological Studies in Costa Rica

The first marine invertebrate species described for Costa Rica belongs to the phylum Mollusca; it was the polyplacophora *Chaetopleura scabricula* Sowerby, 1832. The first species described from Costa Rica came from Puerto Potrero (now known as Bahía Potrero), Guanacaste, and from the Golfo de Nicoya and Golfo Dulce, Puntarenas (Cortés 1998). Other mollusc species were described, thanks to the collections done by Hugh Cuming between 1827 and 1830, along the west coast of the Americas.

The majority of the molluscs collected in Costa Rica are the product of expeditions organized by foreigners, especially from the USA, from the late 19th century to the mid-20th century. The main objective of these expeditions was to explore the coasts of the Americas. One of the most visited sites by these expeditions was Isla del Coco, since it is en route to the Galapagos Islands, and the focus of attention of many naturalists (Hertlein 1963; Cortés & Jiménez 2003b).

The Caribbean coast was not studied until the 1960s. Bakus (1968a) did a comparative study of the littoral zonation of gastropods collected in 1962 at three sites on the Caribbean: Portete, Puerto Limón, and Isla Uvita; he reported 29 species of gastropods. Houbrick (1968a) studied the marine molluscs in 1966 at two localities (Portete and Barra del Colorado), in shallow waters and in the littoral zone; he collected 250 species of molluscs of which 181 were gastropods. Wellington (1974) studied the fauna of Cahuita in 1972 and reported 70 species of gastropods. Robinson & Montoya (1987) collected materials between 1982 and 1986 from the intertidal to 5 m depth in ten locations distributed along the coast: Barra del Colorado, Río Matina, Moín, Portete, Puerto Limón, mouth of Río Banano, Río Estrella, Cahuita National Park, Puerto Viejo, and Manzanillo. They collected 395 species of which 288 were gastropods. Espinosa & Ortea (2001) collected from Cahuita to Gandoca between 1999 and 2001, and listed 484 molluscs of which 340 were gastropods. Finally, Rodríguez-Sevilla *et al.* (2003) published a compedium of the marine gastropods of the Caribbean coast of Costa Rica; they reported 384 species, plus 106 additional in two appendixes, for a total of 490 species.

Many papers have been published on molluscs of the Pacific of Costa Rica, including Isla del Coco. We can divide the publications into three periods. The first publications are descriptions of species based on specimens collected by ship captains or by people who collected for museums during the early part of the 19th century (e.g., Sowerby 1832a, b). The second period, from the late 19th to the mid-20th century, produced a large series of papers based on scientific expeditions to the eastern Pacific, e.g., Dall (1900) with the US Fish Commission Expedition of

1888 on board the steamer *Albatross*. Some of the most productive expeditions were organized by the New York Zoological Society, described by Beebe (1938). An example of one of many papers on molluscs is by Hertlein & Strong (1943). Another productive series of expeditions were the Allan Hancock Expeditions to the eastern Pacific, presented by Fraser (1943). For an example of a paper, among many, on molluscs see Hanna & Hertlein (1938). The third period is from the 1950s to the present date when individuals have done collections, e.g., Jung (1989) studied *Strombina* and published on them. Others collected as part of expeditions; e.g., Høisæter (1998) participated in the Costa Rican–German Expedition and published on the gastropods of the Golfo Dulce. Many papers have been published on the molluscs of Isla del Coco as a result of expeditions between 1983 and 1987 (e.g., Montoya & Shasky 1987; D'Attilio *et al.* 1987; Montoya 1988; Montoya & Kaiser 1988; Kaiser 1998, 2001a, 2002; Kaiser & Hertz 2001).

During the review of this part, Carol Skoglund sent us a list of 84 species in 59 genera and 20 families of gastropods (Table 28.1) collected during three dredging

Table 28.1 List of species collected during three dredging trips (depth: 10–37 m) to Playa del Coco, Guanacaste, by Carol Skoglund and her late husband Paul. These are now, with a few exceptions, in the Skoglund collection, Phoenix, Arizona. This collection will go to the Santa Barbara Museum of Natural History. The species from the late Robert Koch's collection are now at the Santa Barbara Museum of Natural History. The Philadelphia Academy of Sciences (ANSP) has duplicate specimens of all turrids listed, plus the six species shown as ANSP which are not in the Skoglund collection. The list and information was provided by Carol Skoglund (C. Skoglund, personal communication 2005)

Family FISSURELLIDAE	
Lucapinella eleanorae McLean, 1967	Family CALYPTRAEIDAE
Leurolepas roseola McLean, 1970	*Crucibulum castellum* Berry, 1963
Family TROCHOIDAE	Family NATICIDAE
Calliostoma leanum (C.B. Adams, 1852)	*Natica othello* Dall, 1908
Calliostoma rema Strong & Hertlein, 1933	*Polinices caprae* (Philippi, 1852)
	Sinum debile (Gould, 1853)
Family TURBINIDAE	
Arene balboai (Strong & Hertlein, 1939) – SBMNH	Family TRIVIIDAE
	Hespererato columbella (Menke, 1847) – SBMNH
Family RISSOIDAE	Family MURICIDAE
Rissoina axeliana Hertlein & Strong, 1951 – SBMNH	*Pterynotus pinniger* (Broderip, 1833)
	Dermomurex indentata (Carpenter, 1857)
Folinia ericana (Hertlein & Strong, 1951)	*Favartia perita* (Hinds, 1844)
Folinia insignis (de Folin, 1867)	*Trachypollia lugubris* (C.B. Adams, 1852)
	Coralliophila orcuttiana Dall, 1919
Family CAECIDAE	
Caecum parvum C.B. Adams, 1852 – SBMNH	Family BUCCINIDAE
Caecum lohri (Strong & Hertlein, 1939) – SBMNH	*Caducifer nigricostatus* (Reeve, 1846)

(continued)

Table 28.1 (continued)

	Monostiolum crebristriatum (Carpenter, 1856)
	Phos veraguensis Hinds, 1843
Family TURRITELLIDAE	*Phos dejanira* (Dall, 1919)
Turritella mariana Dall, 1908	*Phos cumingii* Reeve, 1846
Family VERMETIDAE	Family COLUMBELLIDAE
Serpulorbis oryzata (Mörch, 1862)	*Columbella haemastoma* Sowerby, 1832
	Mitrella harfordi Strong & Hertlein, 1937
Family EPITONIIDAE	*Nassarina helenae* Keen, 1971 – SBMNH
Epitonium minuticostatum (Deboury, 1912)	*Nassarina perita* Keen, 1971
Epitonium togatum Hertlein & Strong, 1951	*Steironepion tincta* (Carpenter, 1864)
Epitonium hindsii (Carpenter, 1856)	*Strombina carmencita* Lowe, 1935
Opalia spongiosa Carpenter, 1864	*Strombina elegans* (Sowerby, 1832)
	Strombina maculosa (Sowerby, 1832)
Family EULIMIDAE	*Strombina recurva* (Sowerby, 1832)
Scalenostoma rangii (de Folin, 1867)	*Sincola gibberula* (Sowerby, 1832)
Sincola dorsata (Sowerby, 1832) – SBMNH	*Cochlespira cedonulli* (Reeve, 1843)
	Pyrgospira obeliscus (Reeve, 1843)
Family NASSARIIDAE	*Crassispira cortezi* Shasky and Campbell, 1964
Nassarius corpulentus (C.B. Adams, 1852)	*Crassispira brujae* Hertlein & Strong, 1951
Nassarius pagodus (Reeve, 1844)	*Crassispira epicasta* Dall, 1919
Nassarius nucleolus (Philippi, 1846)	*Crassispira adana* (Bartsch, 1950) – ANSP
	Crassispira nigerrima (Sowerby, 1834) – ANSP
Family OLIVIDAE	*Crassispira tepocana* Dall, 1919
Oliva kaleontina Duclos, 1835	*Hindsiclava militaris* (Reeve, 1843 ex Hinds, MS) – SBMNH and ANSP
Olivella gracilis (Broderip & Sowerby, 1829)	
	Miraclathurella bicanalifera (Sowerby, 1834)
Family CANCELLARIIDAE	*Carinodrillia dichroa* Pilsbry & Lowe, 1932
Aphera tessellata (Sowerby, 1832)	*Carinodrillia halis* (Dall, 1919) – ANSP
Narona exopleura (Dall, 1908) – SBMNH	*Strictispira ericana* (Hertlein & Strong, 1951)
Tritonoharpa siphonata (Reeve, 1844)	*Compsodrillia alcestis* (Dall, 1919)
	Compsodrillia duplicata (Sowerby, 1834)
Family TEREBRIDAE	*Compsodrillia haliplexa* (Dall, 1919)
Terebra puncturosa Berry, 1961	*Compsodrillia jaculum* (Pilsbry & Lowe, 1932)
Terebra argosyia Olsson, 1971	*Compsodrillia olssoni* McLean & Poorman, 1971 – ANSP
Family TURRIDAE	*Nannodiella fraternalis* (Dall, 1919)
Calliclava albolaqueata (Carpenter, 1865)	*Glyphostoma candida* (Hinds, 1843)
Imaclava unimaculata (Sowerby, 1834)	*Kurtziella cyrene* (Dall, 1919) – ANSP
Agladrillia pudica (Hinds, 1843)	*Tenaturris verdensis* (Dall, 1919)
Drillia acapulcana (Lowe, 1935)	*Daphnella bartschi* Dall, 1919 – SBMNH
Cerodrillia cybele (Pilsbry & Lowe, 1932) – SBMNH	*Philbertia doris* Dall, 1919 – SBMNH
Gemmula hindsiana Berry, 1958	*Microdaphne trichodes* (Dall, 1919) – SBMNH

Fig. 28.1 *Prunum holandae* from the Caribbean coast of Costa Rica (Photo: Paul V. Scott)

Fig. 28.2 *Jenneria pustulata* from Pacific Costa Rica (Photo: Yolanda Camacho)

trips to Playa del Coco (depth: 10–37 m) by herself and her late husband Paul, which were not included in her publication (Skoglund 1990a).

Herein we report 1,198 benthic, shelled gastropod (prosobranch and pulmonate) species present in both coasts, 389 from the Caribbean (Fig. 28.1) and 818 from the Pacific (Fig. 28.2), with nine species in common. In terms of families, the Caribbean has 63, while the Pacific has 75. Including the species in Table 28.1, the total of benthic, shelled gastropods for Costa Rica is 1,282, and the number of species from the Pacific coast increases to 902. The systematics used in the present part are based on Abbott (1974), Skoglund (1992, 2002), and Beesley *et al.* (1998).

Comparative Species Richness

A recent report by Díaz Merlano and Puyana Hegedus (1994) on the Caribbean and western Atlantic of Colombia revealed 653 species of benthic, shelled gastropods. The number of species reported herein for the Caribbean constitutes 60% of the above-mentioned number from Colombia. Keen (1971) and Skoglund (2002) reported 2,682 for the Panamic province; our number for the Pacific coast comprised 31% of the Panamic species.

The difference in number of species between both Costa Rican coasts and between Costa Rica and the other regions sampled could be explained by several reasons:

1. The Caribbean coast is about five times shorter than the Pacific coast.
2. The collecting efforts have been greater in the Pacific.
3. The morphological differences between both coasts make the Pacific a more heterogeneous region with more diversity in habitats and, hence, more biodiversity.
4. The majority of the sites sampled in the Caribbean coast are located in the central and southern sections of the coast; the northern part has been poorly sampled. However, diversity must be higher in the southern and central areas of the coast, due not only to more intense sampling, but also to the larger variety of habitats in the south.
5. The majority of the collections were made on the shore or by diving in shallow waters; thus, the deep water fauna remains unknown.
6. Most species reported in the literature are macromolluscs. Micromolluscs have received little attention due to the difficulties associated with their identification. In other areas, however, the study of the micromolluscs has received more interest lately, revealing a large number of new species and families.
7. Sampling of the Pacific coast has been patchy and mostly in shallow waters, there are few samples of waters deeper than 100 m.

For all these reasons it is concluded that more sampling is needed on both sides of Costa Rica. Therefore, the number of species presented in this part is probably an underestimation of what really exists. Species Lists 28.1 and 28.2, included on the CD-Rom, are based on the literature and museum collections in Costa Rica. In the collection of the Museo de Zoología of the Universidad de Costa Rica, there is a number of specimens that have not been identified, and some might be new species; these have not been included in the lists.

Specialists

Henry Chaney: Director of Collection and Research, Santa Barbara Museum of Natural History, Santa Barbara, California, USA – Eastern Pacific. hchaney@sbnature2.org

Kirstie L. Kaiser: Museum Associate, Los Angeles County Museum of Natural History, Los Angeles, California, 90007, USA – Eastern Pacific. 76511.3463@compuserve.com

José Espinosa: Instituto de Oceanología de Cuba, La Habana, Cuba – Caribbean. espinosa@oceano.inf.cu
Michel Montoya: San José, Costa Rica – Eastern Pacific. michelmontoya@correo.co.cr
Carol Skoglund: Santa Barbara Museum of Natural History, Santa Barbara, California, USA – Eastern Pacific. carolskoglund@msn.com

Collections

- Museo de Zoología, Escuela de Biología, Universidad de Costa Rica: http://museo.biologia.ucr.ac.cr/
- Academy of Natural Sciences, Philadelphia: http://www.acnatsci.org/
- California Academy of Sciences, San Francisco: http://www.calacademy.org/
- Instituto de Oceanología, Cuba: http://www.cuba.cu/ciencia/citma/AMA/p_a.html
- Los Angeles County Museum of Natural History, California: http://www.nhm.org/
- National Museum of Natural History, Smithsonian Institution, Washington, DC: http://www.mnh.si.edu/
- Natural History Museum, London: http://www.nhm.ac.uk/
- Naturhistorisches Museum, Basel, Switzerland: http://www.nmb.bs.ch/
- San Diego Natural History Museum, California: http://www.sdnhm.org/
- Santa Barbara Museum of Natural History, California: http://www.sbnature.org/

Acknowledgments We are grateful to José H. Leal for his help during the preparation of this part. José H. Leal and David G. Robinson identified and revised specimens. J. Michel Montoya (Costa Rica), Kay Hale and Helen D. Albertson (previously at the RSMAS Library, University of Miami), Marien Faber (Museum of Zoology in Amsterdam), and Emilio Rolán (Sociedad Española de Malacología) provided many papers that contained information on Costa Rican gastropods. Useful comments by anonymous reviewers and by Carol Skoglund are greatly appreciated.

References

1. Abbott RT (1974) American Seashells. Van Nostrand Reinhold, New York, 663 p
2. Abbott RT, Dance SP (1982) Compendium of Seashells. E.P. Dutton, New York, 411 p
3. Adams CB (1852) Catalogue of shells collected at Panama, with notes on synonymy, station, and geographical distribution. Ann Lyc Nat Hist New York 5:229–548
4. Allmon WD (1988) Ecology of recent turritelline gastropods (Prosobranchia, Turritellidae): current knowledge and paleontological implications. Palaios 3:259–284
5. Allmon WD, Jones DS, Vaughan N (1992) Observations on the biology of *Turritella gonostoma* Valenciennes (Prosobranchia: Turritellidae) from the Gulf of California. Veliger 35:52–63
6. Andrews J (1977) Shells and Shores of Texas. University of Texas, Austin, Texas, USA, 365 p
7. Avilés MC (1984) Moluscos de la ensenada de Santa Catalina, distrito de Sona, provincia de Veraguas. 2. Gasterópodos y poliplacóforos. Donax Panamensis, Soc Panam Malacol 40:45–48

8. Baker F (1926) Expedition of the California Academy of Sciences to the Gulf of California in 1921. Mollusca of the family Triphoridae. Proc Calif Acad Sci 15:223–239
9. Baker F, Hanna GD, Strong AM (1928) Some Pyramidellidae from the Gulf of California. Proc Calif Acad Sci 17:205–246
10. Baker F, Hanna GD, Strong AM (1930) Some rissoid Mollusca from the Gulf of California. Proc Calif Acad Sci 19:23–40
11. Bakus GJ (1968a) Quantitative studies on the cowries (Cypraeidae) of the Allan Hancock Foundation collections. Veliger 11:93–97
12. Bakus GJ (1968b) Zonation in marine gastropods of Costa Rica and species diversity. Veliger 10:207–211
13. Bakus GJ (1975) Marine zonation and ecology of Cocos Island, off Central America. Atoll Res Bull 179:1–8
14. Baldazzi S, Di Cecco C, Tiberi F (1986) Contributions to the knowledge of the shells of Ecuador. Conchiglia 18:4–7
15. Bandel K (1974) Spawning and development of some Columbellidae from the Caribbean Sea of Colombia (South America). Veliger 16:271–282
16. Bartsch P (1911) The Recent and fossil mollusks of the genus *Cerithiopsis* from the west coast of America. Proc US Natl Mus 40:327–367
17. Bartsch P (1915) The recent and fossil mollusks of the genus *Rissoina* from the west coast of America. Proc US Natl Mus 49:33–62
18. Bartsch P, Rehder HA (1939) Mollusks collected on the Presidential Cruise of 1938. Smithson Misc Collec 98 10:1–18
19. Bayer FM (1971) Biological results of the University of Miami deep-sea expeditions. 79 New and unusual mollusks collected by R/V John Elliott Pillsbury and R/V Gerda in the tropical western Atlantic. Bull Mar Sci 21:111–236

Beebe W (1938) Eastern Pacific expeditions of the New York Zoological Society, XIV. Introduction, itinerary, list of stations, nets and dredges of the eastern Pacific Zaca expedition, 1937–1938. Zoologica 23:287–298

Beesley PL, Ross GJB Wells A (eds) (1998) Mollusca: the Southern Synthesis. Fauna of Australia, Volume 5. CSIRO, Melbourne. Part A, pp 1–563; Part B, pp 565–1234

20. Berry SS (1964) Notices of new eastern Pacific Mollusca-VI. Leaflets Malacol 1 24:147–154
21. Beu AG (1985) A classification and catalogue of living world Ranellidae (=Cymatiidae). Conchol Am Bull 13:55–66
22. Beu AG, Cernohorsky WO (1986) Taxonomy of gastropods of the families Ranellidae (=Cymatiidae) and Bursidae. Part 1. Adoption of Ranellidae, and a review of Linatella Gray, 1857. New Zealand J Zool 13:241–266
23. Beu AG, Kay EA (1988) Taxonomy of gastropods of the families Ranellidae (=Cymatiidae) and Bursidae. Part IV The Cymatium pileare complex J Roy Soc New Zealand 18:185–223
24. Beu AG, Knudsen J (1987) Taxonomy of gastropods of the families Ranellidae (=Cymatiidae) and Bursidae. Part 3 A review of the trifid-ribbed species of Cymatium (Turritriton) J Roy Soc New Zealand 17:73–91
25. Bieler R (1993) Architectonicidae of the Indo-Pacific (Mollusca: Gastropoda). Abhandl Naturwiss Vereins, Hamburg, 286 p
26. Bieler R, Mikkelsen PM (1988) Anatomy and reproductive biology of two western Atlantic species of Vitrinellidae, with a case of protandrous hermaphroditism in the Rissoacea. Nautilus 102:1–29
27. Birkeland DK, Meyer DL, Stames JP, Buford CL (1975) Subtidal communities of Malpelo Island. *In*: Graham JB (ed) The Biological Investigation of Malpelo Island, Colombia. Smithson Contrib Zool 176:55–68
28. Bishop JA (1992) T*ectarius muricatus* (Linnaeus, 1758) from the northern Gulf of California, México. Festivus 24:81–82

29. Bishop JA (1997) New distributional data on three eastern Pacific gastropod species. Festivus 29:39–40
30. Bishop MJ, Bishop SJ (1973) A census of marine prosobranch gastropods at San Diego, California. Veliger 16:143–152
31. Bouchet P, Tillier S (1978) Two new giant epitoniids (Mollusca: Gastropoda) from West Africa. Veliger 20:345–348
32. Bradshaw-Hawkins VI, Sander F (1981) Notes on the reproductive biology and behavior of the West Indian fighting conch, *Strombus pugilis* Linnaeus in Barbados, with evidence of male guarding. Veliger 24:159–164
33. Bratcher T (1986) Range extensions of five terebrid species. Festivus 18:127–129
34. Bratcher T, Burch RD (1970) Five new species of *Terebra* from the eastern Pacific. Veliger 12:295–300
35. Bratcher T, Burch RD (1971) The Terebridae (Gastropoda) of Clarion, Socorro, Cocos and Galápagos islands. Proc Calif Acad Sci 37:537–566
36. Bratcher T, Cernohorsky WO (1987) Living Terebras of the World. A Monograph of the Recent Terebridae of the World. American Malacologists, Melbourne, Florida, 240 p
37. Broderip WJ (1832) Characters and descriptions of new species of Mollusca and Conchifera, collected by Mr. H. Cuming in 1827–1830. Proc Zool Soc London, for 1832–1833:25–33 (Feb 28, 1832), 50–61 (Feb 28, 1832). (Some of the species cited in these pages were described by G.B. Sowerby, I.)
38. Broderip WJ (1833) Characters and descriptions of new species of Mollusca and Conchifera, collected by Mr. H. Cuming in 1827–1830. Proc Zool Soc London, for 1832:173–179 (Jan 14, 1833), 194–202 (May 13, 1833). (Some of the species cited in these pages were described by G.B. Sowerby, I.)
39. Broderip WJ (1833) Characters and descriptions of new species of Mollusca and Conchifera, collected by Mr. H. Cuming in 1827–1830. Proc Zool Soc London, for 1833:4–8 (May 17, 1833)
40. Broderip WJ (1836) Characters and descriptions of new species of Mollusca and Conchifera, collected by Mr. H. Cuming in 1827–1830. Proc Zool Soc London, for 1835–1836:192–197 (Apr 8, 1836)
41. Bullock RC (1974) A contribution to the systematics of some West Indian *Latirus* (Gastropoda: Fasciolariidae). Nautilus 88:69–79
42. Burch JQ (1949) A new *Trigonostoma* from Central America. Minutes, Conchol Club South Calif 94:2–3
43. Burguess CM (1985). Cowries of the World. Gordon Verhoef Seacomber Publications, South Africa, 289 p
44. Cantera JR (1991) First record of the Indo-Pacific gastropod *Cypraea caputserpentis* (Linnaeus, 1758) at Isla de Gorgona, Colombia. Veliger 34:85–87
45. Cantera K, JR, Rubio AER, Borrero FJ, Contreras R, Zapata F, Buttkus E (1979) Taxonomía y distribución de los moluscos litorales de la isla Gorgona. *In*: Prahl H, Guhl F, Grögl M (eds) Gorgona. Universidad de Los Andes, Bogotá, Colombia, pp 141–167
46. Cardin C, Walls JG (1980) A new false cowry from Pacific Panamá (Ovulidae: Simniini). Pariah 8:1–2
47. Catarius L (1988) *Cymatium gibbosum* at Palos Verdes. Festivus 20:10
48. Cate CN (1969) A revision of the eastern Pacific Ovulidae. Veliger 12:95–102
49. Cate CN (1969) The cowrie species living at Guam. Veliger 12:120–131
50. Cate CN (1969) The eastern Pacific cowries. Veliger 12:103–119
51. Cate CN (1973) A systematic revision of the recent cypraeid family Ovulidae (Mollusca: Gastropoda). Veliger 15(Suppl):1–116 p
52. Cate CN (1976) Five new species of Ovulidae (Mollusca: Gastropoda). Veliger 19:159–162
53. Cate CN (1977) A review of the Eratoidae. Veliger 19:341–366
54. Cate CN (1979) A review of the Triviidae (Mollusca: Gastropoda). San Diego Soc Nat Hist Mem 10:1–126
55. Cernohorsky WO (1969) List of type specimens of Terebridae in the British Museum (Natural History). Veliger 11:210–222

56. Cernohorsky WO (1969) The Muricidae of Fiji. Veliger 11:293–315
57. Cernohorsky WO (1975) The taxonomy of some west American and Atlantic Nassariidae based on their type-specimens. Rec Auckland Inst Mus 12:121–173
58. Cernohorsky WO (1975) Supplementary notes on the taxonomy of buccinid species of the subfamily Pisaniinae (Mollusca: Gastropoda). Rec Auckland Inst Mus 12:175–211
59. Cernohorsky WO (1976) The Mitridae of the world. Part I The subfamily Mitrinae Indo-Pacific Moll 3:273–528
60. Cernohorsky WO (1978) New species of Mitridae, Costellariidae and Turridae from the Hawaiian Islands with notes on *Mitra sphoni* in the Galápagos Islands. Nautilus 92:61–67
61. Cernohorsky WO (1985) Taxonomy of some Indo-Pacific Mollusca. Part 12 With remarks on two American gastropod species. Rec Auckland Inst Mus 22:47–67
62. Cernohorsky WO (1991) The Mitridae of the world. Part 2 The subfamily Mitrinae concluded and subfamilies Imbricariinae and Cylindromitrinae. Mon Mar Moll 4:1–164
63. Chaney HW (1992) Interesting gastropods from the 1992 Cocos Island expedition. Festivus 24:86–91
64. Chaney HW (1996) Additional records of *Polinices simiae* (Gastropoda, Naticidae) in the eastern Pacific. Festivus 28:82
65. Christiaens J (1989) The phenacolepadidea gastropoda: neritoidea part 2. Gloria Maris, Belgian Soc Malacol 28:17–48
66. Clench WJ, Pérez I (1945) The genus *Murex* in the Western Atlantic. Johnsonia 1(17):1–56

Coll M, Fonseca AC, Cortés J (2001) Caracterización del manglar de Laguna Gandoca, Refugio Nacional de Vida Silvestre Gandoca-Manzanillo, Limón, Costa Rica. Rev Biol Trop 49(Suppl 2):321–329

67. Collin R (2000) The development of three heterobranch mollusks from California, USA. Nautilus 114:117–119
68. Coomans HE, Moolenbeek RC, Wils E (1982) Alphabetical revision of the (sub)species in the recent Conidae 5. *Baccatus* to *byssinus* including *Conus brettinghami* nomen novem. Basteria 46:3–67
69. Coomans HE, Moolenbeek RC, Wils E (1983) Alphabetical revision of the (sub)species in the recent Conidae 6. *Cabritii* to *cinereus*. Basteria 47:67–143
70. Coomans HE, Moolenbeek RC, Wils E (1985) Alphabetical revision of the (sub)species in the recent Conidae 8. *Dactylosus* to *dux*. Basteria 49:145–196

Cortés J (1998) El estudio de los invertebrados marinos en Costa Rica. Bionet 4:20–23

71. Cortés J, Guzmán HM (1985) Arrecifes coralinos de la costa Atlántica de Costa Rica. Brenesia 23:275–292

Cortés J, Jiménez CE (2003a) Corals and coral reefs of the Caribbean of Costa Rica: past, present and future. *In*: Cortés J (ed) Latin American Coral Reefs. Elsevier Science B.V., Amsterdam, pp 223–239

Cortés J, Jiménez CE (2003b) Corals and coral reefs of the Pacific of Costa Rica: history, research and status. *In*: Cortés J (ed) Latin American Coral Reefs. Elsevier Science B.V., Amsterdam, pp 361–385

72. Cruz RA (1996) Annotated checklist of marine molluscs collected during the R.V. Victor Hensen Costa Rica Expedition 1993/1994. Rev Biol Trop 44(Suppl 3):59–67
73. Cruz RA (1996) Costa Rican marine gastropods: range extensions. Rev Biol Trop 44(Suppl 3):237–238
74. Cruz RA, Jiménez JA (1994) Moluscos asociados a las áreas de manglares de la costa pacífica de América Central: Guía. Fundación UNA, Heredia, Costa Rica. 182 p

Dall WH (1900) Additions to the insular land-shell faunas of the Pacific Coast, especially of the Galápagos and Cocos islands. Proc Acad Nat Sci Phila 52:88–106

75. Dall WH (1909) Report on a collection of shells from Peru, with a summary of the littoral marine Mollusca of the Peruvian Zoological Province. Proc US Natl Mus 37:147–294
76. Dall WH (1910) Summary of the shells of the genus *Conus* from the Pacific coast of America in the U.S. National Museum. Proc US Natl Mus 38:217–228.
77. D'Attilio A (1979) *Tripterotyphis lowei* from the Solomon Islands. Festivus 11:51–53

78. D'Attilio A (1981) The radula of *Neoteron ariel* (Pilsbry and Lowe, 1932) (Gastropoda: Buccinidae). Festivus 13:104–105
79. D'Attilio A (1987) *Typhisopsis coronatus* and dwarf *Typhisala grandis* at Costa Rica. Festivus 19:32–35
80. D'Attilio A, Hertz CM (1979) The eastern Pacific species of *Murex* sensu stricto. Festivus 11:60–63
81. D'Attilio A, Myers BW (1988) A new species of *Favartia* from the eastern Pacific (Gastropoda: Muricidae). Nautilus 102:106–109
82. D'Attilio A, Myers BW, Shasky DR (1987) A new species of *Phyllonotus* (Muricidae: Muricinae) from Isla del Coco, Costa Rica. Nautilus 101:162–165
83. Dautzenberg P (1932) Mollusques testacés marins de Madagascar. J Conchyl 76:5–119
84. de Jong KM, Coomans HE (1988) Marine Gastropods from Curaçao, Aruba and Bonaire. E.J. Brill, Leiden, The Netherlands, 261 p
85. Demond J (1957) Micronesian reef-associated gastropods. Pac Sci 11:275–341
86. Dexter DM (1974) Sandy-beach fauna of the Pacific and Atlantic coasts of Costa Rica and Colombia. Rev Biol Trop 22:51–66
87. Díaz JM, Mittnacht P (1989) A new species of *Anachis* from the Caribbean coast of Colombia (Prosobranchia: Columbellidae). Arch Moll 120:149–154
88. Díaz JM, Escobar LA, Velásquez LE (1992) Reef associated molluscan fauna of the Santa Marta area, Caribbean coast of Colombia. An Inst Invest Mar Punta Betín 19–20:173–196
89. Díaz Merlano JM, Puyana Hegedus M (1994) Moluscos del Caribe Colombiano. Un Catálogo Ilustrado. Colciencias y Fundación Natura Colombia, Santafé de Bogotá, Colombia. 291 p
90. Dixon J (1981) Evidence of gregarious settlement in the larvae of the marine snail *Collisella strigatella* (Carpenter). Veliger 24:181–184
91. Draper BC (1972) Checklist of shells found in scrapings from *Spondylus princeps* taken at a depth of 40 feet, southeast of Santa Rosalía, Baja California, 1969. Festivus 3(4):5–7
92. Draper BC (1974) Minute shells – part 6. Tabulata 7:17–22
93. Draper BC (1974) Minute shells – part 7. Tabulata 7:44–48
94. Draper BC (1974) Minute shells – part 8. Tabulata 7:69–77
95. Draper BC (1974) Minute shells – part 9. Tabulata 7:100–114
96. Draper BC (1975) Checklist of shells collected at Cholla Bay, Sonora, México. Festivus 6:67
97. DuShane H (1974) The panamic-galapagan epitoniidae. Veliger 16:1–84
98. DuShane H (1982) Notes on living *Casmaria vibexmexicana* (Stearns, 1894). Veliger 24:336–338
99. DuShane H (1985) *Cyclothyca corrugata* Stearns, 1890. Festivus 17:11
100. DuShane H (1985) The wandering wentletrap, *Epitonium (Asperiscala) billeeana* (dushane and Bratcher, 1965). Festivus 17:38–39
101. DuShane H (1985) The family Epitoniidae of Panama Bay. Festivus 17:68–75
102. DuShane H (1988) Pliocene Epitoniidae of the Esmeraldas beds northwestern Ecuador (Mollusca: Gastropoda). Tulane Stud Geol Paleont 21:51–58
103. DuShane H, Brennan E (1969) A preliminary survey of mollusks for Consag rock and adjacent areas, Gulf of California, México. Veliger 11:351–363
104. DuShane H, Draper BC (1975) The genus *Seila* in the eastern Pacific. Veliger 17:335–345
105. DuShane H, Poorman R (1967) A checklist of mollusks for Guaymas, Sonora, México. Veliger 9:413–441
106. Emerson WK (1968) A new species of the gastropod genus *Morum* from the eastern Pacific. J Conchybiol CVII:53–57
107. Emerson WK (1968) A record of the Indo-Pacific cone, *Conus ebraeus*, in Guatemala. Veliger 11:33
108. Emerson WK (1983) New records of prosobranch gastropods from Pacific Panamá. Nautilus 97:119–123

109. Emerson WK (1989) On the occurrence of *Charonia tritonis* (Linnaeus) in the eastern Pacific (Ranellidae: Cymatiinae). Festivus 21:13–15
110. Emerson WK (1991) First records for *Cymatium mundum* (Gould) in the eastern Pacific Ocean, with comments on the zoogeography of the tropical trans-Pacific tonnacean and non-tonnacean prosobranch gastropods with Indo-Pacific faunal affinities in west American waters. Nautilus 105:62–80
111. Emerson WK (1993) On the first records of *Cypraea moneta* Linnaeus and *Cymatium mundum* (Gould) on the west American mainland (Cypraeidae: Ranellidae) with zoogeographic implications. Festivus 25:70–74
112. Emerson WK (1994) A zoogeographic summary of the marine mollusks of Clipperton Island (tropical eastern Pacific Ocean). Festivus 26:62–71
113. Emerson WK (1995) A zoogeographic summary of the marine molluscs of the Revillagigedo Islands (tropical eastern Pacific Ocean). Festivus 27:3–18
114. Emerson WK, d'Attilio A (1969) A new species of *Strombina* from the Galápagos Islands. Veliger 11:195–197
115. Emerson WK, d'Attilio A (1970) Three new species of muricacean gastropods from the eastern Pacific. Veliger 12:270–274.
116. Emerson WK, Chaney HW (1995) A zoogeographic review of the Cypraeidae (Mollusca: Gastropoda) occuring in the eastern Pacific Ocean. Veliger 38:8–21
117. Emerson WK, Hertlein LG (1964) Invertebrate megafossils of the Belvedere expedition to the Gulf of California. Trans San Diego Soc Nat Hist 13:333–368
118. Emerson WK, Jacobson MK (1976) The American Museum of Natural History Guide to Shells. Land, Freshwater, and Marine, from Nova Scotia to Florida. Alfred A. Knopf, New York, 482 p
119. Emerson WK, Old WE Jr (1962) Results of the Puritan-American Museum of Natural History expedition to western México. 16 The recent mollusks: Gastropoda, Conidae Am Mus Novitates 2112:1–44
120. Emerson WK, Old WE Jr (1963) Remarks on *Cassis (Casmaria) vibexmexicana*. Nautilus 76:143–145
121. Emerson WK, Old WE Jr (1963) Results of the Puritan-American Museum of Natural History expedition to western México. 19. The Recent mollusks: Gastropoda, Strombacea, Tonnacea, and Cymatiacea. Am Mus Novitates 2153, 38 p
122. Emerson WK, Old WE Jr (1965) New molluscan records for the Galápagos Islands. Nautilus 78:116–120
123. Emerson WK, Old WE Jr (1968) An additional record for *Cypraea teres* in the Galápagos Islands. Veliger 11:98–99
124. Emerson WK, Piech BJ (1992) Remarks on *Distorsio constricata* (Broderip, 1833) and related species in the eastern Pacific Ocean, with the description of a new species (Gastropoda: Personidae). Veliger 35:105–116
125. Emerson WK, Radwin GE (1969) Two new species of Galapagan turrid gastropods. Veliger 12:149–156
126. Emerson WK, Sage WE III (1988) *Conus baccatus* G.B. Sowerby III, 1877: a Panamic faunal constituent Nautilus 102:110–114
127. Espinosa J, Ortea J (1999) Descripción de nuevas marginelas (Mollusca: Neogastropoda: Marginellidae) de Cuba y del Caribe de Costa Rica y Panamá. Avicennia 10/11:165–176
128. Espinosa J, Ortea J (1999) Dos nuevas especies del género *Hyalina* Schumacher, 1817 (Mollusca: Neogastropoda: Marginellidae) del mar Caribe de Costa Rica y costas de Cuba. Avicennia 10/11:177–183
129. Espinosa J, Ortea J (2000) Descripción de un género y once especies nuevas de Cystiscidae y Marginellidae (Mollusca: Neogastropoda) del Caribe de Costa Rica. Avicennia 12/13:95–114
130. Espinosa J, Ortea J (2001) Moluscos del mar Caribe de Costa Rica: desde Cahuita hasta Gandoca. Avicennia (Suppl 4):1–77
131. Everson G (1984) In search of *Thais planospira*. Festivus 16:113–114

132. Faber MJ (1988) Studies on West Indian marine molluscs 13. The malacological taxa of Gordon W Nowell-Usticke De Kreukel 24(4–5):67–102
133. Faber MJ (1990) Studies on West Indian marine molluscs 19. On the identity of *Turbo bryereus* Montagu, 1803, with the description of a new species of Rissoina (Gastropoda Prosobranchia: Rissoidae). Basteria 54:115–121
134. Faber MJ, Moolenbeek RG (1987) On the doubtful records of *Alvania platycephala*, *Alvania pagodula* and *Alvania didyma*, with the description of two new rissoid species (Mollusca; Gastropoda: Rissoidae). Beaufortia 37:67–71
135. Faber MJ, Moolenbeek RG (1991) Two new shallow water triphorids and a new name in *Metaxia* from Florida and the West Indies. Apex 6:81–85
136. Faber MJ, Moolenbeek RG (1999) *Hummelinckiella borinquensis*, a new genus and species in the subfamily Stephopominae (Caenogastropoda: Siliquariidae) with notes on the genera *Caporbis* and *Stephopoma* and the "Blasian subregion." Bull Zool Mus 17:41–45
137. Fair RH (1976) The Murex Book: An Illustrated Catalogue of Recent Muricidae (Muricinae, Muricopsinae, Ocenebrinae). Sturgis, Honolulu, 138 p
138. Finet Y (1985) Preliminary Faunal List of the Marine Molluscs of the Galápagos Islands. Documents de Travail 20. Inst Roy Sci Nat Belgique, 50 p
139. Finet Y (1991) The marine mollusks of the Galápagos Islands. *In*: James MJ (ed) Galápagos Marine Invertebrates. Plenum, New York, pp 253–280
140. Finet Y (1994) Marine molluscs of the Galápagos. Gastropods. A monograph and revision of the families Haliotidae, Scissurellidae, Fissurellidae and Lottiidae. L'Informatore Piceno, Italy, 110 p
141. Finet Y (1994) The Marine Mollusks of the Galápagos Islands: A Documented Faunal List. Editions du Muséum d'Histoire Naturelle de Genève, 180 p
142. Fortunato H, Penchaszadeh PE, Miloslavich P (1998) Observations on the reproduction of *Bifurcium bicanaliferum* (Sowerby, 1832) (Gastropoda: Columbellidae: *Strombina*-group) from the Pacific coast of Panama. Veliger 41:208–211

Fraser CM (1943) General account of the scientific work of the *Velero III* in the eastern Pacific, 1931–1941, Part I: historical introduction, *Velero III*, Personnel. Allan Hancock Pac Exp 1:1–48

143. García-Cubas A, Reguero M (1987) Caracterización ecológica de moluscos en lagunas costeras de Sonora y Sinaloa. Mem Soc Mex Malacol 3:1–16
144. Gemmell J, Hertz CM, Myers BW (1980) Sea star predation on mollusks in the San Felipe Bay area, Baja California, México. Festivus 12:24–55
145. Gemmell J, Hertz CM, Myers BW (1995) The Epitoniidae of San Felipe and environs, Gulf of California, from the Gemmell collection. Festivus 27:74–80
146. Gibson-Smith J, Gibson-Smith W (1982) The subfamily Melampinae (Pulmonata: Basommatophora) in Venezuela, with descriptions of two new species. Nautilus 96:116–120
147. Greene KW (1953) "The genus *Conus* in the Hawaiian area." (Unpublished mimeographed list.) Board of Public Parks and Recreation, Honolulu, Hawaii
148. Groves LT (1992) California cowries (Cypraeacea): past and present, with notes on recent tropical eastern Pacific species. Festivus 24:101–107
149. Haderlie EC, Donat W III (1978) Wharf piling fauna and flora in Monterey Harbor, California. Veliger 21:45–69

Hanna GD, Hertlein LG (1938) Land and brackish water Mollusca of Cocos Island. Allan Hancock Pac Exp 2:(8): 123–135

150. Harasewych MG, Petit RE (1982) Notes on the morphology of *Cancellaria reticulata* (Gastropoda: Cancellariidae). Nautilus 96:104–113
151. Hardison LK, Kitting CL. (1985) Epiphytic algal browsing by *Bittium varium* (Gastropoda) among *Thalassia testudinum* turtlegrass. J Phycol 21:13
152. Hemmen J, Hemmen C. (1980) Range extension for *Pterotyphis fimbriatus* (A. Adams, 1854) Veliger 23:96
153. Hernández MA (1979) Range extensions of mollusk species found on the tropical coast of El Salvador. Veliger 22:204–205

154. Hernández MA (1992) Inventario preliminar de moluscos marinos de la costa de El Salvador. Trabajo de investigación presentado para la Secretaría Ejecutiva del Medio Ambiente (SEMA), El Salvador, 172 p
155. Hertlein LG (1937) A note on some species of marine mollusks occuring in both Polynesia and the western Americas. Proc Am Phil Soc 78:303–312
Hertlein LG (1963) Contribution to the biogeography of Cocos Island, including a bibliography. Proc Calif Acad Sci 4th Ser 32:123–235
156. Hertlein LG, Allison EC (1968) Descriptions of new species of gastropods from Clipperton Island. Occ Pap Calif Acad Sci 66:1–13
157. Hertlein LG, Emerson WK (1953) Mollusks from Clipperton Island (eastern Pacific) with the description of a new species of gastropod. Trans San Diego Soc Nat Hist 11:345–364
Hertlein LG, Strong AM (1943) Eastern Pacific Expeditions of the New York Zoological Society. XXXII Mollusks from the west coast of Mexico and Central America Part II. Zoologica 28:149–168
158. Hertlein LG, Strong AM. (1951) Eastern Pacific expeditions of the New York Zoological Society. XLIII Mollusks from the west coast of México and Central America Part X. Zoologica 36:67–120
159. Hertlein LG, Strong AM. (1955) Marine mollusks collected during the "Askoy" expedition to Panama, Colombia, and Ecuador in 1941. Bull Am Mus Nat Hist 107:159–318
160. Hertz CM (1989) Two *Turritella* species in San Felipe, Baja California Norte, México. Festivus 21:50–51
161. Hertz CM (1995) A second species of *Haustellotyphis* (Gastropoda: Typhidae) from Costa Rica. Veliger 38:158–161
162. Hertz CM, d'Attilio A (1985) *Murexiella diomedaea* (Dall, 1908), from the Gulf of California. Festivus 17:90–93
163. Hertz CM, Hertz J (1982) A new eastern Pacific species of *Eulimostraca* (Gastropoda: Eulimidae). Veliger 25:72–76
164. Hertz CM, Kaiser KL (2002) Four species of eastern Pacific Columbellidae (Mollusca) with egg masses covering their shells. Festivus 34:61–64
165. Hertz CM, Myers BW, Gemmell J (1992) Two new vitrinellid species from the Gulf of California, México (Gastropoda: Vitrinellidae). Veliger 35:70–73
166. Hertz CM, Myers BW, Gemmell J (1992) Review of the Vitrinellidae from the San Felipe area, Baja California, México, from the Gemmell collection (1965–1976). Festivus 24:68–80
167. Hertz CM, Myers BW, Gemmell J (1994) The Caecidae of San Felipe and environs, Gulf of California, from the Gemmell collection (1965–1976). Festivus 26:89–94
168. Hertz J (1976) Minute shells. Festivus 7:50
169. Hertz J (1977) Minute shells. Festivus 8:3
170. Hertz J (1977) Minute shells. Festivus 8:23
171. Hertz J (1977) Minute shells. Festivus 8:40
172. Hertz J (1977) Minute shells. Festivus 9:80
173. Hertz J (1978) Minute shells. Festivus 10:27
174. Hertz J (1979) Minute shells. Festivus 11:2–4
175. Hertz J (1979) Minute shells. Festivus 11:32–34
176. Hertz J (1983) *Fissurella rugosa* Sowerby, 1835 and *Fissurella microtrema* Sowerby, 1835. Festivus 15:72–79
177. Hertz J (1995) Range extension for *Thais melones* (Duclos, 1832). Festivus 27:88
178. Hertz J, Hertz CM (1984) Collecting in La Paz. Festivus 16:66–70
179. Hertz J, Kaiser KL (1998) *Colubraria (Colubraria) ochsneri* and *Colubraria (C.) Lucasensis* from Isla del Coco, Costa Rica. Festivus 30:35–39
180. Hertz J, Kaiser KL (1998) *Gastrochaena (Rocellaria) ovata* Sowerby, 1834, and *Scalenostoma subulata* (Broderip, 1832), sharing the same niche. Festivus 30:85–88
181. Hoagland KE (1977) Systematic review of fossil and recent *Crepidula* and discussion of the evolution of the Calyptraeidae. Malacologia 16:353–420

182. Hoagland KE (1986) Patterns of encapsulation and brooding in the Calyptraeidae (Prosobranchia: Mesogastropoda). Am Malacol Bull 4:173–183.
183. Hoffman DL, Weldon PJ (1978) Flight responses of two intertidal gastropods (Prosobranchia: Trochidae) to sympatric predatory gastropods from Barbados. Veliger 20:361–366
184. Hoffman DL, Homan WC, Swanson J, Weldon PJ (1978) Flight responses of three congeneric species of intertidal gastropods (Prosobranchia: Neritidae) to sympatric predatory gastropods from Barbados. Veliger 21:293–296
185. Høisæter T (1998) Preliminary check-list of the marine, shelled gastropods (Mollusca) of Golfo Dulce, on the Pacific coast of Costa Rica. Rev Biol Trop 46(Suppl 6):263–270
186. Hollmann M (1996) *Polinices (Mamilla) simiae* on Isla del Coco, Costa Rica: another Indo-Pacific invader into the Panamic Province. Festivus 28:24–29
187. Houbrick JR (1968a) A survey of the litoral marine molluscs of the Caribbean coast of Costa Rica. Veliger 11:4–23
188. Houbrick JR (1968b) New record of *Conus ebraeus* in Costa Rica. Veliger 10:292
189. Houbrick RS (1978) The family Cerithiidae in the Indo-Pacific Part 1: The genera *Rhinoclavis, Pseudovertagas* and *Clavocerithium*. Monogr Mar Moll 1:1–130
190. Houbrick RS (1987) Anatomy, reproductive biology, and phylogeny of the Planaxidae (Cerithiacea: Prosobranchia). Smithson Contrib Zool 445:1–57
191. Houbrick RS (1990) Anatomy, reproductive biology and systematic position of *Fossarus ambiguus* (Linné) (Fossarinae: Planaxidae; Prosobranchia). Açoreana (Suppl):59–73
192. Houbrick RS (1993) Phylogenetic relationships and generic review of the Bittiinae (Prosobranchia: Cerithioidea). Malacologia 35:261–313
193. Hutsell KC (1990) *Casmaria vibexmexicana* (Stearns, 1894), a new resident of Escondido Bay? Festivus 22:37–38
194. Hutsell KC (1993) *Conus mahogani* (Reeve, 1943) the first specimen from Cocos Island, Costa Rica. Festivus 25:20
195. Hutsell KC (1993) *Haliotis roberti* at Cocos Island, Costa Rica. Festivus 25:28
196. Hutsell KC (1994) *Persicula pulchella* (Kiener, 1834): "A numbing experience." Festivus 26:56–57
197. Hutsell KC (1994) *Xenophora robusta* (Verrill, 1870). Festivus 25:72–73
198. Ingham RE, Zischke JA (1977) Prey preferences of carnivorous intertidal snails in the Florida Keys. Veliger 20:49–51
199. Ingram WM (1951) The living Cypraeidae of the western hemisphere. Bull Amer Paleont 33:5–46
200. Jung P (1989) Revision of the *Strombina*-Group (Gastropoda: Columbellidae), Fossil and Living. Distribution, Biostratigraphy, Systematics. Schweiz Paläont Abhandl 111:1–298
201. Jung P, Heitz A (2001) The subgenus *Lentigo* (Gastropoda: Strombidae) in tropical America, fossil and living. Veliger 44:20–53
202. Kaiser KL (1993) An overview to the known Recent micromolluscan marine fauna of the Islas Galápagos including microfaunal list. Festivus 25:90–109
203. Kaiser KL (1995) New distributional record for *Chicoreus (Phyllonotus) peratus* (Keen, 1960). Festivus 27:91–92
204. Kaiser KL (1997) The Recent molluscan marine fauna of the Islas Galápagos. Festivus 29:1–67
205. Kaiser KL (1998) *Cypraea talpa* on eggs and a live *Cymatium (Gutturnium) muricinum*, both at Isla del Coco, Costa Rica. Festivus 30:28–29
206. Kaiser KL (1999a) Growth series of common molluscan species (Family: Muricidae). Festivus 31:98–100
207. Kaiser KL (1999b) *Littoraria (Littoraria) variegata* (Souleyet, *in* Eydoux & Souleyet, 1852) (Gastropoda: Littorinidae) found at Barra de Navidad, Colima, México. Festivus 31:109
208. Kaiser KL (1999c) *Oliva foxi* (Gastropoda: Olividae) at Isla Montuosa, Golfo de Chiriquí, Panamá. Festivus 31:51
209. Kaiser KL (2001a) Comments on four muricoidean (Mollusca) species formerly endemic to Isla del Coco found at Isla de Malpelo. Festivus 33:3–7

210. Kaiser KL (2001b) New distributional records for eight eastern Pacific molluscan species at Panamá. Festivus 33:69–74
211. Kaiser KL (2002) *Coralliophila neritoidea* (Lamarck, 1816) (Gastropoda: Muricidae) reported at Panamá and Costa Rica. Festivus 34:99–100
212. Kaiser KL, Bryce CW (2001) The Recent molluscan marine fauna of Isla de Malpelo, Colombia. Festivus 33(Occ Pap):1–149
213. Kaiser KL, Hertz CM (2001) Notes on *Rhizochilus antipathum* Steenstrup, 1850 (Gastropoda: Coralliophilidae) with new records from Île Clipperton and Isla del Coco (tropical eastern Pacific). Festivus 33:87–94
214. Kay EA (1979) Hawaiian Marine Shells. Reef and Shore Fauna of Hawaii. Section 4: Mollusca. Bernice Pauahi Bishop Mus Spec Publ 64(4). Bishop Museum, Honolulu, Hawaii, 653 p
215. Kay EA (1991) The marine molluscs of the Galápagos. Determinants of insular marine faunas. *In*: James MJ (ed) Galápagos Marine Invertebrates. Plenum, New York, pp 235–252
216. Keen AM (1971) Sea Shells of Tropical West America: Marine Molluscs from Baja California to Peru. Stanford University, Stanford, California, 1064 p
217. Keen AM (1983) Vermetidae of the tropical eastern Pacific. Ann Rep Western Soc Malacol (for 1982) 15:10
218. Keen AM, Coan EV (1975) "Sea shells of tropical west America": additions and corrections to 1975. West Soc Malacol Occ Pap 1:66
219. Kerstitch A (1979) *Conus*. The Conus bartschi/brunneus controversy now over Shell Collector 2:38–41
220. Kosuge S, Suzuki M (1985) Illustrated catalogue of *Latiaxis* and its related groups. Family Coralliophilidae Inst Malacol, Tokyo, Spec Publ 1:1–83
221. Kronenberg GC (1994) *Columbella sonsonatensis* (Mörch, 1860) from Cocos Island, Costa Rica (Gastropoda: Columbellidae). Festivus 26:101
222. Lahmann EJ, González W (1982) Observaciones sobre la distribución espacial y el comportamiento de *Siphonaria gigas* Sowerby, 1825 en la costa Pacífica de Costa Rica (Gastropoda: Siphonariidae). An Inst Cienc Mar Limnol Univ Autón México 9:101–110
223. Leal JH (1991) Marine Prosobranch Gastropods from Oceanic Islands off Brazil. Backhuys, Oegstgeest, The Netherlands, 418 p
224. Lightfoot J (1993a) Caecidae of the Panamic province. Part one of Sea and Shore 16:13–26
225. Lightfoot J (1993b) Caecidae of the Panamic province. Part two of Sea and Shore 16:75–87
226. Lindberg DR, McLean JH (1981) Tropical eastern Pacific limpets of the family Acmaeidae (Mollusca, Archaeogastropoda); generic criteria and description of six new species from the mainland and the Galápagos Islands. Proc Calif Acad Sci 42:323–339
227. Llera Gonzales EM, Ortea Rato JA (1981) Marine molluscs from Isabel Island, México. Conchiglia 13:3–5
228. López AL, Montoya M (1985) More on the validity of *Cypraea aequinoctialis* (Schilder, 1933). Festivus 17:113–114
229. López AL, Montoya M (1985) *Agaronia murrha* Berry: an enigma in black and white. Hawaiian Shell News 33(12):1–10
230. López A, Montoya M, López J (1988) A review of the genus *Agaronia* in the Panamic province and the description of two new species from Nicaragua. Veliger 30:295–304
231. Lowe HN (1932) Shell collecting in West Central America. Nautilus 45:73–82
232. Lowe HN (1935) New marine Mollusca from west México, together with a list of shells collected at Punta Peñasco, Sonora, México. Trans San Diego Soc Nat Hist 8 6:15–34
233. Macdonald KB (1969) Molluscan faunas of Pacific coast salt marshes and tidal creeks. Veliger 11:399–405
234. Main KL (1982) The early development of two ovulid snails, *Simnia aequalis* and *Simnia barbarensis*. Veliger 24:252–258
235. Marincovich L Jr (1973) Intertidal mollusks of Iquique. Chile Nat His Los Angeles County Sci Bull 16:1–49

236. Marincovich L Jr (1977) Cenozoic Naticidae (Mollusca: Gastropoda) of the northeastern Pacific. Bull Am Paleont 70:1–494
237. McLean JH (1969) New species of tropical eastern Pacific Gastropoda. Malacol Rev 2:115–130
238. McLean JH (1970) Descriptions of a new genus and eight new species of eastern Pacific Fissurellidae, with notes on other species. Veliger 12:362–367
239. McLean JH (1970) New species of Panamic gastropods. Veliger 12:310–315
240. McLean JH (1970) Notes on the deep water calliostomas of the Panamic province with descriptions of six new species. Veliger 12:421–426
241. McLean JH (1978) Marine shells of southern California. Nat Hist Mus Los Angeles Sci Ser 24:1–104
242. McLean JH (1984) The world of marine micromollusks. Terra 22(6):26–30
243. Merrill AS, Petit RE (1969) Mollusks new to South Carolina: II. Nautilus 82:117–122
244. Mikkelsen PM, Mikkelsen PS, Karlen DJ (1995) Molluscan biodiversity in the Indian River Lagoon, Florida. Bull Mar Sci 57:94–127
245. Miller AC (1983) A comparison of the species richness and trophic roles of gastropods and chitons on rocky shores of temperate and tropical west America. Veliger 26:62–68
246. Montoya M (1983) A 3000-mile range extension: *Charonia tritonis tritonis* in the Eastern Pacific. Hawaiian Shell News 31:8
247. Montoya M (1983) Los moluscos marinos de la isla del Coco, Costa Rica. I Lista anotada de especies Brenesia 21:325–353

Montoya M (1988) Molluscan fauna of Cocos Island, Costa Rica, based on visits from 1983–1987: preliminary report. West Soc Malacol, Ann Rep 1987 20:30

248. Montoya M, Kaiser KL. (1988) Biogeographical notes on the genus *Terebra* (Gastropoda: Terebridae) at Isla del Coco, Costa Rica. Rev Biol Trop 36:569–574
249. Montoya M, López A, López J. (1987) Range extensions of several Panamic molluscs based on new records made in Nicaragua. Part I Festivus 19:43–45
250. Montoya M, Shasky DR (1987) The marine Mollusca of Cocos Island, Costa Rica. 106 p (unpublished)

Morton JE (1960) Molluscs. An introduction to their form and function. Harper & Brothers, New York, 232 p

251. Moolenbeek RG (1996) New skeneiform species of the genus *Lodderena* Iredale, 1924 from the Sultanate of Oman (Gastropoda: Skeneidae). Vita Mar 44:21–28
252. Moolenbeek RG, Faber MJ (1991) A new deepwater species of *Benthonellania* from the Caribbean (Gastropoda: Rissoidae). Apex 6:51–53
253. Moolenbeek RG, Faber MJ, Iliffe TM (1989) Two new species of the genus *Caecum* (Gastropoda) from marine caves on Bermuda. Studied in honour of Dr Pieter Wagenaar Hummelinck Found Sci Res Surinam Neth Antilles 123:209–216
254. Mörch OAL (1860) Etudes sur la famille des vermets (suite). J Conchyl 8:27–48
255. Morris PA (1975) A Field Guide to Shells of the Atlantic and Gulf Coasts and the West Indies. Houghton Mifflin, Boston, MA, 330 p
256. Morton JE, Keen AM (1960) A new species of *Stephopoma* (Siliquariidae: mesogastropoda) from the eastern Atlantic Ocean. Proc Malacol Soc London 34:27–35
257. Myers BW (2000) New distributional record for *Favartia (Murexiella) mildredae* Poorman, 1980 (Gastropoda: Muricidae). Festivus 32:9–10
258. Myers BW, d'Attilio A (1990) Three new muricacean species from Cocos Island, Costa Rica (Muricidae and Coralliophilidae). Venus, Jap J Malacol 49:281–292
259. Myers BW, Hertz CM, Gemmell J (1989) Preliminary report on the Vitrinellidae from San Felipe, Baja California, México, from the Gemmell collection. Ann Rep West Soc Malacol (for 1988) 21:17 (abstract)
260. Myhre D (1972) Range extension for *Columbella sonsonatensis* Mörch, 1860. Veliger 14:321
261. Nesbitt EC, Pitt W (1986) *Nassarius* (Gastropoda: Neogastropoda) from the Galápagos Islands. Veliger 28:294–301
262. Nybakken JW (1968) Notes on the food of *Conus dalli* Stearns, 1873. Veliger 11:50

263. Nybakken JW (1971) Biological results of the University of Miami deep-sea expeditions. 78 The Conidae of the Pillsbury expedition to the Gulf of Panama. Bull Mar Sci 21:93–110
264. Nybakken JW (1988) Possible ontogenetic change in the radula of *Conus patricius* of the eastern Pacific. Veliger 31:222–225
265. Olsson AA (1971) Biological results of the University of Miami deep-sea expeditions. 77 Mollusks from the Gulf of Panama collected by R/V John Elliott Pillsbury, 1967 Bull Mar Sci 21:35–92
266. Olsson AA, Bayer FM (1972) American metulas (Gastropoda: Buccinidae). Bull Mar Sci 22:900–926
267. Olsson AA, McGinty TH (1958) Recent marine mollusks from the Caribbean coast of Panama with the description of some new genera and species. Bull Am Paleontol 39:1–52
268. Ortega S (1985) Competitive interactions among tropical intertidal limpets. J Exp Mar Biol Ecol 90:11–25
269. Ortega S (1986) Fish predation on gastropods on the Pacific coast of Costa Rica. J Exp Mar Biol Ecol 97:181–191
270. Ortega S (1987) Habitat segregation and temporal variation in some tropical intertidal populations. J Exp Mar Biol Ecol 113:247–265
271. Ostergaard JM (1950) Spawning and development of some Hawaiian marine gastropods. Pac Sci 4:75–115
272. Peña GM (1970) Zonas de distribución de los gastrópodos marinos del Perú. Anales Cient Univ Nac Agraria 8:154–170
273. Peña GM (1971) Descripción de los gasterópodos de los manglares del Perú. Anales Cient Univ Nac Agraria 9:46–55
274. Perrin B (1977) Two days on Clipperton – or getting high on boobies. Festivus 8:24–26
275. Petuch EJ, Sargent DM (1986) Atlas of the Living Olive Shells of the World. The Coastal Education and Research Foundation, Fort Lauderdale, Florida, 253 p
276. Pilsbry HA, Lowe HN (1932) West Mexican and Central American mollusks collected by H.N. Lowe, 1929–31. Proc Acad Nat Sci Philadelphia 84:33–144
277. Pilsbry HA, Olsson AA (1945) Vitrinellidae and similar gastropods of the Panamic province. I Proc Acad Nat Sci Philadelphia 97:249–278
278. Pilsbry HA, Olsson AA (1952) Vitrinellidae of the Panamic province. II Proc Acad Nat Sci Philadelphia 104:35–88
279. Pilsbry HA, Vanatta EG (1902) Papers from the Hopkins Stanford Galápagos expedition, 1898–1899. XIII Marine Mollusca Proc Wash Acad Sci 4:549–560
280. Pitt W, Kohl R (1979) A new Panamic *Mitrella* (Mollusca: Gastropoda). Veliger 21:467–468
281. Ponder WF (1983) A revision of the recent Xenophoridae of the world and of the Australian fossil species (Mollusca: Gastropoda). Australian Mus Mem 17:1–126
282. Ponder WF (1985) A review of the genera of the Rissoidae (Mollusca: Mesogastropoda: Rissoacea). Rec Australian Mus Suppl 4:221
283. Ponder WF (1985) The anatomy and relationships of *Elachisina* Dall (Gastropoda: Rissoacea). J Moll Stud 51:23–34
284. Poorman FL, Poorman LH (1978) Additional molluscan records from Bahía de los Angeles, Baja California Norte. Veliger 20:369–374
285. Poorman FL, Poorman LH (1988) A report on the molluscan species in the San Carlos rectangle, Sonora, México, collected by Forrest L. and Leroy H Poorman from December 1953 to December 1983 Festivus 20:47–63
286. Poorman LH (1977) A diary of retirees. Festivus 8:36–39
287. Poorman LH (1981) Comments on two misunderstood fusinids (Gastropoda: Fasciolariidae) from the tropical eastern Pacific. Veliger 23:345–347
288. Poorman LH, Poorman FL (1981) Comments on a lovely *Murexiella* (Gastropoda: Muricidae) from the Gulf of California. Festivus 13:61–62
289. Porter HJ, Safrit G (1981) The marine faunal zones of North Carolina, with new molluscan records. Nautilus 95:127–130

290. Radwin GE (1977) The family Columbellidae in the western Atlantic. Veliger 19:403–417
291. Radwin GE (1977) The family Columbellidae in the western Atlantic. Part IIa-The Pyreninae Veliger 20:119–133
292. Radwin GE (1978) The family Columbellidae in the Western Atlantic. Part IIb The Pyreninae (continued) Veliger 20:328–344
293. Radwin GE, d'Attilio A (1972) The systematics of some new world muricid species (Mollusca: Gastropoda), with descriptions of two new genera and two new species. Proc Biol Soc Wash 85:323–352
294. Radwin GE, d'Attilio A (1976) Murex Shells of the World. An Illustrated Guide to the Muricidae. Stanford University Press, Stanford, California, 284 p
295. Reed JK, Mikkelsen PM (1987) The molluscan community associated with the scleractinian coral *Oculina varicosa*. Bull Mar Sci 40:99–131
296. Reguero M, García-Cubas A (1987) Sistemática y ecología de moluscos bentónicos de Nayarit. Mem Soc Mex Malacol 3:31–47
297. Rehder HA (1973) The family Harpidae of the world. Indo-Pacific Moll 3:207–274
298. Reid DG (1999) The genus *Littoraria* Griffith & Pidgeon, 1834 (Gastropoda: Littorinidae) in the tropical eastern Pacific. Veliger 42:21–53
299. Reid DG (2002) The genus *Nodilittorina* von Martens, 1897 (Gastropoda: Littorinidae) in the eastern Pacific Ocean, with a discussion of biogeographic provinces of the rocky-shore fauna. Veliger 45:85–170
300. Reid DG, Kaiser KL (2001) Three new records of Indo-West Pacific Littorinidae (Gastropoda) in the tropical eastern Pacific. Veliger 44:381–388
301. Ríos EC (1975) Brazilian Marine Molluscs Iconography. Fundaçao Univ Rio Grande, Centro Ciênc Mar, Mus Oceanogr, Rio Grande, RS, Brasil, 331 p
302. Ríos EC (1985) Seashells of Brazil. Fundaçao Cidade do Rio Grande, Fundaçao Univ Rio Grande, Mus Oceanogr, Rio Grande, RS, Brasil, 328 p
303. Rivadeneira-Giuria VM (1989) Dos gasterópodos registrados por primera vez en el mar tropical del Perú. Bol Lima 11:73–75
304. Rivadeneira-Giuria VM, Cardoso F, Juscamaita J (1990) Especies peruanas del género *Natica* y una clave de identificación. Bol Lima 12:39–46
305. Robertson R (1979) *Philippia (Psilaxis) radiata*: another Indo-West-Pacific architectonicid newly found in the eastern Pacific (Colombia). Veliger 22:191–193
306. Robinson DG (1991) The Systematics and Paleoecology of the Prosobranch Gastropods of the Pleistocene Moín Formation of Costa Rica. Ph.D. dissertation, Tulane University, New Orleans, Louisiana, 748 p
307. Robinson DG, Montoya M (1987) Los moluscos marinos de la costa Atlántica de Costa Rica. Rev Biol Trop 35:375–400
307a. Rodríguez-Sevilla L, Vargas R, Cortés J (2003) Biodiversidad marina de Costa Rica: Gastrópodos (Mollusca: Gastropoda) de la costa Caribe. Rev Biol Trop 51(Suppl 3):305–399
308. Rolán E, Crúz-Abrego FM (1998) Las familias Barleeiidae Gray, 1857 y Anabathridae Coan, 1964 (Mollusca, Gastropoda, Rissooidea) en Quintana Roo, México. Bol Malacol Roma 33(1–4):1–10
309. Rolán E, Espinosa J (1995) The family Cerithiopsidae (Mollusca: Gastropoda) in Cuba 3. The genus Cerithiopsis sl, species with brown shells Iberus 13:129–147
310. Rolán E, Espinosa J (1999) El complejo *Brachycythara biconica* (C.B. Adams, 1850) (Mollusca: Gastropoda: Turridae) en Cuba, con la descripción de una nueva especie Bol Malacol Roma 34(1–4):43–49
311. Rolán E, Fernández Garcés R (1992) La familia Triphoridae (Mollusca, Gastropoda) en la isla de Cuba 1: el género *Metaxia* Monterosato, 1884. Boll Malacol 28:169–176
312. Rolán E, Espinosa J, Fernández-Garcés R (1990) Los géneros *Sansonia* y *Mareleptopoma* en la isla de Cuba (Mollusca, Gastropoda). Thalassas 8:85–91

313. Román Contreras R, Cruz Abrego FM, Ibánez Aguirre AL (1991) Observaciones ecológicas de los moluscos de la zona intermareal rocosa de la bahía de Chamela, Jalisco, México. Anales Inst Biol Univ Nac Aut Méx, Ser Zool 62:17–32
314. Roth B (1978) New species and records of tropical west American Marginellidae (Mollusca: Neogastropoda). Nat His Mus Los Angeles County, Contr Sci 292:1–18
315. Roth B, Coan EV (1971) Marginellidae (Mollusca: Neogastropoda) from the Galápagos Islands and Cocos Islands. Proc Calif Acad Sci 37:575–584

Ruppert EE, Barnes RD (1994) Invertebrate Zoology. Saunders College, Orlando, FL, 1056 p

316. Sander F, Lalli CM (1982) A comparative study of mollusk communities on the shelf-slope margin of Barbados, West Indies. Veliger 24:309–318
317. Seapy RR, Hoppe WJ. (1973) Morphological and behavioral adaptations to desiccation in the intertidal limpet *Acmaea (Collisella) strigatella*. Veliger 16:181–188
318. Shasky DR (1966) Range and bathymetric extensions for *Olivella inconspicua* and *Nassarius limacinus*. Nautilus 80:35–36
319. Shasky DR (1970) New gastropod taxa from tropical western America. Veliger 13:188–195
320. Shasky DR (1975) Range extensions for two tropical west American gastropods. Veliger 18:217
321. Shasky DR (1976) Marine mollusks of Panama Bay. Bull Amer Malacol Union 1975:47
322. Shasky DR (1983) New records of Indo-Pacific Mollusca from Cocos Island, Costa Rica. Nautilus 97:144–145
323. Shasky DR (1983) Update on mollusks with Indo-Pacific faunal affinities in the tropical eastern Pacific – II. Festivus 15:109–110
324. Shasky DR (1984) A preliminary checklist of marine molluscs from Manabí province, Ecuador. Ann Rep West Soc Malacol (for 1983) 16:25–32
325. Shasky DR (1984) Mollusks of Cocos Island – I: *Olivella cocosensis*. Shells and Sea Life 16:151
326. Shasky DR (1984) Update on molluscs with indopacific faunal affinities in the tropical eastern Pacific – III. Festivus 16:123–124
327. Shasky DR (1986) Update on mollusks with Indo-Pacific faunal affinities in the tropical eastern Pacific – IV. Ann Rep West Soc Malacol (for 1985) 18:27 (abstract)
328. Shasky DR (1987) *Cantharus fragarius* (Wood, 1828) at Cocos Island. Festivus 19:30
329. Shasky DR (1987) Update on the mollusks with Indo-Pacific faunal affinities in the tropical eastern Pacific – V. Festivus 19:48–50
330. Shasky DR (1987) Update on the mollusks with Indo-Pacific faunal affinities in the tropical eastern Pacific – VI. Festivus 19:100–101
331. Shasky DR (1988) Update on the mollusks with Indo-Pacific faunal affinities in the tropical eastern Pacific – VIII. Festivus 20:104–105
332. Shasky DR (1988) Wanderings of an itinerant malacologist II. La Plata Island, Ecuador Amer Conchol 16:8–9
333. Shasky DR (1989) An update on the mollusks from the Galápagos Islands as listed in "Preliminary faunal list of the marine mollusks of the Galápagos Islands" by Yves Finet. Festivus 21:7–10
334. Shasky DR (1989) Addition to "An update of the mollusks from the Galápagos Islands as listed in 'Preliminary faunal list of the marine mollusks of the Galápagos Islands' by Yves Finet." Festivus 21:18
335. Shasky DR (1989) Wanderings of an itinerant malacologist III. My highest high Amer Conchol 17:6
336. Shasky DR (1989) My last seven years – at Cocos Island. Festivus 21:72–75
337. Shasky DR (1996) Distributional records of interesting and rarely collected marine gastropods from the tropical eastern Pacific. Festivus 28:35–45
338. Shasky DR (1996) Additional distributional records of interesting and rarely collected marine gastropods (Epitoniidae) from the tropical eastern Pacific. Festivus 28:85–86

339. Shasky DR, Campbell GB (1964) New and otherwise interesting species of mollusks from Guaymas, Sonora, México. Veliger 7:114–120
340. Shimek RL (1983) The biology of the northeastern Pacific Turridae. III The habitat and diet of Kurtziella plumbea (Hinds, 1843) Veliger 26:10–17
341. Skoglund C (1974) Intertidal Marine Mollusca of Cape Tepoca, Sonora, México. M.Sc. thesis, Arizona State University, Arizona, USA, 70 p
342. Skoglund C (1988) Deep water shells from off Isla Smith, Bahía de los Angeles, Baja California, México. Festivus 20:110–116
343. Skoglund C (1989) Range correction for *Agaronia propatula* (Conrad, 1849). Veliger 32:232
344. Skoglund C (1990a) Panamic province molluscan range extensions from Bahia Cocos, Costa Rica. Festivus 22:100
345. Skoglund C (1990b) Molluscan range extensions to Playas de Villamil, Guayas, Ecuador. Festivus 22:119–121
346. Skoglund C (1991) Panamic range extensions for *Drillia* (Gastropoda: Turridae). Festivus 23:47
347. Skoglund C (1991) Small boat dredging with resulting Panamic province molluscan range extensions. Ann Rep West Soc Malacol (for 1990) 23:21 (abstract)
348. Skoglund C (1992) Additions to the Panamic province gastropod (Mollusca) literature 1971 to 1992. Festivus 24(Suppl):169 p

Skoglund C (2002) Panamic province Molluscan literature: additions and changes from 1971 through 2001. III Gastropoda. Festivus 33(Suppl):286 p

349. Skoglund C, Koch R (1993) New distributional information for Panamic province Archaeogastropoda (Mollusca). Festivus 25:116–118
350. Skoglund C, Koch R (1995) New distributional information for Panamic province Rissooidea (Mollusca: Gastropoda). Festivus 27:23–28
351. Small M (1994) Marine gastropod habitats of western Panama. Festivus 26:107–122
352. Sowerby GB (1832a) Characters and descriptions of new species of Mollusca and Conchifera, collected by Mr. H. Cuming in 1827–1830. Proc Zool Soc London, for 1832–1833:25–33 (Feb 28, 1832), 50–61 (Feb 28, 1832). (Some of the species cited in these pages were described by W.J. Broderip)
353. Sowerby GB (1832b) Characters and descriptions of new species of Mollusca and Conchifera, collected by Mr. H. Cuming in 1827–1830. Proc Zool Soc London, for 1832–1833:113–120 (June 12, 1832)
354. Sowerby GB (1833) Characters and descriptions of new species of Mollusca and Conchifera, collected by Mr. H. Cuming in 1827–1830. Proc Zool Soc London, for 1832:173–179 (Jan 14, 1833), 194–202 (May 13, 1833). (Some of the species cited in these pages were described by W.J. Broderip)
355. Sowerby GB (1833) Characters and descriptions of new species of Mollusca and Conchifera, collected by Mr. H. Cuming in 1827–1830. Proc Zool Soc London, for 1833:52–56 (May 24, 1833). (Some of the species cited in these pages were described by W.J. Broderip)
356. Sowerby GB (1834) Characters and descriptions of new species of Mollusca and Conchifera, collected by Mr. H. Cuming in 1827–1830. Proc Zool Soc London, for 1834:6–8 (Jan 28, 1834)
357. Sphon GG (1961) Notes on the Mitridae of the eastern Pacific, I. *Mitra fultoni* E.A. Smith Veliger 4:32–36
358. Sphon GG (1969) Notes on the Mitridae of the eastern Pacific II: the genus *Thala*, with the description of a new species. Veliger 12:84–88
359. Sphon GG (1976) The Mitridae of the Galápagos Islands. Nautilus 90:63–64
360. Spight TM (1976) Censuses of rocky shore prosobranchs from Washington and Costa Rica. Veliger 18:309–317
361. Spight TM (1978) Temporal changes in a tropical rocky shore snail community. Veliger 21:137–143
362. Stingley D (1984) A new *Oliva* from the eastern Pacific (Gastropoda: Olividae). Conchiglia 16:178–179

363. Strong AM (1949) Additional Pyramidellidae from the Gulf of California. Bull South Calif Acad Sci 48:71–93
364. Sutherland JP (1987) Recruitment limitation in a tropical intertidal barnacle: *Tetraclita panamensis* (Pilsbry) on the Pacific coast of Costa Rica. J Exp Mar Biol Ecol 113:267–282
365. Sutherland JP, Ortega S (1986) Competition conditional on recruitment and temporary escape from predators on a tropical rocky shore. J Exp Mar Biol Ecol 95:155–166
366. Sutton B (1975) *Conus ebraeus* Linné. Of Sea and Shore 6:79
367. van Montfrans J, Orth RJ, Vay S (1982) Preliminary studies of grazing by *Bittium varium* on eelgrass periphyton. Aquat Bot 14:75–90
368. van Regteren Altena CO (1975) The marine Mollusca of Suriname (Dutch Guiana) Holocene and Recent. Part III Gastropoda and Cephalopoda. Zool Verh 139:1–104

Vargas R, Cortés J (1999a) Biodiversidad marina de Costa Rica: Crustacea: Decapoda (Penaeoidea, Sergestoidea, Stenopodidea, Caridea, Thalassinidea, Palinura) del Caribe. Rev Biol Trop 47:877–885

Vargas R, Cortés J (1999b) Biodiversidad marina de Costa Rica: Crustacea: Decapoda (Penaeoidea, Sergestoidea, Caridea, Astacidea, Thalassinidea, Palinura) del Pacífico. Rev Biol Trop 47:887–911

369. Vokes EH (1970) The west American species of *Murexiella*, (Gastropoda: Muricidae) including two new species. Veliger 12:325–329
370. Vokes EH (1974) Notes on *Chicoreus* (Mollusca: Gastropoda) from the Cenozoic of the Western Atlantic region, with the description of new species. Tulane Stud Geol Paleontol 11:81–94
371. Vokes EH (1980) Review of the muricid genus *Acanthotrophon* (Mollusca: Gastropoda). Veliger 23:10–14
372. Vokes EH (1985) Review of the west coast aspelloids *Aspella* and *Dermomurex* (Gastropoda: Muricidae), with the descriptions of two new species. Veliger 27:430–439
373. Vokes EH (1988) Muricidae (Mollusca: Gastropoda) of the Esmeraldas beds, northwestern Ecuador. Tulane Stud Geol Paleont 21:1–50
374. Vokes EH (1996) Cenozoic Muricidae of the Western Atlantic region. Part XI – The subfamily Ergalataxinae Tulane Stud Geol Paleontol 29:27–44
375. Vokes EH, d'Attilio A (1980) A new species of *Favartia (Caribiella)* from the Galápagos Islands. Veliger 23:15–18
376. Vokes EH, Houart R (1986) An evaluation of the taxa *Muricopsis* and *Risomurex* (Gastropoda: Muricidae), with one new species of *Risomurex*. Tulane Stud Geol Paleontol 19:63–88
377. Vokes HE, Vokes EH (1983) Distribution of Shallow-Water Marine Mollusca, Yucatan Peninsula, Mexico. Mesoamerican Ecology Institute, Monograph 1. Mid Am Res Inst, Publ. 54, Tulane University, New Orleans, 183 p
378. von Cosel R (1977) First record of *Mitra mitra* (Linnaeus, 1758) (Gastropoda: Prosobranchia) on the Pacific coast of Colombia, South America. Veliger 19:422–424
379. von Cosel R (1984) Moluscos marinos de la isla de Gorgona (costa del Pacífico Colombiano). An Inst Invest Mar Punta de Betín 14:175–257
380. Walls JG (1979) Cone shells. A synopsis of the living Conidae. THF, Neptune City, NJ, 1011 p
381. Warén A (1980) Revision of the genera *Thyca, Stilifer, Scalenostoma, Mucronalia* and *Echineulima* (Mollusca, Prosobranchia, Eulimidae). Zool Scripta 9:187–210
382. Warén A (1992) Comments on and descriptions of eulimid gastropods from tropical west America. Veliger 35:177–194
383. Warmke GL, Abbott RT (1975) Caribbean Seashells. A Guide to the Marine Molluscs of Puerto Rico and other West Indian Islands, Bermudas and the Lower Florida Keys. Dover, New York, 348 p
384. Wellington GM (1974) An Ecological Description of the Marine and Associated Environments to the Monumento Nacional Cahuita. Subdirección de Parques Nacionales, Ministerio de Agricultura y Ganadería, San José, Costa Rica, 82 p

385. Whitney RA (1978) A new species of *Anachis* (Gastropoda: Columbellidae) from the eastern Pacific. Veliger 20:295–296
386. Wicksten MK (1978) Checklist of marine mollusks at Coyote Point Park, San Francisco Bay, California. Veliger 21:127–130
387. Wiley GN, Circé RC, Tunnell JW Jr (1982) Mollusca of the rocky shores of east central Veracruz State, México. Nautilus 96:55–61

Part 29
Pelagic Gastropods

Eduardo Suárez-Morales, Rebeca Gasca, and Iván Castellanos

Atlanta inclinata, a pelagic gastropod from Pacific Costa Rica (Photo: Iván Castellanos)

E. Suárez-Morales, R. Gasca, and I. Castellanos
El Colegio de la Frontera Sur (ECOSUR) Unidad Chetumal. Av. Centenario Km 5.5. C.P. 77900. A.P. 424. Chetumal, Quintana Roo 77000. México.
e-mails: esuarez@ecosur.mx; rgasca@ecosur.mx; invancast@ecosur.mx

Abstract Here we deal with two different lines of evolutionary adaptation of holoplanktonic gastropods. Only a small proportion of gastropods spend their entire life cycle as part of the plankton; these include forms contained in two subclasses of Gastropoda: the Prosobranchia and the Opisthobranchia. The former are represented by the Superfamily Heteropoda, and opisthobranchs by the orders Thecosomata and Gymnosomata. Of these two orders, we considered only the two suborders of Thecosomata: Euthecosomata and Pseudothecosomata, both widely known under the informal name of Pteropoda. This revision updates the knowledge about the taxonomic composition and distribution of the species of pteropods and heteropods recorded in both the Pacific and Atlantic coasts of Costa Rica. In general, it was found that investigations on the Costa Rican pelagic gastropods have been quite scarce and the literature on adjacent areas is scattered. The number of heteropod species effectively recorded in the Pacific coast of Costa Rica (16 spp.) versus the worldwide known species represents about 44%; the figure is 60% with respect to the Eastern Pacific. There are only three records of heteropods in the Caribbean Sea off Costa Rica, the corresponding proportional figures are 10% worldwide, 12% Northwestern Atlantic Ocean. As for pteropods, the percentage of species effectively recorded in the Pacific coast of Costa Rica versus the worldwide known species is low: 7%. The proportion is higher (36%) with reference to the Eastern Pacific. Only nine species of pteropods have been reported in the Costa Rican Caribbean (17% of the Northwestern Atlantic; 4.7% worldwide). Comments are also provided on the relevance of the group and the methodological problems involved in the taxonomic study of these organisms.

Introduction

Pelagic mollusks represent a successful adaptive and evolutionary trend of the Phylum Mollusca to colonize the planktonic environment. Probably the most conspicuous molluscan forms in the marine realm are the cephalopods (Class Cephalopoda). However, the gastropods (Class Gastropoda) are the only other mollusks that include pelagic species. Several adaptations are present in these zooplanktonic forms and are mainly related to feeding, swimming, and buoyancy in the water column. In this part we deal with two different lines of evolutionary adaptation in holoplanktonic gastropods. These are represented by two subclasses of Gastropoda: the Prosobranchia and the Opisthobranchia. Both groups are mostly benthic, and only a few taxa are pelagic; in fact, out of the approximately 40,000 known species of marine gastropods, only about 140 spend their entire life cycle as part of the plankton (Lalli & Gilmer 1989). The prosobranchs are represented in the plankton by the Superfamily Heteropoda, and opisthobranchs by the orders Thecosomata and Gymnosomata. Of these two orders, we consider only the two suborders of Thecosomata: Euthecosomata and Pseudothecosomata, both widely known under the informal name of Pteropoda.

Heteropoda

Heteropods are holoplanktonic mesogastropods widely distributed in oceanic tropical and subtropical waters. Some species of *Atlanta* can penetrate into neritic environments. Heteropods are not particularly abundant and/or frequent in zooplankton samples and this is probably related to (1) their capacity to avoid plankton nets and (2) their moderately low overall abundance. In situ observations and submersible collections have proved to be good methods to investigate their biology and behavior. Several interesting adaptations of these zooplankters that allow them to dwell in the water column are described by Richter & Seapy (1999).

As a group, they are represented by a wide range of morphological types which constitute three well-defined taxonomic groups at the family level: Atlantidae, Carinariidae, and Pterotracheidae. Heteropods are carnivorous, visual predators armed with a protrusible radula to capture and handle their preys. Partly because of limitation by light, on which they depend to feed, most species dwell in surface waters (0–200 m depth). Some mesopelagic forms, living at depths between the 300 and 700 m, exhibit extended migrational patterns (Van der Spoel 1996a). Species living in the lower part of this epipelagic vertical range tend to be more active migrators than the surficial forms (Michel & Michel 1991). Van der Spoel (1996a) stated that the general horizontal distribution of the heteropod mollusks shows a certain degree of avoidance of the central areas of the oceans; their distributional pattern is therefore assignable to the distant neritic type.

Status of taxonomical knowledge: Besides the pteropods, the heteropods are the most diverse group of holoplanktic gastropods. Probably because of their peculiar beauty and size, they have been studied for more than two centuries (Van der Spoel 1996a). Most of the species currently recognized as valid taxa were described before the beginning of the 20th century. The more recent contributions by Richter (1961, 1968, 1972, 1974, 1986, 1987), Taylor & Berner (1970), Van der Spoel (1972, 1976, 1996a), and Lalli & Gilmer (1989) provided an important amount of information on the heteropod taxonomy, biogeography, and biology. This information has allowed a reasonably well-developed taxonomy, with well-defined taxa at both the supraspecific and specific levels.

There are several sympatric taxa which are close in morphological terms; these were considered as sibling species by Van der Spoel (1976, 1996a). Some of them are the following pairs: *Atlanta lesueuri* – *A. oligogyra*, *A. peroni* – *A. gaudichaudi*, *A. inclinata* – *A. tokiokai*. Some of the subspecific forms described by Van der Spoel (1996a) may be sibling forms also.

Heteropods have been classified as part of the subclass Mesogastropoda although their relations within this taxon are still obscure (Van der Spoel 1996a). A problem related to the taxonomy of some of the heteropod taxa is the effect of acidic preservation media on their thin shells. Just as the pteropods, representatives of the Heteropoda secrete calcium carbonate shells. These are composed of aragonite, which is both lighter and more soluble than calcite. Shells can be dissolved after long- and medium-term storage in formalin solutions. It is recommended that these shelled taxa be promptly transferred into an ethanol solution (70–85%); however, this is a good medium for atlantids, but it is not as good for the gelatinous carinariid and pterotracheid heteropods.

Species richness and comparison: The number of species recognized before the works of Tesch (1909, 1949) exceeded 100. He revised the taxonomy of the group and recognized synonymies, thus reducing drastically the previous number of species. According to Van der Spoel (1976, 1996a) the number of heteropod species currently known is 35 plus several (7–10) subspecific forms. Richter and Seapy's (1999) figure is 36, without recognition of subspecific taxa; they stated that there are 33 species known to be distributed in the Pacific Ocean, 32 in the Indian Ocean, and 23–25 species in the Atlantic. Most species are true cosmopolitan forms (Richter & Seapy 1999); only 3–4 species seem to be confined to a particular ocean. This explains why figures of species richness are not significantly different when comparing both the Pacific and the Atlantic coasts of Costa Rica. There are few works dealing with this group in waters adjacent to Costa Rica; this information is analyzed separately for the Pacific and the Atlantic waters.

Pacific Ocean: In the California Current area, which is probably the most studied oceanic zone of the eastern Pacific, up to 15 species of heteropod mollusks have been recorded (Dales 1953, McGowan 1967; Seapy 1974; Seapy & Richter 1993). Sánchez-Hidalgo y Anda (1989) found four species in the western coast of the Baja California. Seapy (1990) recorded 13 species of Atlantidae from Hawaiian waters. In a survey of the Panama Province mollusks, Keen (1971) recorded 20 species of heteropods, but only 4 were effectively recorded in Panama, none off Costa Rica; however, at least the other 16 species are expected to occur in Costa Rican waters. Van der Spoel (1976) included the Pacific area of Costa Rica in the regional distribution of several species (Species List 29.2 is included on the CD-Rom). In the area off Ecuador we found an isolated report of heteropods by Cruz (1998), but the number or identity of the species he recorded is unavailable. South of Central America, Dall (1909) recorded two species of atlantiids off the coasts of Peru. There is only one published taxonomic work on heteropod mollusks in the Pacific waters of Costa Rica, from the oceanic upwelling system known as the Costa Rica Dome which includes oceanic waters of different countries: El Salvador, Nicaragua, and Costa Rica. Sánchez-Nava (1984) recorded seven species of Heteropoda from the analysis of samples collected during two oceanographic cruises carried out in this area in 1979. He found three species of Atlantidae, two of Pterotracheidae, and two of Carinariidae. He reported *A. lesueuri*, *Oxygyrus keraudreni*, and *Firoloida desmaresti* as the most abundant species in the Dome area. The same records were published later on by Sánchez-Nava and Segura-Puertas (1987). Castellanos-Osorio (personal observation) found species of *Atlanta* and a pterotracheid heteropod while examining zooplankton samples collected in Bahía Culebra.

Therefore, considering Van der Spoel's (1996a) and Richter and Seapy's (1999) heteropod species richness, the number of species effectively recorded in the Pacific coast of Costa Rica (16 spp.) versus the worldwide known species (35–36) represents about 44%. The figure is higher (ca. 60%) when the Costa Rican records are compared with the number of species known in the Pacific. Records of heteropod mollusks from Costa Rican waters of the Pacific Ocean are presented in Species List 29.2. Based on the literature, the Costa Rican Pacific heteropod fauna is expected to be much richer, the number of species dwelling in this area is probably higher than that presented herein.

Atlantic Ocean: Only a few studies have been published on the abundance and distribution of the heteropod mollusks in the northwestern tropical Atlantic (Castellanos & Suárez-Morales 2001). These studies include those of: Taylor and Berner (1970) in the northern Gulf of Mexico (16 species), Michel & Michel (1991) in the Straits of Florida (8 species), Gasca (1992) and Castellanos & Suárez-Morales (2001) in the western Caribbean Sea (6 species), and González & Princz (1979) in the Venezuelan Caribbean (5 species). Michel & Foyo (1976) recorded eight species of the three families in the Caribbean.

Records of heteropod mollusks from the Atlantic Ocean sum up to 25 species (Richter & Seapy 1999). In the Caribbean region the number of species is 12 (Michel & Michel 1991; Gasca 1992; Castellanos & Suárez-Morales 2001). Overall, the heteropod fauna of the Caribbean is dominated by *A. peroni*, *A. inclinata*, *Pterotrachea coronata*, *P. hippocampus*, and *F. desmaresti* (Michel & Foyo 1976; González & Princz 1979; Michel & Michel 1991). There are three records of heteropods in the Caribbean Sea off Costa Rica (Van der Spoel 1976, 1996a; Species List 29.1 is included on the CD-Rom).

Recommendations: Heteropod mollusks are an appealing group of the zooplankton community. Besides their beauty, they have been tested as bioindicators of oceanographic conditions (Cruz 1998) and they are part of the food available for several other zooplankters; the study of their adaptations to the pelagic realm is interesting. Taxonomic and distributional surveys of this group for both Costa Rican coasts should be supported in order to evaluate the distributional patterns of these mollusks and relate them to oceanographic processes in waters of Costa Rica.

Pteropoda

The informal term "pteropod" includes the shelled (Thecosomata) and the shell-less (Gymnosomata) pteropods (Lalli & Gilmer 1989). The name derives from the shape of part of the foot, which forms paired wing-like swimming structures (*pteron*). Only the thecosomes (suborders Euthecosomata and Pseudothecosomata) will be included in this part. As part of the holoplankton, the pteropod mollusks spend their entire life cycle in the water column. They are the most widely distributed pelagic mollusks. Pteropods can be abundant and frequent in zooplankton samples of neritic and oceanic environments. For instance, *Creseis acicula*, a common species in coastal-neritic areas has been reported to form highly dense swarms.

As a group, they represent an adaptative tendency towards the reduction of the molluscan shell to dwell in the water column (Van der Spoel 1996b). Thecosomata are mainly herbivorous forms, and feed upon microplankton and diatoms (Lalli &

Gilmer 1989). Most species dwell in surface waters, some mesopelagic forms exhibit large migration patterns from depths of 1,000–2,000 m, others are bathypelagic. Some species are useful as hydrological, geological, and ecological indicators (Biekart 1989).

Status of taxonomical knowledge: Most of the species currently recognized as valid taxa were described at the beginning of the 20th century. It is known that several nominal species that were described are now considered to be subspecies or formae (Van der Spoel 1996b). With the application of geographic speciation concepts, morphologic variations within several taxa had to be reevaluated (Van der Spoel 1967). Therefore, species with a wide latitudinal distribution tend to reflect the climatic influence, and frequently, species are restricted to a water mass. It has been necessary to revise most of the original species and examine specimens from different parts of the world in order to determine the morphological variation and the real distributional ranges of the subspecific forms (Van der Spoel & Boltovskoy 1981; Van der Spoel 1996b).

One of the main problems related to the complex taxonomy of some of the pteropod taxa is the effect of acidic preservation media on their thin shells. Shells are one of the main structures used in the taxonomy of this group and they can be dissolved after long- and medium-term storage in formalin solutions. In samples stored for more than 3–4 years in formalin preservation it is common that only the soft parts of both pteropod and heteropod mollusks remain. While it is possible to identify many of these species from the shape and features of their soft parts alone, it is frequently difficult, if not impossible, to provide reliable identifications in such conditions (McGowan 1967). It is recommended that these shelled taxa be promptly transferred into an ethanol solution (70%); well-preserved samples can be identified after 200 years.

The analysis of the shell structure and shape can be made easily with a stereomicroscope. However, the current taxonomic standards and the discovery of new characters to separate species of pteropod mollusks require the observation of smaller structures.

Species richness: Currently, up to 188 species and subspecific forms of pteropod mollusks have been recognized worldwide. A key for the identification of all these species was published recently by Van der Spoel (1996b). However, Lalli and Gilmer (1989) did not recognize subspecific forms and they considered several species invalid; thus, their list is much reduced when compared with that of Van der Spoel (1996b), including just 94 valid species.

Bé and Gilmer (1977) stated that the general composition of the circumglobal pteropod mollusk assemblages in the Atlantic and Pacific oceans is quite similar between 40°N and 40°S. Therefore, it is not expected that figures of species richness will be significantly different when comparing both the Pacific and the Atlantic coasts of Costa Rica. There are several works dealing with this group in tropical waters adjacent to Costa Rica; this information is analyzed for the Pacific and the Atlantic waters separately.

Pacific Ocean: In the eastern Pacific Ocean, the early surveys by Meisenheimer (1905), who presented five records, and later by Tesch (1948) and Tokioka (1955) constitute the basic information sources for this group in this region. In the California Current area, which is probably the most studied area of the eastern Pacific, McGowan (1967, 1968) reported 29 species of pteropod mollusks. Sánchez-Hidalgo (1989) found 11 species in the western coast of the Baja California, and Fernández-Alamo (1989) recorded 13 species of Pteropoda from the gulf. Montoya (1983) recorded *Cavolinia longirostris* from the area of Isla del Coco, Costa Rica. In the Costa Rica Dome, Sánchez-Nava (1984) and Sánchez-Nava & Segura-Puertas (1987) recorded ten species from the analysis of two oceanographic cruises carried out in 1979. He found *C. acicula clava*, *Limacina inflata*, *Hyalocylis striata*, *Desmopterus papilio* (as *D. pacificus*), and *Diacavolinia longirostris longirostris* as the most abundant species in the Dome area. Beese (1995) recorded *C. virgula* from samples collected in the Golfo Dulce during a German-Costa Rican oceanographic expedition. Keen (1971) found nine species of pteropods. Rodríguez *et al.* (2003) recorded *D. longirostris longirostris* in Costa Rican waters. Van der Spoel (1996b) found *Diacria schmidti schmidti* in the Pacific waters of Costa Rica. Vicencio-Aguilar & Fernández-Álamo (1996) reported the species previously recorded by Sánchez-Nava (1984) and Sánchez-Nava & Segura-Puertas (1987). South of Central America, in Peruvian Pacific waters, Dall (1909) recorded 18 species of pteropods. The number of species recorded from the Pacific of Costa Rica is 13. Therefore, considering Van der Spoel's (1996b) species list (190 species and subspecies), the percentage of species effectively recorded in the Pacific coast of Costa Rica versus the worldwide known species is low: 6,8%. The figure is much higher (37%) with reference to the species richness known in the eastern Pacific. It must be noted here that the genus *Diacria* contains taxonomically complex species groups in which *D. quadridentata* is an Indo-Pacific form, not present in the Atlantic Ocean (see Van der Spoel & Dadón 1999; Species List 29.2), where a closely related form, *D. danae* is widely distributed.

Atlantic Ocean: in the Northwestern Atlantic Ocean pteropod mollusks have been surveyed mainly in the Gulf of Mexico and along the eastern coast of the United States (Wormelle 1962; Chen & Hillman 1970; Matsubara 1975; Michel & Michel 1991; Suárez-Morales & Gasca 1992). Studies have been carried out in the Caribbean Sea also (Wells 1975, 1976; Haagensen 1976; Velázquez & Lozano 1987; Gasca & Suárez-Morales 1992; Suárez-Morales & Gasca 1998). A general overview of the pteropod records in the Gulf of Mexico and adjacent zones was presented by Suárez-Morales (1994).

Records of thecosomatous pteropods from the tropical northwestern Atlantic yield up to 51 species and subspecific forms (Wells 1975, 1976; Haagensen 1976; Michel & Michel 1991; Suárez-Morales 1994). Most of these records are from the Gulf of Mexico and the central and eastern portions of the Caribbean Sea. A total of 17 species and subspecific forms were recorded by Suárez-Morales & Gasca (1998) in the western Caribbean. In a bay-type coastal system of the Mexican Caribbean, only six species have been recorded during a year-round sampling program (Gasca & Suárez-Morales 1992). The survey by Haagensen (1976) included one sampling site off the Caribbean coasts of Costa Rica; at this station (sta. 11 of cruise 6811) he recorded up to seven species of thecosomatous pteropods plus at least two species of the pseudothecosomatous *Corolla* (Species List 29.1). Also, Robinson & Montoya (1987) recorded *Cavolinia longirostris* from this area.

Only a few pteropods are distributed near the coastal areas. In contrast, a much higher number of taxa is expected to be found in oceanic areas of the Caribbean coast of Costa Rica.

Overall, the pteropod fauna of tropical surface waters of the Caribbean is dominated by a group of several species (Haagensen 1976), which include *C.s acicula*, *L. trochiformis*, *C. virgula*, and *Cavolinia longirostris*. However, in the western part of the Caribbean the main species group is even more similar to that reported by Suárez-Morales & Gasca (1992) from the southern Gulf of Mexico (*C. acicula acicula*, *L. inflata*, *L. trochiformis*, *C. longirostris longirostris*). These four species are of course expected to be found as abundant or dominant forms in the neritic and oceanic areas of the Atlantic side of Costa Rica. In fact, Haagensen (1976) reported *L. inflata*, *L. trochiformis*, and *C. acicula* as the most abundant species off Costa Rica.

Most species have been reported as common in neritic and oceanic waters; however, *C. acicula acicula* is considered one of the few species able to penetrate farther into coastal environments than any other thecosome. In the Caribbean the most abundant populations have been recorded near the shore (Haagensen 1976). In a coastal-estuarine system of the western Caribbean, species such as *C. acicula acicula*, *L. trochiformis*, and *C. longirostris longirostris* have been recorded well inside this system (Suárez-Morales & Gasca 1992). These species are thus expected to be found also in the coastal systems of Costa Rica.

Recommendations: Pteropod mollusks are an interesting group of the zooplankton community. They have been known for long as excellent indicators of water masses and oceanographic/environmental conditions (Chen & Hillman 1970; Biggs *et al.* 1997) and they are part of the food available for several fish species with commercial value (Lewis 1962) whales and marine turtles. These are two strong reasons to start taxonomic and distributional surveys of this group in both Costa Rican coasts. Of course, these potential studies should have a sound hydrological background in order to evaluate the distributional patterns of the pteropods and relate them to oceanographic processes in waters of Costa Rica.

A reliable identification of this group is an essential step to follow either biogeographical or oceanographic interpretations. This work is supported by having well-preserved specimens with complete shells. Hence, it is recommended to transfer the specimens collected to 70% ethanol after fixation with buffered formalin. This will allow a better preservation of the aragonite shells for taxonomical examination. A reference collection should be formed for consulting by regional specialists.

The study of these mollusks is always satisfactory; even in areas which have been surveyed previously, new records are constantly being reported. Suárez-Morales & Gasca (1998) found at least three species known in the southern Gulf of Mexico but not previously recorded in the western Caribbean Sea.

Specialists

There are few specialists working on these taxa of pelagic mollusks and currently none is studying the Costa Rican fauna. Dr. S. Van der Spoel (University of Amsterdam), who produced benchmark studies about this group in the last three decades, is now retired. This list includes active colleagues familiar with these mollusks in tropical regions of the Atlantic and the Pacific adjacent to Costa Rica

Heteropoda

IVÁN CASTELLANOS-OSORIO: El Colegio de la Frontera Sur (ECOSUR), Unidad Chetumal, Chetumal, Quintana Roo, Mexico – Caribbean Sea and Gulf of Mexico. ivancast@ecosur.mx

GOTTHARD RICHTER: Forschunginstitut Senckenberg, Frankfurt a.M., Germany – Worldwide. sbaszio@sng.uni-frankfurt.de

MARCO SÁNCHEZ-HIDALGO Y ANDA: CICIMAR-IPN, La Paz, Baja California Sur, Mexico – Baja California and Mexican Tropical Pacific. mhidalgo@ipn.mx

ROGER SEAPY: Dept. Biological Science, California State University at Fullerton, California, USA – Worldwide. rseapy@fullerton.edu

Pteropoda

JOSÉ DADÓN: Depto. Ciencias Biológicas, Fac. Ciencias Exactas y Naturales, Universidad de Buenos Aires – Southwestern Atlantic. dadon@bg.fcen.uba.ar

MARCO SÁNCHEZ HIDALGO Y ANDA: CICIMAR-IPN, La Paz, Baja California Sur, Mexico – Baja California and Mexican Tropical Pacific. mhidalgo@ipn.mx

EDUARDO SUÁREZ-MORALES: El Colegio de la Frontera Sur (ECOSUR), Unidad Chetumal, Chetumal, Quintana Roo, Mexico – Caribbean Sea and Gulf of Mexico. esuarez@ecosur.mx

Collections

Heteropoda: There are no collections of heteropod mollusks in Costa Rica. Probably the geographically closest collection is that of El Colegio de la Frontera Sur (ECOSUR), one which contains 22 specimens that represent five genera and six species, most of them probably present in Costa Rican waters. This collection is formed by specimens from coastal, neritic, and oceanic waters of the western

Caribbean Sea. The Laboratorio de Invertebrados of the Facultad de Ciencias, UNAM in Mexico City holds a large collection of zooplankton samples from the Costa Rica Dome; the pelagic mollusks (including Heteropoda) were studied by Sánchez-Nava (1984), and specimens, either sorted out or unsorted are presumed to be deposited in this collection. Also, a large collection of Caribbean zooplankton is deposited in the Instituto de Ciencias del Mar y Limnología of the Universidad Nacional Autónoma de México at Puerto Morelos, Quintana Roo, Mexico. Heteropods of those samples have not been studied. The Centro de Investigación en Ciencias del Mar y Limnología (CIMAR) in Costa Rica holds a large collection of zooplankton samples collected mainly in several systems along the Pacific coast; the pelagic mollusks remain unstudied.

Pteropoda: There are no collections of pteropod mollusks in Costa Rica. Probably the closest collection is that of El Colegio de la Frontera Sur (ECOSUR) which contains 14 specimens, which represent 6 genera and 11 species, several of them present in Costa Rican waters. Most of this collection is formed by specimens from coastal, neritic, and oceanic waters of the western Caribbean Sea. The Laboratorio de Invertebrados of the Facultad de Ciencias, UNAM in Mexico City holds a large collection of zooplankton samples from the Costa Rica Dome; the pelagic mollusks were studied by Sánchez-Nava (1984), and the specimens are presumed to be deposited in this collection. The CIMAR in Costa Rica holds a large collection of zooplankton samples collected mainly in several systems along the Pacific coast, but the pelagic mollusks have not been analyzed.

Acknowledgments We thank the editors for the opportunity to participate in this effort. Leonora Rodríguez Sevilla kindly provided us some of her records of heteropod mollusks. We also thank Álvaro Morales-Ramírez for allowing us to look at his samples during a visit to the CIMAR (Universidad de Costa Rica). As well as the comments of two anonymous reviewers that greatly improved this contribution.

References

Abbott RT (1974) American Seashells. Van Nostrand Reinhold, New York, 663 p

Bé AW, Gilmer RW (1977) A zoogeographic and taxonomic review of Euthecosomatous Pteropoda. *In*: Ransay ATS (ed) Micropaleontology. Academic Press, London, pp 733–808

1. Beese D (1995) Zur Rekonstruktion des Paläoklimas im Golfo Dulce, Costa Rica, anhand der stabilen Isotopenverhältnisse in Foraminiferen- und Pteropodenschalen. Diplomarbeit Teil 2 (Laborarbeit), Fachbereich Geowissenschaften, Univ. Bremen, 41 p

Biekart JW (1989) Euthecosomatous pteropods as paleo-hydrological and paleoecological indicators in a Tyrrhenian deep sea core. Palaeogeogr, Palaeoclim, Palaeoecol 71:205–224

Biggs DC, Zimmerman RG, Gasca R, Suárez-Morales E, Castellanos I (1997) Note on plankton and cold-core rings in the Gulf of Mexico. Fish Bull 95:369–375

Castellanos I, Suárez-Morales E (2001) Heteropod molluscs (Carinariidae and Pterotracheidae) of the Gulf of Mexico and the western Caribbean Sea. An Inst Biol, Univ Nac Autón Méx, Ser Zool 72:221–232

Chen C, Hillman N (1970) Shell bearing pteropods as indicators of water masses off Cape Hatteras, North Carolina. Bull Mar Sci Gulf Carib 20:350–367

Cruz M (1998) Gastrópodos planctónicos (Pteropoda y Heteropoda) como bioindicadores de los eventos "El Niño" 1992 y 1997–98 en la estación fija "La Libertad", Ecuador. Acta Oceanogr Pacíf 9:12–18

Dales RP (1953) The distribution of some heteropod mollucks off the Pacific coast of North America. Proc Zool Soc London 122:1007–1015

Dall WH (1909) Report on a collection of shells from Peru, with a summary of the littoral marine Mollusca of the Peruvian Zoological Province. Proc US Nat Mus 37:147–294

Fernández-Alamo MA (1989) Una aportación al conocimiento del zooplancton del Golfo de California. Mem. VII Simp Int Biol Mar: 17–22

Gasca R (1992) Heterópodos (Mollusca: Gastropoda; Atlantidae) de la Bahía de la Ascensión, Reserva de la Biosfera de Sian Ka'an, Quintana Roo, México. *In*: Navarro D, Suárez-Morales E (eds) Diversidad Biológica en la Reserva de la Biosfera de Sian Ka'an, Quintana Roo, México. II. CIQRO/SEDESOL, México, pp 115–122

Gasca R, Suárez-Morales E (1992) Pterópodos (Mollusca: Gastropoda; Thecosomata) de la Bahía de la Ascensión, Reserva de la Biosfera de Sian Ka'an, Quintana Roo, México. *In*: Navarro D, Suárez-Morales E (eds) Diversidad Biológica en la Reserva de la Biosfera de Sian Ka'an, Quintana Roo, México. II. CIQRO/ SEDESOL, México, pp 123–127

González LA, Princz D (1979) Moluscos heterópodos y pterópodos de la zona central del Caribe venezolano. Bol Inst Oceanogr, Univ Oriente 18:99–107

2. Haagensen DA (1976) Caribbean zooplankton. Part II. Thecosomata. Office of Naval Research, Dept. Navy, Washington, DC, pp 551–712

3. Keen AM (1971) Sea shells of tropical West America: marine mollusks from Baja California to Peru. Stanford University Press, Stanford,1064 p

Lalli CM, Gilmer RW (1989) Pelagic snails. The biology of holoplanktonic Gastropod mollusks. Stanford University Press, Stanford, 259 p

Lewis JB (1962) Notes on the Barbados 'sea egg'. J Barbados Mus Hist Soc 30:1–3

Matsubara JA (1975) Sistemática, distribución, abundancia y relaciones ambientales de los pterópodos tecosomados de la Bahía de Campeche, México. Tesis Prof., Fac. Ciencias, UNAM, México, 52 p

McGowan JA (1967) Distributional atlas of pelagic mollusks in the California Current region. CalCOFI Atlas 6:1–218

McGowan JA (1968) The Thecosomata and Gymnosomata of California. Veliger 3:103–129

Meisenheimer J (1905) Pteropoda. Wiss Ergebn dt Tiefsee-Exped "Valdivia" 9:1–314

Michel HB, Foyo M (1976) Caribbean zooplankton. Part. 1. Siphonophora, Heteropoda, Copepoda, Euphausiacea, Chaetognatha and Salpidae. Off. Naval Res., Washington DC, 549 p

Michel HB, Michel JF (1991) Heteropod and thecosome (Mollusca: Gastropoda) macroplankton in the Florida Straits. Bull Mar Sci 49:563–574

Richter G (1961) Die Radula der Atlantiden (Heteropoda, Prosobranchia) und ihre Bedeutung für die Systematik und Evolution der Familie. Zeitschr Morphol Ökol Tiere 50:163–238

Richter G (1968) Heteropoden und Heteropodenlarven im Oberflächenplankton des Golfes von Neapel. Publ Staz Zool Napoli 36:347–400

Richter G (1972) Zur Kenntnis der Gattung *Atlanta* (Heteropoda: Atlantidae). Arch Molluskenkd 102:85–91

Richter G (1974) Die Heteropoden der "Meteor" Expedition in der Indischen Ozean 1964/65. "Meteor" Forschungsergeb. (D) 17:55–78

Richter G (1986) Zur Kenntnis der Gattung *Atlanta*, 2. *Atlanta lesueuri* Souleyet und *Atlanta oligogyra* Tesch (Prosobranchai: Heteropoda). Arch Molluskenkd 117:19–31

Richter G (1987) Zur Kenntnis der Gattung *Atlanta*, 3. *Atlanta inflata, Atlanta helicinoides, A. echinogyra* und *Atlanta plana* (Prosobranchai: Heteropoda). Arch Molluskenkd 117:177–201

Richter G, Seapy RR (1999) Heteropoda. *In*: Boltovskoy D (ed) South Atlantic Zooplankton. Backhuys, Leiden, The Netherlands, pp 621–647

Ríos EC (1985) Seashells of Brazil. Fundaçao Cidade do Rio Grande, Fundaçao Universidade do Rio Grande, Museu Oceanográfico, Rio Grande, RS, 328 p

4. Rodríguez SL, Vargas R, Cortés J (2003) Biodiversidad marina de Costa Rica: gastrópodos de la costa Caribe (Mollusca: Gastropoda). Rev Biol Trop 51(Suppl 3):305–399
Sánchez-Hidalgo y Anda M (1989) Gasterópodos holoplanctónicos de la costa occidental de Baja California Sur, en mayo y junio de 1984. Inv Mar CICIMAR 4:1–14
5. Sánchez-Nava S (1984) Sistemática, distribución y abundancia de los moluscos pelágicos (Gastropoda: Pteropoda y Heteropoda) en el Domo de Costa Rica. Tesis Prof., Fac. Ciencias, UNAM, 180 p
6. Sánchez-Nava S, Segura-Puertas L (1987) Los moluscos pelágicos (Gastropoda: Heteropoda y Pteropoda) recolectados en el domo de Costa Rica y regiones adyacentes. Mem Soc Mex Malacol 3:232–240
Seapy RR (1974) Distribution and abundance of the epipelagic mollusk *Carinaria japonica* in waters off southern California. Mar Biol 24:243–250
Seapy RR (1987) The heteropod fauna of oceanic waters off Hawaii. Western Soc Malacol Ann Rep 19:9–12
Seapy RR (1990) The pelagic Family Atlantidae (Gastropoda: Heteropoda) from Hawaiian waters: a faunistic survey. Malacologia 32:107–130
Seapy RR, Richter G (1993) *Atlanta californiensis*, a new species of atlantid heteropod (Mollusca: Gastropoda) from the California Current. Veliger 36:389–398
Suárez-Morales E (1994) Distribución de los pterópodos (Gastropoda: Thecosomata y Pseudothecosomata) del Golfo de México y zonas adyacentes. Rev Biol Trop 42:523–530
Suárez-Morales E, Gasca R (1992) Pterópodos (Gastropoda: Thecosomata y Pseudothecosomata) del estrato superficial (0–50 m) del sur del Golfo de México. An Inst Cienc Mar Limnol, UNAM 19:199–207
Suárez-Morales E, Gasca R (1998) Thecosome pteropod (Gastropoda) assemblages of the Mexican Caribbean Sea (1991). Nautilus 112:43–51
Taylor D, Berner L (1970) The Heteropoda (Mollusca: Gastropoda). *In*: Pequegnat WE, Chace FA (eds) Contributions on the Biology of the Gulf of Mexico. Texas A&M Univ. Oceanogr. Stud. 1, pp 231–244
Tesch JJ (1909) Systematic monograph of the Atlantidae (Heteropoda) with enumeration of the species in the Leiden Museum. Not Leiden Mus 30:1–30
Tesch JJ (1948) The thecosomatous pteropods. II. The Indo-Pacific. Dana Rep 30:1–44
Tesch JJ (1949) Heteropoda. Dana Rep 34:1–55
Tokioka T (1955) On some plankton animals collected by the Syunkoto Maru in May-Jun, 1954. Thecosomattous Pteropoda. Publ Seto Mar Biol Lab 5:59–74
Van der Spoel S (1967) Euthecosomata, a group with remarkable developmental stages (Gastropoda, Pteropoda). Gorichem. J. Noorduijn, 375 p
Van der Spoel S (1972) Notes on the identification and speciation of Heteropoda (Gastropoda). Zool Meded, Leiden 47:545–560
7. Van der Spoel S (1976) Pseudothecosomata, Gymnosomata and Heteropoda (Gastropoda). Bohn, Scheltema and Holkema, Utrecht,484 p
8. Van der Spoel S (1996a) Heteropoda. *In*: Gasca R, Suárez-Morales E (eds) Introducción al Estudio del Zooplancton Marino. ECOSUR/ CONACYT, México, pp 407–457
9. Van der Spoel S (1996b) Pteropoda. *In*: Gasca R, Suárez-Morales E (eds) Introducción al Estudio del Zooplancton Marino. ECOSUR/ CONACYT, México, pp 459–528
Van der Spoel S, Boltovskoy D (1981) Pteropoda. *In*: Boltovskoy D (ed) Atlas del zooplancton del Atlántico Sudoccidental y métodos de trabajo con el zooplancton marino. Publ. Esp. INIDEP, Mar del Plata, Argentina, pp 493–531
10. Van der Spoel S, Dadón JR (1999) Pteropoda. *In*: Boltovskoy D (ed) South Atlantic Zooplankton. Backhuys, Leiden, The Netherlands, pp 650–706
Velázquez LE, Lozano F (1987) Algunos pterópodos Euthecosomata (Mollusca: Opistobranchia) de la Bahía de Gaira, Santa Marta (Caribe Colombiano). An Inst Inv Punta Betín 17:81–85
11. Vicencio-Aguilar ME, Fernádez-Alamo MA (1996) Zooplancton del Domo de Costa Rica: taxonomía y biogeografía. Rev Biol Trop 44:631–642

Wells FE (1975) Comparison of euthecosomatous pteropods in the plankton and sediments off Barbados, West Indies. Proc Malac Soc London 41:503–509
Wells FE (1976) Seasonal patterns of abundance and reproduction of euthecosomatous pteropods off Barbados, West Indies. Veliger 18:241–248
Wormelle RL (1962) A survey of the standing crop of plankton of the Florida Current. VI. A study of the distribution of the pteropods of the Florida Current. Bull Mar Sci Gulf Carib 12:95–136

Part 30
Benthic Opisthobranchs

Yolanda E. Camacho-García

Aeolidiella indica from Pacific Costa Rica, with parasitic copepods (Photo: Y. Camacho-García)

Abstract National and international experts are working together to complete the inventory of benthic opisthobranchs in Costa Rica. The most extensive and comprehensive collection of these organisms (including holotypes and paratypes) from the country is housed at the Museo de Zoología at the Universidad de Costa Rica. A total of 250 species (127 only found in the Pacific, 111 only found in the Caribbean, and 12 species common to both coasts) are reported for Costa Rica. Seven species

Y.E. Camacho-García
Museo de Zoología, Escuela de Biología, Universidad de Costa Rica, San Pedro de Montes de Oca, 11501-2060 San José, Costa Rica.
e-mail: ycamacho_99@yahoo.com

from the Pacific and nine from the Caribbean are reported in this part for the first time. The number of species represented in Costa Rica is compared to other adjacent faunistic regions. The distribution of the species is discussed in a biogeographical context. Recommendations are given for future inventories in adjacent areas.

Introduction

Opisthobranch molluscs, also known as sea slugs, are some of the most colorful and beautiful marine organisms found along the Pacific and Caribbean coasts of Costa Rica. With more than 6,000 documented species worldwide, sea slugs are also one of the most morphologically and ecologically diverse groups of organisms in the world's oceans.

Sea slugs are almost exclusively marine animals which can be found from the intertidal zone down to a depth of 4,000 m (Valdés 2001). These molluscs are not commonly found in extreme environments; however, one species has been described from hydrothermal vents (Valdés & Bouchet 1998). Many species are cryptically colored and resemble the substrate on or in which they are normally found. Others have bright colors, which advertise potentially toxic defense chemicals to predators. Specialized structures such as radula, jaw elements, gills, rhinophores (and other sensory organs such as cephalic tentacles), and other appendages are all characteristics of opisthobranchs, but they may or may not be present in the same clade. These molluscs exhibit an evolutionary trend towards the reduction, internalization, or complete loss of the shell (Thompson 1976; Gosliner 1981b).

Most opisthobranchs are simultaneous hermaphroditic animals, and partners interchange sperm during copulation (Gosliner 1981b). In the wild, they typically live for 1 year or less, but in captivity some may live up to 4 years (Camacho-García *et al.* 2005). They feed on a variety of sessile organisms such as sponges, algae, anemones, hydroids, and bryozoans (Thompson 1976).

Opisthobranchs are important study animals in different fields of biology. For example, several species (i.e., *Aplysia*) have been used in neurobiological studies because of the large size of their individual nerve cells (Kandel 1979). Additionally, the secondary metabolites used by opisthobranchs as chemical defenses are being surveyed for potential pharmaceutical use (Yamada & Kigoshi 1997).

The highest diversity of opisthobranch molluscs occurs in the tropical regions, especially in the diversity triangle formed by Papua New Guinea, the Philippines, and Indonesia (Gosliner 1992). The number of species decreases dramatically in temperate and cold waters, but a few species have been recorded from polar environments in Antarctica and the Arctic Ocean (Camacho-García *et al.* 2005). For additional information on their ecology, morphology, and systematics, see Todd (1981), Gosliner (1994a), and Wägele & Willan (2000), respectively.

Until recently, the opisthobranch diversity of Costa Rica was poorly known. Today, this group of marine molluscs of the Pacific and Caribbean coasts of Costa Rica is better known. Houbrick (1968) and Robinson & Montoya (1987) provided a few citations of opisthobranchs among a general list of the marine molluscs from the Caribbean coast of Costa Rica. However, the most relevant faunal studies in this region have been conducted recently by Ortea & Espinosa, who documented about 80 species between 1995 and 2003 (Ortea 2001a, b; Ortea & Espinosa 2000a, b, 2001a, b, c). In the Pacific, the following studies provide taxonomic revisions of groups, new records, or descriptions of new species: Bertsch *et al.* (1973), Bertsch & Ferreira (1974), Bertsch (1978a, b), Skoglund (1991, 2002), Mulliner (1993), Høisæter (1998), Camacho-García & Ortea (2000), Valdés & Camacho-García (2000), García & Troncoso (2001), Atjai *et al.* (2003), Camacho-García & Valdés (2003), Fahey & Gosliner (2003), Valdés & Camacho-García (2004), Camacho-García & Gosliner (2004), Gosliner *et al.* (2004), and Camacho-García & Valdés (2005).

This part focuses on the benthic opisthobranchs found in Costa Rica and does not include information for pelagic, interstitial, and other groups of opisthobranchs. However, it does include a list of higher heterobranchs that were once considered cephalaspidean opisthobranchs. The National Biodiversity Institute (INBio) in Costa Rica initiated a substantial collecting effort throughout the country from 1995 to 2004. The entire collection of opisthobranchs (as well as other molluscs) was recently donated to the Museo de Zoología at the Universidad de Costa Rica (UCR) after INBio closed its Malacology Department in September 2004. The majority of the information on the presence of the species provided in this part is the result of this effort and current sampling conducted as part of my work as the curator of molluscs of the Museo de Zoología at UCR. Other data have been added from the literature. Species lists are intended to contain all the species of benthic opisthobranchs found in Costa Rica (Species Lists 30.1 and 30.2 which are included on the CD-Rom). Classifications reflect current systematic arrangements and the present understanding of synonymy (Rudman 1998; Wägele & Willan 2000; Valdés 2002a).

Results

There are 250 species included in this part; 127 are found only on the Pacific side of Costa Rica, 111 species are restricted to the Caribbean (Table 30.1), and 12 are common to both coasts. Seven records from the Costa Rican Pacific and nine from the Caribbean are reported for the first time.

Skoglund (1991) reported 187 species of benthic opisthobranchs for the Panamic Province. Later, Skoglund (2002) indicated a total of 194 species for this biogeographic province, increasing the number of species by about 4% (Table 30.1). This is the currently accepted total count for the province. Thus, 139 total species reported in this part for the Pacific represents 72% of all the known benthic species in the Panamic Province. Forty of these species (29%) are new to science and have been described in the past few years or are currently being described.

Table 30.1 Number of opisthobranch species in Costa Rica and adjacent faunistic regions

Taxa	Panamic Province[a]	Costa Rican Pacific	Caribbean[b]	Costa Rican Caribbean
Cephalaspidea	44	22[c]	37	22[c]
Anaspidea	11	8	15	9
Sacoglossa	15	17[d]	43	28
Notaspidea	8	7	8	3
Nudibranchia	116	83	129	61
Doridacea	64	42	65	27
Cryptobranchia	38	29	42	15
Porostomata	7	3	6	4
Suctoria	7	2	0	1
Nonsuctoria	12	8	17	4
Dendronotacea	10	9	19	14
Arminacea	4	5	3	1
Aeolidacea	38	29	42	19
Totals	194	139[e]	232	123[e]

[a] Data from Skoglund (2002)
[b] Data from Gosliner (1992)
[c] Number includes higher heterobranchs for comparison purposes only
[d] Numbers are higher than those of the Panamic Province because they include new undescribed species
[e] Number includes species common to both coasts

Fig. 30.1 *Hypselodoris ruthae* from the Caribbean coast of Costa Rica (Photo: Y. Camacho-García)

There are 232 species reported for the entire Caribbean (Gosliner 1992). Therefore, 123 total species reported here represents 53% of all the known Caribbean species. Twenty-five of these (20%) have been described since 1999, or are new to science. One common representative of the opisthobranch fauna of the Caribbean coast of Costa Rica is shown in Fig. 30.1.

Biogeography

Costa Rica has two clearly differentiated opisthobranch faunas: one in the Caribbean and the other in the Pacific. The Pacific coast of Costa Rica belongs to the Panamic Province, which extends from 25° N (Baja California, Mexico) to 6° N (northern Peru). This region is characterized by reef habitats distributed along a continuous continental coastline (with a few offshore islands) and an open basin with unstable currents, cold water intrusions, and extensive mixing due to El Niño events and other Pacific phenomena, all of which influence the distributional patterns of organisms (Coates 1997).

It is likely that the same species composition will be found throughout the Panamic Province, but there are some regional variations, especially near the boundaries of the region. Variations are more pronounced during abnormal years such as those with El Niño events (Coates 1997).

Within the two biogeographical areas, the Caribbean supports a slightly greater diversity of opisthobranchs, as indicated in Table 30.1. For example, Gosliner (1992) has shown that the Caribbean has a much higher diversity of sacoglossans than the Pacific. In addition, nudibranch dorids and aeolids are more abundant in the Caribbean than in the Pacific (although in Costa Rica, the opposite trend is seen: the Pacific coast shows a higher number of nudibranch dorids and aeolids than the Caribbean; see Table 30.1). However, overall, there is little difference in species diversity between these regions.

Dorids are normally divided into four groups, according to ecological specialization: cryptobranchs and porostomes, which feed on sponges; and non-suctorians and suctorians, which feed on bryozoans and tunicates (Gosliner 1992). In both the Pacific and the Caribbean, the sponge-eaters dominate.

There has not been any contact between opisthobranchs from the different sides of Costa Rica since the closure of the Isthmus of Panama about 3.1 million years ago (Coates 1997). Yet, there are several species in common between the Pacific and the Caribbean coasts. Most of these species, such as *Stylocheilus striatus* and *Aplysia parvula*, are also found worldwide in tropical areas. But a few, such as *Flabellina marcusorum* and *Glossodoris sedna* (Fig. 30.2), are restricted to these two bodies of water. There are also clear sister relationships in many other groups of opisthobranchs that suggest a common ancestry for many lineages.

There are several circumtropical species present in Costa Rica: a sacoglossan (*Lobiger souverbii*; Fig. 30.3), a notaspidean (*Berthella stellata*), three aolidaceans (*Spurilla neapolitana*, *Aeolidiella indica*, and *Phidiana lascrucensis*), five anaspideans (*A. dactylomela*, *Dolabrifera dolabrifera*, *S. striatus*, *Bursatella learchii*, and *Dolabella auricularia*), and one dendronotacean (*Crosslandia daedalli*). All of these except *B. learchii* and *P. lascrucensis* have been found on both coasts of Costa Rica, along with *Polybranchia viridis*, *A. parvula*, *Cadlina evelinae*, *F. marcusorum*, *Pleurobranchus areolatus*, and *Navanax aenigmaticus*.

Although the nudibranch *G. sedna* and the aeolid *Fionna pinnata* have been reported for the Caribbean coast of Panama, they have not yet been found in the Costa Rican Caribbean.

Fig. 30.2 *Glossodoris sedna* with eggs; Isla Murciélago, Pacific Costa Rica (Photo: Ingo S. Wehrtmann)

Fig. 30.3 *Lobiger souverbii* from Pacific Costa Rica (Photo: Y. Camacho-García)

It is possible that species found in Costa Rica have a geographic range much larger than originally thought. For example, the species *Cyerce orteai*, which was only known from the Costa Rican Pacific, has recently been found in Hawaii and Mexico (Camacho-García *et al.* 2005). The presence of many species that appear to occur exclusively in Costa Rica could be due to a lack of adequate sampling in other areas of the Panamic region and the Caribbean.

In the Caribbean, Espinosa & Ortea (2001) indicated that the area from Cahuita to Gandoca, based on the amount of new species of gastropods found, contained high endemism and biological speciation. They also suggested that the Sixaola River could be a biogeographical barrier for several species that were not present in the

Panamanian Caribbean coast. In order to test this hypothesis, more sampling should be conducted in this area and in more localities along the Caribbean coast of Costa Rica as well as in adjacent areas in Central America, such as Bocas del Toro, Panama, and Nicaragua. Other marine taxa should be included to test whether this area is a true area of high endemism and not an artifact of sampling.

While human activities always have the potential to alter the natural distributional patterns of opisthobranchs, there are no known artificially introduced species in Costa Rica.

Specialists

The following specialists have worked with opisthobranchs in Costa Rica and other countries in Central America:

YOLANDA E. CAMACHO-GARCÍA: Museo de Zoología, Escuela de Biología, Universidad de Costa Rica – Panamic Province, Caribbean. ycamacho_99@yahoo.com

JOSÉ ESPINOSA: Instituto de Oceanología de la Habana, Cuba – Instituto de Oceanología, Ave. 1ra No. 18406 e/184 y 186, Rpto. Flores, Playa, C. Habana, CP. 12100, Cuba-Caribbean

TERRENCE GOSLINER: California Academy of Sciences (CAS), San Francisco, U.S.A. – Indo-Pacific, Panamic Province. tgosliner@calacademy.org

ALICIA HERMOSILLO: Universidad de Guadalajara, México – Panamic Province. gueri25@hotmail.com

JESÚS ORTEA: Universidad de Oviedo, Spain – Western Atlantic, Caribbean. kankel@teleline.es

ÁNGEL VALDÉS: Department of Biological Sciences, California State Polytechnic University, U.S.A. – Worldwide. aavaldes@csupomona.edu

Collections

The Instituto Nacional de Biodiversidad (INBio) closed its Malacology Department in September 2004. The collections of opisthobranchs and other marine molluscs are now located in the Museo de Zoología at the Universidad de Costa Rica. The opisthobranch collection is composed of a total of 5,440 specimens from both coasts, where 4,632 specimens (1,475 taxonomic lots) are from the Pacific and 811 specimens (316 taxonomic lots) are from the Caribbean (information updated to June 2006).

The specimens were collected using direct and indirect methods between 0 and 30 m. Three web pages are available for the public:

- http://www.inbio.ac.cr/papers/babosasmarinas/index.html
- http://www.inbio.ac.cr/papers/moluscoscaribe/index.html
- www.slugophile.org

Outside of Costa Rica, there are two main collections:

(a) California Academy of Sciences. Main geographical area: Indo-Pacific, Panamic Province. Web-pages:

- http://www.calacademy.org/research/izg/
- http://www.calacademy.org/research/izg/dorids/peet.htm

(b) Natural History Museum of Los Angeles County. Main geographical area: Panamic Province. Web-pages:

- http://www.nhm.org/research/malacology/index.html
- http://www.nhm.org/research/malacology/avaldes/dorids/index.html

Scientific Value

The collection of the Museo de Zoología at the Universidad de Costa Rica currently hosts 23 holotypes and 37 paratypes of opisthobranch molluscs from Costa Rica. However, some of the type material described from the Caribbean was either previously lost at INBio (*Noumea regalis*) or not deposited in that institution and neither at the museum at the Universidad de Costa Rica (*Cratena piutaensis*, *Millereolidia ritmica*, and *P. adiuncta*). Some paratypes were deposited at the Instituto de Oceanología de La Habana, Cuba, Museo de Ciencias Naturales de Tenerife, and the California Academy of Sciences (CAS). Twenty-three scientific papers have been published since 1999 describing 36 new species from Costa Rica and other areas: *Jorunna osne* Camacho-García & Gosliner, 2008; *J. tempisquensis* Camacho-García & Gosliner, 2008; *Thordisa nieseni* Chan & Gosliner 2007; *Cuthona destinyae*, Hermosillo & Valdés, 2007; *C. millenae*, Hermosillo & Valdés, 2007; *Janolus anulatus* Camacho-García & Gosliner, 2006; *P. adiuncta* Ortea *et al.*, 2004, *Mexichromis tica* Gosliner *et al.*, 2004; *Cylichnella goslineri* Valdés & Camacho-García, 2004; *Okenia academica* Camacho-García & Gosliner, 2004; *C. piutaensis* Ortea *et al.,* 2003; *M. ritmica* (Ortea *et al.,* 2003); *Phestilla hakunamatata* Ortea *et al.,* 2003; *Hoplodoris bramale* Fahey & Gosliner, 2003; *Elysia eugeniae* Ortea & Espinosa, 2002; *E. zuleicae* Ortea & Espinosa, 2002; *Eubranchus convenientis* Ortea & Caballer, 2002; *N. regalis*, Ortea *et al.,* 2002; *Ancula espinosai* Ortea, 2001; *Dendrodoris magagnai* Ortea & Espinosa, 2001; *Doto awapa* Ortea, 2001; *D. cabecar* Ortea, 2001; *D. curere* Ortea, 2001; *D. duao* Ortea, 2001; *D. iugula* Ortea, 2001; *D. kekoldi* Ortea, 2001; *D. proranao* Ortea, 2001; *Ercolania selva* Ortea & Espinosa, 2001; *E. leopoldoi* Caballer *et al.,* 2001; *Philine caballeri* Ortea *et al.* 2001; *Philinopsis aeci* Ortea & Espinosa, 2001; *C. orteai* Valdés & Camacho-García, 2000; *Janolus costacubensis* Ortea & Espinosa, 2000; *Trapania inbiotica* Camacho-García & Ortea, 2000; *Thuridilla mazda* Ortea & Espinosa, 2000, and *Polycera manzanilloensis* Ortea *et al.*, 1999. Manuscripts describing several new species belonging to the genera *Atagema*, *Cadlina*, and *Marionia* are currently being prepared. A guide to the opisthobranchs of the Tropical Eastern Pacific has recently been published (Camacho-García *et al.* 2005).

Information about type material and other material collected in the field is being stored in a database at INBio (ATTA, http://atta.inbio.ac.cr/). Through this database, it is possible to obtain basic and advanced reports for all the animals collected. Following the terms of use, the user has access to locality, depth, collector, date, number of specimens, and catalog number among other details for any specimen.

A high percentage of the opisthobranchs deposited at the Universidad de Costa Rica has been identified by international experts. Photographs have been taken, and descriptions and drawings have been completed for the living animals in most species. INBio has published several Basic Information Units (Unidades Básicas de Información [UBIS]), electronic publications that summarize the basic information on the taxonomy, ecology, and potential uses of species from Costa Rica. This information can be accessed through the following address: http://darnis.inbio.ac.cr/ubis/.

More than 90% of the specimens collected were fixed in formalin or Bouin, which make the extraction of DNA for molecular work difficult or impossible. However, tissue samples from a few species have been taken in the field in order to support future molecular studies.

Recommendations

So far, most collecting efforts have been focused in the Osa, Tempisque, and Amistad Caribe conservation areas. Little sampling has been conducted in the Guanacaste Conservation Area (but see www.slugophile.org; proyecto Guanacaste Atjai *et al.*). Future sampling should include more areas along both the Pacific and Caribbean coasts of Costa Rica in order to compare the fauna between localities. Although past efforts have been concentrated between 0 and 30 m depth, where the richest biological diversity of these animals can be expected, deep water dredging should be carried out along the coasts (especially in those areas that present unique conditions such as Golfo Dulce), since the deep water fauna of many localities is poorly known. Sampling should also be done near oceanic islands such as Isla del Coco, and inventories around Islas Murciélago and Isla del Caño should be intensified.

Different substrates, such as red and green algae, seagrass, bryozoans, hydroids, sponges, and soft coral should be sampled to collect associated species. However, these indirect methods can be highly destructive and should only be used by well-trained people, and only when strictly necessary and when there is evidence of the presence of opisthobranchs in a determined substrate (i.e., visible egg masses or animals). Since many of the taxonomically important characters of opisthobranchs are found in their external morphology, it is critical that collected specimens are studied and photographed alive.

Despite the scientific importance of opisthobranch molluscs, there is limited knowledge of their systematics and biogeography due to the lack of taxonomic expertise in this group, particularly in tropical countries. The Pacific and Caribbean coasts of most Central American countries have been poorly sampled, while in other countries of the region collections do not exist at all. Consequently, there is an urgent need to sample these areas to improve our knowledge and understanding of the biology and taxonomy of these organisms.

In addition to improved specimen collections from throughout Central America, a library or libraries with comprehensive collections of basic, up-to-date literature about opisthobranchs in this region is vital. Scientists, in coordination with administrators, should take into consideration that good collections are not as valuable if researchers do not have access to appropriate literature (guides, original descriptions, journals, and taxonomic keys) that allow them to identify the specimens to the genus level at least. PDF files of Panamic opisthobranch literature should be available for users through the internet.

Institutions in Costa Rica should be encouraged to work together and contribute to the urgent need to train additional local taxonomists. In the future, a main goal should be to increase the number of national experts in this group and target areas of biological interest with high levels of biodiversity and endemism. The training of Costa Rican students should be further strengthened with graduate training as well as field and laboratory experience, which will effectively help Costa Rica understand, preserve, and wisely use its species richness.

Additionally, future collaborations with international museums such as the California Academy of Sciences and the Natural History Museum of Los Angeles County, which together have the most extensive representative collection of opisthobranchs from the Panamic Province in the world, should be encouraged.

Opisthobranchs in general are potential important sources for pharmacological investigations. Many different chemicals have been isolated for the first time from some species (Avila 1995). Key species with sufficiently abundant populations should be identified in order to select those that are more likely to be found throughout the year and that can provide enough dry weight for future extractions.

Future collections should include tissue samples of Costa Rican opisthobranchs preserved in 95% alcohol suitable for molecular work, thus facilitating future investigations about the relationships between taxa.

Most species of opisthobranchs are not in danger of extinction, but collections must be done for scientific study and for educational purposes only. Since most opisthobranch species are associated with coral reefs, the destruction of these environments could cause the extinction of some species, as well as entire lineages. Therefore, it is in the best interest of biologists, conservationists, and the pharmacological industry to preserve these critical ecosystems.

Acknowledgments I am very grateful to the curators, technicians, parataxonomists, park rangers, and national and international students who assisted during the fieldwork in Costa Rica. I am indebted to Ángel Valdés and Terrence Gosliner for their guidance in conducting the inventory of opisthobranchs along both the Pacific and Caribbean coasts of Costa Rica. Fieldwork was funded by the INBio through the Netherlands Government and the project "Development of Biodiversity Knowledge and Sustainable Use in Costa Rica." This research was made possible thanks to the Cooperation Agreement between the Ministerio del Ambiente y Energía de Costa Rica (MINAE) and INBio. This part has also been supported in part by the Lakeside Foundation through a scholarship to the author at the California Academy of Sciences and the PEET grant DEB-9978155, "Phylogenetic systematics of dorid nudibranchs," to Terrence Gosliner and Ángel Valdés. David Butvill and Terry Erwin made constructive comments on the manuscript.

References

1. Abbott T (1974) American Seashells. Nostrand Reinhold, New York, 663 p
2. Atjai P, Camacho Y, Wehrtmann IS (2003) Range extension of *Crosslandia daedali* Poorman and Mulliner, 1981 (Nudibranchia: Scyllaeidae). Rev Biol Trop 51:274

Avila C (1995) Natural products of opisthobranch molluscs: a biological review. Oceanogr Mar Biol Annu Rev 33:487–559

3. Baba K (1949) Opisthobranchia of Sagami Bay collected by His Majesty The Emperor of Japan. Iwanami Shoten, Tokyo, 194 p
4. Baba K (1959) The family Stiligeridae from Japan (Opisthobranchia-Sacoglossa). Publ Seto Mar Biol Lab 7:327–334
5. Baba K (1974) Some comments of *Lobiger souverbii* Fischer, 1856, re-identified, of Japan (Opisthobranchia; Sacoglossa; Lobigeridae). Veliger 16:253–257
6. Behrens D (1980) Pacific Coast Nudibranchs: A Guide to the Opisthobranchs of the Northeastern Pacific. Sea Challengers, Los Osos, CA, 112 p
7. Behrens D, Henderson R (1982) *Taringa aivica timia* Marcus and Marcus, 1967 (Nudibranchia: Doridacea) in California. Veliger 24:197–199
8. Behrens D, Gosliner T (1988) The first record of *Polycerella* Verrill, 1881, from the Pacific, with the description of a new species. Veliger 30:319–324
9. Bertsch H (1972) Two additions to the opisthobranch fauna of the southern Gulf of California. Veliger 15:103–106
10. Bertsch H (1976) A new species of *Chromodoris* (Opisthobranchia: Nudibranchia) from tropical west America. Veliger 19:156–158
11. Bertsch H (1978a) The Chromodoridinae nudibranchs from the Pacific coast of America. Part II. The genus *Chromodoris*. Veliger 20:307–327
12. Bertsch H (1978b) The Chromodoridinae nudibranchs from the Pacific coast of America. Part III. The genera *Chromolaichma* and *Mexichromis*. Veliger 21:70–86
13. Bertsch H (1978c) The Chromodoridinae nudibranchs from the Pacific coast of America. Part IV. The genus *Hypselodoris*. Veliger 21:236–250
14. Bertsch H (1979a) The eastern Pacific chromodorids (Opisthobranchia: Chromodoridinae). Festivus 11:44–46
15. Bertsch H (1979b) Tropical faunal affinities of opisthobranchs from the Panamic province (eastern Pacific). Nautilus 93:57–61
16. Bertsch H (1988) Anatomy and zoogeography of *Glossodoris sedna* and *Chromodoris grahami* (Opisthobranchia: Nudibranchia) in the tropical western Atlantic and Caribbean. Veliger 30:395–399
17. Bertsch H, Smith A (1973) Observations on three opisthobranchs (Mollusca: Gastropoda) of the La Paz area, Baja California, Mexico. Southwest Nat 18:165–176
18. Bertsch H, Ferreira A (1974) Four new species of nudibranchs from tropical west America. Veliger 16:343–353
19. Bertsch H, Gosliner T (1984) *Tritonia pickensi* (Nudibranchia: Tritoniidae) from Baja California, Mexico. Shells Sea Life 16:138–139
20. Bertsch H, Kerstitch A (1984) Distribution and radular morphology of various nudibranchs (Gastropoda: Opisthobranchia) from the Gulf of California, México. Veliger 26:264–273

21. Bertsch H, Mozqueira O (1986) A new species of *Tritonia* (Nudibranchia) from southern California and Baja California. Nautilus 100:46–49
22. Bertsch H, Angulo O, Arreola J (2000) New distributional record of opisthobranchs from the Punta Eugenia Region of the Baja California Peninsula: a report based on 1997–1998 CONABIO-sponsored expeditions. Festivus 32:99–104
23. Bertsch H, Ferreira A, Farmer W, Hayes T (1973) The genera *Chromodoris* and *Felimida* (Nudibranchia: Chromodorididae) in tropical west America: distributional data, description of a new species, and scanning electron microscopic studies of radulae. Veliger 15:287–294
24. Brodie G, Willan R, Collins J (1997) Taxonomy and occurrence of *Dendrodoris nigra* and *Dendrodoris fumata* (Nudibranchia: Dendrodorididae) in the Indo-West Pacific region. J Moll Stud 63:411–427
25. Bullis HJ Jr (1956) The genus *Scaphander* in the Gulf of Mexico and notes on the Western Atlantic species. Bull Mar Sci 6:1–17
26. Caballer M, Ortea J, Espinosa J (2001) Descripción de una nueva especie de *Eubranchus* Forbes, 1834. Avicennia Suppl 4:55–56
27. Camacho-García Y, Ortea J (2000) A new species of *Trapania* (Nudibranchia: Goniodorididae) from the Pacific coast of Central America. Rev Biol Trop 48:317–322
28. Camacho-García Y, Valdés A (2003) Caryophyllidia-bearing dorid nudibranchs (Mollusca, Nudibranchia, Doridacea) from Costa Rica. Proc Calif Acad Sci 54:65–79
28a. Camacho-García Y, Gosliner T (2004) A new species of *Okenia* (Gastropoda: Nudibranchia) from the Pacific coast of Costa Rica. Proc Calif Acad Sci 55:431–438
28b. Camacho-García Y, Gosliner T (2006) New species of zephyrinid genus *Janolus* Bergh, 1884 (Opisthobranchia: Nudibranchia) from North America and Costa Rica. Rev Biol Trop 54:1295–1305
28c. Camacho-García Y, Gosliner T, Valdés A (2005) Field guide to the sea slugs of the tropical eastern Pacific. Calif Acad Sci, California, USA, 129 p
28d. Camacho-Garcia Y, Gosliner T (2008) Systematic revision of *Jorunna* Bergh, 1876 (Nudibranchia: Discodorididae) with a morphological phylogenetic analysis. J. Moll Stud 74: 143–181
28e. Chan J, Gosliner T (2007) Preliminary phylogeny of *Thordisa* (Nudibranchia: Discodorididae) with descriptions of five new species. Veliger 48: 284–308
29. Cleveland P Jr, Finet Y (1999) A Field Guide to Marine Molluscs of Galapagos. An Illustrated Guidebook to the Common Intertidal and Shallow-water Snails, Bivalves and Chitons of the Galapagos Islands. Galapagos Marine Life Series, Sugar Spring, Lexington, VA, 150 p

Coates A (1997) The Forging of Central America. *In*: Coates AG (ed) Central America: A Natural and cultural History. Yale University Press, New Haven, CT, pp 1–37

30. De Jong K, Coomans H (1988) Marine Gastropods from Curaçao, Aruba and Bonaire. Stud Fauna Curaçao Carib Is. E.J. Brill, Leiden, Netherlands, 69. 261pp.
31. Debelius H (1996) Nudibranchs and Sea Snails: Indo-Pacific Field Guide. IKAN-Unterwasserarchiv, Frankfurt, Germany, 321 p
32. D'Orbigny A (1841) Mollusques. *In*: de la Sagra R(ed) Histoire Physique, Politique et Naturelle de l'île de Cuba 1: Arthus Bertrand, Paris, pp 1–240
33. Espinosa J, Ortea J (2001) Moluscos del Mar Caribe de Costa Rica: desde Cahuita hasta Gandoca. Avicennia Suppl 4:1–77
33a.Espinosa J, Ortea J (2004) Adiciones al Catálogo de Moluscos del Mar Caribe de Costa Rica: desde Cahuita hasta Gandoca. Avicennia 16:113–120
34. Fahey S, Gosliner T (2003) Mistaken identities: on the Discodorididae genera *Hoplodoris* Bergh, 1880 and *Carminodoris* Bergh, 1889 (Opisthobranchia, Nudibranchia). Proc Calif Acad Sci 54:169–208
35. Farmer W (1975) Anatomical and distributional observations of some opisthobranchs from the Panamic faunal province. Veliger 17:323–330
36. Farmer W (1978) *Tambja* and *Roboastra* (Mollusca: Opisthobranchia from the Gulf of California and the Galapagos Islands. Veliger 20:375–385
37. Farmer W (1980) Sea-slug Gastropods. Wesley M. Farmer Enterprises, Tempe, AZ, 177 p

38. Ferreira A, Bertsch H (1975) Anatomical and distributional observations of some opisthobranchs from the Panamic faunal province. Veliger 17:323–330
39. García F, Troncoso J (2001) *Favorinus elenalexiae*, a new species (Opisthobranchia: Aeolidiidae) from the eastern Pacific Ocean. Nautilus 115:55–61
40. Gosliner T (1979) The systematics of the Aeolidacea (Nudibranchia: Mollusca) of the Hawaiian Islands, with descriptions of two new species. Pac Sci 33:37–77
41. Gosliner T (1980) Systematics and phylogeny of the Aglajidae (Opisthobranchia: Mollusca). Zool J Linn Soc 68:325–360
42. Gosliner T (1981a) A new species of tergipedid nudibranch from the coast of California. J Moll Stud 47:200–205

Gosliner T (1981b) Origins and relationships of primitive members of the Opisthobranchia (Mollusca: Gastropoda). Biol J Linn Soc 16:197–225

43. Gosliner TM (1982) The genus *Janolus* (Nudibranchia: Arminacea) from the Pacific Coast of North America, with a reinstatement of *Janolus fuscus* O'Donoghue, 1924. Veliger 24:219–226
44. Gosliner T (1985) The aeolid nudibranch family Aeolidiidae (Gastropoda, Opisthobranchia) from tropical southern Africa. Ann S Afr Mus 95:233–267
45. Gosliner T (1991a) Four new species and a new genus of opisthobranch gastropods from the Pacific coast of North America. Veliger 34:272–290
46. Gosliner T (1991b) The Opisthobranch gastropod fauna of the Galápagos Islands. *In*: James MJ (ed) Galápagos Marine Invertebrates: Taxonomy, Biogeography, and Evolution in Darwin's Islands. Plenum, New York, pp 281–305

Gosliner T (1992) Biodiversity of tropical opisthobranch gastropod faunas. Proc 7th Int Coral Reef Symp Guam 2:702–709

Gosliner T (1994a) Gastropoda: Opisthobranchia. *In*: Harrison FW, Kohn AJ (eds) Microscopic Anatomy of Invertebrates, vol 5, Mollusca I. Wiley-Liss, New York, pp 253–355

47. Gosliner T (1994b) New records of Flabellinidae (Opisthobranchia: Aeolidacea) from the Tropical Americas, with descriptions of two new species. Proc Calif Acad Sci 48:171–183
48. Gosliner T, Griffiths R (1981) Description and revision of some South African aeolidacean Nudibranchia (Mollusca, Gastropoda). Ann S Afr Mus 84:105–150
49. Gosliner T, Behrens D (1986) Two new species and genera of aeolid nudibranchs from the tropical eastern Pacific. Veliger 29:101–113
50. Gosliner T, Bertsch H (1988) A review of the genus *Berthella* (Opisthobranchia: Notaspidea) from the Pacific coast of North America. Veliger 31:46–67
51. Gosliner T, Kuzirian A (1990) Two new species of Flabellinidae (Opisthobranchia: Aeolidacea) from Baja California. Proc Calif Acad Sci 47:1–15
52. Gosliner T, Williams G (1972) A new species of *Chelidonura* from Bahia San Carlos, Gulf of California, with a synonymy of the family Aglajidae. Veliger 14:424–36

52a. Gosliner T, Ortea J, Valdés A (2004) New data on Tropical Eastern Pacific Chromodorididae (Nudibranchia: Doridina) with description of a new species of *Mexichromis* Bertsch, 1979. Proc Calif Acad Sci 55:590–599

53. Harry H (1967) A review of the living tectibranch snails of the genus *Volvulella*, with descriptions of a new subgenus and species from Texas. Veliger 10:133–147

53a. Hermosillo A (2004) Opisthobranch molluscs of Parque Nacional de Coiba, Panamá (tropical eastern Pacific). The Festivus 36:105–117

53b. Hermosillo A, Camacho-García Y (2006) A note on the opisthobranchs of Parque Nacional de Coiba, Panamá (Tropical Eastern Pacific). The Festivus 38:95–98

53c. Hermosillo A, Valdés A (2007) Five new species of aeolid nudibranchs (Mollusca, Opisthobranchia) from the tropical eastern Pacific. Am Malacol Bull 22:119–137

53d. Hermosillo A, Behrens DW, Jara ER (2006) Opistobranquios de México: guía de babosas marinas del Pacífico, Golfo de California y las islas oceánicas. Editorial Impre-Jal, México, 143 p

54. Høisæter T (1998) Preliminary check-list of the marine, shelled gastropods (Mollusca) of Golfo Dulce, on the Pacific coast of Costa Rica. Rev Biol Trop 46(Suppl 6):263–270
55. Houbrick J (1968) A survey of the litoral marine molluscs of the Caribbean coast of Costa Rica. Veliger 11:4–23

56. Humann P, Deloach N (2002) Reef Creature Identification: Florida, Caribbean, Bahamas. 2nd ed. New World Publications, Jacksonville, FL, 448 p
57. Jensen K (1992) Anatomy of some Indo-Pacific Elysiidae (Opisthobranchia, Sacoglossa (=Ascoglossa), with a discussion of the generic division and phylogeny. J Moll Stud 58:257–296
Kandel E (1979) Behavioral Biology of *Aplysia*: A Contribution to the Comparative Study of Opisthobranch Molluscs. W.H. Freeman, San Francisco, CA, 463 p
58. Keen A (1971) Seashells of Tropical West America: Marine Molluscs from Baja California to Peru. 2nd ed. Stanford University Press, Stanford, 1064 p
59. Kerstitch A (1989) Sea of Cortez Marine Invertebrates: A Guide for the Pacific Coast, Mexico to Ecuador. Sea Challengers, Monterey, 214 p
60. Lance J (1971) Observations on the sea hare *Aplysia parvula* (Gastropoda: Opisthobranchia) from the Gulf of California. Veliger 14:60–63
61. MacFarland F (1966) Studies of opisthobranch molluscs of the Pacific coast of North America. Mem Cal Acad Sci 6:546
62. Marcus E (1955) Opisthobranchia from Brazil. Bol Fac Filosofia, Ciencias e Letras, Univ São Paulo, Zool 207:89–261
63. Marcus E (1957) On Opisthobranchia from Brazil (2). Zool J Linn Soc 43:390–486
64. Marcus E (1958) On western Atlantic opisthobranchiate gastropods. Am Mus Novitates (1906): 1–82
65. Marcus E (1958) (for 1956) Notes on Opisthobranchia. Bol Inst Oceanogr, Univ São Paulo 7:31–78
66. Marcus E (1964) A new species of *Polycera* (Nudibranchia) from California. Nautilus 77:128–131
67. Marcus E (1965) Some Opisthobranchia from Micronesia. Malacologia 3:263–286
68. Marcus E (1970) Opisthobranchs from northern Brazil. Bull Mar Sci 20:922–951
69. Marcus E (1971) On some euthyneuran gastropods from the Indian and Pacific Oceans. Proc Malac Soc Lond 39:355–369
70. Marcus E (1972) On the Anaspidea (Gastropoda: Opisthobranchia) of the warm waters of the Western Atlantic. Bull Mar Sci 22:841–874
70a. Marcus Ev (1976) On *Kentrodoris* and *Jorunna* (Gastropoda: Opisobranquia) Bolm. Zool, Univ São Paulo 1:11–68
71. Marcus E (1977) An annotated checklist of the Western Atlantic warm water opisthobranchs. J Moll Stud Suppl 4:1–23
72. Marcus E (1980) Review of western Atlantic Elysiidae (Opisthobranchia: Ascoglossa) with a description of a new *Elysia* species. Bull Mar Sci 30:54–79
73. Marcus E (1984) The western Atlantic warm water Notaspidea (Gastropoda, Opisthobranchia), Part 2. Bol Zool 8:43–76
73a. Marcus E, Marcus E (1960) Opisthobranchs from American Atlantic warm waters. Bull Mar Sci Gulf Carib 10:129–203
74. Marcus E, Marcus E (1963) Opisthobranchs from the Lesser Antilles. Stud Fauna Curaçao Carib Isl 19:1–76
75. Marcus E, Marcus E (1967) American opisthobranch molluscs. Stud Trop Oceanogr 6:3–137
76. Marcus E, Marcus E (1968) *Flabellina engeli*, a new nudibranch from Curaçao. Beaufortia 15:139–142
77. Marcus E, Marcus E (1970) Opisthobranchs from Curacao and faunistically related regions. Stud Fauna Curaçao Carib Isl 33:1–129
78. Marcus E, Hughes H (1974) Opisthobranch molluscs from Barbados. Bull Mar Sci 24:498–532
79. Meyer K (1977) Dorid nudibranchs of the Caribbean coast of the Panama Canal Zone. Bull Mar Sci 27:299–307
80. Mikkelsen P, Mikkelsen P (1984) Comparison of *Acteocina canaliculata* (Say, 1826), A. *candei* (d'Orbigny, 1841), and *A. atrata* spec. nov. (Gastropoda: Cephalaspidea). Veliger 27:164–192

81. Mulliner D (1993) Opisthobranchs of Cocos Island. West Soc Malacol Ann Rep [for 1992] 25:2–3
82. Nunn J (1994) *Cuthona genovae* (O'Donoghue, 1926) (Mollusca: Opisthobranchia) from the North Coast of Ireland. Irish Nat J 24:416–417
83. Ortea J (1981) Moluscos opistobranquios de las Islas Canarias. Primera parte: Ascoglosos. Bol Inst Español Oceanogr 6(Part 4):180–199
84. Ortea J (2001a) Descripción de una nueva especie de *Ancula* Loven, 1846. Avicennia Suppl 4:49
85. Ortea J (2001b) El género *Doto* Oken, 1815 (Mollusca: Nudibranchia) en el mar Caribe: historia natural y descripción de nuevas especies. Avicennia Suppl 3:1–46
86. Ortea J, González L (1981) Un nuevo dórido (Mollusca: Nudibranchiata) de la Isla Isabel, Nayarit, México. Iberus 1:47–52
87. Ortea J, Templado J (1988) Una nueva especie de *Cyerce* Bergh, 1871 (Opisthobranchia: Ascoglossa) de la isla de Cuba. Iberus 8:11–14
88. Ortea J, Espinosa J (2000a) Nueva especie del género *Janolus* Bergh, 1884 (Mollusca: Nudibranchia) de Cuba y Costa Rica. Avicennia 12/13:79–83
89. Ortea J, Espinosa J (2000b) Nueva especie del género *Thuridilla* Bergh, 1872 (Mollusca: Sacoglossa) de Cuba y Costa Rica. Avicennia 12/13:87–90
90. Ortea J, Espinosa J (2001a) Descripción de una nueva especie de *Dendrodoris* Ehrenberg, 1831. Avicennia Suppl 4:52–53
91. Ortea J, Espinosa J (2001b) Descripción de una nueva especie de *Ercolania* Trinchese, 1872. Avicennia Suppl 4:45–47
92. Ortea J, Espinosa J (2001c) Descripción de una nueva especie de *Philinopsis* Pease, 1860. Avicennia Suppl 4:41
93. Ortea J, Caballer M (2002) Nuevos datos sobre el género *Eubranchus* Forbes, 1838 (Mollusca: Nudibranchia) en aguas templadas del Atlántico Oeste. Avicennia 15:77–90
94. Ortea J, Espinosa J (2002) Nuevas especies del género *Elysia* Risso, 1818 (Mollusca: Sacoglossa) con caracteres singulares. Avicennia 15:129–140
95. Ortea J, Pérez J, Llera E (1982) Moluscos Opistobranquios recolectados durante el Plan de Bentos Circumcanario. Doridacea: primera parte (1). Cuadernos del Crinas No. 3. Consejerías de Agricultura y Pesca del Consejo de Gobierno del Principado de Asturias y de la Junta de Canarias. 48 p
95a. Ortea J, Luque A, Templado J (1990) Contributions to the knowledge of the genus *Aegires* Lovén, 1844 (Opisthobranchia: Doridoidea: Aegiretidae) in the North Atlantic, with descriptions of two new species. J Mollusc Stud 56:333–337
96. Ortea J, Bacallado J, Valdés A (1992) Chromodorididae (Mollusca, Nudibranchia). *In*: Bacallado JJ, Ortea J (eds) Resultados científicos del proyecto Galápagos, patrimonio de la humanidad I. Museo de iencias Naturales, Tenerife, pp 31–70
97. Ortea J, Espinosa J, Camacho Y (1999) Especies del género *Polycera* Cuvier, 1816 (Mollusca: Nudibranchia) recolectadas en la epifauna de algas rojas del Caribe de Costa Rica, Cuba. Avicennia 10/11:157–164
98. Ortea J, Caballer M, Moro L (2001) Nuevos datos acerca de *Cuthona iris* Edmunds & Just, 1983. Avicennia Suppl 4:57–58
99. Ortea J, Caballer M, Moro L (2001) Una nueva especie del género *Noumea* Risbec, 1928 (Mollusca: Nudibranchia) del Caribe arrecifal de Costa Rica, descrita con motivo del Quinto Centenario de su descubrimiento. Avicennia 14:16
100. Ortea J, Espinosa J, Moro L (2001) Descripción de una nueva especie de *Philine* Ascanius, 1772. Avicennia Suppl 4:38–40
101. Ortea J, Caballer M, Espinosa J (2003) Nuevos Aeolidaceos (Mollusca: Gastropoda) de Costa Rica. Avicennia 16:129–142
101a. Ortea J, Caballer M, Moro L (2004) Dos aeolidáceos con ceratas rojos de la región macaronésica y el mar Caribe (Mollusca: Nudibranchia). Vieraea 32:83–96
102. Pease W (1868) Descriptions of marine Gasteropodae, inhabiting Polynesia. Am J Conch 4:71–80

102a. Pola M Cervera J, Gosliner T (2005) Review of the systematics of the genus *Roboastra* Bergh, 1877 (Nudibranchia, Polyceridae, Nembrothinae) with the description of a new species from the Galápagos Islands. Zool J Linn Soc 144:67–189
103. Poorman L, Mulliner D (1981) A new species of *Crosslandia* (Nudibranchia: Dendronotacea) from the Gulf of California. Nautilus 95:96–99
104. Redfern C (2001) Bahamian Seashells: A Thousand Species from Abaco, Bahamas. BahamianSeashells.com, Inc., Boca Raton, Florida, 280 p
105. Radwin G (1969) A recent molluscan fauna from the Caribbean coast of southeastern Panama. San Diego Soc Nat Hist Trans 15:229–236
106. Robinson D, Montoya M (1987) Los moluscos marinos de la costa Atlántica de Costa Rica. Rev Biol Trop 35:375–400
107. Rochebrune A (1881) Materiaux pour la faune del'Archipel du Cap Vert. Nouv Arch Mus Nat d'Hist Natur, Paris, Ser 2:215–340
108. Rodríguez L, Vargas R, Cortés J (2003) Biodiversidad marina de Costa Rica: Gastrópodos (Mollusca: Gastropoda) de la costa Caribe. Rev Biol Trop 51(Suppl 3):293–304
Rudman W (1998) Opisthobranchia. *In*: Beesley PL, Ross GJB, Wells A (eds) Mollusca, The Southern Synthesis, Part B: Australian Biological Resources Study, Canberra, pp 915–1036
109. Skoglund C (1991) Additions to the Panamic Province Opisthobranchia (Mollusca): Literature 1971 to 1990. Festivus 22(Suppl 1):1–27
110. Skoglund C (2002) Panamic Province molluscan literature: additions and changes from 1971 through 2001. III Gastropoda Festivus 32:201–286
111. Sphon G, Mulliner D (1972) A preliminary list of known opisthobranchs from the Galápagos Islands collected by the Ameripagos Expedition. Veliger 15:147–152
112. Thompson T (1976) Biology of Opisthobranch Molluscs I. Ray Society, London, 206 p
113. Thompson T (1977) Jamaican opisthobranch molluscs I. J Moll Stud 43:93–140
114. Thompson T (1980) Jamaican opisthobranch molluscs II. J Moll Stud 46:74–99
Todd CD (1981) The ecology of nudibranch molluscs. Oceanogr Mar Biol Ann Rev 19:141–234
115. Vagel R (1971) The biology and a redescription of the opisthobranch mollusc *Hermaea cruciata* Gould, from Chesapeake Bay (Sacoglossa: Hermaeidae). Veliger 14:155–157
Valdés A (2001) Deep-sea cryptobranch dorid nudibranchs (Mollusca, Opisthobranchia) from the Tropical West Pacific, with descriptions of two new genera and eighteen new species. Malacologia 43:237–311
Valdés A (2002a) A phylogenetic analysis and systematic revision of the cryptobranch dorids (Mollusca, Nudibranchia, Anthobranchia). Zool J Linn Soc 136:535–626
116. Valdés A (2002b) How many species of *Hexabranchus* (Opisthobranchia: Dorididae) are there? Moll Res 22:289–301
Valdés A, Bouchet P (1998) Naked in toxic fluids: a nudibranch mollusc from hydrothermal vents. Deep-Sea Research Part II. Trop Stud Oceanogr 45:319–327
117. Valdés A, Angulo O (2000) Redescription and reassessment of *Cadlina luarna* (E. Marcus and E. Marcus, 1967), *comb. nov*. (Mollusca, Opisthobranchia, Doridina). Proc Calif Acad Sci 52:77–85
118. Valdés A, Camacho-García Y (2000) A new species of *Cyerce* Bergh, 1871 (Mollusca, Sacoglossa, Polybranchiidae) from the Pacific coast of Costa Rica. Bull Mar Sci 66:445–456
119. Valdés A, Camacho-García Y (2004) "Cephalaspidean" heterobranchs (Gastropoda) from the Pacific coast of Costa Rica. Proc Calif Acad Sci 55:459–497
Wägele H, Willan R (2000) Phylogeny of the Nudibranchia. Zool J Linn Soc 130:83–181
120. Williams G, Gosliner T (1973) A new species of anaspidean opisthobranch from the Gulf of California (Mollusca: Gastropoda). Veliger 16:216–232
Yamada K, Kigoshi H (1997) Bioactive compounds from the sea hares of two genera: *Aplysia* and *Dolabella*. Bull Chem Soc Japan 70:1479–1489.

Part 31
Bivalves

Julio A. Magaña and José Espinosa

A representative of the genus *Tellina*, known to occur on both coasts of Costa Rica (Photo: Alcides Berrocal)

J.A. Magaña
Universidad Estatal a Distancia. Programa de Manejo de Recursos Naturales, Sabanilla, San José Costa Rica.
e-mail: eubranchus@yahoo.com

J. Espinosa
Instituto de Oceanología de la Habana, Cuba. Ave 1[a], N 18406, E. 184 y 186, Playa, La Habana, Cuba.
e-mail: espinosa@oceano.inf.cu

Abstract Here we report the presence of 521 species of marine bivalves from Costa Rica, including Isla del Coco. A total of 151 species are from the Caribbean and 375 from the Pacific, including 5 species present on both coasts. From the Pacific, 274 species are new records for Costa Rica. The species distribution, habitat, depth, and relevant references are indicated.

Introduction

Bivalves are mollusks with two valves joined together by a hinge (Keen 1971). They are the second largest class within the Phylum Mollusca, and the number of species may range from about 8,000 (Boss 1982) to as many as 20,000 living species (Brusca & Brusca 2003). All are exclusively aquatic, although some intertidal species can survive many days out of water, and others can live in ephemeral lakes or can survive in moist sediments for months (Coan *et al.* 2000). Even though bivalves have a wide geographic and bathymetric distribution, the majority of the species of this class inhabit the neritic zone of tropical seas (Espinosa *et al.* 1994).

The earliest known bivalves were from the protobranch and pteriomorph lineages from which it is possible to derive the other major groups. Fossil records of bivalves are scarce from the Middle to Late Cambrian, but they are more abundant since the Early Ordovician, 480 million years ago. Mollusks are considered to share a common ancestor with the annelid worms and sipunculans (Coan *et al.* 2000).

Few studies have dealt exclusively with the marine bivalves of Costa Rica. There have been a variety of reports that focus on the eastern Pacific and western Atlantic areas, or that deal with specific groups of bivalves from one or both coasts. The first papers describing new species were published in a series of reports authored by G.B. Sowerby I; G.B. Sowerby II and W.J. Broderip (e.g., Sowerby I 1833; Sowerby II 1865–1878; Broderip 1835). These descriptions were based on collections made by Hugh Cuming from 1828 to 1830. Cuming traveled as far north as the Golfo de Fonseca (Dance 1986), and it is known that he also collected around Puntarenas, Pacific Costa Rica.

The next major collection from Costa Rica was made by A.S. Oersted, who sampled mollusks near Puntarenas on the Pacific coast and Limón on the Caribbean (Keen 1966). An account of the material he collected, mostly gastropods, was published in several parts by O.A.L. Mörch (1859–1861); later, Keen (1966) reviewed and illustrated his type material. Mörch (1860, 1861) described seven new species of bivalves collected from the Golfo de Nicoya. The first account on the bivalves of Isla del Coco was published by Biolley (1907), based on a collection done by the author in 1902. However, the first collection was obtained during the 1891 "Albatross" expedition (Dall 1908).

The first extensive studies related to biodiversity of bivalves were based on the collections of expeditions to the eastern Pacific by the New York Zoological Society aboard the *Zaca* (Hertlein & Strong 1943, 1946a, b, 1947, 1949a, b, 1950) and the Allan Hancock Foundation (Rost 1955; Soot-Ryen 1955; Grau 1959; Bernard 1974) during the first half of the 20th century. Keen (1958, 1971), in her compilation of mollusks from California to Perú, included all the papers in which species from Costa Rica were reported. Skoglund (1991, 2001) updated Keen's species list.

More recent research (mostly on ecology) was done by scientists of both Universidad de Costa Rica and Universidad Nacional. Papers were published on the mollusks from Isla del Coco based on collections made between 1982 and 1992 (Montoya 1983; Chaney 1992), but most are on gastropods. Cruz & Jiménez (1994) described the mollusks associated to mangrove forests in Central America. One of the most complete collections of mollusks in Costa Rica was made by the Instituto Nacional de Biodiversidad (INBio). The Malacology Department was closed in September 2004, and the entire collection is now at the Museo de Zoología, Universidad de Costa Rica.

The first paper on bivalves from the Caribbean coast of Costa Rica was by Houbrick (1969) and was based on a collection he made in 1966. He reported 250 species of mollusks of which 35 were bivalves. Robinson & Montoya (1987) published an inventory of the mollusks of this area, including 100 species of bivalves. Associated with the work developed by INBio, Espinosa & Ortea (2001) published a catalogue of the mollusks of the Caribbean coast of Costa Rica that reported 41 additional species of bivalves, including one of only three new bivalves described from Costa Rican waters (*Adrana elizabethae*).

Bivalves played an important role in the economy of aboriginal Costa Rican cultures, evidenced by the number of shells found around places where indigenous people had settled (Solís Del Vecchio 2005; Herrera 2005). Studies on the importance of bivalves as a human food source were published in the 1980s, beginning with a report about the potential and characteristics of the Costa Rican species (Glude 1981). Published studies on Arcidae or "chuchecas" (*Anadara grandis*), "pianguas" (*A. tuberculosa*), and "boludos" (*A. similis*) concerned reproductive aspects (Cruz 1984a, b, Fournier & de la Cruz 1987), growth (Cruz 1982, 1986; Villalobos & Baéz 1984), size (Cruz & Palacios 1983), estimation of the population size in the most important mangrove areas (Campos *et al.* 1990), and size, biometric characteristics, sexual dimorphism, and meat yield (Cruz & Palacios 1983).

Oysters (Ostreidae) comprise another important group frequently used as food. The following aspects of the Caribbean mangrove oyster, *Crassostrea rhizophorae*, have been studied: commercial size (Cabrera-Peña *et al.* 1983), growth rates in cultivation systems (Pacheco-Urpí *et al.* 1983), filtration rates under different salinity conditions (Alfaro *et al.* 1985), population structure (Madrigal *et al.* 1985a, b), experimental transfer to the Pacific coast (Alfaro 1985a, Quesada *et al.* 1985), and finally the development of a technique to induce egg laying (Alfaro 1985b). Oysters are also the subject of studies on the experimental cultivation of the Japanese oyster *C. gigas* on the Costa Rican Pacific coast (Villalobos 1982). Coquina or wedge clams (*Donax*) comprise the third most important group for human consumption; many studies have been done on several species, e.g., ecological analysis of one of the most-used clams *Donax dentifer* (Palacios *et al.* 1982).

The present list of Costa Rican marine bivalves is based mainly on the literature and on the records from the former Malacological Collection of the INBio, now deposited at the Museo de Zoología, Universidad de Costa Rica. Although we found many instances that could represent important extensions of species distribution ranges, we decided not to include these records, because they could be based on misidentifications. Unfortunately, verification of these doubtful records was not possible; future studies need to verify the identification of these species.

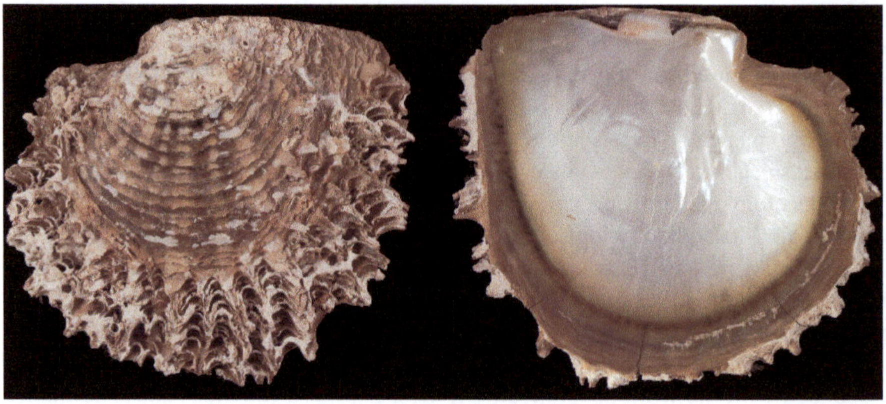

Fig. 31.1 External and internal view of *Pinctada mazatlanica*, one of the largest bivalve shells from the Pacific shore (©2008, Santa Barbara Museum of Natural History)

Additionally, species in the collections classified only to Family or genus were not considered herein.

The number of living marine bivalve species reported from Costa Rica is 521; 151 species from the Caribbean, or 29% of the total (Species List 31.1 is included on the CD-Rom), and 375 species from the Pacific (Fig. 31.1), 71% of the total (Species List 31.2 is included on the CD-Rom); five species occur in both oceans. The marine bivalves of Costa Rica are classified in four subclasses, 10 orders, 6 suborders, 36 superfamilies, 57 families, 48 subfamilies, 189 genera, and 96 subgenera (Table 31.1). Two hundred and seventy-four species from the Pacific are new records for Costa Rica.

The most specious families in Costa Rica are: Subclass Heterodonta: Veneridae (64 spp.), Tellinidae (59), Cardiidae (22), Semelidae (18), Mactridae (17), Lucinidae (20), Chamidae (15), and Corbulidae (12); Subclass Pteriomorphia: Arcidae (37), Mytilidae (28), and Pectinidae (17); the remaining families have less than 17 species. Sixty-four species of bivalves are recorded from Isla del Coco (17% of the Pacific species of bivalves) with 18 species having Isla del Coco as the only known locality in Costa Rica.

So far, three species have been reported to occur only in Costa Rica: *Adrana elizabethae* from the Caribbean, and from the Pacific the recently discovered *Calyptogena costaricana*; *Solen* (*Solen*) *oerstedii*. These four species are considered endemic to Costa Rica, but more taxonomic research is necessary to verify this status. Five species are recorded from both the Pacific and Caribbean shores: *Modiolus* (*Modiolus*) *americanus*, *Malleus* (*Malvufundus*) *regulus*, *Martesia* (*Martesia*) *striata*, *Sphenia fragilis* and *Crenella decussata*.

Caribbean Sea: The number of marine bivalves of the Costa Rican Caribbean coast is 151 species. This is a relatively high number of species compared to the 80 species from the Panamian areas of the central Caribbean (Radwin 1969) and 138

Table 31.1 Taxonomic and geographic distribution of Costa Rican bivalve species

Subclass	Protobranchia	Pteriomorphia	Heterodonta	Anomalodesmata	Total
Orders	1	5	2	2	10
Superfamilies	2	11	19	4	36
Families	2	17	31	7	57
Genera	3	51	121	14	189
Species	19	150	334	23	521*
Caribbean	4	51	93	3	151
Pacific	15	99	241	20	375
Isla del Coco	-	37	25	2	64

*Five species are present along both coasts of Costa Rica.

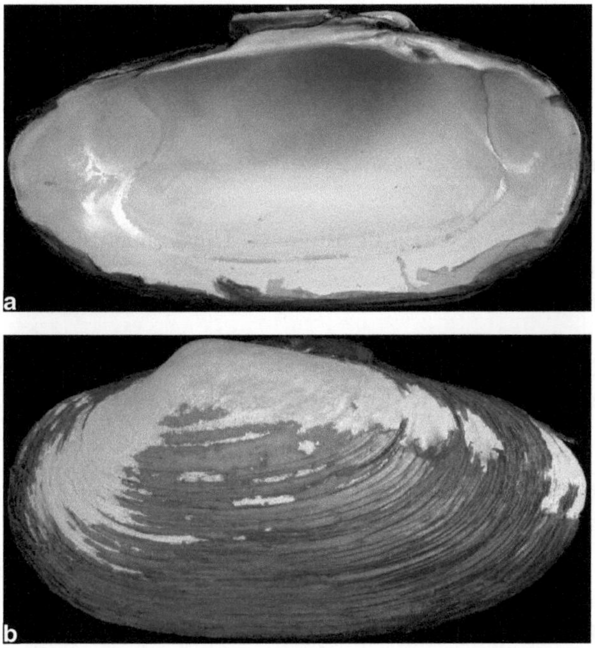

Fig. 31.2 *Calyptogena costaricana*, type material collected at a depth of 2,263 m off Pacific Costa Rica (Photos: Heiko Sahling)

species from Bocas del Toro (Olsson & McGinty 1958). However, it is low in relation to the bivalve diversity of Cuba (320 spp.; Espinosa *et al.* 1994) and Colombia (315 spp.; Díaz & Puyana 1994). These differences may be a result of more intense collecting efforts in some countries than in others.

Pacific Ocean: The Costa Rican inventory of marine bivalves comprises 375 species from the Pacific coast, or 44.6% of the species recorded in the Panamic Zoogeographic Province (ca. 856 species; Coan & Valentich-Scott personal communication, 2006). Here we report for the first time the presence of 274 species from Pacific Costa Rica, including a recently described species (*Calyptogena costaricana*; Fig. 31.2) from deepwaters (Krylova & Sahling 2006).

Specialists

Rüdiger Bieler: Department of Zoology, The Field Museum, Chicago. Bivalve and gastropod evolution, comparative anatomy, zoogeography and reproduction biology – Worldwide. bieler@fieldmuseum.org

Eugene V. Coan: California Academy of Sciences/Natural History Museum of Los Angeles County. Bivalve systematics – Worldwide. gene.coan@sierraclub.org

Paula M. Mikkelsen: American Museum of Natural History. Systematics, phylogeny, and the evolution of diversity in marine molluscs – Caribbean. mikkel@amnh.org

Paul Valentich-Scott: Santa Barbara Museum of Natural History. Bivalve systematics – Worldwide. pvscott@sbnature2.org

Collections

Caribbean coast

- Museo de Zoología, Escuela de Biología, Universidad de Costa Rica: http://museo.biologia.ucr.ac.cr/
- National Museum of Natural History, Smithsonian Institution: http://www.mnh.si.edu/
- American Museum of Natural History, New York: http://www.amnh.org/
- Field Museum of Natural History, Chicago: http://www.fieldmuseum.org/

Pacific coast

- Museo de Zoología, Escuela de Biología, Universidad de Costa Rica: http://museo.biologia.ucr.ac.cr/
- National Museum of Natural History, Smithsonian Institution: http://www.mnh.si.edu/
- Academy of Natural Sciences of Philadelphia: http://www.acnatsci.org/
- Rosenstiel School of Marine and Atmospheric Science, University of Miami, Florida: http://www.rsmas.miami.edu
- Paleonotological Research Institute, Cornell University
- Natural History Museum of Los Angeles County, California: http://www.nhm.org/
- Santa Barbara Museum of Natural History, California: http://www.sbnature.org/

Acknowledgments Our special thanks go to Eugene V. Coan and Paul Valentich-Scott who provided advice on the systematics and bivalve research in Costa Rica. Paula Mikkelsen and Jesús Ortea Rato did a general revision of an earlier draft of the manuscript. Jorge Cortés (CIMAR) and Ingo Wehrtmann, collaborated with the final edition of this work. Frank González helped us with geographic information about the species included in the INBio database and acquiring literature. We appreciate the goodwill of "Chumi" and Socorro Ávila for trying to identify doubtful material in the previous INBio collection.

Priscilla Hurtado, Alejandro Herrera, Gilberth Solis, Andrea Zumora, Alejandra Mijangos, and Manuel Caballer provided literature and technical support. Anayensy Herrera (Museo Nacional) helped with information on archaeological uses of bivalves. Raul Fernández Garcés "Cacho" assisted in the communication between us. Heiko Sahling provided pictures for a recent discovered specie.

References

1. Abbott RT (1974) American Seashells; the Marine Mollusca of the Atlantic and Pacific Coasts of North America, 2nd ed. Van Nostrand Reinhold, New York, 663 p

Alfaro J (1985a) Cultivo experimental del ostión de mangle, *Crassostrea rhizophorae* (Guilding 1828) en el estero Vizcaya, Limón y Golfo de Nicoya, Costa Rica. Uniciencia 2:81–84

Alfaro J (1985b) Una técnica desarrollada para inducir el desove del ostión de mangle, *Crassostrea rhizophorae*, Guilding 1828. Uniciencia 2:51–52

Alfaro J, Quesada R, Zamora E, Madrigal E, Pacheco O (1985) Análisis comparativo de sustratos para recolectar juveniles del ostión de manglar (*Crassostrea rhizophorae* Guilding 1828) en Estero Vizcaya, Limón, Costa Rica. Rev Biol Trop 33:1–6

2. Bakus GJ (1975) Marine zonation and ecology of Cocos Island, off Central America. Atoll Res Bull 179:1–9
3. Bernard FR (1974) Septibranchs of the eastern Pacific (Bivalvia: Anomalodesmata). Allan Hancock Monogr Mar Biol 8:1–279.
4. Bernard FR (1976) Living Chamidae of the eastern Pacific (Bivalvia: Heterodonta). Nat Hist Mus Los Angeles County, Contr Sci 278:1–43
5. Bernard FR (1979) Bivalve mollusks of the western Beaufort Sea. Nat Hist Mus Los Angeles County, Contr Sci 313:1–80
6. Bernard FR (1983) Catalogue of the living Bivalvia of the eastern Pacific Ocean: Bering Strait to Cape Horn. Canadian Spec Publ Fish Aquat Sci 61:102
7. Bernard FR (1989) Living Periplomatidae of the Pacific and Indo-Pacific regions (Bivalvia: Anomalodesmata). Venus 48:1–11
8. Bernard FR, McKinnell SM, Jamieson GS (1991) Distribution and zoogeography of the Bivalvia of the eastern Pacific Ocean. Canadian Spec Publ Fish Aquat Sci 112:60

Biolley P (1907) Mollusques de l'Isle del Coco. Museo Nacional de Costa Rica, Costa Rica, 30 p

Boss KJ (1982) Mollusca, pp 946–1166. *In*: Parker SP (ed) Synopsis and Classification of Living Organisms, Vol. 1, and pp 1092–1096 in Vol. 2. McGraw-Hill, New York

Broderip WJ (1835) Description of some species of *Chama*. Zool Soc London, Proc 1834 [2] (25):148–151

9. Britton JC Jr (1972) Two new species and a new subgenus of Lucinidae (Mollusca: Bivalvia) with notes on certain aspects of lucinid phylogeny. Smithson Contr Zool 129:1–18

Brusca RC, Brusca GJ (2003) Invertebrates, 2nd ed. Sinauer Assoc., Sunderland, MA, 936 p

10. Buck GL (1994) *Kellia suborbicularis* (Montagu 1803) an unusual and interesting habitat. Festivus 25:38
11. Bullock RC, Boss KJ (1971) *Lithophaga aristata* in the shell-plates of chitons (Mollusca). Breviora 369:1–10

11a. Campbell MR, Steiner G, Campbell LD, Dreyer H (2004) Miocene to Recent Chamidae (Bivalvia) from the Western Atlantic Ocean. Malacologia 46:381–415

Cabrera-Peña J, Zamora-Madriz E, Pacheco-Urpí O (1983) Determinación del tamaño comercial de la ostra de manglar *Crassostrea rhizophorae* (Guilding 1828) en sistema de cultivo suspendido en Estero Vizcaya, Limón, Costa Rica. Rev Biol Trop 31:257–262

Campos JA, Fournier ML, Soto R (1990) Estimación de la población de *Anadara tuberculosa* (Bivalvia: Arcidae) en Sierpe-Térraba, Costa Rica. Rev Biol Trop 38:477–480

Chaney HW (1992) Interesting gastropods from the 1992 Cocos Island Expedition. Festivus 24:86–89

12. Coan EV (1971) The northwest American Tellinidae. Veliger 14(Suppl):63 p
13. Coan EV (1973a) The northwest American Semelidae. Veliger 15:314–329
14. Coan EV (1973b) The northwest American Psammobiidae. Veliger 16:40–57
15. Coan EV (1979) Recent eastern Pacific species of the crassatellid bivalve genus *Crassinella*. Veliger 22:1–11
16. Coan EV (1983a) The eastern Pacific Donacidae. Veliger 25:273–298
17. Coan EV (1983b) A *Semele* story (Bivalvia: Semelidae). Nautilus 97:132–134
18. Coan EV (1984) The Recent Crassatellinae of the eastern Pacific, with some notes on *Crassinella*. Veliger 26:153–169
19. Coan EV (1984b) The Bernardinidae of the eastern Pacific (Mollusca: Bivalvia). Veliger 27:227–237
20. Coan EV (1985) Some additional notes on the distribution of eastern Pacific Donacidae. Veliger 28:217
21. Coan EV (1987) Concerning the type material of *Lasaea subviridis* Dall 1899. Veliger 29:445
22. Coan EV (1988a) The eastern Pacific species of *Semele*. Western Soc Malac Ann Rep [for 1987] 20:26
23. Coan EV (1988b) Recent eastern Pacific species of the bivalve genus *Semele*. Veliger 31:1–42
24. Coan EV (1990) The Recent eastern Pacific species of the bivalve family Thraciidae. Veliger 33:20–55
25. Coan EV (1996) Recent species of the genus *Petricola* in the eastern Pacific (Bivalvia: Veneroidea). Festivus 28:118–124
26. Coan EV (1997a) Recent species of the genus *Petricola* in the eastern Pacific (Bivalvia: Veneroidea). Veliger 40:298–340
27. Coan EV (1997b) The eastern Pacific Sportellidae (Bivalvia: Cyamioidea). Festivus 29:107–112
28. Coan EV (1999a) The eastern Pacific Sportellidae (Bivalvia). Veliger 42:132–151
29. Coan EV (1999b) The eastern Pacific species of *Sphenia* (Bivalvia: Myidae). Nautilus 113:103–120
30. Coan EV (2000a) The eastern Pacific recent species of the bivalve genus *Gari* (Tellinoidea: Psammobiidae), with notes on western Atlantic fossil taxa. Malacologia 42:1–29
31. Coan EV (2000b) The eastern Pacific recent species of the Corbulidae (Bivalvia). Malacologia 44:47–105
32. Coan EV (2003) The tropical eastern Pacific species of the Condylocardiidae (Bivalvia). Nautilus 117:47–61
33. Coan EV, Scott PH (1997) Checklist of the marine bivalves of the northeastern Pacific Ocean. Santa Barbara Mus Nat Hist Contr Sci 1:1–28
33a. Coan EV, Valentich-Scott, P (2008a) Three nomenclatural notes on Panamic bivalves Festivus, 40(4):49–54
33b. Coan EV, Valentich-Scott, P (2008b) Bivalvia, pp 807–859. *In*: Carlton JP (ed) Light's Manual Intertidal Invertebrates from the Central California Coast. Berkeley, California: University of California, 1002p
34. Coan EV, Scott PV, Bernard F (2000) Bivalve seashells of western North America. Marine mollusks from Arctic Alaska to Baja California. Santa Barbara Mus Nat Hist Monogr 2:764
35. Crisp DJ, Standen A (1988) *Lassaea rubra* (Montagu) (Bivalvia: Erycinacea), an apomictic crevice-living bivalve with clones separated by tidal level preference. J Exp Mar Biol Ecol 117:27–47

Cruz RA (1982) Variación mensual del índice de condición del molusco *Anadara tuberculosa* (Pelecypoda: Arcidae) en Punta Morales, Puntarenas, Costa Rica. Rev Biol Trop 30:1–4

Cruz RA (1984a) Algunos aspectos de la reproducción en *Anadara tuberculosa* (Pelecypoda: Arcidae) en Punta Morales, Puntarenas, Costa Rica. Rev Biol Trop 32:45–50

Cruz RA (1984b) Algunos aspectos reproductivos y variación mensual de índice de condición de *Anadara similis* (Pelecypoda: Arcidae) de Jicaral, Puntarenas, Costa Rica. Brenesia 22:95–105

Cruz RA (1986) Caracteres generales, edad y crecimiento de *Anadara grandis* (Pelecypoda: Arcidae). Uniciencia 3:25–29

36. Cruz RA (1996) Annotated checklist of marine molluscs collected during the R.V. Victor Hensen Costa Rica Expedition 1993/1994. Rev Biol Trop 44(Suppl 3):59–67

Cruz RA, Jiménez JA (1994) Moluscos asociados a las áreas de manglar de la costa Pacífica de América Central: guía. Editorial Fundación UNA, Heredia, 182 p

Cruz RA, Palacios JA (1983) Biometría del molusco *Anadara tuberculosa* (Pelecypoda: Arcidae) en Punta Morales, Puntarenas, Costa Rica. Rev Biol Trop 31:175–179

36a. Dall WH (1908) Reports on the dredging operations off the west coasts of Central America to the Galapagos, to the west coast of Mexico, and in the Gulf of California, in charge of Alexander Agassiz, carried on by the U.S. Fish Commission steamer "Albatross", during 1891, Lieut. Commander Z.L. Tanner, U.S.N., commanding. XXXVII, and Reports on the scientific results of the expedition to the eastern tropical Pacific, in charge of Alexander Agassiz, by the U.S. Fish Commission steamer "Albatross," from October, 1904, to March, 1905, Lieut. Commander L.M. Garrett, U.S.N., commanding. XIV. The Mollusca and Brachiopoda. Bull Mus Comp Zoöl Harvard Coll XLIII:205–487

Dance SP (1986) A History of Shell Collecting, 2nd ed. Brill, Leiden, 265 p

Díaz JM, Puyana M (1994) Moluscos del Caribe Colombiano. Un catálogo ilustrado. COLCIENCIAS-Fundación Natura-INVEMAR, Bogotá, 291 p

37. Dodge H (1950) *Hiatella* Daudin versus *Saxicaba* Bellevue. Nautilus 63:138–139
38. Emerson WK (1978) Mollusks with Indo-Pacific faunal affinities in the eastern Pacific Ocean. Nautilus 92:91–96
39. Emerson WK (1994) A zoogeographic summary of the marine mollusks of Clipperton Island (tropical eastern Pacific Ocean). Festivus 26:62–71
40. Emerson WK (1995) A zoogeographic summary of the marine mollusks of Revillagigedo Islands (tropical eastern Pacific Ocean). Festivus 27:3–18
41. Espinosa J, Ortea J (2001) Moluscos del Mar Caribe de Costa Rica: desde Cahuita hasta Gandoca. Avicennia (Suppl 4): 1–77
42. Espinosa J, Ortea J (2003) Adiciones al Catálogo de Moluscos del Mar Caribe de Costa Rica: desde Cahuita hasta Gandoca. Avicennia 16:113–120
43. Espinosa J, Ortea J, Valdés A (1994) Catálogo de los moluscos bivalvos recientes del Archipiélago cubano. Avicennia 2:109–129
44. Fischer PH II (1941) Sur l'habitat de *Lasaea rubra* Mtg. Mus Natl Hist Nat Bull (2) 12(7): 433–435

Fournier ML, de la Cruz E (1987) Reproduction of the cockle *Anarara grandis* in Costa Rica. NAGA 10:6

45. Gallardo SC, Osorio C R (1978) *Hiatella solida* (Sowerby 1834) (Mollusca: Hiatellidae) on *Concholepas concholepas* (Brugière 1789) and other substrates. Veliger 20:274–278

Glude J (1981) The Potential for Mollusks Farming in Costa Rica. Final Report, FAO, Seattle, Washington 28 p

46. Grau G (1959) Pectinidae of the eastern Pacific. Allan Hancock Pac Exp 308 p
47. Harte MH (1993) A review of *Pitar* (*Hyphantosoma*) Dall 1902 (Veneridae: Pitarinae) and a description of *Pitar* (*H.*) *festoui* sp. nov. Veliger 36:343–350

Herrera A (2005) Reconstrucción del procedimiento precolombino para la extracción de los moluscos de sus conchas. Vínculos 27:35–62

48. Hertlein LG, Strong AM (1943) Eastern Pacific expeditions of the New York Zoological Society. XXXII. Mollusks from the west coast of Mexico and Central America. Part II. Zoologica 28:149–168
49. Hertlein LG, Strong AM (1946a) Eastern Pacific expeditions of the New York Zoological Society. XXXIV. Mollusks from the east coast of Mexico and Central America. Part III. Zoologica 31:53–76
50. Hertlein LG, Strong AM (1946b) Eastern Pacific expeditions of the New York Zoological Society. XXXV. Mollusks from the west coast of Mexico and Central America. Part IV. Zoologica 31:93–120
51. Hertlein LG, Strong AM (1947) Eastern Pacific expeditions of the New York Zoological Society. XXXVI. Mollusks from the west coast of Mexico and Central America. Part V. Zoologica 31:129–150
52. Hertlein LG, Strong AM (1949a) Eastern Pacific expeditions of the New York Zoological Society. XL. Mollusks from the west coast of Mexico and Central America. Part VII. Zoologica 34:63–97

53. Hertlein LG, Strong AM (1949b) Eastern Pacific expeditions of the New York Zoological Society. XLI. Mollusks from the west coast of Mexico and Central America. Part VIII. Zoologica 34:239–258
54. Hertlein LG, Strong AM (1950) Eastern Pacific expeditions of the New York Zoological Society. XLII. Mollusks from the west coast of Mexico and Central America. Part IX. Zoologica 35:217–252
55. Hertz CM, Gemmell J, Meyers BW (1982) A review of *Chama buddiana* C.B. Adams 1852 (Bivalvia: Chamidae). Festivus 14:15–24
56. Hertz CM, Myers BW, Gemmell J (1984) A comparison of two related species of *Tellina: Tellina hiberna* Hanley 1844 and *Tellina amianta* Dall 1900. Festivus 16:28–34
57. Hertz CM, Meyers BW, Gemmell J (1985) *Pandora* (*Clidiophora*) *cornuta* C.B. Adams 1852, and new information on its distribution. Festivus 17:101–110
58. Hickman CS (1994) The genus *Parvilucina* in the eastern Pacific: making evolutionary sense of a chemosymbiotic species complex. Veliger 37:43–61
59. Hodgkin NM (1962) Limestone boring by the mytilid *Lithophaga*. Veliger 4:123–129
59a. Houbrick JR (1969) A survey of the littoral marine mollusks of the Caribbean coast of Costa Rica. Veliger 11:4–23
60. Isham LB, Moore HB, Walton Smith FG (1951) Growth rate measurement of shipworms. Bull Mar Sci Gulf Carib 1:136–147
61. Kay EA (1979) Hawaiian marine shells. Reef and shore fauna of Hawaii Section 4: Mollusca. Bishop Mus Spec Publ 64(4):654

Keen AM (1958) Sea Shells of Tropical West America; Marine Mollusks from Lower California to Colombia, 1st ed. Stanford University Press, Stanford, CA, 624 p

Keen AM (1966) Moerch's West Central American molluscan types with proposal of a new name for a species of *Semele*. Occ Pap Calif Acad Sci 59:1–33

62. Keen AM (1971) Sea Shells of Tropical West America; Marine Mollusks from Baja California to Peru, 2nd ed. Stanford University Press, Stanford, CA, 1064 p
63. Keen AM (1980) The pelecypod family Cardiidae: a taxonomic summary. Tulane Studies Geol Paleontol 16:1–44
64. Keen AM, Coan EV (1975) "Sea Shells of Tropical West America": additions and corrections to 1975. West Soc Malacol Occ Pap 1:66
65. Kemperman TCM (1990) *Jouannetia* des Moulins 1828 (Bivalvia: Pholadidae) from Costa Rica. Basteria 54(1–3):122

Krylova EM, Sahling H (2006) Recent bivalve molluscs of the genus *Calyptogena* (Vesicomyidae). J Moll Stud 72:359–395

66. Lamprell K (1987) *Spondylus*. Spiny Oyster Shells of the World. Brill-Backhuys, Leiden, 84 p
67. Lebour MV (1938) The life history of *Kellia suborbicularis*. J Mar Biol Ass UK 22:447–451

Madrigal E, Alfaro J, Quesada R, Pacheco O, Zamora E (1985a) Estructura de la población y distribución de talla del ostión de manglar (*Crassostrea rhizophorae*, Guilding 1828) en el estero Vizcaya, Limón, Costa Rica. Rev Biol Trop 33:61–62

Madrigal E, Pacheco O, Zamora E, Quesada R, Alfaro J (1985b) Tasa de filtración del ostión de manglar (*Crassostrea rhizophorae*, Guilding 1828) a diferentes salinidades y temperaturas. Rev Biol Trop 33:77–79

67a. Mikkelsen PM, Bieler R (2001) *Varicorbula* (Bivalvia: Corbulidae) of the western Atlantic: taxonomy, anatomy, life habits, and distribution. Veliger 44:271–293
67b. Mikkelsen PM, Bieler R (2003) Systematic revision of the western Atlantic file clams, *Lima* and *Ctenoides* (Bivalvia: Limoidae: Limidae). Invert Syst 17:667–710
67c. Mikkelsen PM, Bieler R (2007) Seashells of Southern Florida Living Marine Mollusks of the Florida Keys and Adjacent Regions, Bivalves. Princeton Univ Press, Princeton, New Jersey, 703 p
68. Montoya M (1983) Los moluscos marinos de la Isla del Coco, Costa Rica. I. Lista anotada de especies. Brenesia 21:325–353

Mörch OAL (1859) Beiträge zur Molluskenfauna Central Amerika's. Malakozool Blät 6(4):102–126

Mörch OAL (1860a) Beiträge zur Molluskenfauna Central Amerika's. Malakozool Blät 7(3):97–106
68a. Mörch OAL (1860b) Beiträge zur Molluskenfauna Central Amerika's. Malakozool Blät 7(4):170–192
Mörch OAL (1861) Beiträge zur Molluskenfauna Central Amerika's. Malakozool Blät 7(5):193–213
69. Myers BW, D'Attilio A (1979) Range extension for *Isongomon janus* Carpenter 1857 (Bivalvia). Festivus 11:79–81
70. Nicol D, Jones DS (1984) *Bellaxinaea,* a new subgenus of glycymeridids (Pelecypoda) from the western hemisphere. Nautilus 98:126–128
71. Nichols-Driscoll J (1976) Benthic invertebrate communities in Golfo Dulce, Costa Rica, an anoxic basin. Rev Biol Trop 24:281–297
72. Norrid CC (2000) Range extensions and size records of bivalves from the Panamic Province in the Norrid Collection. Festivus 32:151–152
Olsson AA, McGinty TL (1958) Recent marine mollusks from the Caribbean coast of Panama with the description of some new genera and species. Bull Am Paleont 177:1–59
73. Ortea J, Espinosa J (2001) Descripción de una nueva especie de *Adrana* H. y A. Adams 1858. Avicennia (Suppl 4):61–64
Pacheco-Urpí O, Cabrera-Peña J, Zamora-Madriz E (1983) Crecimiento y madurez sexual de *Crassostrea rhizophorae* (Guilding 1828) cultivada en sistema suspendido en Estero Vizcaya, Limón, Costa Rica. Rev Biol Trop 31:277–282
Palacios JA, Cruz RA, Pacheco O (1982) Contribución al análisis ecológico y biológico de *Donax dentifer* Hanley 1843 (Pelecypoda: Donacidae) en Playa Garza, Puntarenas, Costa Rica. Bol Infor UNA, Heredia 6:33–41
Palacios JA, Rodríguez JA, Cruz RA, Bolaños J (1986a) Estudios sobre la biología de *Protothaca asperrima* (Pelecypoda: Veneridae). I. Distribución y cuantificación. Brenesia 25/26:1–11
74. Pilsbry HA, Lowe HN (1932) West Mexican and Central American mollusks collected by H. N. Lowe 1929–31. Proc Acad Nat Sci Philadelphia 84:33–144
Quesada R, Madrigal E, Alfaro J, Pacheco O, Zamora E (1985) Crecimiento y supervivencia del ostión de manglar (*Crassostrea rhizophorae*, Guilding 1828), trasladado de Estero Vizcaya, costa del Caribe, a estanques de cultivo de camarones en Chomes, costa pacífica de Costa Rica. Rev Biol Trop 33:7–12
Radwin GE (1969) A recent molluscan fauna from the Caribbean coast of southeastern Panama. Trans San Diego Soc Nat Hist 15:229–236
75. Redfern C (2001) Bahamian Seashells. A Thousand Species from Abaco, Bahamas. Bahamian seashells.com, Inc., Boca Ratón, FL, 280 p
76. Robinson DG, Montoya M (1987) Los moluscos marinos de la costa Atlántica de Costa Rica. Rev Biol Trop 35:375–400
Rojas J, Villalobos CE, Chartier F, Villalobos CR (1988) Tamaño, densidad y reproducción de la barba de hacha, *Tagelus peruvianus* (Bivalvia: Solecurtidae) en el estero de Puntarenas, Costa Rica. Rev Biol Trop 36:479–483
77. Rombouts A (1991) Guidebook to Pecten Shells: Recent Pectinidae & Propeamussiidae of the World. W. Backhuys, Oegstgeest, Netherlands, 157 p
78. Rost H (1955) A report on the family Arcidae (Pelecypoda). Allan Hancock Pac Exp 20:173–248
79. Salas C, Gofas S (1998) Description of four new species of *Neolepton* Monterosato 1875 (Mollusca: Bivalvia: Neoleptonidae), with comments on its affinity with the Veneracea. Ophelia 48:35–70
80. Salvat B, Salvat F (1972) Geographic distribution of *Pinna rugosa* Sowerby 1835 (Mollusca: Bivalvia). Veliger 15:43–44
81. Shasky DR (1986a) Update on mollusks with Indo-Pacific faunal affinities in the tropical eastern Pacific IV. [abstract]. West Soc Malac Ann Rep [for 1985] 18:27
82. Shasky DR (1986b) Update on mollusks with Indo-Pacific faunal affinities in the tropical eastern Pacific IV. Festivus 18:3–5
83. Shasky DR (1987) Update on mollusks with Indo-Pacific faunal affinities in the tropical eastern Pacific – Part V. Festivus 19:48–50
Sibaja W (1986a) Madurez sexual en el mejillón chora *Mytella guyanensis* Lamark 1819 (Bivalvia: Mytillidae) del manglar de Jicaral, Puntarenas, Costa Rica. Rev Biol Trop 34:151–155

Sibaja W (1986b) Rendimiento de la carne fresca en el mejillón *Mytella guyanensis* (Bivalvia: Mytillidae) en el Golfo de Nicoya, Costa Rica. Ciencia y Tecnología (Costa Rica) 10:25–30

Sibaja W, Villalobos CR (1986) Crecimiento del mejillón chora *Mytella guyanensis* (Bivalvia: Mytillidae) en el Golfo de Nicoya, Costa Rica. Rev Biol Trop 34:231–236

84. Skoglund C (1990) Panamic Province molluscan range extensions from Bahía Cocos, Guanacaste, Costa Rica. Festivus 22:100

Skoglund C (1991) Additions to the Panamic Province bivalves (Mollusca) literature 1971 to 2000. Festivus 23:(Suppl):1–74

85. Skoglund C (2000) *Spondylus liguafelis* Sowerby 1847 (Bivalvia: Spondylidae) in the Panamic Province, with notes on the occurrence of *Spondylus nicobaricus* Schreibers 1793. Festivus 32:3–7

Skoglund C (2001) Panamic Province molluscan literature: additions and changes from 1971 through 2000. I. Bivalvia. Festivus 33(Suppl):1–119

86. Skoglund C, Mulliner DK (1996) The genus *Spondylus* (Bivalvia: Spondylidae) of the Panamic Province. Festivus 28:91–107

87. Smith LC (1945) Notes on the family Cardiidae. Mus Comp Zool Occ Pap Moll 1:29–32

Solís Del Vecchio F (2005) El criterio "facilidad de obtención" para explicar las frecuencias relativas de moluscos en los depósitos de concha precolombinos de la Península de Nacascolo. Vínculos 27(1–2):63–79

88. Soot-Ryen T (1955) A report on the family Mytilidae (Pelecypoda). Allan Hancock Pac Exp 20:1–261

Sowerby I GB (1833) Characters of new species of Mollusca and Conchifera, collected by Hugh Cuming. Proc Zool Soc London (May 17):16–22

Sowerby II GB (1866) Monograph of the genus *Donax*, Linnaeus. In: GB Sowerby II. ed. Thesaurus conchyliorum or, monograph of genera of shells 3(24–25):305–317

89. Tevesz M (1977) Taxonomy and ecology of the Philobryidae and Limopsidae (Mollusca: Pelecypoda). Postilla 171:1–64

90. Turner RD (1954) The family Pholadidae in the western Atlantic and the eastern Pacific. Part I – Pholadininae. Johnsonia 3(33):1–63

91. Turner RD, Boss KJ (1962) The genus *Lithophaga* in the western Atlantic. Johnsonia 4(41):81–116

92. Valentich-Scott PH (1998) Class Bivalvia, pp 97–113. *In*: Valentich-Scott PH, Blake JA (eds.). Taxonomic Atlas of the Benthic Fauna of the Santa Maria Basin and the Western Santa Barbara Channel. Volume 8. The Mollusca Part 1. The Aplacophora, Polyplacophora, Scaphopoda, Bivalvia, and Cephalopoda. Santa Barbara Mus Nat Hist, California, 250 p

93. Vidal J (1999) Taxonomic review of the elongated cockles, genera *Trachycardium*, *Vasticardium* and *Acrosterigma* (Mollusca, Cardiidae). Zoosystema 21:259–335

Villalobos CR (1982) Cultivo experimental del ostión japonés *Crassostrea gigas* en una región estuarina del Pacífico de Costa Rica. Mem II Congr Nac Biol:1–6

Villalobos CR, Baéz AL (1984) Tasa de crecimiento y mortalidad en *Anadara tuberculosa* (Bivalvia: Arcidae) bajo dos sistemas de cultivo. Rev Latinoamer Acuicult 17:9–19

94. Vokes HE, Vokes H (1983) Distribution of shallow-water marine Mollusca, Yucatan Peninsula, Mexico. Middle Amer Res Inst Publ 54:1–183

95. Wagner HP (1985) Notes on type material of the family Pectinidae (Mollusca: Bivalvia). 3. On the identity of *Pecten solidulus* Reeve 1853, and *Pecten commutatus* Monterosato 1875. Basteria 49:81–84

96. Warmke GL, Abbott RT (1961) Caribbean Seashells. Livingstone, Pennsylvania, 346 p

97. Wilson BR, Tait R (1984) Systematics, anatomy and boring mechanisms of the rock-boring mytilid bivalve *Botula*. Proc. Royal Soc Victoria 96:113–125

Part 32
Squids and Octopuses

Frederick G. Hochberg and Yolanda E. Camacho-García

Unidentified squid collected in Cahuita, Caribbean coast of Costa Rica (Photo: Leslie Harris)

F.G. Hochberg
Department of Invertebrate Zoology, Santa Barbara Museum of Natural History, 2559 Puesta del Sol, Santa Barbara, CA 93105, USA.
e-mail: fghochberg@sbnature2.org

Y.E. Camacho-García
Museo de Zoología, Escuela de Biología, Universidad de Costa Rica, San Pedro de Montes de Oca, 11501-2060 San José, Costa Rica.
e-mail: ycamacho_99@yahoo.com

I.S. Wehrtmann, J. Cortés (eds.) *Marine Biodiversity of Costa Rica, Central America*,

Abstract The squid and octopus fauna (Cephalopoda) in the waters of Costa Rica is poorly known. In this part we summarize the species of cephalopods known to occur along the coasts of Costa Rica. Due to the lack of distributional and presence data in Costa Rican waters, we do not include small oceanic squids and octopuses nor do we include new shallow-water benthic octopuses that have not yet been described. Our review indicates a total of 46 cephalopod species in seven families, where 26 species are exclusively from the Caribbean and 18 only from the Pacific; two species are common to both coasts. No cephalopod species are known to be endemic to Costa Rica. We recommend that future investigations of the cephalopod fauna of Costa Rica and Central America consider existing museum collections, apply different collecting methods, and include records of paralarval diversity as well as collections of females brooding eggs.

Introduction

The Cephalopoda is a class of molluscs that are widely distributed in oceanic and benthic marine habitats around the world. Cuttlefishes, squids, and octopuses typically are dominant predators wherever present in marine environments. Many of the large, schooling, neritic, and oceanic squids, as well as a number of the large shallow-water octopuses are the targets for commercial or artisanal fisheries (Hochberg & Couch 1971). Although diversity in terms of numbers of taxa is relatively low compared with other molluscan classes, based on our observations of specimens present in museum collections, this group of molluscs is well represented in the marine fauna of Costa Rica.

At present no published accounts accurately summarize the taxonomic diversity, distribution, and biology of cephalopods that live in the waters of Costa Rica. A few papers provide annotated lists of marine molluscs found on both coasts of Costa Rica and Isla del Coco; however, the names for the cephalopod species listed typically are not correct (Montoya 1983a, b; Robinson & Montoya 1987).

Several reviews have been published that treat cephalopod species encountered in Central America, but in general the information available is out of date, or based on old reports. The following taxonomic or fishery overviews are especially helpful: Robson (1929, 1932), Pickford (1945, 1946), Voss (1956, 1971, 1973), Voss & Toll (1998), Voss *et al.* (1973), Rathjen *et al.* (1979), Reid & Jereb (2005), Roper (1978), Roper *et al.* (1984, 1995), Voss & Brakoniecki (1985a, b), Nesis (1987), Voss *et al.* (1998a, b). Several publications that illustrate body patterns of living cephalopods with extensive color photographs are helpful for identification of squid and octopus species (see especially: Hanlon 1982, 1988; Norman 2000). Publications that document the taxonomy and biology of the cephalopod fauna are also available for the following regions adjacent to Costa Rica: Cuba (Voss 1955); Mexico (Lipka 1975; Hochberg 1980; Roper *et al.* 1995); Panama (Voss 1971); or Venezuela (Arocha & Urosa 1982, 1983, 1991; Arocha & Robaina 1984; Arocha 1989; Arocha *et al.* 1991).

In the present summary of the cephalopods known to occur in Costa Rica (Species Lists 32.1 and 32.3 which are included on the CD-Rom), we have chosen to mostly focus on the larger taxa in the families that have potential economic or fishery potential, namely: Sepiolidae, Loliginidae, Lepidoteuthidae, Onychoteuthidae, Ommastrephidae, Thysanoteuthidae, Octopodidae, and Argonautidae. We have not included many small oceanic squids and octopuses because we know too little about their specific presence and distribution in Costa Rica. Nor have we listed a number of species of small, shallow-water benthic octopuses that currently are recognized as "undescribed" in a review of the octopuses in the eastern central Pacific (Hochberg & Sweeney, unpublished data).

A distinct and nonoverlapping fauna of cephalopods is present on both coasts with similar numbers of taxa represented in each family. New species of shallow-water octopuses from the eastern central Pacific are currently being described; once these descriptions are published, the numbers of taxa on both coasts will be almost identical. At least two, and possibly three, additional undescribed genera of octopuses are also endemic to both coasts of Central America (Hochberg & Norman, unpublished data). At present no cephalopod species are known to be endemic to Costa Rica.

Several genera included in the cuttlefish family Sepiidae (e.g., *Sepia*, *Metasepia*, etc.) have never been found in the Americas (Voss 1974; Reid *et al.* 2005). The octopus genus *Eledone* (Octopodidae) is unknown in the Caribbean and northwestern Atlantic although members of this Atlantic genus have been described from off South America (southern Brazil). Trans-isthmian connections are discussed by Nesis (1975) and Voight (1988).

The taxonomy of shallow-water octopuses is currently in a state of flux. In the past the genus *Octopus* served as a catch-all where any octopus-looking animal was placed. Recently, the genus has been redefined and the majority of species removed (Norman & Hochberg 2005). However, until new genera are erected, a number of species cannot be properly allocated. Consequently, in the tables that follow, the genus for these unallocated taxa is indicated as "*Octopus*."

Specialists

At present, relatively few researchers have experience with cephalopods that live along the coasts of Costa Rica or elsewhere in Central America. Scientists that are actively working on either basic taxonomy or biology in this geographic area are especially rare. We have listed the key researchers below.

A. Cephalopods in general

Roger T. Hanlon: Marine Resources Center, Marine Biological Laboratory, Woods Hole, Massachusetts, USA – Biology & behavior. rhanlon@mbl.edu

Manuel Haimovici: Departamento de Oceanografia, Fundação Universidade, Federal do Rio Grande, Rio Grande, Brazil – Biology & taxonomy. docmhm@furg.br

Jennifer Mather: Department of Psychology & Neuroscience, University of Lethbridge, Lethbridge, Alberta, Canada – Behavior. mather@uleth.ca

Mark N. Norman: Department of Marine Invertebrates, Museum of Victoria, Melbourne, Victoria, Australia – Taxonomy & biology. mnorman@museum.vic.gov.au

B. Squids and bobtail squids

Alexander I. Arkhipkin: Fisheries Department, Falkland Islands, Government, Stanley, Falkland Islands – Biology & taxonomy. aarkhipkin@fisheries.gov.fk

Roxana de Silva Dávila: Departamento de Plancton y Ecología Marina, Instituto Politechnico Nacional, Centro Interdisciplinario de Ciencias Marinas, La Paz, Baja California Sur, México – Taxonomy & biology. rdesilva@ipn.mx

Unai Markaida: Laboratorio de Pesquerías Artesanales, El Colegio de la Frontera Sur, Unidad Campeche, Campeche, México – Biology. umarkaida@camp.ecosur.mx

Michael Vecchione: Systematics Laboratory, National Marine Fisheries, Smithsonian Institution, Washington, DC, USA – Taxonomy. vecchiom@si.edu

C. Octopuses

Franz Cardoso: Laboratorio de Biología y Sistemática de Invertebrados Marinos, Departamento de Zoología, Universidad Nacional Mayor de San Marcos, Lima, Perú – Biology & taxonomy. fcardosop@unmsm.edu.pe

John A. Cigliano: Department of Biological Sciences, Cedar Crest College, Allentown, Pennsylvania, USA – Behavior and ecology. cigliano@bu-bio.bu.edu

F.G. Hochberg: Department of Invertebrate Zoology, Santa Barbara Museum of Natural History, Santa Barbara, California, USA – Taxonomy, biology, paralarvae & parasites. fghochberg@sbnature2.org

Janet R. Voight: Department of Zoology, Field Museum, Chicago, Illinois, USA – Biology & morphology. jvoight@fmnh.org

Collections

Relatively few preserved specimens of cephalopods obtained from the coasts of Costa Rica are available in museum collections. At present significant collections of cephalopods are not known to be housed in any museum or university in Costa Rica. The following museums in the USA and Europe have minor but important collections of cephalopods from Central America. Cephalopods are also represented in a few private collections. In all cases less than 100 specimens or specimen lots specifically from Costa Rica are present in each collection.

A. North America (USA)

NMNH: Department of Systematic Biology, Division of Molluscs, National Museum of Natural History, Smithsonian Institution, Washington, DC (collections of shallow- and deepwater cephalopods from the Pacific coast of Costa Rica, collected by private collectors).

LACM: Mollusc Section, Natural History Museum of Los Angeles County, Los Angeles, California (collections of shallow- and deepwater cephalopods from the Pacific coast of Costa Rica, mainly collected by the R/V *Velero IV* and R/V *Searcher* expeditions).

CAS: Department of Invertebrates and Geology, California Academy of Sciences, San Francisco, California (collections of shallow-water cephalopods from the Pacific coast of Costa Rica, including Isla del Coco, mainly collected by individual collectors and the CAS-Crocker Expedition in 1932].

SBMNH: Department of Invertebrate Zoology, Santa Barbara Museum of Natural History, Santa Barbara, California (synoptic marine mollusc collections, especially from Isla del Coco and expeditionary material originally obtained by the Allan Hancock Foundation on R/V *Velero* expeditions).

UMML: Division of Marine Biology & Fisheries, Rosenstiel School of Marine & Atmospheric Science, University of Miami, Miami, Florida (expeditionary material of shallow-water cephalopods from the Caribbean coast of Costa Rica obtained by 1971 Central American Cruise of R/V *John Elliott Pillsbury* [RSMAS, University of Miami]).

B. Europe

MNHN: Département Milieux et Peuplements Aquatiques, Muséum National d'Histoire Naturelle, Paris, France.

BMNH: Department of Zoology, Malacology Section, The Museum of Natural History, London, England.

Recommendations

The cephalopod fauna of Costa Rica is poorly known especially with regard to the diversity of benthic octopuses and small oceanic squids. Based on unpublished information, at least six new species of shallow- and deepwater octopuses are recognized as being present in the waters off both coasts of Central America (Hochberg, unpublished data). Many of these undescribed species are likely to be present in Costa Rica.

Existing museum holdings need to be systematically and critically analyzed to document the taxonomic diversity of cephalopods not only from Costa Rica, but also from neighboring countries in Central America. A concerted effort needs to be undertaken to adequately sample a diversity of marine habitats for cephalopods. Collections need to be made by hand, trawl, and plankton net. The latter is especially important to record the diversity of paralarval stages present in surface waters (see Hochberg *et al.* 1992). Whenever females are encountered, brooding eggs should be collected in order to determine egg size for each octopus species present in the region. Eggs should be reared until the embryos hatch to confirm paralarval diversity obtained in plankton samples.

Collection of cephalopods should not only result in the fixation and preservation of whole animals of a variety of sizes and ages, but should be utilized also for the preparation of tissue for molecular analysis that compliments modern taxonomic studies.

In addition, a concerted effort should be made to collect parasites of cephalopods as they often provide useful host-specific information that can be used to confirm or support baseline taxonomy. In particular, the following parasitic phyla from cephalopod hosts are totally unknown for Costa Rica: Apicomplexa, Ciliophora, Dicyemida, Platyhelminthes (Digenea, Cestoda), Nematoda, and Arthropoda (Copepoda) (see Hochberg, 1990).

Acknowledgments We are thankful to Lindsey Groves and Angel Valdés (LACM), Linda Cole, Michael Sweeney, and Michael Vecchione (NMNH), and Nancy Voss (UMML) for providing valuable information about the cephalopod material from Costa Rica housed at their institutions.

References

1. Adam W (1936) Notes sur les Céphalopodes. VI. Une nouvelle espece d'Octopus (*Octopus hummelincki* sp. nov.) des Indes Occidentales Néerlandaises. Bull Mus R Hist Nat Belgique 12:1–3

Arocha F (1989) Cephalopod resources of Venezuela. Mar Fish Rev 51:47–51

Arocha F, Robaina G (1984) First record of *Octopus defilippi* Verany, 1851 in Venezuelan coastal waters. Nautilus 98:25–28

Arocha F, Urosa LJ (1982) Cefalópodos del género *Octopus* en el área nororiental de Venezuela. Bol Inst Oceanogr Venez Univ Oriente 21:167–189

Arocha F, Urosa LJ (1983) Notes on two Octopoda from Venezuelan waters. Bull Mar Sci 33:941–942

Arocha F, Urosa LJ (1991) Some biological studies of *Loligo plei* and occurrence of *Loligo pealii* (Cephalopoda, Myopsida) in northeastern Venezuela. Biol Mar 42:145–152

Arocha F, Marcano L, Cipriani R (1991) Cephalopods trawled from Venezuelan waters by the R/V *Dr. Fridtjof Nansen* in 1988. Bull Mar Sci 49:231–234

2. Berry SS (1954) *Octopus penicillifer*, new species. Leafl Malac 1(11):66
3. Brakoniecki TF (1980) *Lolliguncula tydeus*, a new species of squid (Cephalopoda; Myopsida) from the Pacific coast of Central America. Bull Mar Sci 30:424–430
4. Brakoniecki TF (1996) A revision of the genus *Pickfordiateuthis* Voss, 1953 (Cephalopoda; Myopsida). Bull Mar Sci 58:9–28
5. Brakoniecki TF, Roper CFE (1985) *Lolliguncula argus*, a new species of loliginid squid (Cephalopoda: Myopsida) from the tropical eastern Pacific. Proc Biol Soc Wash 98:47–53
6. Burgess LA (1966) A study of the morphology and biology of *Octopus hummelincki* Adam, 1936 (Mollusca: Cephalopoda). Bull Mar Sci 16:762–813
7. Cohen A (1976) The systematics and distribution of *Loligo* (Cephalopods: Myopsida) in the western North Atlantic, with description of two new species. Malacologia 15:299–367
8. Forsythe JW, Toll RB (1991) Clarification of the western Atlantic Ocean pygmy octopus complex: the identity and life history of *Octopus joubini* (Cephalopoda: Octopodinae). Bull Mar Sci 49:88–97

Gleadall IG (2004) Some old and new genera of octopus. Interdisciplinary Information Series 10(2):99–112

8a. Goldman DA (1995) A juvenile of the scaled squid, *Pholidoteuthis adami* Voss, 1956 (Cephalopoda: Oegopsida), from the Florida Keys. Proc Biol Soc Wash 108:136–146

9. Hanlon RT (1982) The Functional Organization of Chromatophores and Iridescent Cells in the Body Patterning of *Loligo plei* (Cephalopoda: Myopsida). Malacologia 23:89–119
10. Hanlon RT (1983) *Octopus briaeus*. *In*: Boyle PR (ed) Cephalopod Life Cycles, vol I. Species Accounts. Academic Press, London, pp 251–266
11. Hanlon RT (1988) Behavioral and Body Patterning Characters Useful in Taxonomy and Field Identification of Cephalopods. Malacologia 29:247–265
12. Hanlon RT, Hixon RF (1980) Body patterning and field observations of *Octopus burryi* Voss, 1950. Bull Mar Sci 30:749–755
13. Hanlon RT, Wolterding MR (1989) Behavior, body patterning, growth and life history of *Octopus briareus* cultured in the laboratory. Am Malacol Bull 7:21–45
14. Hanlon RT, Forsythe JW, von Boletzky S (1985) Field and laboratory behavior of "*Macrotritopus* larvae" reared to *Octopus defilippi* Verany, 1851 (Mollusca: Cephalopoda). Vie Milieu 35:237–242

Hochberg FG (1980) Class Cephalopoda. *In*: Brusca RC (ed) Common Intertidal Invertebrates of the Gulf of California, 2nd ed. University of Arizona Press, Tucson, AZ, pp 201–204

Hochberg FG (1990) Diseases of Mollusca: Cephalopoda. Diseases caused by protistans and metazoans. *In*: Kinne O (ed) Diseases of Marine Animals, vol III. Biologische Anstalt Helgoland: Hamburg, Germany, pp. 47–227

15. Hochberg FG, Couch JA (1971) Biology of cephalopods. Tektite II, Scientists in the Sea, Mission 8–50. Department of Interior, pp VI-221–228

Hochberg FG, Nixon M, Toll RB (1992) Order Octopoda Leach, 1818. *In*: Sweeney MJ, Roper CFE, Mangold K, Clarke MR, Boletzky Sv (eds) "Larval" and Juvenile Cephalopods: a Manual for their Identification. Smithson. Contrib Zool 513:213–280

15a. Lipka DA (1975) The systematics and zoogeography of Cephalopods from the Gulf of Mexico. Ph.D. dissertation, Texas A&M University, College Station, Texas, 351 p

Montoya M (1983a) Los moluscos marinos de la Isla del Coco, Costa Rica. I. Lista anotada de especies. Brenesia 21:325–353

Montoya M (1983b) Marine molluscs of Cocos Island, Costa Rica. I. Bibliographic compilation of species. Ann Rep West Soc Malac 16:1–56 [translated by H. Bertsch]

16. Moynihan MH, Rodaniche AF (1982) The behavior and natural history of the Caribbean reef squid *Sepioteuthis sepioidea*. Adv Ethol 25:1–150

Nesis KN (1975) [Cephalopods of the American Mediterranean Sea]. Trudy Inst Okeanol Akad Nauk SSSR 100:259–288 [In Russian]

Nesis KN (1987) Cephalopods of the World – Squids, Cuttlefishes, Octopuses and Allies. T.F.H. Publications, Neptune City, NJ, 351 p

17. Nesis KN, Nikitina IV (1981) [*Macrotritopus*, a planktonic larva of the benthic octopus *Octopus defilippi*: identification and distribution]. Zool Mag 60:835–846 [In Russian]
18. Nesis KN, Nikitina IV (1991) Larvae of the benthic octopus, *Euaxoctopus panamensis* (Cephalopoda, Octopodidae), in the plankton of the open waters of the eastern Pacific. Zool Zhur 70:30–35

Norman M (2000) Cephalopods: a World Guide. ConchBooks, Hackenheim, Germany, 320 p

Norman MD, Hochberg FG (2005) The current state of octopus taxonomy. Phuket Mar Biol Cent Res Bull 66:127–154

19. Pickford GE (1945) Le Poulpe Américaine: a study of the littoral Octopoda of the western Atlantic. Trans Connecticut Acad Arts Sci 36:701–811
20. Pickford GE (1946) A review of the littoral Octopoda from central and western Atlantic stations in the collections of the British Museum. Ann Mag Nat Hist (ser 11) 13:412–429

Rathjen WF, Hixon RF, Hanlon RT (1979) Squid fishery resources and development in the northwest Atlantic and Gulf of Mexico. Proc Ann Gulf Carib Fish Inst 31:145–157

Reid M, Jereb P (2005) Family Sepiolidae Leach, 1817. *In*: Jereb P, Roper CFE (eds) Cephalopods of the World: an Annotated and Illustrated Catalogue of the Cephalopod Species known to Date, vol 1. Chambered Nautiluses and Sepiolids (Nautilidae, Sepiidae, Sepiolidae, Sepiadariidae, Idiosepiidae and Spirulidae). FAO Species Catalogue for Fishery Purposes No. 4. Food and Agriculture Organization of the United Nations: Rome, pp 153–203

Reid M, Jereb P, Roper CFE (2005) Family Sepiidae Keferstein, 1866. *In*: Jereb P, Roper CFE (eds) Cephalopods of the World: an Annotated and Illustrated Catalogue of the Cephalopod Species known to date, vol 1. Chambered Nautiluses and Sepiolids (Nautilidae, Sepiidae, Sepiolidae, Sepiadariidae, Idiosepiidae and Spirulidae). FAO Species Catalogue for Fishery Purposes No. 4. Food and Agriculture Organization of the United Nations: Rome, pp 57–152

21. Robaina G (1983) Sobre el cultivo y el mantenimiento de *Octopus briareus* (Cephalopoda: Octopoda). Inf Tech Invest Pesq 106:3–20
22. Robaina G, Voglar J (1986) Maduración sexual de las hembras de *Sepioteuthis sepioidea* en las costas venezolanas. Invest Pesq 50:57–68

Robinson DG, Montoya M (1987) Los moluscos marinos de la costa Atlántica de Costa Rica. Rev Biol Trop 35:375–400

Robson GC (1929) A Monograph of the Recent Cephalopoda. Pt. I. Octopodinae. London, England, British Museum (Natural History), 236 p

Robson GC (1932) A Monograph of the Recent Cephalopoda. Pt. II. The Octopoda, Excluding the Octopodinae. British Museum (Natural History), London, England. 359 p

23. Rodaniche AF (1984) Iteroparity in the Lesser Pacific Striped Octopus, *Octopus chierchiae* (Jatta, 1889). Bull Mar Sci 35:99–104
24. Roper CFE (1978) Cephalopods. *In*: W. Fischer (ed.). FAO Species Identification Sheets for Fishery Purposes, vol VI. Western Central Atlantic (Fishing Area 31). FAO, Rome, pp 277
25. Roper CFE, Sweeney MJ, Nauen C (1984). FAO Species Catalogue, vol 3. Cephalopods of the World: an Annotated and Illustrated Catalogue of Species of Interest to Fisheries. FAO Fisheries Synopsis 125. FAO, Rome, 277 p
26. Roper CFE, Sweeney MJ, Hochberg FG (1995). Cefalópodos. *In*: Fischer W, Krupp F, Schneider W, Sommer C, Carpenter KE, Niem VH (eds) Guía FAO para la identificación de especies para los fines de la pesca. Pacifico Centro-oriental, vol I. Plantas e Invertebrados. FAO, Rome, pp 305–353
27. Squires HJ, Barragan JH (1979) *Lolliguncula panamensis* (Cephalopoda: Loliginidae) from the Pacific coast of Colombia. Veliger 22:67–74
28. Summers WC (1983) *Loligo pealei*. *In*: Boyle PR (ed) Cephalopod Life Cycles, vol I. Species Accounts. Academic Press, London, pp 115–142
29. Toll RG (1981) *Benthoctopus oregonae*, a new species of octopod (Mollusca: Cephalopoda) from the southern Caribbean with a redescription of *Benthoctopus januarii* (Hoyle, 1885). Bull Mar Sci 31:83–95
30. Van Heukelem WF (1983) *Octopus maya*. *In*: Boyle PR (ed) Cephalopod Life Cycles, vol I. Species accounts. Academic Press, London, pp 311–323

Voight JR (1988) Trans-Panamanian geminate octopods (Mollusca: Octopoda). Malacologia 29:289–294

Voss GL (1950) Two new species of cephalopods from the Florida Keys. Rev de la Soc Malacol "Carlos de la Torre" 7(2):73–79

31. Voss GL (1951) Further description of *Octopus burryi* Voss, with a note on its distribution. Bull Mar Sci Gulf Carib 1:231–240
32. Voss GL (1953) A new family, genus and species of myopsid squid from the Florida Keys. Bull Mar Sci Gulf Carib 2:602–609

Voss GL (1955) The Cephalopoda obtained by the Harvard-Havana expedition off the coast of Cuba in 1938–39. Bull Mar Sci Gulf Carib 5:81–115

33. Voss GL (1956) A review of the cephalopods of the Gulf of Mexico. Bull Mar Sci Gulf Carib 6:85–178
34. Voss GL (1957) A first record of *Octopus macropus* Risso from the United States with notes on its behavior, color, feeding and gonads. Quart J Flor Acad Sci 20:223–232
35. Voss GL (1958) Observations on *Ornithoteuthis antillarum* Adam, 1957, an ommastrephid squid from the West Indies. Bull Mar Sci Gulf Carib 7:370–378
36. Voss GL (1964) *Octopus defilippi* Verany, 1851. An addition to the cephalopod fauna of the western Atlantic. Bull Mar Sci Gulf Carib 14:554–560

37. Voss GL (1968) Octopods from the R/V Pillsbury southwestern Caribbean cruise, 1966, with a description of a new species, Octopus zonatus. Bull Mar Sci 18:645–659

Voss GL (1971) The cephalopod resources of the Caribbean Sea and adjacent regions. In: Symposium on Investigations and Resources of the Caribbean Sea and Adjacent Regions. Pap Fish Resources, FAO, Rome, vol 1, pp 307–323

38. Voss GL (1971) Cephalopods collected by the R/V John Elliott Pillsbury in the Gulf of Panama in 1967. Bull Mar Sci 21:1–34

Voss GL (1973) Cephalopod resources of the world. FAO Fish Circ 149:1–75

Voss GL (1974) On the absence of cuttlefish in the western Atlantic. Veliger 16:367–369

39. Voss GL (1975) Euaxoctopus pillsburyae, new species (Mollusca: Cephalopoda) from the southern Caribbean and Surinam. Bull Mar Sci 25:346–352

Voss GL, Brakoniecki TF (1985a) Squid resources of the Gulf of Mexico and the southeast Atlantic coasts of the United States. NAFO Sci Coun Stud 9:27–37

Voss GL, Brakoniecki TF (1985b) The distribution and numbers of the commercial or potentially commercial squid of the Gulf of Mexico and southeastern Atlantic Ocean. Gulf & South Atlantic Fisheries Development Foundation, Tampa, Florida, 61 p

40. Voss GL, Solis-Ramirez M (1966) Octopus maya, a new species from the Bay of Campeche, Mexico. Bull Mar Sci 16:615–625

Voss GL, Toll RB (1998) The systematics and nomenclatural status of the Octopodinae described from the Western Atlantic Ocean. In: Voss NA, Vecchione M, Toll RB, Sweeney MJ (eds). Systematics and Biogeography of Cephalopods, vol II. Smithson Contrib Zool No 586, pp 457–474

Voss GL, Opresko L, Thomas R (1973) The potentially commercial species of octopus and squid of Florida, the Gulf of Mexico and the Caribbean Sea. Sea Grant Field Guide Ser 2:1–33

Voss NA, Vecchione M, Toll RB, Sweeney MJ (1998a) Systematics and Biogeography of Cephalopods, vol I. Smithson Contr Zool No. 586, 586 p

Voss NA, Vecchione M, Toll RB, Sweeney MJ (1998b) Systematics and Biogeography of Cephalopods, vol II. Smithson Contr Zool No. 586, 599 p

Part 33
Phoronids

Christian C. Emig

Phoronis hippocrepia: possibly present on the Pacific coast of Costa Rica (Photo: Christian C. Emig)

C.C. Emig
Centre d'Océanologie, 20, Rue Chaix, F-13007 Marseille, France.
e-mail: emig@free.fr

I.S. Wehrtmann, J. Cortés (eds.) *Marine Biodiversity of Costa Rica, Central America*,

Abstract Worldwide there are ten valid species of Phoronida. All are known from the Pacific Ocean. Five species, viz. *Phoronis hippocrepia*, *P. muelleri*, *P. psammophila*, *Phoronopsis harmeri*, and *Phoronopsis albomaculata*, have been cited along the Pacific coast of Central America, and the former three also occur in the Caribbean. However, so far none have been recorded from the coasts of Costa Rica.

Introduction

Phoronids are exclusively marine animals with only two genera (*Phoronis* and *Phoronopsis*) and ten species. The Phoronida Hatschek, 1888 are considered a phylum or a class according to various authors, and have no intermediate hierarchical level until the genus. They are sedentary infaunal benthic suspension-feeders, with a vermiform body enclosed in a slender chitinous tube in which it moves freely, anchored by the ampulla, which is the end-bulb of the body. The specific diagnostic features are: presence of a lophophore, defined as "a tentaculated extension of the mesosome, and its cavity, the mesocoelom, which embraces the mouth but not the anus" (Emig 1976); U-shaped digestive tract; nervous center between mouth and anus, a ring nerve at the basis of the lophophore, one or two giant nerves fibers; one pair of nephridia acting also as gonoducts; closed circulatory system with red blood corpuscles; hermaphroditic or dioecious, some species have brooding patterns; egg cleavage is radial and total, with three types of developments. The larva is called *Actinotrocha* (or actinotroch). Asexual reproduction by transverse fission will also occur in most of the species.

Separate names for larval and adult forms are still used in taxonomy. Despite the priority of the larval name *Actinotrocha*, the International Commission of Zoological Nomenclature accepted the name *Phoronis* as also valid (see Silén 1952). Consequently, the actinotroch keeps a separate "generic" name considered as a technical term under *Actinotrocha*, which is sometimes still different from the adult species name.

Summary and Comments

Ten valid species of Phoronida are known worldwide, but none have been recorded along the Costa Rican coast. From the nearest records located along the coasts of southern California and of Panama (Emig 1982; Emig & Golikov 1990; Emig & Roldán 1992; Bailey-Brock & Emig 2000) five species have been cited, of which three occur also in the Caribbean (Species List 33.1 is included on the CD-Rom). However, all phoronid species have occurrences in the Pacific Ocean (Species List 33.2 is included on the CD-Rom).

Phoronis hippocrepia has mainly been reported from the Atlantic. This species is known only from two locations in the eastern Pacific (Hawaii, west coast of Panama) and in the Caribbean in Veracruz (Mexico).

P. muelleri is a cosmopolitan species which occurs in the eastern Pacific in Hawaii, along the west coast of Panama, and off the coast of Ecuador at a depth of 156 m (7°45′S 80°05′W), collected during cruise n° 34 of the R/VAkademik Kurtchatov (Emig 1984). In the Caribbean this species has been recorded from the coast of Florida (USA).

P. psammophila has been sampled in the eastern Pacific in Hawaii, California (USA) and west coast of Panama, as well as along the coast of Florida (USA) in the Caribbean where this species is generally described under its synonymous name *Phoronis architecta*.

For *Phoronopsis albomaculata*, the only known occurrence in the eastern Pacific is on the west coast of Panama; however, this species has a large distribution in the western Pacific and in other oceans.

Phoronopsis harmeri is a well-known species in the Pacific, in particular along the coast from Canada to California, and on the west coast of Panama, as well as in the Cook Island. Along the US Pacific coast, this species is often described under its synonymous name *Phoronopsis viridis*. During the last decade this species was recorded in several locations in the Atlantic Ocean and in the Mediterranean Sea.

Updated information on Phoronida taxonomy is available at the Phoronida web site: http://paleopolis.rediris.es/phoronida and at the mirror site http://emig.free.fr/Phoronida/.

Specialist and Collections

Christian C. Emig (other email address: brachnet@aliceadsl.fr)

The collection of the phoronid specimens collected from Central America is housed at the Smithsonian Institution, National Museum of Natural History, Washington DC (USA).

References

1. Bailey-Brock JH, Emig CC (2000) Hawaiian Phoronida (Lophophorata) and their distribution in the Pacific region. Pac Sci 54:119–126

Emig CC (1976) Le lophophore - structure significative des Lophophorates (Brachiopoda, Bryozoa, Phoronida). Zool Scripta 5:133–137

2. Emig CC (1982) The biology of Phoronida. Adv Mar Biol 19:1–89
3. Emig CC (1984) New data on the phoronids, Lophophorata, collected by Soviet expeditions (in Russian). Biol Mor Vladivostok 4:65–67
4. Emig CC, Golikov AN (1990) On phoronids of the Far Eastern Seas of the USSR and their distribution in the Pacific Ocean (in Russian). Zool Zh 69:22–30
5. Emig CC, Roldán C (1992) The occurrence in Australia of three species of Phoronida (Lophophorata) and their distribution in the Pacific area. Rec S Aust Mus 26:1–8

Silén L (1952) Researches on Phoronida of the Gullmar Fiord area (west coast of Sweden). Ark Zool 4:95–140

Part 34
Bryozoans

Jorge Cortés, Vanessa Nielsen, and Amalia Herrera-Cubilla

Two unidentified species of the Order Cyclostomata plus one Cheilostomata from Bahía Salinas, north Pacific of Costa Rica (Photo: Andrea Bernecker)

J. Cortés
Centro de Investigación en Ciencias del Mar y Limnología (CIMAR), and Escuela de Biología, Universidad de Costa Rica, 11501–2060 San José, Costa Rica.
e-mail: jorge.cortes@ucr.ac.cr

A. Herrera-Cubilla
Center for Tropical Paleoecology and Archeology, Smithsonian Tropical Research Institute, P.O. Box 2072. Balboa/Ancón, Panamá, República de Panamá.
e-mail: herreraa@ancon.si.edu

V. Nielsen
Unidad Investigación Pesquera y Acuicultura (UNIP) of the Centro de Investigación en Ciencias del Mar y Limnología (CIMAR), Universidad de Costa Rica. 11501–2060 San José, Costa Rica.
e-mail: vnielsen@cariari.ucr.ac.cr

I.S. Wehrtmann, J. Cortés (eds.) *Marine Biodiversity of Costa Rica, Central America*,

Abstract Sixty-one species of bryozoans have been reported in the literature from Costa Rica and are distributed as follows: Caribbean: 13 species in 11 genera, 10 families, 1 order, and 1 class; Pacific: 49 species in 41 genera, 31 families, 3 orders, and 2 classes. One species is present in both coasts. The 61 species are in 49 genera, 36 families, and 3 orders in 2 classes. Only one site in the Caribbean (Portete) has been sampled, and from the Pacific most of the collections were carried out along the shore or by dredging in shallow waters. More studies are needed to obtain a more complete picture of the diversity and ecology of bryozoans in Costa Rica and Central America.

Introduction

The bryozoans (Phylum Bryozoa) are known as moss animals or sea mats because of their fuzzy appearance; they are also called Phylum Ectoprocta. They consist of individual zooids that are boxlike, the small calcareous chambers often less than one millimeter in length and form colonies (zoaria) of interconnected individuals (Soule *et al.* 1975; Brusca & Brusca 1990). Colonies vary greatly in size and are formed from few to thousands of zooids. They are usually encrusting over hard substrates like rocks, corals, and shells, and even over soft organisms like algae and seagrasses. The colonies growth forms vary from encrusting to erect zoaria that form lacy or fan-like colonies, some of which are extremely fragile. Encrusting colonies range from a few millimeters to as much as 50 cm^2 (McKinney & Jackson 1989), most being between 5–25 cm^2. Erect colonies could reach up to 1.5 m in diameter. Their oceanic distribution is worldwide and can be found from the intertidal zone to 6,000 m and from the tropics to the polar seas; but they are most abundant in the littoral and neritic zones down to approximately 200 m (Soule *et al.* 1975). Although not a single compilation of bryozoan taxa exists, the fossil record of bryozoans in current listing up to the Tertiary includes over 17,000 names of which 14,700 were accepted in the relatively recent review on the species diversity of the bryozoan fossil record, done by Horowitz & Pachut (2000).

Bryozoans are benthic organisms that feed on microorganisms using their ciliated feeding structures called lophophore (Brusca & Brusca 1990), and usually have to compete with other benthic organisms for space (Jackson & Hughes 1985).

The first and the most extensive lists of bryozoans from Costa Rica were published by Raymond Carroll Osburn with material collected from several localities during the Allan Hancock Pacific Expeditions (Osburn 1950, 1952, 1953; Osburn & Soule 1953). He reported 41 species collected in Costa Rica. The next paper published was by R.J. Cuffey (1971) and it is a report of a bryozoan, *Electra* cf. *angulata*, living on a sea snake, *Pelamis platurus*, collected in Golfo Dulce. Finally, William C. Banta and Renate J.M. Carson reported the presence of 24 species of Bryozoa, collected in 1964, 22 of the Class Gymnolaemata, and 2 of the Class Cyclostomata, from two localities in Costa Rica (Banta & Carson 1977). They found 13 species at the Caribbean site (Portete), 12 species from the Pacific (Playas del Coco), and 1 species, *Hippopodina feegeensis*, from both sites, but the Pacific specimens may be a misidentification. There are no other published reports of bryozoans from Costa Rica.

We present lists of the reported bryozoans from the Caribbean coast (Species List 34.1 included on the CD) and from the Pacific Ocean (Species List 34.2 included on the CD) of Costa Rica in published accounts. Those collections should be checked for changes in names, wrong identifications, and possible new species. When a change in name is known, the new name is followed by the name used in the paper after the connotation ("in ref.:"). We compiled a list of 61 species, 54 in the Order Cheilostomata, 6 in the Order Cyclostomata, and 1 in the Order Ctenostomata, in 36 families, and 49 genera; 13 species from the Caribbean and 49 from the Pacific, with one from both coasts, *H. feegeensis*, which might be a misidentification on the Pacific. In Panama, however, the most recent studies, which included only encrusting species of the Order Cheilostomata, 50 species were reported for the Caribbean and 37 species for the Pacific (Hughes & Jackson 1992; Jackson & Herrera-Cubilla 2000).

Few samples have been collected from the Caribbean, only from Portete, and from the Pacific from a few localities. Most of the Pacific collections were done by dredging in shallow waters or from the shore (see papers by Osburn). Of the rest of the Central American countries, little has been published with the exception of Panama (Hughes & Jackson 1992; Jackson & Herrera-Cubilla 2000) and Belize (Winston 1984). More research is needed on this group not only in Costa Rica but also in the rest of Central America.

Specialists

AMALIA HERRERA-CUBILLA: Smithsonian Tropical Research Institute, Panama – Tropical America. herreraa@ancon.si.edu

JO ANN SANNER: Department of Paleobiology, National Museum of Natural History Smithsonian Institution, Washington, DC 20560 – Worldwide. Sanner.Joann@nmnh.si.edu

Collections

There is a small collection of Costa Rican bryozoans at the Museo de Zoología of the Universidad de Costa Rica.

Recommendations

More work is needed on both coasts of Costa Rica and in the other Central American countries on the species richness, distribution, and ecology of the bryozoans.

Acknowledgments We appreciate the support with the literature by K. Hale and H. Albertson, RSMAS Library, University of Miami. Funding to study the marine biodiversity of Costa Rica was provided by the Vicerrectoría de Investigación of the Universidad de Costa Rica (Projects 808-92-237; 808-96-601) and by CONICIT (Project 90-326-BID). We appreciate the review by Dr. R. Petersen.

References

1. Banta WC, Carson RJM (1977) Bryozoa from Costa Rica. Pac Sci 31:381–424.

Brusca RC., Brusca GJ (1990) Invertebrates. Sinauer Assoc, Sunderland, MA, 922 p

2. Cuffey RJ (1971) Pacific sea snakes – a highly mobile newly recognized substrate for bryozoans. Prog. 2nd Conf. Int. Bryozoology Assoc., Durham, England. Abstract, pp 32

Horowitz AS, Pachut JF (2000) The fossil record of bryozoan species diversity. *In*: Herrera-Cubilla A, Jackson JBC (eds) Proc. 11th Int. Bryozool. Assoc. Conf., STRI, Panamá, pp 245–248.

Hughes DJ, Jackson JBC (1992) Distribution and abundance of cheilostome bryozoans on the Caribbean reefs of central Panama. Bull Mar Sci 51:443–465

Jackson JBC, Herrera-Cubilla A (2000) Adaptation and constraint as determinants of zooid and ovicell size among encrusting ascophoran Cheilostome Bryozoa from opposite sides of the Isthmus of Panama. *In*: Herrera-Cubilla A, Jackson JBC (eds) Proc. 11th Int. Bryozool. Assoc. Conf., STRI, Panamá, pp 249–258

Jackson JBC, Hughes TP (1985) Adaptive strategies of coral reef invertebrates. Am Sci 73:265–274

McKinney FK., Jackson JBC (1989) Bryozoan evolution. University of Chicago Press, Chicago, 252 p

3. Osburn RC (1950) Bryozoa of the Pacific coast of America. Part I, Cheilostomata-Anasca. Allan Hancock Pac Exp 14:1–269
4. Osburn RC (1952) Bryozoa of the Pacific coast of America. Part 2, Cheilostomata-Ascophora. Allan Hancock Pac Exp 14:271–611
5. Osburn RC (1953) Bryozoa of the Pacific coast of America. Part 3, Cyclostomata, Ctenostomata, Entoprocta, and Addenda. Allan Hancock Pac Exp 14:613–725
6. Osburn RC, Soule JD (1953) Bryozoa of the Pacific coast of America. Part 3, Order Ectoprocta. Allan Hancock Pac Exp 14:726–758

Soule JD, Soule DF, Pinter PA (1975) Phyllum Ectoprocta (Bryozoa). *In*: Smith RI, Carlton JT (eds) Light's Manual—Intertidal Invertebrates of the Central California Coast. University of California Press, Berkeley, pp 579–608

Winston JD (1984) Shallow-water bryozoans of Carrie Bow Cay, Belize Am Mus Novitates No. 2799:1–38

Part 35
Brachiopods

Christian C. Emig

Glottidia audebarti, a species from Pacific Costa Rica (Photo: Christian C. Emig)

C.C. Emig
Centre d'Océanologie, Rue de la Batterie-des-Lions, F-13007 Marseille, France.
Present address: 20, rue Chaix, F-13007 Marseille France
e-mail: brachnet@alideads.fr

I.S. Wehrtmann, J. Cortés (eds.) *Marine Biodiversity of Costa Rica, Central America*,

Abstract Ten brachiopod species have been recorded in the waters of Central America, and eight of these species occur on the Pacific coast of Costa Rica. Among the Linguliformea, two *Glottidia* species, *G. albida*, and *G. audebarti*, live in the Golfo de Nicoya (Costa Rica). Two discinid species occur in the Pacific waters, viz. *Discradisca strigata* and the deep-sea *Pelagodicus atlanticus*. Among the Rhynchonelliformea, all recorded species are living in the deep-sea, in the bathyal and abyssal zones: *Neorhynchia strebeli, Liothyrella clarkeana, L. moseleyi, Macandrevia diamantina, M. americana, and M. craniella.*

Introduction

Brachiopods, or lamp shells, are an exclusively marine group of lophophorate animals. They are sessile benthic suspension-feeders (Emig 1997a, b), bilaterally symmetrical, and they are solitary coelomates. They are enclosed within a shell formed by a dorsal and a ventral valve, and fixed to or into the substrate by a pedicle, lacking in some taxa, and cemented to the substratum by one of the valves. The pedicle has the capacity to adjust the position of the organism in relation to its surroundings (Emig 1997a; Richardson 1997). The lophophore of brachiopods varies in complexity, and is usually supported by the brachidium (Emig 1992). Larvae are either planktotrophic or non-planktotrophic.

Following the classification established in the "Treatise on Invertebrate Paleontology" (Kaesler 2000–2007), brachiopods are divided into three subphyla: the Linguliformea, the Craniiformea, and the Rhynchonelliformea. There are, at least, 114 extant brachiopod genera represented by 401 species. Representatives are found from littoral waters (generally subtidal) to the abyssal zone, and are generally epifaunal on hard substrata; only the lingulides are exclusively infaunal in soft substrata. In the waters of Central America, ten brachiopod species belonging to five genera have been recorded, only one species of which is known for the Caribbean (Species List 35.1 is included on the CD-Rom).

Summary and Comments

Glottidia albida and *G. audebarti* are the only brachiopod species recorded in the Golfo de Nicoya (see Species List 35.2, which is included on the CD-Rom). Both species have been redescribed by Emig (1983) and by Emig & Vargas (1990), respectively. They also occur on the Pacific coast of Mexico with *G. palmeri. G. audebarti* has also been sampled on the Panama and Ecuador coasts. In the Caribbean and Atlantic waters another *Glottidia* species, *G. pyramidata*, occurs (Emig 1983). The genus *Glottidia* is restricted to the American coasts, while in the other tropical and temperate areas the genus *Lingula* occurs (Emig 1997a). Two discinid species occur in Central America: *D. strigata* has been studied in the intertidal zone of Panama (La Barbera 1985) and *P. atlanticus* off the Mexican and Peruvian coasts (Zezina 1961).

Neorhynchia strebeli (= *N. profunda*) has been collected off Isla del Coco at a depth of 2,150 m on muddy bottom (type-locality, Species List 35.2 is included on the CD-Rom), and SW Galapagos at 3,800 m, as well as off California and in the SE Pacific (Dall 1908; Cooper 1972) in the bathyal and abyssal zones (2,000–4,500 m depth), and in the Antarctic where this species occurs from the immediate subtidal to a depth of several hundred meters (Barnes & Peck 1997).

Liothyrella clarkeana has only been recorded from the two localities cited above, off the Gulf of Panama: off Isla del Coco at 2,150 m on muddy bottom (type-locality) and SW Galapagos at 3,724 m (Dall 1908). *L. moseleyi* is also a poorly known species collected near Isla del Coco at 250 m and in several other Pacific locations (from about 250–4,000 m deep). This species occurs also in the Atlantic off Martinique, at a depth of 310 m in the Caribbean (Species List 35.1 is included on the CD-Rom), and in the Indian Ocean (type-locality is west of Kerguelen Island at 384 m) (Dall 1908).

Two *Macandrevia* species occur along the west coast of North and South America, extending from San Diego (California) to the Antarctic (Dall 1908, 1921; Cooper 1972, 1973, 1982): *M. diamantina* (= *Notorygmia abyssa*), off Isla del Coco at a depth of 2,150 m on muddy bottom (type-locality), and off the west coast of Colombia at a depth of 3,250–3,260 m. Depth distribution is known until 4,600 m. This species commonly has homeomorphic features with *N. strebeli*. The other species, *M. americana* (= *M. vanhoeffeni*, = *M. lata*) extends off Isla del Coco at a depth of 3,000 m on muddy bottom. The bathymetric range of this species is between 100 and 4,000 m.

The third species, *M. craniella*, has only been recorded off Isla del Coco at a depth of 2,150 m on muddy bottom (type-locality) in the Gulf of Panama (Dall 1908).

Specialists and Collections

Information on Brachiopoda, including directories of the specialists and references, is available at http://paleopolis.rediris.es/BrachNet/ (or in the mirror sites at http://emig.free.fr/BrachNet/ and at http://www.marinespecies.org/brachiopoda/). Except the present author on Lingulides, there is no specialist on the other extant brachiopods cited herein. Because most of the experts are retired, the directories at the web site Br@chnet remain the best way to contact a specialist (webmaster is Christian C. Emig, email: emig@free.fr and brachnet@aliceads.fr). The largest collection of American brachiopod specimens is located at the Smithsonian Institution, National Museum of Natural History, Washington DC (USA).

References

Barnes DKA Peck LS (1997) An Antarctic shelf population of the deep-sea, Pacific brachiopod *Neorhynchia strebeli*. J Mar Biol Ass UK 77:399–407

Cooper GA (1972) Homeomorphy in recent deep-sea brachiopods? Smithson Contr Paleobiol 11:1–25
1. Cooper GA (1973) *Vema's* Brachiopoda (Recent). Smithson Contr Paleobiol 17:1–51
Cooper GA (1982) New Brachiopoda from the Southern hemisphere and *Cryoptopora* from Oregon (Recent). Smithson Contr Paleobiol 41:1–43
2. Dall WH (1908). Reports on the Mollusca and the Brachiopoda. Bull Mus Comp Zool Harvard Univ 43(6):205–487
3. Dall WH (1921) Annotated list of the recent Brachiopoda in the collection of the United States National Museum, with descriptions of thirty-three new forms. Proc US Nat Mus 57(2314):261–377
Emig CC (1983) Taxonomie du genre *Glottidia* (Brachiopodes Inarticulés). Bull Mus Nat Hist Nat Paris (Sér. 4) 5 (Sect.4) (n° 2):469–489
Emig CC (1992) Functional disposition of the lophophore in living Brachiopoda. Lethaia 25:291–302
Emig CC (1997a) Ecology of inarticulated brachiopods. Biogeography of inarticulated brachiopods. *In*: Kaesler RL (ed) Treatise on Invertebrate Paleontology. Part H, Revised, Brachiopoda, Vol. 1. Geol Soc Am and Univ. Kansas Press. Boulder, Colorado & Lawrence, Kansas, pp 473–502
Emig CC (1997b) Les Lophophorates constituent-ils un embranchement? Bull Soc Zool Fr 122:279–288
4. Emig CC Vargas JA (1990) Notes on *Glottidia audebarti* (Broderip) (Brachiopoda, Lingulidae) from the Gulf of Nicoya, Costa Rica. Rev Biol Trop 38(2A):251–258
Kaesler RL (ed) 2000–2007 Treatise on Invertebrate Paleontology. Part H, Revised, Brachiopoda. Geol. Soc. Am. and Univ. Kansas Press, Boulder, Colorado/Lawrence, Kansas, Vol. 2–7, 1–3226
5. Kowalevski M, Dryreson E, Marcot JD, Vargas JA, Flessa KW & Hallman DP (1997). Phenetic discrimination of biometric simpletons: paleobiological implications of morphospecies in the lingulide brachiopod *Glottidia*. Paleobiology 23:444–469
6. La Barbera M (1985) Mechanisms of spatial competition of *Discinisca strigata* (Inarticulata: Brachiopoda) in the intertidal of Panama. Biol Bull 168:91–105
7. Owen R (1835) Mémoire sur l'anatomie des Mollusques Brachiopodes (Cuv.) et plus spécialement des Térébratules et Orbicules. Ann Sci Nat (2) 3:52–77
Richardson J (1997) Ecology of articulated brachiopods. Biogeography of articulated brachiopods. *In*: Kaesler RL (ed) Treatise on Invertebrate Paleontology. Part H, Revised, Brachiopoda, Vol. 1. Geol Soc Am. and Univ. Kansas Press. Boulder, Colorado & Lawrence, Kansas, pp 441–472
8. Zezina ON (1961) Distribution of the deepwater brachiopod *Pelagodiscus atlanticus* (King) (in Russian). Okeanologiya 5:354–358

Part 36
Echinoderms

Juan José Alvarado and Jorge Cortés

Nidorellia armata, a common species from Pacific Costa Rica (Photo: Jorge Cortés)

J.J. Alvarado
Centro de Investigación en Ciencias del Mar y Limnología (CIMAR), Universidad de Costa Rica, San Pedro, San José 11501-2060, Costa Rica.
e-mail: juanalva76@yahoo.com

J. Cortés
Centro de Investigación en Ciencias del Mar y Limnologia (CIMAR), and
Escuela de Biología, Universidad de Costa Rica, 11501-2060 San José, Costa Rica
e-mail: jorge.cortes@ucr.ac.cr

I.S. Wehrtmann, J. Cortés (eds.) *Marine Biodiversity of Costa Rica, Central America*,

Abstract Two hundred and twenty nine species of echinoderms (Phylum Echinodermata), including feather stars, sea stars, brittle stars, sea urchins, and sea cucumbers, are reported for Costa Rica. The Class Crinoidea is represented by six species from five genera, four families and one order; the Class Holothuroidea by 49 species, 24 genera, 15 families, and 6 orders; and the Asteroidea by 50 species, in 38 genera, 19 families, and 6 orders. The most diverse classes are Ophiuroidea (69 species, 31 genera, 11 families, and 2 orders) and Echinoidea (55 species, 33 genera, 17 families, and 11 orders). The genus *Holothuria* is the largest with 20 species. Of the 229 species reported, 187 are from the Pacific (103 from the Pacific mainland and 124 from Isla del Coco), 44 are from the Caribbean, and there were two species in common to both coasts (*Ophioderma appresum* and *Ophiactis savigny*). Most of the taxonomic research on echinoderms was done in the first half of the twentieth century. Here, we report 50 new records from Costa Rica. Due to the lack of collections from some areas of the country and from deepwaters, it is expected that more species will be found in the future, possibly including undescribed species.

Introduction

The echinoderms (feather stars, sea stars, brittle stars, sea urchins, and sea cucumbers) are exclusively marine animals. The Phylum Echinodermata (literally: spiny skinned animals) is divided into five living classes: Crinoidea, Asteroidea, Ophiuroidea, Echinoidea, and Holothuroidea. About 6,500 species of living echinoderms and over 13,000 fossil species in many extinct classes have been described (Hendler *et al.* 1995).

The echinoderms have important ecological roles such as the control of algal growth in Caribbean coral reefs by the black sea urchin, *Diadema antillarum* (Carpenter 1997). An example from the Pacific is the coral-eating sea star, *Acanthaster planci*, which reaches population densities in the Pacific that can devastate entire coral reefs (Moran 1986).

The echinoderms from Costa Rica have been studied in a sporadic way. The first work dates from the late nineteenth and early twentieth century by the expeditions of the US Fish Commission Steamer "Albatross" in charge of Alexander Agassiz (Ludwig 1894, 1905; Agassiz 1898, 1904; Lütken & Mortensen 1899; Clark 1917). In the 1930s and the 1940s, as a result of the expeditions to the eastern Pacific by the Allan Hancock Foundation and the New York Zoological Society, there has been a significant amount of research on echinoderms (Ziesenhenne 1937, 1940, 1942, 1955; Clark 1939, 1940, 1948; Deichmann 1941, 1958). Among the numerous species reported in these papers, 11 species of echinoderms have been described as new. Many of these new species had been collected at Isla del Coco (summary in Hertlein 1963). One of these species recorded from Isla del Coco is *Stichopus horrens* (Fig. 36.1).

Fig. 36.1 *Stichopus horrens*, a common species around Isla del Coco, Pacific Costa Rica (Photo: Graham Edgar)

More recent publications are ecologically oriented. Lawrence (1967) studied the lipid reserves in the gut of three species of sea urchins (*Echinometra lucunter*, *Tripneustes ventricosus*, and *Lytechinus variegatus*) collected in Portete, Limon. Bakus (1974) experimented with the toxicity of holothurians to fishes at Isla del Coco, finding that six of seven species tested were toxic. The density, distribution, and massive death of the black sea urchin, *D. antillarum*, at Parque Nacional Cahuita, were studied in 1977, 1980, 1983, 1992, 1999, and 2003 (Valdez & Villalobos 1978; Cortés 1981, 1994; Murillo & Cortés 1984; Alvarado *et al.* 2003). Guzmán (1988) and Guzmán & Cortés (1992) provided information on the density of the corallivorous sea star, *A. planci*, at Isla del Caño and Isla del Coco, respectively. Guzmán (1988), moreover, presented densities of the sea urchin *D. mexicanum* from Isla del Caño and Guzmán & Cortés (1992) from Isla del Coco. Rojas (1990) and Rojas *et al.* (1998) analyzed seven heavy metals in the sea cucumber, *Holothuria mexicana*, from Cahuita, and found that the concentrations of lead, iron, and copper are higher than in other organisms. Moreover, they suggested that the holothurians are organisms, suitable for the detection of metal pollution, because they feed on bottom sediment. Cortés *et al.* (1992) published on the massive death of organisms due to the uplift of the Caribbean coast during the Limón 1991 earthquake. Among the most affected was the boring sea urchin, *E. lucunter*. During the evaluation of the effects of the earthquake they collected a great number of skeletons of the irregular sea urchins, *Meoma ventricosa* and *Echinoneus cyclostomus*. The great abundance of skeletons of these sea urchins was used to reconstruct their population structure (Soto *et al.* in preparation). Lessios *et al.* (1996) reported the

Fig. 36.2 *Echinothrix diadema*, in Costa Rica known exclusively from Isla del Coco (Photo: Graham Edgar)

presence of two Indo-Pacific diadematid sea urchins (*Echinothrix diadema*, Fig. 36.2, and *E. calamaris*) at Isla del Coco and proposed that their arrival was enhanced by the North Equatorial Counter Current during the 1982–1983 El Niño. The same authors suggest that the echinoderm fauna of Coco, Revillagigedo, and Clipperton Island, is a mixture of both east- and west-Pacific faunas. Lessios *et al.* (1998) postulated that Isla del Coco is a stepping stone between both sides of the Pacific Ocean, and found evidence of genetic flow across the largest marine biogeographic barrier of the world (the Eastern Pacific Barrier). At the end of the twentieth century and beginnings of the twenty-first century several researchers have being centering their attention on the biogeographic relationships of sea urchin populations between both sides of the Pacific by molecular analysis (Palumbi 1996; Lessios *et al.* 1999, 2001, 2003; McCartney *et al.* 2000; Zigler & Lessios 2003, 2004; Palumbi & Lessios 2005). Alvarado & Cortés (2004) reviewed the state of knowledge on echinoderms of Costa Rica and Central America until 2003 pointed out the needs for research on the area. The diversity, abundance, and distribution of echinoderms from Parque Nacional Marino Ballena, on the Pacific coast of Costa Rica was studied by Alvarado & Fernández (2005), and they suggested that sedimentation is having a negative effect on echinoderm diversity. Meanwhile, Bolaños *et al.* (2005) compared the diversity and distribution of echinoderms in two environments (seagrass and coral reefs) at the reef lagoon of Parque Nacional Cahuita. More recently, Hendler (2005) described two new species of brittle star of the genus *Ophiothrix* collected from Cahuita: *O. stri* and *O. cimar*.

Furthermore, there are nine publications on fossil echinoderms of Costa Rica (Gabb 1881; Jackson 1917; Durham 1961; Aguilar 1978, 1997, 1999; Fisher 1985; Aguilar & Cortés 2001; Alvarado *et al.* 2006). Alvarado *et al.* (2006) made a review on the research done on that area in Costa Rica. The fossil fauna is composed of one crinoid and 45 species of echinoids. Most of the specimens are from the

Cretaceous and Miocene, found in 11 geological formations throughout the country (Alvarado *et al.* 2006).

In this part we present lists of the echinoderms of Costa Rica. To compile these lists we used the information in the literature, collections in the National Museum of Natural History, Smithsonian Institution (USNM), Colección Nacional de Equinodermos, Instituto de Ciencias del Mar y Limnología, Universidad Nacional Autónoma de Mexico (UNAM), and Museo de Zoología, Escuela de Biología, Universidad de Costa Rica (UCR). New records to Costa Rica are indicated in the list and in some cases specimens have been deposited at the Museo de Zoología (these specimens are indicated by UCR and a collection number) or in the National Museum of Natural History. We follow the taxonomy used by Maluf (1988).

Discussion

The information in this part indicates that six species of crinoids, 50 asteroids, 69 ophiuroids, 55 echinoids, and 49 holothurians live in Costa Rica, comprising a total of 229 species (Species Lists 36.1 and 36.2 are included on the CD-Rom, Table 36.1). Forty four species are from the Caribbean, 185 from the whole Pacific (103 from the Pacific mainland and 124 from Isla del Coco) (Table 36.1), and there are two species occurring on both coasts in Costa Rica (*Ophioderma appresum* and *Ophiactis savigny*) (Fig. 36.3). Fifty species reported herein are new records for Costa Rica (Species Lists 36.1 and 36.2).

Eleven species have been described from material collected in Costa Rica: two asteroids *Pauliella aenigma* (Albatross Expedition of 1891, in charge of Alexander Agassiz), and *Tamaria obstipa*, both from Isla del Coco. Two ophiuroids, *Amphiodia vicina*, collected in Bahía Ballena, Golfo de Nicoya during the Eastern Pacific Expeditions of the New York Zoological Society lead by William Beebe, and *Ophionereis dictyota*, collected at Chatham Bay, Isla del Coco in 1938, during the Allan Hancock Pacific Expeditions on the Velero III. Three echinoids, *Hesperocidaris panamensis*, and *Centrocidaris doederleini*, from material collected in Isla del Coco during the Albatross Expedition of 1891 (Clark 1948), and *Encope cocosi*, collected at Wafer Bay, Isla del Coco, on a sandy bottom between 4 and 8 m deep. Finally, four holothurians: *Psolus diomedeae*, Isla del Coco, collected during the Albatross Expedition in 1891, *Pentamera beebi*, Bahía Ballena, and *Labidodemas americanum*, Isla Tortuga, Golfo de Nicoya, both collected during the Eastern Pacific Expeditions of the New York Zoological Society (Deichmann 1938); and *Euthyonidium veleronis*, from Playa Blanca, Guanacaste, collected by the Allan Hancock Pacific Expedition of 1935 on the Velero III. Three species are endemic: the irregular sea urchin, *E. cocosi* and the sea stars *Astropecten benthophilus* and *Persephonaster armiger* from Isla del Coco.

The sites with the highest number of species are Isla del Coco (124 species), Golfo de Nicoya (26), Playas del Coco (12), and Santa Elena (11), which are also

Table 36.1 Number of species, genera, families, and orders of echinoderms from Costa Rica

Class	Crinoidea	Asteroidea	Ophiuroidea	Echinoidea	Holothuroidea	Total
Orders	1	6	2	11	6	26
Families	4	19	11	17	15	66
Genera	5	38	31	33	24	131
Species	6	50	69	55	49	229
Caribbean	4	7	17	12	4	44
Whole Pacific	2	43	55	42	45	187
Pacific Mainland (PM)	0	14	36	25	28	103
Isla del Coco (IC)	2	34	30	31	27	124
% species in common between PM and IC	0.00	14.71	40.00	3.65	37.04	32.26

Fig. 36.3 *Ophiactis savigny* from the Parque Nacional Marino Ballena, Pacific Costa Rica (Photo: Ingo Wehrtmann)

the sites visited by the expeditions of the 1930s. The number of species from other sites is expected to increase when more extensive collections will be carried out.

Panamá and Belize are the Central American countries from which more species of echinoderms have been reported at 276 and 110, respectively (Table 36.2). These are also the best studied countries in the region. Four hundred and twenty seven species of echinoderms have been reported for Central America, but this certainly is an underestimation of the true diversity of echinoderms of the region. This

Table 36.2 Number of species of echinoderms reported for each Central American country. CA = number of species for the entire Central American region

	Belize	Guatemala	Honduras	El Salvador	Nicaragua	Costa Rica	Panamá	CA
Crinoidea	5	2	9	0	3	6	13	20
Asteroidea	11	2	3	7	9	50	47	86
Ophiuroidea	55	11	7	13	12	69	71	135
Echinoidea	20	11	18	13	8	55	89	98
Holothuroidea	19	7	8	12	6	49	56	88
Total	110	33	45	45	38	229	276	427
Species in common with Costa Rica	37 (33%)	27 (82%)	26 (57%)	36 (80%)	26 (76%)		119 (43%)	
References	[24, 33, 34, 36, 40, 41, 51]	[9, 13, 22, 28, 37, 49, 52, 54]	[9, 13, 21, 22, 23, 32, 36, 37, 41, 44, 49, 52, 54]	[6, 9, 13, 14, 22, 23, 37, 49, 52, 54, 60]	[9, 13, 22, 23, 36, 42, 49, 52, 54]		[9, 10, 12–14, 21, 23, 25, 31, 34, 35, 37, 41–43, 45, 49, 52–54, 57, 63, 67]	

number represents 6.5% of the total diversity of described species of echinoderms. There are few studies from the other countries in the region, and many habitats and sites have not yet been sampled.

The number of species in common between Costa Rica and the other Central American countries is presented in Table 36.2. Compared to the countries with low species diversity reported so far, the percentage of species in common is high, probably due to the fact that in those countries only the most abundant and widespread species have been reported. The number of species between Costa Rica (229) and Panama (276) is not substantially different, but only 43% of the species occur in both countries. This indicates, either that the faunas of each country are localized and differ widely, or that more species will be found in both countries as more areas and environments are studied. In conclusion, more studies are needed on the biodiversity of the echinoderms in Central America.

Specialists

Gordon Hendler: Curator of Echinoderms, Natural History Museum of Los Angeles County, Los Angeles, California, USA – Worldwide. hendler@nhm.org

Francisco A. Solís-Marín: Colección Nacional de Equinodermos, Laboratorio de Sistemática y Ecología de Equinodermos, Instituto de Ciencias del Mar y Limnología, Universidad Nacional Autónoma de México, México-Caribbean and tropical eastern Pacific. fasolis@icmyl.unam.mx

Sabine Stöhr: Curator of Echinoderms, Swedish Museum of Natural History, Department of Invertebrate Zoology, Box 50007 SE-10405, Stockholm, Sweden-Worldwide. sabine.stohr@nrm.se

Collections

There are collections of Costa Rican echinoderms in the Museo de Zoología of the Universidad de Costa Rica, Los Angeles County Museum, the National Museum of Natural History at the Smithsonian Institution, Colección Nacional de Equinodermos from the Instituto de Ciencias del Mar y Limnología at the Universidad Nacional Autónoma de México, the Natural History Museum of Santa Barbara, and in the Museum of Comparative Zoology, Harvard University.

Recommendations

More studies are needed in deepwaters, in the Caribbean intertidal zone and in some sections of the coasts, for example, Bahía Salinas on the North Pacific. More research is needed in the rest of the Central American countries.

Acknowledgments We appreciate the help received over the years by R. Vargas and A. Villalobos, Museo de Zoología, Escuela de Biología, Universidad de Costa Rica; Francisco Solís-Marín, Colección Nacional de Equinodermos, Instituto de Ciencias del Mar y Limnología, Universidad Nacional Autónoma de México; Cinthya Ahearn, National Museum of Natural History, Smithsonian Institution; and the support with the literature by K. Hale and H. Albertson, RSMAS Library, University of Miami. We appreciate that Graham Edgar allowed us to use his excellent photos for this part. Funding to study the marine biodiversity of Costa Rica was provided by the Vicerrectoría de Investigación of the Universidad de Costa Rica (Projects 808-92-237; 808-96-601) and by CONICIT (Project 90-326-BID). Comments by anonymous reviewers improved the paper.

References

Agassiz A (1898). Preliminary report on the echini. Dredging operations of the Albatross. Bull Mus Comp Zool 32:71–86.

1. Agassiz A (1904) Reports on an exploration off the west coast of Mexico, Central and south America, and off Galapagos Islands, in charge of Alexander Agassiz, by the U.S. Fish Commission Steamer "Albatross", during 1891, Lieut. Commander Z.L. Tanner, U.S. Commanding. XXXII. The Panamic deep sea echini. Mem Mus Zool Harvard Coll 31:1–246

Aguilar T (1978) Fauna de un perfil de la formación Térraba (Oligoceno, Costa Rica). Tesis de grado, Universidad de Costa Rica, San Pedro, Costa Rica. 84 p

Aguilar T (1997) Parches arrecifales de Jesús María (Oligoceno-Mioceno Inferior): una de las primeras comunidades del actual Caribe de Costa Rica. Tesís de Maestría, Universidad de Costa Rica, San Pedro, Costa Rica. 61 p

Aguilar T (1999) Organismos de un arrecife fósil (Oligoceno Superior-Mioceno Inferior), del Caribe de Costa Rica. Rev Biol Trop 47:453–474

Aguilar T, Cortés J (2001) Arrecifes coralinos del Oligoceno Superior-Mioceno Inferior, de Turrialba, Costa Rica. Rev Biol Trop 49:203–213

2. Alvarado JJ (2004) Abundancia del erizo de mar *Centrostephanus coronat*us (Echinoidea: Diadematidae) en el Pacífico de Costa Rica. Rev Biol Trop 52:911–913

Alvarado JJ, Cortés J (2004) The state of knowledge on echinoderms of Costa Rica and Central America. Proceedings of the 11th International Echinoderm Conference, Münich, Germany, vol 1, pp 149–155

3. Alvarado JJ, Fernández C (2005) Equinodermos del Parque Nacional Marino Ballena, Pacífico, Costa Rica. Rev Biol Trop 53(Suppl 3):275–284

Alvarado JJ, Cortés J, Salas E (2003) Population densities of *Diadema antillarum* Philippi at Cahuita National Park (1977–2003), Costa Rica. Carib J Sci 40:257–259

Alvarado JJ, Zeledón LA, Boyd R (2006) Notas sobre equinodermos fósiles de Costa Rica. Rev Biol Trop 54(Suppl 1):287–299

4. Bakus GJ (1974) Toxicity in holothurians: a geographical pattern. Biotropica 6:229–236
5. Bakus GJ (1975) Marine zonation and ecology of Cocos Island, off Central America. Atoll Res Bull 179:1–11
6. Barraza JE, Hasbún CR (2005) Los equinodermos (Echinodermata) de El Salvador. Rev Biol Trop 53(Suppl 3):139–146
7. Bolaños N, Bourg A Gómez J, Alvarado JJ (2005) Diversidad de equinodermos en la laguna arrecifal del Parque Nacional Cahuita, Caribe de Costa Rica. Rev Biol Trop 53(Suppl 3):285–290
8. Bradley GE, Ireland J (1993) Such is Life in Costa Rica, Underwater Photo Guide. Printex Int., San José, 93 p
9. Brusca RC (1980) Common intertidal invertebrates of the Gulf of California, 2nd ed. University of Arizona Press, Phoenix, 513 p

Carpenter RC (1997) Invertebrate predators and grazers. *In*: Birkeland C (ed) Life and Death of Coral Reefs. Chapman & Hill, New York, pp 198–229
10. Chesher RH (1972). The status of knowledge of Panamanian Echinoids, 1971, with comments on other Echinoderms. Bull Biol Soc Wash 2:139–157
11. Clark AH (1939) Echinoderms (other than Holothurians) collected on the Presidential Cruise of 1938. Smithson Miscell Collec 98(11):1–22
12. Clark AH (1946) Echinoderms from the Pearl Islands, Bay of Panama, with a revision of the Pacific species of the genus *Encope*. Smithson Miscell Collec 106(5):1–11
Clark HL (1917) Reports on the scientific results of the expedition to the tropical Pacific in charge of Alexander Agassiz, on the U.S. Fish Commission Steamer "Albatross" from August 1899 to March 1900, commander Jefferson F. Mooser, U.S.N Commanding. XVIII. Reports on the scientific results of the expedition to the tropical Pacific in charge of Alexander Agassiz, on the U.S. Fish Commission Steamer "Albatross" from October 1904 to March 1905, Lieut. Commander L.M. Garret, U.S.N. Commanding. Bull Mus Comp Zool Harvard Coll 61:429–466
13. Clark HL (1940) Eastern Pacific Expeditions of the New York Zoological Society. XXI Notes on Echinoderms from the west coast of Central America Zoologica 25:331–352
14. Clark HL (1948) A report on the Echini of the warmer eastern Pacific, based on the collections of the Velero III. Allan Hancock Pac Exp 8(5):225–352
15. Cortés J (1981) The Coral Reef at Cahuita, Costa Rica: A Reef Under Stress. M.Sc. thesis, McMaster University, Ontario, Canada, 176 p
16. Cortés J. 1991. Ambientes y organismos marinos del Refugio Nacional de Vida Silvestre Gandoca-Manzanillo, Limón, Costa Rica. Geoistmo 5(1–2):61–68
17. Cortés J (1994) A reef under siltation stress: a decade of degradation. *In*: Ginsburg RN (Compiler) Proceedings of the Colloquium on Global Aspects of Coral Reefs: Health, Hazards and History, 1993. RSMAS, University of Miami, Miami, Florida, pp 240–246
18. Cortés J, Guzmán HM (1985) Arrecifes coralinos de la costa Atlántica de Costa Rica. Brenesia 23:275–292
19. Cortés J, León A (2002) Arrecifes coralinos del Caribe de Costa Rica. INBio, Santo Domingo de Heredia, Costa Rica, 140 p
20. Cortés J, Soto R, Jiménez C, Astorga A (1992) Earthquake associated mortality of intertidal and coral reef organisms (Caribbean of Costa Rica). Proc 7th Int Coral Reef Symp, Guam 1:235–240
21. Deichmann E (1938) Holothurians from the western coast of Lower California and Central America, and from the Galápagos Islands; Eastern Pacific Expeditions of the New York Zoological Society XVI. Zoologica 23:361–187
22. Deichmann E (1941) The Holothuroidea collected by the Velero III during the years 1932 to 1938. Part I Dendrochirota Allan Hancock Pac Exp 8(3):61–195
23. Deichmann E (1958) The Holothurioidea collected by the Velero III and IV during the years 1932 to 1954. Part II Aspidochirota Allan Hancock Pac Exp 11(2):253–349
24. Devaney DM (1974) Shallow water echinoderms from British Honduras with a description of a new species of *Ophiocoma* (Ophiuroidea). Bull Mar Sci 24:122–164
25. Dexter DM (1977) A natural history of the sand dollar *Encope stokesi* L. Agassiz in Panama Bull Mar Sci 27:544–551
Durham JW (1961) Miocene echinoids from the Valle Central, Costa Rica. J Paleont 35:480–488
Fischer R (1985) La fauna de la formación Turrúcares (Mioceno, Valle Central, Costa Rica). Geol Paleontol 19:191–225
26. Fischer R (1990) Biogenetic and biogenetically determined morphologies of the Costa Rican Pacific coast. Z Geomorph NF 34:313–321
27. Fisher WK (1928) Sea stars from the Arcturus oceanographic expedition. Zoologica 8:487–493
28. Fonseca AC, Arrivillaga A (2003) Coral reefs of Guatemala. *In*: Cortés J (ed) Latin American Coral Reefs. Elsevier Science B.V., Amsterdam, pp 159–169

Gabb WM (1881) Description of Caribbean Miocene fossils. J Acad Nat Sci Philadelphia 8:337–348

29. Guzmán HM (1988) Distribución y abundancia de organismos coralívoros en los arrecifes coralinos de la Isla del Caño, Costa Rica. Rev Biol Trop 36:191–207
30. Guzmán HM, Cortés J (1992) Cocos Island (Pacific of Costa Rica) coral reefs after the 1982–83 El Niño disturbance. Rev Biol Trop 40:309–324
31. Guzmán HM, Guevara CA (2002) Annual reproductive cycle, spatial distribution, abundance, and size structure of *Oreaster reticulatus* (Echinodermata: Asteroidea) in Bocas del Toro, Panama. Mar Biol 141:1077–1084
32. Hasbún CR, Lawrence AJ (2002) An annotated description of shallow water holothurians (Echinodermata: Holothuroidea) from Cayos Cochinos, Honduras. Rev Biol Trop 50:669–678
33. Hendler G (1988) Western Atlantic *Ophiolepsis* (Echinodermata: Ophiuroidea): a description of *O*. pawsoni new species, and a key to the species Bull Mar Sci 42:265–272
34. Hendler G (1995) New species of brittle stars from the Western Atlantic, *Ophionereis vittata, Amphioplus sepultus,* and *Ophiostigma siva,* and the designation of a neotype for *Ophiostigma isocanthum* (Say) (Echinodermata: Ophiuroidea). Nat Hist Mus Los Angeles County, Contr Sci 458:1–19
35. Hendler G (2005) Two new brittle star species of the genus *Ophiothrix* (Echinodermata: Ophiuroidea: Ophiotrichidae) from coral reefs in the southern Caribbean sea, with notes in their biology. Carib J Sci 41:583–599
36. Hendler G, Miller JE, Pawson DL, Kier PM (1995) Sea Stars, Sea Urchins, and Allies: Echinoderms of Florida and the Caribbean. Smithsonian Institution Press, Washington, DC, 390 p
37. Hendrickx, ME (1995) Equinodermos. *In*: Fischer W, Krupp F, Schneider W, Sommer C, Carpenter KE, Niem VH (eds) Guía FAO para la identificación de especies para los fines de las pesca. Pacifico Centro-Oriental, vol I. Plantas e Invertebrados. Organización de la Naciones Unidas para la Agricultura y la Alimentación (FAO), Rome, pp 637–646
38. Hertlein LG (1963) Contribution to the biogeography of Cocos Island, including a bibliography. Proc Cal Acad Sci 32(4th ser):123–235
39. Hickman CP Jr (1998) A field guide to sea stars and other echinoderms of Galápagos. Sugar Spring Press, Lexington, VA, 83 p
40. Hotchkiss FHC (1982) Ophiuroidea (Echinodermata) from Carrie Bow Cay, Belize. *In*: Rützler K, Macintyre, IG (eds) The Atlantic Barrier Reef Ecosystem at Carrie Bow Cay, Belize, I. Structure and Communities. Smithsonian Institution Press, Washington, DC, pp 387–412

Jackson RT (1917) Fossil echinoids of the Panama coastal zone and Costa Rica. US Nat Mus Bull 103:103–116

41. Kaplan EH (1988) Southeastern and Caribbean Seashores. Peterson Field Guides, Houghton Mifflin Company, New York, 425 p
42. Kerstitch A (1989) Sea of Cortez Marine Invertebrates. A Guide for the Pacific Coast, Mexico to Ecuador. Sea Challengers, Monterrey, California, 114 p

Lawrence JM (1967) Lipid reserves in the gut of three species of tropical sea urchins. Carib J Sci 7:65–68

43. Lessios HA (1987) Temporal and spatial variation in egg size of 13 Panamanian echinoids. J Exp Mar Biol Ecol 114:217–239
44. Lessios HA (1998) Shallow water echinoids of Cayos Cochinos, Honduras. Rev Biol Trop 46(Suppl 4):95–101
45. Lessios HA (2005) Echinoids of the Pacific waters of Panama: status of knowledge and new records. Rev Biol Trop 53(Suppl 3):147–170
46. Lessios HA, Kessing BD, Wellington GM, Graybeal A (1996) Indo-Pacific echinoids in the tropical eastern Pacific. Coral Reefs 15:133–142

Lessios HA, Kessing BD, Robertson DR (1998) Massive gene flow across the world's most potent marine biogeographic barrier. Proc R Soc Lond B 265:583–588

47. Lessios HA, Kessing BD, Robertson DR (1999) Phylogeography of the pantropical sea urchin *Eucidaris* in relation to land barriers and ocean currents. Evol 53:806–817

Lessios HA, Kessing BD, Pearse JS (2001) Population structure and speciation in tropical seas: global phylogeography of the sea urchin *Diadema*. Evol 55:955–975

Lessios HA, Kane J, Robertson DR (2003) Phylogeography of the pantropical sea urchin *Tripneustes*: contrasting patterns of population structure between oceans. Evol 57:2026–2036

48. Ludwig H (1894) Reports on an exploration off the west coast of Mexico, Central America and South America, and off the Galápagos Islands, in charge of Alexander Agassiz, by the US Fish Commission steamer "Albatross," during 1891. XII. The Holothurioidea. Mem Mus Comp Zool 17(3):1–183

49. Ludwig H (1905) Asteroidea in: Explorations of "Albatross" in Tropical Pacific, 1891 and 1899–1900. Mem Mus Comp Zool 22(7):1–290

50. Lütken CF, Mortensen T (1899) Reports on an exploration off the west coast of Mexico, Central America and South America, and off the Galápagos Islands, in charge of A. Agassiz, during 1891 on the "Albatross." XXV. The Ophiuridae. Mem Mus Comp Zool 23(2):93–208

51. Macurda DB Jr (1982) Shallow-water Crinoidea (Echinodermata) from Carrie Bow Cay, Belice. *In*: Rützler K, Macintyre IG (eds) The Atlantic Barrier Reef Ecosystem at Carrie Bow Cay, Belize, I. Structure and Communities. Smithsonian Institution Press, Washington, DC, pp 413–416

52. Maluf LY (1988) Composition and distribution of the central eastern Pacific echinoderms. Nat Hist Mus Los Angeles County, Tech Rep 2:1–242

53. Maluf LY (1991) Echinoderm fauna of the Galápagos Islands. *In*: James MJ (ed) Galápagos Marine Invertebrates: Taxonomy, Biogeography and Evolution in Darwin's Islands. Plenum, New York, pp 345–367

McCartney MA, Keller G, Lessios HA (2000) Dispersal barriers in tropicals oceans and speciation in Atlantic and eastern Pacific sea urchins of the genus *Echinometra*. Mol Ecol 9:1391–1400

54. Miller JE, Pawson DL (1984) Memoirs of the Hourglass Cruises, vol VII. Holothurians (Echinodermata: Holothuroidea). Florida Department of Natural Resources, St. Petersburg, 79 p

55. Miskelly, A (2002) Sea urchins of Australia and the Indo-Pacific. Capricornia Publications, Sydney, 180 p

Moran PJ (1986) The *Acanthaster* phenomenon. Oceanogr Mar Biol Annu Rev 24:379–480

56. Murillo MM, Cortés J (1984) Alta mortalidad en la población del erizo de mar *Diadema antillarum* en el Parque Nacional Cahuita, Limón, Costa Rica. Rev Biol Trop 32:167–170

57. Nielsen E (1932) Papers from Dr. Th. Mortensen's Pacific Expedition 1914–16. LIX. Ophiurans from the Gulf of Panama, California and the Strait of Georgia. Vidensk Medd Dansk naturh Forem Bd 91:240–346

58. Palumbi SR (1996) What can molecular genetics contribute to marine biogeography' An urchin's tale. J Exp Mar Biol Ecol 203:75–92

Palumbi SR, Lessios HA (2005) Evolutionary animation: how do molecular phylogenies compare to Mayr's reconstruction of speciation patterns in the sea? PNAS 102:6566–6572

59. Reyes-Bonilla H (1995) Asteroidea and Echinoidea (Echinodermata) of Isla San Benedicto, Revillagigedo Archipelago, México. Rev Inv Cient, Ser Cienc Mar UABCS 6:29–38

60. Reyes-Bonilla H, Barraza JE (2003) Corals and associated marine communities from El Salvador. *In*: Cortés J (ed) Latin American Coral Reefs. Elsevier Science B.V., Amsterdam, pp 351–360

61. Rojas G.MT (1990) Determinación de cadmio, cromo, cobre, hierro, manganeso, plomo y zinc en el pepino de mar, *Holothuria* sp. (Echinodermata) del arrecife coralino del Parque Nacional Cahuita, Costa Caribe, Costa Rica. Tesis de Licenciatura en Química, Universidad de Costa Rica, 69 p.

62. Rojas MT, Acuña JA, Rodríguez OM (1998) Metales traza en el pepino de mar *Holothuria (Halodeima) mexicana* del Caribe de Costa Rica. Rev Biol Trop 46(Suppl 6):215–220

63. Solís-Marín FA, Laguarda-Figueras A (1999) *Cucumaria flamma*, a new species of sea cucumber from the central eastern Pacific (Echinodermata: Holothuroidea). Proc Biol Soc Wash 112:778–786

64. Solís-Marín FA, Reyes-Bonilla H, Herrero-Pérezrul MD, Arizpe-Covarrubias O, Laguarda-Figueras A (1997) Sistemática y distribución de los equinodermos de la Bahía de La Paz. Cienc Mar 23:249–263
65. Valdez MF, Villalobos CR (1978) Distribución espacial, correlación con el substrato y grado de agregación en *Diadema antillarum* Phillipi (Echinodermata: Echinoidea). Rev Biol Trop 26:237–245
66. Vargas JA (1989) A three year survey of the macrofauna of an intertidal mud flat in the Gulf of Nicoya. *In*: Magoon O, Converse M, Miner D, Tobin LT, Clark D (eds) Coastal Zone 89. American Society of Civil Engineers, New York, pp 1905–1919
67. Verrill AE (1867) Notes on the echinoderms of Panama and west coast of America, with description of new genera and species. Trans Connect Acad Arts Sci 1(2):251–322
68. Villalobos SC (1982) Animales y plantas comunes de las costas de Costa Rica. Ed. Univ. Estatal a Distancia, San José, Costa Rica, 145 p
69. Wellington GM (1974) An Ecological Description of the Marine and Associated Environments at Monumento Nacional Cahuita. Subdirección de Parques Nacionales, Ministerio de Agricultura y Ganadería (unpublished)
70. Ziesenhenne FC (1937) The Templeton Crocker Expedition. X Echinoderms from the West coast of Lower California, the Gulf of California and Clarion Island. Zoologica 22:209–239
71. Ziesenhenne FC (1940) New ophiurans of the Allan Hancock Pacific Expeditions. Allan Hancock Pac Exped 8(2):9–59
72. Ziesenhenne FC (1942) New eastern Pacific sea stars. Allan Hancock Pac Exped 8(4):197–223
73. Ziesenhenne FC (1955) A review of the genus *Ophioderma*. Essays in the Natural Sciences in Honor of Captain Allan Hancock on the Occasion of his Birthday, July 26, 1955. University of Southern California Press, Los Angeles, CA, pp 185–201

Zigler KS, Lessios HA (2003) Evolution of bindin in the pantropical sea urchin *Tripneustes*: comparisons to bindin of other genera. Mol Biol Evol 20:220–231

Zigler, K.S., H.A. Lessios (2004) Speciation on the coasts of the world: phylogeography and the evolution of bindin in the sea urchin genus *Lytechinus*. Evol. 58:1225–1241

Part 37
Chaetognaths or Arrow Worms

Eduardo Suárez-Morales, Rosa Ma. Hernández-Flores, and Álvaro Morales-Ramírez

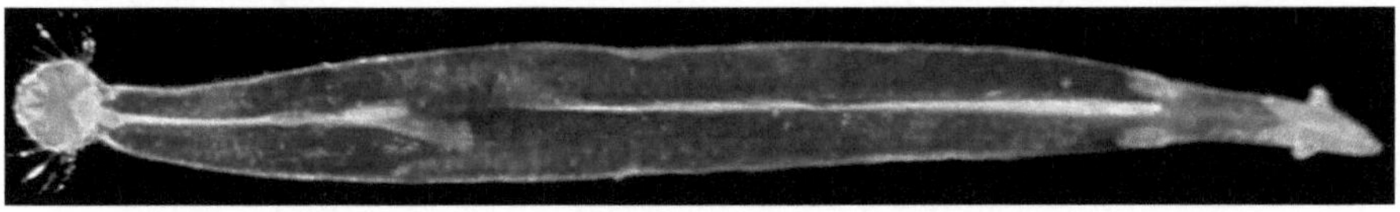

Flaccisagitta enflata reported from Pacific Costa Rica (Photo: Rosa María Hernández)

Abstract The knowledge of chaetognaths in the two Costa Rican coasts is both scarce and asymmetrical. Less than 20% of the number of potential species (115) have been effectively recorded from this country. The Pacific side has been surveyed much more intensely than the Caribbean coast. This is attributed to the large zooplankton surveys in the Eastern Tropical Pacific (ETP) that included sampling sites in Costa Rican waters. These expeditions yielded relevant data to the current knowledge of the ETP region chaetognath fauna, including Costa Rica. Up to 19 species have been effectively recorded in the Costa Rican Pacific waters, about 50% of the known species richness in the ETP. Conversely, only nine species have been reported for the Caribbean waters of Costa Rica, a figure far from the 25 species known to be distributed in the Caribbean Basin. The relevance of developing zooplankton surveys that include a close analysis of the chaetognath fauna is emphasized in terms of the trophic relevance of these predators in the pelagic realm.

E. Suárez-Morales, R.M. Hernández-Flores
El Colegio de la Frontera Sur-Chetumal. A.P. 424. Chetumal, QR. 77000. Mexico.
e-mail: esuarez@ecosur.mx

Á. Morales-Ramírez
CIMAR and Escuela de Biología. Universidad de Costa Rica. 11501-2060 San José, Costa Rica.
e-mail: alvaro.morales@ucr.ac.cr

I.S. Wehrtmann, J. Cortés (eds.) *Marine Biodiversity of Costa Rica, Central America*,

Introduction

Chaetognaths, also called arrow worms, are active zooplanktonic predators. They have a wide distribution in the oceanic, neritic, and even estuarine waters of the world. They are relatively small animals (2–120 mm), all exclusively marine and most of them are holoplanktonic forms, dwelling along the water column (0–1,000 m) during their entire life cycle. A few genera are exclusively benthic (i.e., *Spadella, Paraspadella, Bathyspadella*), and some genera are represented by deep-living forms (i.e., *Hemispadella, Heterokrohnia*) (Casanova 1986, 1994, 1996, 1999). Representatives of this phylum constitute between 5% and 10% of the zooplankton numerical abundance in the world oceans (Bone *et al.* 1991). After copepods, chaetognaths have the most consistent frequency of occurrence: rarely are they conspicuously absent, or unusually abundant. From the oceans to the innermost reaches of bays and coastal systems, and from the cold to the tropical latitudes, chaetognaths represent between 5% and 15% of the zooplankton biomass. Only in estuaries and macrotidal bays with a heavy silt load they are consistently absent (Longhurst 1985). These pelagic predators have been estimated to consume a highly variable percentage of zooplankton standing stock in coastal environments (Feigenbaum & Maris 1984). However, the knowledge of their actual ecological role in oceanic areas is still limited.

Taxonomic status and species richness: Taxonomically, Chaetognatha is considered to be a separate, isolated phylum. For a long time its systematic arrangement was built around a single taxonomic group until Tokioka (1965), and later on Casanova (1985) proposed a new classification of the phylum that has been refined over the years with new taxonomic data (Casanova 1999).

Currently, up to 115 species of Chaetognatha are known to be distributed in the world oceans (Boltovskoy 1981, Bieri 1991, Pierrot-Bults 1996, Casanova 1999); these are contained in 15 genera, some of them described after 1991 (Casanova 1999). Several genera are monotypic (i.e., *Heterokrohnia, Bathyspadella, Pterosagitta, Bathybelos, Xenokrohnia, Archeterokrohnia*) and others (i.e., *Eukrohnia, Krohnitta, Krohnitella*) contain only a few species each. New genera are still being described (i.e., *Pseudeukrohnia*) (Kasatkina 1998). Among chaetognath genera, *Sagitta* is probably the most representative group of the phylum, and with nearly 30 species known, it is clearly the most diverse and widely distributed genus of Chaetognatha.

The taxonomic characters currently used to separate genera have not changed much in decades and they include the presence of transverse muscles in the trunk and/or the tail, the number of lateral fins, the relative size of the tail segment, and the position of the gonads, among other features (Casanova 1999). At the species level, new more reliable and stable characters have been introduced from different studies and have substituted or were added to others previously used. The presence of eyes with photoreceptive or ommatidia-like organs, the structure and size of the hooks, the presence or absence of a collarete, the relative size of the tail, the shape of the head, and the number of anterior teeth are some of the species level characters used in the most recent keys and diagnoses (Pierrot-Bults 1996; Casanova 1999).

Comparison of Species Richness

There are only a few previous works on the Chaetognatha of Costa Rican waters. In order to achieve a better understanding and to facilitate a comparative analysis of the chaetognath fauna richness of the waters adjacent to Costa Rica, we divided our comments on the previous works into two geographic areas: Caribbean Sea and Pacific Ocean.

Caribbean Sea: The chaetognaths of this region have been surveyed by a few authors only. The isolated surveys on Chaetognatha made before the 1970s in this basin have been cited by Legaré & Zoppi (1961) and Owre & Foyo (1972). Out of these, it is noteworthy to mention the work of Alvariño (1968), who provided abundant data about the chaetognath distribution and composition of an area near the easternmost part of the Caribbean, and also in the area of influence of the Amazon River. She recorded *Sagitta enflata*, *S. serratodentata*, and *S. tenuis* as the most abundant species. She found 14 species in the western Caribbean; some of the stations could be inside Costa Rican waters. Hence, a few species (five) can be included safely in the list of chaetognaths dwelling in the Caribbean of Costa Rica. Also, Suárez-Caabro (1955) made the first systematic survey of the Chaetognatha in Cuban waters. Owre & Foyo (1972) found 14 species from isolated collections made in the two geographic extremes of the Caribbean Basin. Later on, Owre (1972) made some observations on the distribution and environmental affinities of *Eukrohnia bathyantarctica* in Caribbean waters. Owre (1973) published new records of *Eukrohnia* for the Gulf of Mexico and the western Caribbean. Michel & Foyo (1976) increased the overall number of records of chaetognaths in this region (24). Michel & Foyo's (1976) station plan included one sampling site in Costa Rican waters (sta. 11 of cruise 6811); up to six species of Chaetognatha were recorded in this station (see Species List 37.1 which is included on the CD-Rom). Based on these works and in several other more recent ones (Michel 1984, Gasca *et al.* 1996, McLelland 1989, Suárez-Morales *et al.* 1990, McLelland *et al.* 1992, Alvarez-Cadena *et al.* 1996, Suárez-Morales 1998, Ramírez-Ávila & Alvarez-Cadena 1999, Hernández *et al.* 2004), it is stated herein that the current number of species recorded in the entire Caribbean Sea is 25. The most abundant species in the Caribbean region, but not necessarily recorded in the Costa Rican coasts of the Caribbean Sea, are: *Sagitta enflata*, *S. serratodentata*, *Pterosagitta draco*, *S. decipiens*, *S. hexaptera*, *Krohnitta pacifica*, *S. bipunctata*, *K. subtilis*, *S. hispida*, *S. lyra*, *S. minima*, and *Eukrohnia bathyantarctica*. This number of species in the Caribbean is low when compared with the figure stated by Casanova (1999) for the South Atlantic (38 species). In this large area richness is particularly high due to the meeting of species from different latitudes (temperate, tropical, and subantarctic).

Following the number of species known worldwide (115) as stated by Bieri (1991) and revised recently by Casanova (1999), the species richness in the Caribbean Basin represents nearly 22% of the total number of known species. The nine species effectively recorded in the Costa Rican Caribbean (Species List 37.1 which is included on the CD-Rom) represent one third of the potential number of species, which are probably distributed in neritic and oceanic waters off the Atlantic coast of Costa Rica. It is also predictable that the most abundant species in Costa Rica will belong to the same group of dominant forms known in the Caribbean.

Eastern Tropical Pacific (ETP): This large region has been surveyed by several authors: Bieri (1959) studied the distribution of the Pacific chaetognaths as related to water masses; he found 19 species in the ETP. Sund (1959, 1961) provided an identification key for the known species and described their distribution. The abundant information published by Alvariño (1961, 1962, 1964, 1972, 1986, 1992) represents a benchmark in the study of Chaetognatha of the entire Eastern Pacific, including the California and Baja California regions. Southward, along the coast of the Eastern Pacific, in Mexico, Laguarda-Figueras (1965) recorded four species in a coastal system; the most abundant in this zone was *S. enflata*. Gómez-Aguirre & Rivero-Beltrán (1987) found *S. euneritica* associated to an estuarine system in Sonora. In southern Baja California and the central Gulf of California, the number of species is around 17–25 (Brinton *et al.* 1986, Franco-Gordo 1997). Bernache-Jiménez (1993) and Arciniega-Flores (1994) studied the abundance and distribution of the chaetognaths collected in the neritic-oceanic area off the coasts of Jalisco and Colima, Mexican Tropical Pacific. These authors reported *S. enflata* as the most abundant species in the area. Further south, along the Mexican Pacific coast, Cambrón (1981) surveyed the chaetognaths collected in the Golfo de Tehuantepec; she recorded ten species, *S. enflata*, *S. pacifica*, and *S. bedoti* being the most abundant. Alvariño (1972) studied plankton samples from a large area including the northwestern Atlantic and the Eastern Tropical Pacific collected during nine oceanographic campaigns. Several stations were located off the Pacific coasts of Costa Rica. She presented distributional maps of some of them; 19 species were mentioned for the area (12 common in the Eastern Tropical Pacific plus 7 circumtropical forms) (for details of records see Species List 37.2 which is included on the CD-Rom). Alvariño (1972) recorded 26 species in the western Caribbean Sea including the zone of Panama and the Gulf of Honduras. The species of Chaetognatha observed were grouped into 3 categories: (1) common to the Gulf of Mexico, the Caribbean and the Pacific adjacent zone: (*Krohnitta subtilis*, *Pterosagitta draco*, *Sagitta bipunctata*, *S. decipiens*, *S. enflata*, *S. hexaptera*, *S. minima*). (2) Tropic-equatorial Atlantic species (*Krohnitta mutabbii*, *S. helenae*, *S. hispida*, *S. serratodentata*, *S. tenuis*, *S. lyra*, and *S. friderici*) (3) Tropic-equatorial Pacific species (*Krohnitta pacifica*, *S. bedoti*, *S. ferox*, *S. neglecta*, *S. pacifica*, *S. peruviana*, *S. popovicci*, *S. pulchra*, *S. regularis*, *S. robusta*, *S. euneritica*, and *S. bierii*. *Sagitta pseudoserratodentata*, a well-known eastern Pacific species was not found by Alvariño (1972). Pineda (1977) recorded 18 species from the Colombian Pacific coast and the Bight of Panama, adding morphological data for two species of *Sagitta*. Casanova (1991) published six additional reports of deep-living chaetognaths in the ETP. Morones (1988) and Segura *et al.* (1992) found 13 species of Chaetognatha in plankton samples collected in the Costa Rican Dome (Pacific Ocean). These authors reported *Sagitta enflata*, *S. decipiens*, *Pterosagitta draco*, and *Ş. pacifica* as the most abundant in this oceanic area. Later on, Vicencio & Fernández (1996) included the same reports in an overall view of the zooplankton of the Costa Rican Dome. Hossfeld (1996) studied the chaetognath fauna of the

Pacific coast of Costa Rica and reported ten species from both the Golfo Dulce and the Golfo de Nicoya. She evaluated the distribution and biomass of this group in two different systems of Costa Rica (Golfo de Nicoya and Golfo Dulce). In terms of species richness, only eight species were found in each system (Species List 37.2; see CD-Rom) chaetognath diversity showed a strong inshore-offshore gradient in both systems surveyed by Hossfeld (1996), with eight species near the oceanic front and only one of them at the innermost reaches of the systems.

Therefore, the analysis of the published records of this taxon yielded an overall estimate of 19 species effectively reported from Costa Rican waters. Clearly, the chaetognath fauna of the Pacific side has been investigated much more intensively than that in the Caribbean coast. The number of species recorded in the ETP is 38 (Alvariño 1972, Pierrot-Bults 1996); this figure represents over 33% of the known species of chaetognaths. Hence, the 19 species effectively reported from Costa Rican waters represent 16% of the total number of known species, and nearly 50% of the species known to be distributed in the Eastern Tropical Pacific region.

Specialists

There are no recognized experts on this group currently working in Costa Rica or in Central America. There are, however, several researchers who have studied the chaetognaths of Costa Rica or adjacent zones and are familiar with the regional fauna. Their affiliation is followed by the area or regions they have worked on.

José N. Alvarez-Cadena: Instituto de Ciencias del Mar y Limnología, Puerto Morelos, Universidad Nacional Autónoma de México, Mexico – Caribbean Sea. inac@mar.icmyl.unam.mx

Jean-Paul Casanova: Université de Provence, Marseille, France – Worldwide. bioplank@up.univ-mrs.fr

Rosa Ma. Hernández-Flores: El Colegio de la Frontera Sur, Chetumal, Mexico – Eastern Tropical Pacific, Caribbean Sea. rosamahe@ecosur.mx

Jerry A. McLelland: University of Southern Mississippi, USA – Western Atlantic. jerry.mclelland@usm.edu

Silvia Mille-Pagaza: Lab. de Ecología, Escuela Nacional de Ciencias Biológicas I.P.N., Mexico – Gulf of Mexico

Annelies C. Pierrot-Bults: University of Amsterdam, The Netherlands – Worldwide. pierrot@science.uva.nl

Collections

There are but a few recognized or institutional collections of American Chaetognatha. In Mexico, El Colegio de la Frontera Sur at Chetumal, holds a collection of chaetognaths with 85 specimens and 10 species, mainly from samples gathered in the western Caribbean Sea. This collection represents 40% of the Caribbean records of this taxon and up to 44% of the species known in the Costa Rican waters of the Caribbean. The Laboratorio de Invertebrados of the Facultad de Ciencias, UNAM in Mexico City, holds the zooplankton collections of the Costa Rican Dome. These fully oceanic samples, which have been studied by several colleagues at different times, have representative specimens of different zooplankton groups (including the ten species of Chaetognatha collected in the Dome). The Centro de Ecología Costera of the Universidad de Guadalajara in western Mexico harbors a collection of more than 1,500 specimens belonging to 11 species from the Mexican Tropical Pacific. Several other regional and local collections of Chaetognatha are personal, not institutional initiatives, and most are not openly available for consultation.

There are several sources of electronic information about chaetognaths on the web; we found that one of the most useful, updated, and accessible is the site by E. Thuesen (http://academic. evergreen.edu/t/thuesene/home.htm). He works on Chaetognath biology and ecology at the Evergreen State College, Washington, USA (thuesene@evergreen.edu). This site contains abundant "information about chaetognath biology including keys, images, species lists and links to other on-line references".

Recommendations

The wide distribution of chaetognaths and their well-known relevance as zooplankton predators (Saito & Kiørboe 2001) are two strong reasons to develop local and regional studies on this group. Their distributional patterns seem to be related to the distribution and abundance of different prey. The hypothesis that chaetognaths can predate heavily on fish larvae and affect the recruitment of fishes at early stages (Alvariño 1985) seems to be still a matter of discussion (Khulman 1977, Feigenbaum & Maris 1984).

There is an evident asymmetry when the knowledge of chaetognaths is compared between the Atlantic and the Pacific coasts of Costa Rica. There are 19 species of Chaetognatha effectively recorded for the Pacific versus less than half this figure of the Atlantic coast. It is recommended to start taxonomic and distributional surveys of this group along and off the Atlantic coasts of Costa Rica. Although both the general composition and the dominant species are predictable in this area to some extent, the overall knowledge of the Costa Rican fauna would be complemented with data from the Caribbean side. This would probably add interesting information to the Central American chaetognath fauna and their particular situation as a region in which the Pacific and the Atlantic pelagic faunas meet (Alvariño 1972).

Chaetognaths are also excellent indicators of water masses or oceanographic/environmental conditions (Alvariño 1977). A basic set of distributional and abundance data on chaetognaths would be useful also in these terms (i.e., interpretation and confirmation of hydrologic conditions and data) thus improving the overall knowledge of the oceanography of Costa Rica.

References

Alvarez-Cadena JN, Suárez-Morales E, McLelland JA (1996) Observations on an isolated population of *Sagitta hispida* Conant (Chaetognatha) in a tropical lagoon system of northeast Yucatan (Mexican Caribbean). Gulf Res Rep 9:197–204

Alvariño A (1961) Two new chaetognaths from the Pacific. Pac Sci 15:67–77

Alvariño A (1962) Two new Pacific chaetognaths, their distribution and relationship to allied species. Bull Scripps Inst Oceanogr 8:1–50

Alvariño A (1964) Zoogeografía de los quetognatos especialmente de la región de California. Cienc Mex 23:51–74

Alvariño A (1968) Los quetognatos, sifonóforos y medusas en la región del Atlántico ecuatorial bajo la influencia del Amazonas. An Inst Biol Univ Nac Autón Méx, Ser Cienc Mar y Limnol 39:41–76

1. Alvariño A (1972) Zooplancton del Caribe, Golfo de México y regiones adyacentes del Pacífico. Memorias del IV Congreso Nacional de Oceanografía. México, D.F., Nov., 1969, pp 223–247

Alvariño A (1977) Indicadores planctónicos y la oceanografía de la región de California. Memorias del V Congreso Nacional de Oceanografía. México, pp 52–78

Alvariño A (1985) Predation in the plankton realm; mainly with reference to fish larvae. Inv Mar CICIMAR 2:1–122

Alvariño A (1986) El zooplancton del Pacífico Ecuatoriano. Inv Mar CICIMAR 3:69–110

Alvariño A (1992) Distribución batimétrica, diurna y nocturna de diez y siete especies de quetognatos durante las cuatro estaciones del año 1969, en aguas de California y Baja California. Inv Mar CICIMAR 1:1–169

Arciniega-Flores J (1994) Distribución y abundancia de los quetognatos de la costa de Jalisco (septiembre de 1990). Tesis Profesional, Universidad de Guadalajara, Mexico. 26 p

Bernache-Jiménez L (1993) Quetognatos de la plataforma continental de Jalisco y Colima (agosto, 1988). Tesis Profesional, Universidad de Guadalajara, México. 57 p

2. Bieri R (1959) The distribution of planktonic Chaetognatha in the Pacific and their relationship to the water masses. Limnol Oceanogr 4:1–28

Bieri R (1991) Systematics of the Chaetognatha. *In*: Bone Q, Kapp H, Pierrot-Bults AC (eds.). The Biology of Chaetognatha. Oxford University Press, Oxford, pp 122–136

Boltovskoy D (1981) Chaetognatha. *In*: Boltovskoy D (ed.). Atlas del zooplancton del Atlántico Sudoccidental. Publ. Esp. INIDEP, Mar del Plata, Argentina, pp 759–791

Bone Q, Kapp H, Pierrot-Bults AC (eds) (1991) The Biology of Chaetognaths. Oxford University Press, Oxford,173 p

Brinton E, Fleminger A, Siegel-Causey D (1986) The temperate and tropical planktonic biotas of the Gulf of California. CalCOFI Rep 27:228–266

Cambrón MM (1981) Estudio preliminar de las especies planctónicas del Phylum Chaetognatha en el Golfo de Tehuantepec, México. Tesis Profesional, Fac. Ciencias, UNAM, Mexico, 78 p

Casanova JP (1985) Description de l'appareil génital primitif du genre *Heterokrohnia* et nouvelle classification des chaetognathes. Comp Rend Acad Sci Paris 301:397–402

Casanova JP (1986) Quatre nouveaux chaetognathes atlantiques abyssaux (genre *Heterokrohnia*). Description, remarques éthologiques et biogéographiques. Ocean Acta 9:469–477

Casanova JP (1991) Chaetognaths from the Alvin dives on the Seamount Volcano 7 (east tropical Pacific). J Plankton Res 13:539–548

Casanova JP (1994) Three new rare *Heterokrohnia* species (Chaetognatha) from deep benthic samples in the Northeast Atlantic. Proc Biol Soc Wash 107:743–750

Casanova JP (1996) A new genus and species of deep-benthic chaetognath from the Atlantic: a probable link between the families Heterokrohnidae and Spadellidae. J Nat Hist 30:1239–1245

Casanova JP (1999) Chaetognatha. *In*: Boltovskoy D (ed) South Atlantic Zooplankton. Backhuys Publ., Leiden, The Netherlands, pp 1353–1374

Fagetti EG (1968) New record of *Eukrohnia bathyantarctica* David, 1958, from the Gulf of Mexico and Caribbean Sea. Bull Mar Sci 18:383–387

Feigenbaum D, Maris RC (1984) Feeding in the Chaetognatha. Oceanogr Mar Biol Ann Rev 22:343–392

Franco-Gordo MC (1997) Estructura de la taxocenosis de quetognatos en la región central del Golfo de California. Tesis de Maestría, CICIMAR-IPN, La Paz, Mexico, 56 p

Gasca R, Alvarez-Cadena JN, Suárez-Morales E (1996) Chaetognath assemblages in the Mexican Caribbean Sea (1991). Carib Mar Stud 5:41–50

Gómez-Aguirre S, Rivero-Beltrán C (1987) Variación estacional de *Sagitta eurenitica* (Chaetognatha) en la Laguna de Agiabampo, México. An Inst Biol Univ Nac Autón Méx, Ser Zool 58:697–706

Hernández RM, Suárez-Morales E, Gasca R (2005) Seasonal distribution of Chaetognatha in a Mexican neotropical bay during a year cycle. J Coast Res 21:186–1

3. Hossfeld B (1996) Distribution and biomass of arrow worms (Chaetognatha) in Golfo de Nicoya and Golfo Dulce, Costa Rica. Rev Biol Trop 44:(Suppl 3): 157–172

Kasatkina AP (1998) A new genus *Pseudeukrohnia* of the family Eukrohniidae (Chaetognatha). Biol Morya 24:359–363 (in Russian)

Khulman D (1977) Laboratory studies on the feeding behaviour of the chaetognaths *Sagitta setosa* Müller and *S. elegans* Verrill with special reference to fish eggs and larvae. Ber Dt Wiss Komm Meeresforsch 25:163–171

Laguarda-Figueras A (1965) Contribución al conocimiento de los quetognatos de Sinaloa. An Inst Biol Univ Nac Autón Méx, Ser Zool 36:215–228

Legaré JEH, Zoppi E (1961) Notas sobre la abundancia y distribución de Chaetognatha en las aguas del oriente de Venezuela. Bol Inst Oceanogr Univ Oriente 1:1–25

Longhurst A (1985) The structure and evolution of plankton communities. Progr Oceanog 15:1–35

McLelland JA (1989) An illustrated key to the Chaetognatha of the northern Gulf of Mexico with notes on their distribution. Gulf Res Rep 8:145–172

McLelland JA, Heard RW, Hardy JD, Jr (1992) Arrow worms (Chaetognatha) from the near-shore waters of Tobago. Carib Mar Stud 3:33–40

Michel HB (1984) Chaetognatha of the Caribbean Sea and adjacent areas. NOAA Tech Rep NMSF 15, US Dept. Commerce, 33 p

7. Michel HB, Foyo M (1976) Caribbean zooplankton. I. Siphonophora, Heteropoda, Copepoda, Euphausiacea, Chaetognatha and Salpidae. Office of Naval Research, Dept. Navy, USA, pp 433–461

4. Morones L (1988) Análisis cualitativo y cuantitativo de los quetognatos (Chaetognatha) en la región del Domo de Costa Rica (otoño, 1981). Tesis Profesional, Fac. Ciencias, UNAM, Mexico, 99 pp

Owre HB (1972) Some temperatures, salinities, and depths of collection of *Eukrohnia bathyantarctica* (Chaetognatha) in the Caribbean Sea. Bull Mar Sci 22:94–99

Owre HB (1973) A new chaetognath genus and species, with remarks on the taxonomy and distribution of others. Bull Mar Sci 23:903–948

Owre HB, Low JK (1969) Methods of collecting net plankton from a series of known depths through the water column. Bull Mar Sci 19:911–921

Owre HB, Foyo M (1972) Studies in Caribbean zooplankton. Description of the program and results of the first cruise. Bull Mar Sci 22:483–521

Pierrot-Bults A (1996) Chaetognatha. *In*: Gasca R, Suárez E (eds) Introducción al estudio del zooplancton marino. El Colegio de la Frontera Sur (ECOSUR)/CONACYT, Mexico, pp 529–596

5. Pierrot-Bults A, Nair VR (1991) Distribution patterns in Chaetognatha. *In*: Bone Q, Kapp H, Pierrot-Bults AC (eds) The Biology of Chaetognaths. Oxford University Press, Oxford, pp 86–116

6. Pineda FHP (1977) Taxonomy of the chaetognaths of the Bight of Panama. An Inst Inv Mar-Punta Betín, Santa Marta 9:225–240
Ramírez-Ávila J, Alvarez-Cadena JN (1999) Chaetognath species composition from a coral reef lagoon in the Mexican Caribbean Sea. Rev Biol Trop 47(Suppl 1):157–163
Saito H, Kiørboe T (2001) Feeding rates in the chaetognath *Sagitta elegans*: effects of prey size, prey swimming behavior and small-scale turbulence. J Plankton Res 23:1385–1398
9. Segura L, Hernández RM, Morones L (1992) Distribución y abundancia de los quetognatos (Chaetognatha) en la región del Domo de Costa Rica. Rev Biol Trop 40:35–42
Suárez-Caabro JA (1955) Quetognatos de los mares cubanos. Mem Soc Cuba Hist Nat 22:125–180
8. Suárez-Caabro JA, Madruga JE (1960) The Chaetognatha of the northeastern coast of Honduras, Central America. Bull Mar Sci Gulf & Caribb 10:421–429
Suárez-Morales E (1998) Quetognatos, Tomo 7. *In*: Xacur JA (comp.). Enciclopedia de Quintana Roo. Edit.Verdehalago. 1ª- Ed. Chetumal, Mexico, pp 160–167
Suárez-Morales E, Hernández RM, Gasca R (1990) Quetognatos (Chaetognatha) de la Bahía de la Ascensión, Reserva de la Biosfera de Sian Ka'an, Quintana Roo, México. *In*: Navarro D, Robinson JG (eds) Diversidad biológica en la Reserva de la Biosfera de Sian Ka'an, Quintana Roo, México. Centro de Investigaciones de Q. Roo/Univ. of Florida, Mexico, pp 137–146
Sund NP (1959) A key to the Chaetognatha of the Tropical Eastern Pacific Ocean. Pac Sci 13:269–285
Sund NP (1961) Some features of the autoecology and distributions of Chaetognatha in the Eastern Tropical Pacific. Inter-Am Trop Tuna Comm Bull 5:307–331
Tokioka T (1965) The taxonomical outline of Chaetognatha. Publ Seto Mar Biol Lab 12:335–357
10. Vicencio ME, Fernández MA (1996) Zooplancton del Domo de Costa Rica: taxonomía y biogeografía. Rev Biol Trop 44:631–642

Part 38
Appendicularians (Urochordata)

Iván A. Castellanos, Álvaro Morales-Ramírez, and Eduardo Suárez-Morales

Bathochordaeus charon, a species from Pacific Costa Rica (Photo: Humberto Bahena)

I.A. Castellanos, E. Suárez-Morales
El Colegio de la Frontera Sur (ECOSUR) Unidad Chetumal. Av. Centenario Km 5.5, C.P. 77900, A.P. 424, Chetumal, Q. Roo, México.
e-mail: ivancast@ecosur.mx; esuarez@ecosur.mx

Á. Morales-Ramírez
Centro de Investigaciones Marinas, Universidad de Costa Rica, 11501-2060 San José, Costa Rica.
e-mail: alvaro.morales@ucr.ac.cr

I.S. Wehrtmann, J. Cortés (eds.) *Marine Biodiversity of Costa Rica, Central America*,

Abstract This revision updates the knowledge about the taxonomic composition and distribution of the species of appendicularians found in both the Pacific and Atlantic coasts of Costa Rica. Comments are provided also on the methodological problems involved in the taxonomic study of these organisms. In general, it was found that investigations on the Costa Rican appendicularians have been scarce. These organisms are widely distributed in the oceanic and neritic realms and although they have not been studied in detail, they are actually frequent in coastal and lagoonal zooplankton communities of Costa Rica. They have been mentioned from different areas of the Costa Rican Pacific such as the Golfo de Nicoya and Golfo Dulce and in the coastal embayments Murciélago and Culebra. On the Caribbean coasts, they have been known to occur off the urban area of Limón and in the reef zone of Cahuita. The regional studies on this group are also quite scarce; there is only one general reference in the Eastern Pacific. Up to six species of appendicularians have been known to occur in the Pacific coast of Costa Rica; this figure represents only about 14% of the total species number (43) recorded in the Pacific Ocean. The current status of knowledge is not better in the Caribbean coast; there are no species records of appendicularians in this zone.

Introduction

Appendicularians are exclusively marine holoplanktonic urochordates. Their body is formed by a trunk measuring between 2 and 4 mm (exceptionally as long as 25 mm in *Bathochordaeus charon*) and a characteristic tail, which is generally several times longer than the trunk. The main morphologic features of this group are discussed in detail by several authors (Esnal 1981, 1996; Fenaux 1998). This group is among the most common members of the zooplankton community, often the second or third most abundant group in plankton samples from estuarine, coastal, and oceanic waters. They are widely distributed in the world ocean, most species are regarded as warm water forms, but several are able to dwell in cold waters (Fenaux *et al.* 1998). They are not regarded as strong vertical migrators along the water column and they can be found mainly in the epipelagic layer. It has been suggested that they are more diverse in deeper waters, and that meso and bathypelagic appendicularian taxa remain undiscovered. The fact that they can easily be damaged during the plankton hauls has made the taxonomical recognition of deep-living forms difficult (Fenaux *et al.* 1998).

Previous Works in Costa Rican Waters and Adjacent Areas

The occurrence of these zooplankters has been reported by several authors in different coastal systems with variable marine influence such as Isla del Caño, the Cahuita reef zone, Golfo Dulce, Golfo de Nicoya, bahías Murciélago and Culebra, and in Puntarenas (Barham 1979; Morales & Murillo 1996; Bednarski 2001; Quesada-Alpízar 2001), all along the Pacific coast of Costa Rica. In the Caribbean side studies on this group are even scarcer: Morales (1996–1997) found relatively high densities of *Oikopleura* spp. in areas related to urban drainage near Ciudad de Limón. However, most of these zooplankton surveys have reported appendicularians as a group, without a more accurate taxonomic analysis. This probably explains the scarce knowledge on the appendicularians of Costa Rican waters, which is restricted currently to a single publication, with the analysis of zooplankton collections from Puntarenas. In order to achieve a better understanding of the Costa Rican and Central American appendicularian fauna, our analysis was divided into two geographic categories: Caribbean Sea and Pacific Ocean.

Caribbean Sea: In this region the earliest surveys on this group started during the 1950s decade. On the first one, Tokioka & Suárez-Caabro (1956) recorded up to 18 species in Cuban oceanic and neritic environments, providing brief descriptions and a distribution map for each species. They found *Oikopleura longicauda* (Vogt, 1854) as the most abundant species in this area. Almost two decades later, Zoppi de Roa (1971) provided a brief description and the density values of each of the 18 species collected in systems on the eastern coast of Venezuela; she recorded several uncommon species such as: *Tectillaria fertilis* (Lohmann, 1896), *F. haplostoma* f. *abjörnseni* Lohmann, 1909 and *F. formica* f. *tuberculata* Lohmann & Bückmann, 1926. These were the first records of these species in the Caribbean Sea. In Jamaican waters, Hopcroft & Roff (1995) surveyed the growth rates of *Oikopleura dioica*, stating that it is ten times higher than that known for copepods. The same authors (Hopcroft & Roff 1998) studied the production rates of the appendicularians in Jamaica; they found that they are an important constituent of the coastal zooplankton biomass. Hopcroft *et al.* (1998) compared the growth rates of several genera such as *Appendicularia*, *Fritillaria*, and *Oikopleura* in Jamaican waters.

In the Mexican Caribbean, the space-time variations of the appendicularians was studied by Castellanos *et al.* (1994) in two coastal systems of the eastern coast of the Yucatan Península, Bahía de la Ascensión, and Bahía de Chetumal. They found *Oikopleura dioica* as the most common species in Chetumal, where it is very abundant, making up to 10% of the total zooplankton biomass.

Out of the ten species recorded in Bahía de la Ascensión, the most abundant was *O. longicauda*, followed by *O. dioica* and *O. rufescens*. *O. longicauda* was the only species present in the samples year-round. It occurred also throughout the entire environmental gradient represented by the bay; the other species were restricted to the oceanic fringe of the bay (Castellanos *et al.* 1994). Castellanos & Gasca (1998a) recorded 8 species in the reef area of Mahahual and 13 in the surface layer of the Mexican Caribbean. In the oceanic waters of the Mexican Caribbean, the most abundant species was *O. longicauda*, followed by *Fritillaria borealis* f. *sargassi* and *O. fusiformis* f. *typica*. The other ten species showed an average density below 25 org/1,000 m^3. The same authors (Castellanos & Gasca 1998b) provided a key for the identification of the 14 species recorded in different areas of the western Caribbean Sea. Castellanos (1998) stated that *O. longicauda*, *O. dioica*, and *Fritillaria borealis* f. *sargassi* are those with the highest frequency in the Mexican

Caribbean. Castellanos (2003) recorded 11 species of *Oikopleura*, *Fritillaria*, and *Stegosoma* in Banco Chinchorro, around an oceanic atoll off the coasts of the Mexican Caribbean. *O. longicauda* was the most abundant species in this system.

Pacific Ocean: The only data available about previous surveys on appendicularians in the Eastern Pacific Ocean are: Barham (1979), who made observations on the houses of *B. charon* Chun, 1900 in mesopelagic waters; the author reported the finding of few appendicularian houses in the area of Puntarenas, Costa Rica. Adame-Rodríguez (1982) studied the distribution and abundance of ten species of Appendicularia in the Golfo de Tehuantepec, in the Mexican Tropical Pacific area. She found *O. longicauda*, *O. rufescens*, and *O. parva* as the most abundant species. Morales (1999–2000) analyzed the filtration rates of the appendicularians in the Golfo Dulce, Costa Rica. Recent observations by Castellanos (2001) from samples collected in Bahía Culebra, Costa Rica, allowed the addition of *O. rufescens*, *O. longicauda*, *O. fusiformis*, and *Fritillaria haplostoma* to the Costa Rican fauna of appendicularians (Species List 38.1 which is included on the CD-Rom).

Number of Species Described

The appendicularians are urochordates which have been divided taxonomically into three well-defined families: Kowalevskiidae, with only one known genus and two species: *Kowalevskia oceanica* Lohmann, 1899 and *K. tenuis* Fol, 1872; Fritillariidae which contains three genera and up to 30 species, and Oikopleuridae, with 11 genera and 38 species (Fenaux 1998). Fenaux (1993a) recognized up to 65 species and 14 genera within the Order Appendicularia. However, with the description of *Mesoikopleura enterospira* Fenaux, 1993, *M. youngbluthi* Fenaux, 1993, and *M. gyroceanis* Fenaux, 1993, plus the amendment and validation of the name given to *Pelagopleura haranti* (Vernières, 1934), which was reallocated in *Mesoikopleura* (Fenaux 1993b), and the recent description of *Mesochordaeus erythrocephalus* Hopcroft & Robison, 1999 in Californian waters (Monterey Bay), the currently recognized number of appendicularian species rises to 70, distributed in 15 genera.

Out of the 43 species described for the Pacific Ocean (Fenaux *et al.* 1998), only 6 have been effectively recorded in Costa Rican waters; this figure represents only 14% of the probable records considering the regional fauna. In the Western Atlantic Ocean the scenario is even more critical. The number of known species of appendicularians in this region is 47 (Fenaux *et al.* 1998) and none of them has been recorded in the Atlantic waters of Costa Rica. According to the reports of authors such as Tokioka & Suárez-Caabro (1956), Zoppi de Roa (1971), Castellanos & Gasca (1998b), Hopcroft *et al.* (1998), in the Caribbean Sea the number of species known to dwell in this area goes up to 23. Therefore, this figure represents more closely the potential number of species to be recorded in the Atlantic coastal and oceanic waters of Costa Rica. It is expected that the faunistic lists of both the

Atlantic and the Pacific coasts of Costa Rica will grow substantially as taxonomic studies on this group are supported and increased. Therefore, the lack of knowledge in this group can be viewed both as a challenge and as an appealing task for Costa Rican zooplankton taxonomists and ecologists.

Specialists

In general, the research on zooplankton in Costa Rican seas has developed amidst less than favorable conditions. The lack of infrastructural and human resources represents a common factor which has characterized, for decades, the marine research not only in Costa Rica but in other Latin American countries. These conditions have favored an emphasis on surveys in coastal areas and on selected taxa. Hence, several interesting areas remain unknown even when they have been sampled. As in other countries, much of the information on the Costa Rican zooplankton fauna has been generated by foreign researchers (Suárez-Morales & Gómez-Aguirre 1996). This is represented somehow in this list, but this tendency is much clear in other taxa. Included herein are those persons currently working on appendicularians in the region or specialists with a worldwide vision.

Iván A. Castellanos Osorio: El Colegio de la Frontera Sur (ECOSUR), Unidad Chetumal, Zona Ind. No. 2, Carr. Chetumal-Bacalar Km 2, Apdo. postal 424 AV. Centenario km 5.5 A.P 424, C.P. 77000 Chetumal, Q. Roo, México – Gulf of Mexico and Caribbean Sea. ivancast@ecosur.mx

Robert Fenaux: Observatoire Océanologique, URA2077, BP 28, 06234, Villefranche-sur-Mer, Cedex, France – Atlantic, Pacific and Indian Oceans and Mediterranean.

Gabriel Gorsky: Laboratoire d'Océanographie Biologique et Ecologie du Plancton Marine. LOBEPM, ESA 7076. Station Zoologique BP28. F-06234. Villefranche-sur-Mer. Cedex. France – Mediterranean. gorsky@obs-vlfr.fr

Evelyn Zoppi de Roa: Instituto de Zoología Tropical, U.C.V., Caracas, Venezuela – Caribbean Sea. ezoppi@strix.ciens.ucv.ve

Collections

There are no collections of appendicularians in Costa Rica. Probably the closest collection is that of El Colegio de la Frontera Sur (ECOSUR) which contains 242 specimens, which represent 51 records of four genera (*Fritillaria*, *Oikopleura*, *Folia*, *Stegosoma*) and 14 species, several of them probably present in Costa Rican waters. In fact, this collection contains now several specimens of appendicularians from Bahía Culebra, Costa Rica (ECO-CHZ01182–01186). More specimens from these collections are being examined and will be deposited eventually in this collection. The Laboratorio de Invertebrados of the Facultad de Ciencias, UNAM in Mexico City holds a large collection of zooplankton samples from the Costa Rica Dome; apparently the appendicularians have not been studied. It is expected that the CIMAR at Costa Rica will hold a collection of appendicularians which will arise from this sampling.

Problems and Recommendations

Despite their importance in the marine trophic webs, the appendicularians have been scarcely studied in Costa Rica and in other Latin American countries. The main reasons that limit the development of taxonomic studies of the appendicularians is their small size (1–2 mm), and the fact that these animals are fragile, and they are often damaged during net sampling.

Another important factor which can generate problems when a researcher intends to obtain complete, representative collection of appendicularians is the net size. Most of the regional information we have on this group has been generated upon material collected with oversized meshes (0.5 mm or more). These nets will allow sampling of the largest species, but many of the smaller forms will be undersampled and their density and relevance underestimated. Therefore, it is recommended here that appendicularians should be collected using a net mesh of no more than 0.200–0.250 mm (Esnal 1981, 1996). However, the standard procedure in most oceanic zooplankton surveys includes hauls using Bongo nets with two mesh sizes (0.3 and 0.5 mm) clearly too large to properly sample appendicularians. Therefore, a general underestimation of the density and species richness of this group is probably related to this kind of sampling in oceanic areas. The scenario changes for coastal, estuarine systems; these are sampled mainly with smaller plankton nets and with tighter meshes. New methodological developments have allowed the observation and study of the distributional patterns of deep-living appendicularians (Gorsky & Fenaux 1998). Fenaux (1993b) and Fenaux and Youngbluth (1990) described two new genera of mesopelagic appendicularians. Gorsky *et al.* (1991) reported *Oikopleura villafrancae* Fenaux, 1992, in the mesopelagic layer of the Mediterranean Sea. Hopcroft & Robison (1999) described a new species collected at a depth of 1,600 m, off Monterey Bay. These specimens were all captured using submersibles.

At the institutional level, it is recommended to increase the exchange with other regional institutes or museums with similar interests. This would be the base of multi-institutional surveys of a given area. For instance, the European Commission supported financially (1998–2001) the project EURAPP: "The impact of Appendicularia in European marine ecosystem"; this is a multidisciplinary survey designed to evaluate the ecological importance of the different species of appendicularians in the marginal seas of Europe, coordinated by Dr. Gabriel Gorsky.

References

Adame-Rodríguez MT (1982) Apendicularias del Golfo de Tehauntepec, México. Tesis Profesional. Fac. de Ciencias, UNAM, México, 87 p

1. Barham EG (1979) Giant larvacean houses: observations from deep submersibles. Science 205: 1129–1131

Bednarski M (2001) Macrozooplankton of Culebra Bay, Costa Rica, with an emphasis on copepods. M.Sc. Thesis. University of Bremen, Germany, 155 p

Castellanos I (1998) Apendicularias. Tomo I, pp 179–184. *In*: Xacur-Maiza J.A (ed) Enciclopedia de Quintana Roo. Chetumal, Q. Roo, México

Castellanos I (2003) Appendicularians (Tunicata) of Banco Chinchorro, Caribbean Sea. Bulletin of Marine Science 73: 133–140

Castellanos I, Gasca R (1998a) Distribución y abundancia de las apendicularias (Appendicularia) del Mar Caribe occidental. Oceánides 13: 13–19

Castellanos I, Gasca R (1998b) Lista faunística y clave de identificación de las apendicularias (Urochordata: Larvacea) del Caribe mexicano. AvaCient 24: 20–29

Castellanos I, Gasca R, Esnal GB (1994) Apendicularias de dos sistemas costeros. Sian Ka'an Serie Documentos 2: 18–22

Esnal GB (1981) Appendicularia. *In*: Boltovskoy D (ed) Atlas del Zooplancton del Atlántico Sudoccidental y métodos de trabajo con el zooplancton marino. Pub. Esp. INIDEP, Mar del Plata, Argentina, pp 809–827

Esnal GB (1996) Thaliacea y Appendicularia. XV. *In*: Gasca R., Suárez-Morales E (eds) Introducción al estudio del zooplancton marino. El Colegio de la Frontera Sur/CONACyT, México, pp 597–630

Fenaux R (1993a) The classification of Appendicularia (Tunicata): History and current state. Memoires de'l Institut Oceanographique 17:104

Fenaux R (1993b) A new genus of midwater appendicularian: *Mesoikoleura* with four species. J Mar Biol Assoc UK 73: 635–646

Fenaux R (1998) The classification of Appendicularia. XVIII. *In*: Bone Q (ed) The Biology of Pelagic Tunicates. Oxford University Press, England, pp 295–306

Fenaux R, Youngbluth MJ (1990) A new mesopelagic appendicularian *Mesochordeus bahamasi* gen. nov. sp. nov. J Mar Biol Assoc UK 70:755–760

Fenaux R, Bone Q, Diebel D (1998) Appendicularian Distribution and Zoogeography. XIV. *In*: Bone Q (ed) The Biology of Pelagic Tunicates. Oxford University Press, England, pp 251–264

Gorsky G, Lins da Silva N, Dallot S, Laval P, Braconnot JC, Prieur L (1991) Midwater tunicates: are they related to the permanent front of the Ligurian Sea (NW Mediterranean)? Mar Ecol Prog Ser 74:195–204

Gorsky G, Fenaux R. (1998) The role of Appendicularia in marine food webs X. Pp. 161–169. *In*: Bone Q (ed) The Biology of Pelagic Tunicates. Oxford University Press, England

Hopcroft RR, Roff JC (1995) Zooplankton growth rates: extraordinary production by the larvacean *Oikopleura dioica* in tropical waters. J Plank Res 17:205–220

Hopcroft RR, Roff JC (1998) Production of tropical larvaceans in Kingston Harbour, Jamaica: are we ignoring an important secondary producer? J Plank Res 20:557–569

Hopcroft RR, Robison BH (1999) A new mesopelagic larvaceans, *Mesochordeus erythrocephalus*, sp. nov., from Monterrey Bay, with a description of its filtering house. J Plank Res 21:1923–1937

Hopcroft RR, Roff JC, Bouman HA (1998) Zooplankton growth rates: the larvaceans *Appendicularia*, *Fritillaria* and *Oikopleura* in tropical waters. J Plank Res 20:539–555

Morales A, Murillo MM (1996) Distribution, abundance and composition of coral reef zooplankton, Cahuita National Park, Limon, Costa Rica. Rev Biol Trop 44:619–630

Quesada-Alpízar MA (2001) Caracterización, composición, abundancia y biomasa del zooplancton en el Golfo Dulce durante el periodo 1997–1998. Tesis de Maestría. Universidad de Costa Rica, San José, Costa Rica, 210 p
Suárez-Morales E, Gómez-Aguirre S (1996) Aspectos históricos y perspectivas del estudio de zooplancton marino en México. Ciencia 47:215–223
Tokioka T, Suárez-Caabro JA (1956) Apendicularias de los mares cubanos. Mem Soc Cubana Hist Nat 23(1):37–89
Zoppi de Roa E (1971) Apendicularias de la región oriental de Venezuela. Studies on the Fauna of Curaçao and other Caribbean Islands 132:76–109

Part 39
Marine Fish

William A. Bussing and Myrna López

Urobatis halleri from Bahía Salinas, Pacific Costa Rica (Photo: Jorge Cortés)

W.A. Bussing and M. López
Museo de Zoología and Centro de Investigación de Ciencias Marinas y Limnología (CIMAR), Universidad de Costa Rica, 11501-2060 San José, Costa Rica.
e-mail: wbussing@biologia.ucr.ac.cr; myrnal@biologia-ucr.ac.cr

Abstract The expected diversity of Pacific marine fishes found in waters between 0 and 200 m depth off the mainland of Costa Rica and the oceanic island Isla del Coco is greater than 800 species. The Costa Rican Caribbean ichthyofauna totals somewhat more than 600 species. Fishes new to science or new to the region are constantly being found in Central American waters wherever intensive collecting takes place. Habitats are diverse and some species apparently have spotty or limited geographic distributions. The broad outlines of lower Central American fish biodiversity are presented, although additions to this fauna are continually being made.

Introduction

The present lists of species include all material in the Museo de Zoología of the Universidad de Costa Rica (UCR) collections plus species whose geographic ranges, as are reported in the scientific literature (principally Fischer *et al.* 1995; Carpenter 2002), include Costa Rica and adjacent countries. Numerous species discovered on the exceptional Atlantic coral reefs and islands found off Belize, but not expected to occur in lower Central America, are not included. Likewise, the San Blas archipelago on the Caribbean coast of Panama presumably harbors additional reef species not expected in Costa Rican waters. The totals to date of expected species (based on the literature) for each coast of Costa Rica and the actual number of species deposited in the University of Costa Rica collections are shown in Table 39.1.

The UCR Atlantic fish holdings include about 50% of the total estimated number of species present along the Caribbean coast of Costa Rica (Species List 39.1 is included on the CD-Rom). Thus the UCR Pacific fish collections represent 85% of the estimated total number of species found between the surface and 200 m depth (Species List 39.2 is included on the CD-Rom). One representative of the Caribbean fish fauna of Costa Rica is shown in Fig. 39.1.

Because of fortuitous circumstances and events, we have placed greater emphasis on the Pacific coast ichthyofauna than on the Caribbean fishes; Figs. 39.2 and 39.3 show two representatives of the Pacific fish fauna of Costa Rica. The Golfo de Nicoya on the Pacific coast is of commercial interest for the country due to its importance as a nursery and major producer of shrimp and fishes. The taxonomic status of the Pacific ichthyofauna too was lesser known and attracted several research expeditions to sample the mainland and Isla del Coco habitats using modern sampling methods. In exchange for numerous opportunities to collect materials aboard research vessels provided by the Natural History Museum of

Table 39.1 Species counts for both coasts of Costa Rica

Ocean	Families	Genera	UCR species	Lit. species
Pacific	119	436	719	838
Atlantic	116	333	332	625

Fig. 39.1 *Lutjanus apodus* from Cahuita, Caribbean coast of Costa Rica (Photo: Andrea Bernecker)

Fig. 39.2 *Diodon holocanthus* from Bahía Salinas, Pacific Costa Rica (Photo: Jaime Nivia)

Los Angeles County (LACM), duplicate specimens have been deposited in both UCR and LACM collections since 1962. Holotypes and paratypes of new species described by us are deposited at LACM; large series of paratypes are also maintained at UCR.

The tropical eastern Pacific ichthyofauna extends from the tip of the Baja California peninsula to northern Peru. Briggs (1974) reviewed the distibution of fishes, crustaceans, and mollusks and discussed the boundaries of the two major zoogeographic provinces, a northern "Mexican Province" and southern "Panamanian Province." Numerous larger schooling fishes inhabit both provinces, although many, especially

Fig. 39.3 *Hippocampus ingens* from Bahía Culebra, Pacific Costa Rica (Photo: Jaime Nivia)

smaller less vagile species, are endemic to each province. A Central American faunal gap of less species diversity forms a wide boundary (Gulf of Tehuantepec to Gulf of Fonseca) between Mexican and Panamanian provinces principally due to the paucity of species typical of rocky shorelines. Robertson & Allen (2002) provide a list of 683 fishes for the Mexican Province, 612 species in the faunal gap, and 890 species in the Panamanian Province.

We have documented over 300 species from Isla del Coco (Lavenberg & Bussing 2000; Bussing & López 2005). Most of these are also present on the Costa Rican mainland, but about 85 species are either endemic to the island (20) or found at both Isla del Coco and Islas Galápagos (the rest). Other species are vagrants from the Indo-West Pacific or from the Central American mainland and not necessarily establish resident breeding populations at Isla del Coco.

The distribution of the Caribbean fish fauna is less clearly defined (Briggs 1974). The northern boundary of the "Caribbean Province" includes the tip of Florida, excludes the northern Gulf of Mexico, and continues from Tampico, Mexico, on the continental coastline of Central and South America to eastern Venezuela. To the south vast stretches of mud bottom extend across the mouths of the Orinoco and Amazon Rivers and delimit the "Brazilian Province." The "West

Indian Province" consists entirely of islands and contains a large diversity of species isolated from the Central American mainland and even to a lesser degree from the close-by Florida Keys. Böhlke & Chaplin (1968) list 507 species from the Bahamas, many of which have never been recorded from the Caribbean mainland.

No estimates of the total species diversity of the Western Atlantic Province are available, but Smith *et al.* (2002) plotted the combined distribution of 987 fish species as well as the composite distribution of 1,172 fish, invertebrates, and tetrapods in the Western Central Atlantic. The resulting maps reveal the area of highest species richness is located in southern Florida, the eastern Bahamas, and northern Cuba. Other secondary centers of diversity in descending order of richness are northern South America, Central America, and the northern Gulf of Mexico. They consider the relatively low endemism and species diversity on the Central American shelf to reflect, in part, its relatively recent reconnection with older North and South American faunas. Abundant evidence of this is shown by the large number of geminate transisthmian species of fish and other organisms. Most fish genera are present on both coasts of the isthmus.

Perhaps the relative paucity of Atlantic species is partly due to the physical nature of the Caribbean coastline of Costa Rica that is principally high-energy sand beach with only a few coral reef habitats in the south. The continental shelf drops off rapidly and provides a minimum diversity of benthic habitats. More collecting by scuba, trawls, and dredges would clearly improve our knowledge of the biodiversity and geographical distribution of the Costa Rican Caribbean ichthyofauna.

Specialists

Names of specialists of most fish families are available in the FAO volumes cited below.

Collections

The main collections of Costa Rican fishes are deposited in the Museo de Zoología, Universidad de Costa Rica, and in the Natural History Museum of Los Angeles County, California, USA.

References

Allen GR, Robertson DR (1994) Fishes of the Tropical Eastern Pacific. Crawford House Press, Bathhurst, Australia, 332 p

Allen GR, Robertson DR (1999) Shorefishes of the Tropical Eastern Pacific. Version 1.0 compact disk. Smithsonian Tropical Research Institute, Panamá

Böhlke JE, Chaplin CCG (1968) Fishes of the Bahamas and adjacent tropical waters. Livingston, Pennsylvania, 771 p

Briggs JC (1974) Marine Zoogeography. McGraw-Hill, New York, 475 p

Bussing WA, López MI S. (1993) Demersal and pelagic inshore fishes of the Pacific coast of lower Central America. An illustrated guide. (Spanish & English). Spec Publ Rev Biol Trop, 164 p

Bussing WA, López MI S (2005) Fishes of Cocos Island and reef fishes of the Pacific coast of lower Central America. An illustrated guide (Spanish & English). Rev Biol Trop 52(Suppl 2):192

Carpenter KE (ed) (2002) The living marine resources of the Western Central Atlantic. FAO species identification guide for fishery purposes and American Society of Ichthyologists and Herpetologists Special Publication No. 5. Vols 1–3. FAO, Rome, 2127 p

Cervigón MF (1966). Los peces marinos de Venezuela. Fundación La Salle, Caracas, 951 p

Cervigón F, Cipriani R, Fischer W, Garibaldi L, Hendrickx M, Lemus AJ, Márquez R, Poutiers JM, Robaina G, Rodríguez B (1993) Field guide to the commercial marine and brackish-water resources of the northern coast of South America. FAO, Rome, 513 p

Fischer, W. (1978) FAO species identification sheets for fishery purposes. Western Central Atlantic (Fishing Area 31). FAO, Rome, Vols 1–6 (no pagination)

Fischer W, Krupp F, Schneider W, Sommer C, Carpenter KE, Niem VH (1995) Guía FAO para la identificación de especies para los fines de la pesca. Pacífico Centro-oriental. FAO, Rome, Vols 1–3, 1813 p

Lavenberg RJ, Bussing WA (2000) Lista de especies del Parque Nacional Isla del Coco. Peces. Appendix. *In*: Garrison G (ed) Peces de la Isla del Coco/Fishes of Cocos Island. INBio, Santo Domingo, Heredia, Costa Rica, pp 334–353

Grove JS, Lavenberg RJ (1997) The Fishes of the Galápagos Islands. Stanford University Press, California, 863 p

Robertson DR, Allen GR (2002) Shorefishes of the Tropical Eastern Pacific. Compact Disk, Version 1.0. ISBN 9962-614-02-3. www.biobase.org/sftep; Smithsonian Tropical Research Institute, Panama

Smith ML, Carpenter KE, Waller RW (2002) An Introduction to the Oceanography, Geology, Biogeography and Fisheries of the Tropical and Subtropical Western Central Atlantic. *In*: K.E. Carpenter (ed) The Living Marine Resources of the Western Central Atlantic. Vol. 1: Introduction, Mollusks, Crustaceans, Hagfishes, Sharks, Batoid Fishes and Chimaeras. FAO Species Identification Guide for Fishery Purposes and American Society of Ichthyologists and Herpetologists Special Publication No. 5, Rome, pp 1–599

Thomson DA, Findley LT, Kerstitch AN (2000) Reef Fishes of the Sea of Cortéz. University of Texas Press, Austin, TX, 353 p

Part 40
Marine Reptiles and Amphibians

Mahmood Sasa, Gerardo A. Chaves, and Lisa D. Patrick

Massive "arribada" of olive ridleys (*Lepidochelys olivacea*) in Ostional, Pacific Costa Rica (Photo: Emel Rodriguez)

M. Sasa
Instituto Clodomiro Picado, Universidad de Costa Rica. San José, Costa Rica.
e-mail: msasamarin@gmail.com

G.A. Chaves
Escuela de Biología, Universidad de Costa Rica, San José, Costa Rica.
e-mail: cachi@biologia.ucr.ac.cr

L.D. Patrick
Department of Biology, Texas Tech University, Lubbock, Texas, USA.
e-mail: lisa-d-patrick@yahoo.com

I.S. Wehrtmann, J. Cortés (eds.) *Marine Biodiversity of Costa Rica, Central America*,

Abstract The herpetofauna of Costa Rica is diverse, and has been the focus of much attention in the last six decades. Despite the extensive general knowledge, little information exists about the species of reptiles and amphibians that inhabit and use the marine and insular environments of the country. Furthermore, practically there is no data on the adaptations that reptiles and amphibians have evolved in order to cope with the rough conditions that characterize these environments. Based on our field observations and previously published reports, we identify 47 species of reptiles and amphibians that occur in five marine environments (including islands). In the Caribbean coast of the country, herpetological studies are centered in sea turtle biology and conservation. In the Pacific coast, marine turtle conservation, dynamics of crocodile populations, and several records on the species inhabiting various mangroves and islands are available. The number of species reported here represents ~12% of the species known to Costa Rica, a percentage that is expected to increase as more research is conducted in those environments. Ecological and physiological studies of the reptiles and amphibians inhabiting marine environments are needed to preserve this important portion of our biodiversity.

Introduction

Modern amphibians and reptiles have radiated extensively throughout terrestrial and freshwater habitats in the tropical and subtropical regions of the world. In contrast, relatively few species of amphibians and reptiles actually inhabit or use marine environments (McCoy 1982), probably due to the severe physical conditions that characterize these ecosystems: high levels of solar radiation, high temperatures and salinity, and an overall lower amount of vegetation cover and shelter.

High salinity is an especially difficult condition, given that amphibians are hypo-osmotic, causing them to lose water and gain ions in marine environments. Due to these fluctuations in water equilibrium, most amphibians are unable to cross even narrow salty water barriers (Duellman & Trueb 1994). Additionally, amphibians lack salt glands, rendering them unable to eliminate high concentrations of salt.

Species of reptiles that tolerate saltwater are more numerous: their impermeable skin is an effective mechanism for protection from desiccation. Furthermore, marine reptiles have specialized glands for excreting excessive salt, mostly in the form of sodium chloride (Peaker & Linzell 1975; Zug 1993). The use of the marine environment by reptiles is more limited during the stages of development when these protective mechanisms are not yet fully evolved.

In a recent, monumental review, Savage (2002) listed 396 species of reptiles and amphibians that are known to occur in Costa Rica. From this figure, only seven species (six sea turtles and a pelagic sea snake) can be considered truly marine, because they spend most of their life cycle in marine waters. However, Savage also described three additional marine environments potentially used by other species of amphibian and reptiles of Costa Rica: salt marshes, mangrove swamps, and coastal beaches and dunes. In the following account, we attempt to expand upon Savage's view and identify the amphibian and reptiles that use suitable marine environments.

We include here species that – to our knowledge – are known to occur in the littoral and supralittoral zones (e.g., mangroves, sand beaches) and those inhabiting off-shore islands. Our account should not be considered a definitive list and is aimed at suggesting further study of the herpetofauna associated with these areas, as well as their biological adaptations to the marine environment.

Diversity of Marine Reptiles and Amphibians

We distinguish five main species groups of amphibians and reptiles, based on location:

1. Pelagic species: organisms that spend most of their life in the open sea.
2. Marine occasional species: organisms that are mainly terrestrial or riparian but occasionally are found in marine waters.
3. Littoral species: organisms that are found in littoral zone, including salt marshes and mangrove swamps.
4. Supralittoral species: organisms that are found in supralittoral zone, including sandy beaches and rocky beaches.
5. Insular species: organisms that have established populations on oceanic or continental islands. We distinguish between oceanic and continental islands based on distance from the mainland, history of geological formation, and origin of species. Oceanic islands are located greater distances from the mainland, are formed through volcanic processes, and have a high level of species endemism. Continental islands, on the other hand, are on the continental shelf, originate from sedimentary processes, were often formerly connected to the mainland, and contain similar species to those on the mainland.

Pelagic species: Included in this category are sea turtles and the pelagic sea snake *Pelamis platurus* (Fig. 40.1). These species can survive in the water column for most of their lives. Sea turtles must return to the beach to lay their eggs, but the pelagic sea snake is viviparous and therefore does not require land for reproduction.

Six species of sea turtles are found in Costa Rican continental waters, and all use the country's beaches as nesting grounds. Breeding season vary among species: on the Caribbean coast, nesting season of the leatherback *Dermochelys coriacea* begins around February and extends through July (Chacón *et al.* 1996), whereas in the same coast the green turtle *Chelonia mydas* nests from July to October (Bjorndal *et al.* 1999). On Pacific beaches, the leatherback nests from October to March, while in the Pacific green turtle *C. agassizi* (Fig. 40.2) nesting occurs from July to October (Chaves, personal observations: 1999–2005). Conversely, the olive ridley *Lepidochelys olivacea* has a widespread nesting season in the Pacific that extends throughout the wet season (June–December, Cornelius 1986), although in recent years it is not rare to see large numbers of females nesting during the dry season (Chaves, personal observations: 2006–2008). In general, marine turtles make

Fig. 40.1 The pelagic sea snake *Pelamis platurus*, a species from Pacific Costa Rica (Photo: Alejandro Solórzano)

Fig. 40.2 *Chelonia agassizi,* photographed at Isla del Coco, Pacific Costa Rica (Photo: Jaime Nivia)

transoceanic migrations early in their life. Except for leatherbacks – that retain such transoceanic journeys – other Costa Rican turtle species migrate along the coastline during adulthood, moving between feeding grounds and reproductive areas (Savage 2002). Mating takes place in the water just before nesting, although sperm storage is known to occur in several species. Females may lay multiple clutches in a single season. The incubation period for Costa Rican species is between 46 and 80 days (Chacón *et al.* 2001) and sex determination is temperature-dependent. Marine turtles are able to remain under water for up to 3 h and surface only to breath. All sea turtles in Costa Rica are currently under protection from trade by the Convention on International Trade in Endangered Species of Wild Flora and Fauna (CITES), given the heavy overexploitation of their eggs for human and animal consumption, and the frequent "accidental" deaths in fishing lines and nets. In addition, some species of sea turtles found in our country are also protected under the United States Endangered Species Act.

The leatherback turtle *D. coriacea* is the largest species of sea turtle (Savage 2002). *Dermochelys* is able to withstand low temperatures, allowing it to reach the cold waters of Alaska, Iceland, and South Africa (Greer *et al.* 1973). This species feeds mainly on jellyfish and other soft-bodied pelagic invertebrates, and its meat is considered toxic due to the retention of jellyfish nematocysts although still is consumed by humans in some countries. Anthropogenic effects seem responsible for the reduction of *D. coriacea* populations at an alarming rate (Spotila *et al.* 1996). In Costa Rica, Spotila *et al.* (1996) and Chacón *et al.* (2001) have studied extensively the biology of the species.

The green turtles *C. agassizi* and *C. mydas* are also large forms (up to 1.3 m in carapace length and close to 200 kg) that feed mainly on sea grass. These two species are allopatric: *C. agassizi* occur in the Pacific Ocean, whereas *C. mydas* inhabits the Caribbean Sea and the Atlantic Ocean. In Costa Rica and neighboring countries, extensive hunting for the meat and eggs of *Chelonia* has put these species in danger of extinction. Nevertheless, conservation efforts conducted at Tortuguero (northeastern Caribbean Costa Rica) result in an increase in *C. mydas* population size there (Bjorndal *et al.* 1999).

Next in body size is the loggerhead, *Caretta caretta*, a rare turtle in Costa Rica found only on the Caribbean coast. This species feeds on a wide variety of invertebrates and occasionally fish (Savage 2002).

The two remaining species of sea turtles found in Costa Rica are relatively small: the hawksbill turtle, *Eretmochelys imbricata*, and the olive ridley, *L. olivacea*. *Eretmochelys* is an uncommon species on both coasts, most likely due to hunting pressure for their carapace shields, used intensively in fine art craft (the famous carey). This species feeds on sponges, tunicates, and mollusks (Meylan 1988, 1990) and nests on solitary rocky beaches. The olive ridley is restricted to the Pacific coast, where it is still frequently observed. These turtles prefer sandy beaches to nest, and the species is well known by their "arribada" or synchronized nesting. During arribada, several thousand females arrive to nest on the same beach for two or three nights, usually once per month during the rainy season. In Costa Rica, this event occurs on the beaches of Ostional and Nancite (Península de Nicoya and Santa Elena).

The pelagic sea snake has a broad oceanic distribution. The species is known from the Indian Ocean throughout the Indo-Australian region across the tropical Pacific Ocean. In America, it is distributed from Peru to California. In Costa Rica, *P. platurus* is known to congregate in large groups that float along drift lines or slicks. Our observations confirm that such congregations can be particularly dense: during a 1 h survey, one of us (Mahmood Sasa) counted 326 individuals just outside Playa del Coco. Apparently, this species does not have any fish predators and studies conducted at the Smithsonian Tropical Research Institute in Panama (Kropach

1975) suggest that large fish predators avoid the snake. Population control may be mostly mediated by the changes in wind pattern and currents during the early dry season: yearly, thousands of individuals are deposited by the tides on the beach, an event that periodically alarms the public. Deposited individuals are incapable of crossing the wave line to return to their slicks, and thus die of exhaustion or thermal shock. Research on their venom characteristics is currently being conducted at the Instituto Clodomiro Picado (University of Costa Rica).

Marine occasional species: We are aware of two species that frequently cross the shoreline to enter saline waters, usually in or around estuaries: the American crocodile, *Crocodylus acutus*, and the green iguana, *Iguana iguana*. Both species are strong swimmers capable of tolerating saline water as adults. As juveniles, however, it seems their distribution is restricted to freshwater areas.

C. acutus is the larger of the two species of crocodilians occurring in Costa Rica, and is distributed in the majority of large rivers on both versants. Populations of this species are particularly dense in Costa Rica (Sasa & Chaves 1992; Bolaños *et al.* 1996–1997) and commonly occur in mangrove swamps and estuaries. Large individuals may be observed swimming along solitary beaches where sea turtles are known to nest. In Playa Nancite, for example, crocodile attacks are one of the main causes of turtle mortality during the *Lepidochelys* arribadas (Arauz & Morera 1990; Maziars *et al.* 1994). The other Costa Rican crocodilian, the spectacled caiman, *Caiman crocodilus*, is occasionally found in brackish water and mangroves, but since it has not been observed along beaches or entering marine water it is excluded from this section.

On the beaches of Carate, Madrigal, and Sirena (Península de Osa) and in Ostional (Península de Nicoya) several visitors and local researchers have witnessed green iguanas crossing the sand and entering the ocean. Recently, two of us (Mahmood Sasa and Gerardo A. Chaves) saw individuals coming from the ocean towards the beach. The animals appeared exhausted while crawling back towards the beach vegetation. There is no definite information on the circumstances in which these animals enter the ocean, but the frequency of observations indicates that their occurrence in marine water is probably not accidental. Closely related iguanas, like the marine iguana (genus *Amblyrhynchus*), enter ocean waters to feed on algae (Zug 1993). Whether the green iguanas enter the ocean to use a food resource is still unknown, but this phenomenon deserves further attention. Green iguanas have been extensively studied by Burghardt and Rand (1982) in Panama, and by Van Devender (1982) in Costa Rica.

Littoral species: Several species of amphibians and reptiles inhabit and use resources in the littoral areas (Species Lists 40.1 and 40.2 are included on the CD-Rom): mangroves and salt marshes. In mangrove swamps of the Caribbean coast, *Tretanorhinus nigroluteus* enters brackish water (W. Lamar, personal observation). This aquatic snake is rare along Costa Rica, but is not uncommon to find in other countries of Central America. In the mangrove swamps of the South Pacific coast, two nocturnal snakes, *Corallus ruschenbergerii* and *Leptodeira rubricata*, can be observed in the vegetation. In Costa Rica, *C. ruschenbergerii* is known to feed mainly on birds and mammals, although juveniles also prey on lizards (Henderson 1993).

L. rubricata feed mainly on small crabs and fishes, but their hunting strategies and behavior have not been formally described. Other snake species, such as *Chironius grandisquamis*, have also been observed along the shoreline, especially during their breeding season (A. Solórzano, personal observation).

Supralittoral species: The supralittoral areas (sandy and rocky beaches) also exhibit a diverse array of amphibians and reptiles. Temporal ponds along mangroves are breeding places for the tree-frogs *Smilisca baudini* and *Trachycephalus venulosus*. Likewise, the toad Chaunus (=*Bufo*) *marinus* and the frog *Leptodactylus melanonotus* have been observed breeding at night in temporal ponds in beaches, at the point where the vegetation line ends. However, no information is available about the salinity tolerance of eggs and tadpoles of these species, or about the physical characteristics of the breeding ponds. Along the Pacific shore of Costa Rica, it is not uncommon to discover the rain-frog Craugastor (=*Eleutherodactylus*) *fitzingeri* escaping the high day temperatures under logs and vegetation on the beach.

Reptiles are better represented than amphibians in supralittoral zones. Several lizards (i.e., *Ameiva quadrilineata*, *Cnemidophorus deppii*, *Ctenosaura similis*, and *Sceloporus variabilis*) are heliophilic, and thus require high levels of solar radiation to be active (Fitch 1973a). These species are active foragers and are in almost constant movement, searching for arthropods in the sand during the hottest hours of the day. Other species are sit-and-wait predators, and prefer to ambush larger prey. For example, the Jesus Christ lizard, *Basiliscus basiliscus*, has been observed hunting insects and crabs on the beaches of Bahía Drake (Península de Osa).

Other species of reptiles are active in the supralittoral zone at night. The New World python, *Loxocemus bicolor*, is found occasionally on sandy beaches, digging up nests of sea turtles and preying upon their eggs (Mora & Robinson 1984). Likewise, the scorpion-eating snake *Stenorrhina freminvillei* is sporadically observed in the beach, in areas covered by vegetation. Also, *Boa constrictor* is commonly found in the beaches of Península de Osa and the Caribbean shore, resting on logs on the beach or moving in the sand (W. Lamar, personal observation).

Insular species: In this section we include species that inhabit continental or oceanic islands (Species List 40.2). The majority of reptiles and amphibians reported here are also known from other marine environments. Furthermore, since marine climate and marine environments heavily influence the physical conditions on islands, a description of the diversity and ecology of insular species is important to understand their capacities to inhabit or to use marine environments.

To our knowledge, no serious attempt has been made to describe the herpetological diversity of insular Costa Rica, although some information is available from our informal visits to the islands, and from material at the Museo de Zoología (Universidad de Costa Rica). The majority of islands are found off the Pacific Coast of the country and are all continental: a series of small islands surround Península de Santa Elena (Islas Murciélago, Isla San José), the Golfo de Nicoya (Chira, Bejuco, Caballo, Venado, San Lucas, and Negritos), and the Península de Osa (Isla del Caño). On the islands of Santa Elena and the Golfo de Nicoya, all of the species observed are typical of dry forest, and are adapted to the strong seasonal rainfall pattern which characterizes the region (Sasa & Solórzano 1996). Isla del Caño, perhaps the best known island herpetologically, is home to 16 species also known to occur in mainland wet forest on the South Pacific coast (Species List 40.2). On the Caribbean side, no information is available for the reptiles and amphibians that inhabit Isla Uvita, or any of the other small islands off shore.

Costa Rica has few oceanic islands; the most important is Isla del Coco, located 532 km off the Pacific coast. Two endemic species of lizards are the only reptiles present on this island: *Norops townsendi* and *Sphaerodactylus pacificus. N. townsendi*, a small, diurnal short-legged anole, is abundant and can be found on rocks, abandoned shelters, or on low herbaceous plants. Most individuals of this species have been observed with broken or regenerated tails, perhaps implying high predation or intraspecific aggression (Savage 2002). *S. pacificus*, one of the largest species in the genus, is nocturnal (Carpenter 1965) and found mainly in lowland wet forests. It is abundant near the shoreline under debris and around the base of coconut palms (Savage 2002), where empty shells and eggs of this species have also been found (Taylor 1956).

To this date, there is no information on the molecular and genetic variation of populations of any reptile or amphibian inhabiting these islands. Thus, the origin and biogeographic events that shape current distributions remains unclear. Research in this direction is strongly recommended.

As shown by this brief summary, Costa Rica is a home to few marine reptiles and amphibians. By expanding our definitions, however, to include environments other than the ocean itself, we are able to demonstrate a higher diversity of species. Unfortunately, most of our information comes from studies done on the Pacific side of the country. Further research, especially on the Caribbean coast, to explore the adaptations and species richness of these taxonomic groups is crucial to our understanding of Costa Rica's marine environment.

Specialists and Collections

The herpetofauna of Costa Rica has been extensively studied in the last 50 years or so. Jay M. Savage (savy1@cox.net) has been working in Costa Rica for more than 40 years. Savage is emeritus professor of biology at the University of Miami, and adjunct at San Diego State University. William W. Lamar (University of Texas at Tyler, lamar@ballistic.com) and Louis Porras have conducted intensive collections from the country. Didiher Chacón (Red Centroamericana de Tortugas Marinas) has studied the populations of leatherback sea turtles in Costa Rica. Federico Bolaños, curator at the Museo de Zoología, Universidad de Costa Rica, has extensive experience with amphibians. Anny Chaves Quirós, now at the Centro de Gestión Ambiental Instituto Costarricense de Electricidad, has been studying intensively the biology and conservation of sea turtles in various beaches in Costa Rica. Alejandro Solórzano (Serpentario Nacional, asolorz@racsa.co.cr) recently published an extensive guide of the snakes of Costa Rica, and is currently studying the ecology of several species of the region, including the pelagic sea snake.

Acknowledgements The authors thank A. Solórzano, J. M. Savage, W.W. Lamar, and an anonymous reviewer for sharing some unpublished information and for making suggestions to improve the manuscript. This study was supported partially by Vicerrectoría de Investigación, Universidad de Costa Rica (Project 741-A1-021) and by a US Student Fulbright Program to LDP.

References

1. Arauz M, Morera R (1990) Monitoreo de las poblaciones de tortuga lora (*Lepidochelys olivacea*) en playa Nancite, Parque Nacional Santa Rosa, Costa Rica. Periodo 1 de agosto al 31 diciembre de 1989. Mimeografiado, 12 p
2. Bolaños JR, Sánchez J, Piedra L (1996–1997) Inventario y estructura poblacional de crocodilos en tres zonas de Costa Rica. Rev Biol Trop 44(3)/45(1):283–287

Bjorndal KA, Wetherall JA, Bolten AB, Mortimer JA (1999) Twenty-six years of green turtle nesting at Tortuguero, Costa Rica: an encouraging trend. Conserv Biol 13:126–134

3. Burghardt GM, Rand AS (1982) Iguanas of the World: Their Behavior, Ecology and Conservation. Noye, Park Ridge, NJ, 472 p
4. Carpenter CC (1965) The display of the Cocos Island anole. Herpetol 21:256–260
5. Carr AF (1983) *Chelonia mydas* (tortuga, tortuga blanca, green turtle). *In*: Janzen DH (ed). Costa Rican Natural History. University of Chicago Press, Chicago, IL, pp 394–396
6. Chacón D, McLarney W, Ampie C, Venegas B (1996) Reproduction and conservation of the leatherback *Dermochelys coriacea* (Testudines: Dermochelyidae) in Gandoca, Costa Rica. Rev Biol Trop 44:853–860
7. Chacón D, Valerín N, Cajiao MV, Gamboa H, Marín G (2001) Manual para mejores prácticas de conservación de las tortugas marinas en Centroamérica. Asociación ANAI, San José, Costa Rica, 139 p
8. Cornelius SE (1986) The Sea Turtles of Santa Rosa National Park. Fundación de Parques Nacionales, San José/INCAFO, Spain, 64 p
9. Dial BE, Grismer LL (1992) A phylogenetic analysis of physiological-ecological character evolution in the lizard genus *Coleonyx* and its implications for historical biogeographic reconstruction. Syst Biol 41:178–195
10. Dodd C (1990b) *Caretta*. Cat Amer Amph Reptl 482:1–2
11. Dodd CK (1990a) *Caretta caretta*. Cat Amer Amph Reptl 483:1–7

Duellman WE, Trueb, L (1994) Biology of Amphibians. John Hopkins University Press, Baltimore, MD, 669 p

12. Fitch HS (1973) A field study of Costa Rican lizards. Univ Kansas Sci Bull 50:39–126
13. Fitch HS (1973) Population structure and survivorship in some Costa Rican lizards. Occas Pap Mus Nat Hist Univ Kansas 18:1–41
14. Greene HW (1983) *Boa constrictor* (boa, bequer). *In*: Janzen DH (ed) Costa Rican Natural History. University of Chicago Press, Chicago, IL, pp 380–382

Greer BJ, Lazell JD, Wright RM (1973) Anatomical evidence for a counter-current heat exchanger in the leatherback turtle (*Dermochelys coriacea*). Nature 244:181

15. Henderson RW (1993) On the diets of some arboreal boids. Herp Nat Hist 1:91–96
16. Hirth HF (1971) Synopsis of biological data on the green turtle, *Chelonia mydas* (Linnaeus). FAO Fish Symp 85:1–8, 19
17. Hirth HF (1980) *Chelonia mydas*. Cat Amer Amph Reptl 249:1–4
18. Kropach C (1975) The yellow bellied sea snakes *Pelamis* in the eastern Pacific. *In*: Dunson WA (ed) The Biology of Sea Snakes. University Park Press, Baltimore, MD, pp 185–213

19. Lieberman SS (1986) Ecology of the leaf litter herpetofauna of a Neotropical rain forest La Selva, Costa Rica. Acta Zool Mex ns 15:1–72

Maziars JD, Bower T, Quesada J (1994) Monitoring of the olive ridley (*Lepidocelys olivacea*) population at Nancite Beach, Santa Rosa National Park, Guanacaste, Costa Rica. Mimeografiado, 13 p

McCoy CJ (1982) Reptiles. *In*: Hurlbert SH, Villalobos-Figueroa A (eds) Aquatic Biota of Mexico, Central America, and the West Indies. San Diego State University, San Diego, CA, pp 515–520

20. Medem F (1983) Los Crocodylia de Sur América. Vol. 2. Venezuela-Trinidad y Tobago-Guyana-Suriname-Guayana Francesa-Ecuador-Perú-Bolivia-Brasil-Paraguay-Argentina-Uruguay. Univ. Nac. Colombia/Fondo Colombiano Invest. Cient. Proyectos Espec. "Francisco José Caldas," Bogota, 271 p

21. Meylan A (1988) Spongivory in hawksbill: a diet of glass. Science 238:393–395

22. Meylan A (1990) Nutritional characteristics of sponges in the diet of the hawksbill turtle *Eretmochelys imbricate*. *In*: Rützler K (ed) New Perspectives in Sponge Biology. Smithsonian Institution Press, Washington, DC, pp 472–477

23. Mora JM, Robinson DC (1984) Predation on sea turtle eggs (*Lepidochelys*) by the snake *Loxocemus*. Rev Biol Trop 32:161–162

Peaker M, Linzell JL (1975) Salt Glands in Birds and Reptiles. Cambridge University Press, Cambridge, 307 p

24. Pritchard PCH, Trebbau P (1984) The Turtles of Venezuela. Contributions to Herpetology 2. Society for the Study of Amphibians and Reptiles: Oxford, Ohio, 403 p

25. Rand SA, Dungan BA (1983) Structure of complex iguana nests. Copeia 3:705–711

26. Sasa M, Solórzano A (1996) The reptiles and amphibians of Santa Rosa National Park, Costa Rica, with comments about the herpetofauna of xerophytic areas. Herpetol Nat Hist 3:113–126

27. Sasa M, Chaves G (1992) Tamaño, estructura y distribución de una población de *Crocodylus acutus* (Crocodylia: Crocodylidae). Rev Biol Trop 40:131–34

28. Savage JM (2002) The Amphibians and Reptiles of Costa Rica. University of Chicago Press, Chicago, IL, 934 p

29. Schwartz A, Henderson RW (1988) West Indian amphibians and reptiles: a checklist. Milwaukee Publ Mus Contr Biol Geol 74:1–264

30. Sites JW, Dixon JR (1982) Geographic variation in *Sceloporus variabilis* and its relationship to *S. teapensis* (Sauria: Iguanidae). Copeia 1:14–27

31. Solórzano A (2004) Serpientes de Costa Rica. Snakes of Costa Rica. Instituto Nacional de Biodiversidad, Santo Domingo de Heredia. Costa Rica, 792 pp

32. Spotila J, Dunham A, Leslie A, Steyermark A, Plotkin P, Paladino F (1996) Worldwide population decline of *Dermochelys coriacea*; are leatherback turtles going extinct? Chelonian Conserv Biol 2:209–222

33. Taylor EH (1956) A review of the lizards of Costa Rica. Univ Kansas Sci Bull 38:3–322

34. Van Devender RW (1978) Growth ecology of a tropical lizard, *Basiliscus basiliscus*. Ecology 59:1031–1038

Van Devender RW (1982) Growth and ecology of spiny-tailed and green iguanas in Costa Rica, with comments on the evolution of herbivory and large body size. *In*: Burghardt GM, Rand AS (eds) Iguanas of the World: Their Behavior, Ecology and Conservation. Noyes, Park Ridge, NJ, pp 162–183

Zug GR (1993) Herpetology: An Introductory Biology of Amphibians and Reptiles. Academic Press, San Diego, CA, 527 p

Part 41
Birds in Coastal and Marine Environments

Gilbert Barrantes and Johel Chaves-Campos

Group of brown pelicans (*Pelecanus occidentalis*) off Pacific Costa Rica (Photo: Ingo S. Wehrtmann)

G. Barrantes
Museo de Zoología, Escuela de Biología, Universidad de Costa Rica, 11501-2060 San José, Costa Rica.
e-mail: gbarrantes@biologia.ucr.ac.cr

J. Chaves-Campos
Department of Biological Sciences, Purdue University, West Lafayette, Indiana 47907, USA.
e-mail: jchavesc@costarricense.cr

Abstract In Costa Rican coastal and marine environments (e.g., estuaries, mudflats, islands, open ocean) 96 bird species have been recorded, 11% of the total avifauna in the country. This high diversity is primarily explained by the complex topography of the coasts, the large variety of habitats available for coastal and marine birds, and Isla del Coco, a volcanic island that includes a particular marine avifauna. The Pacific coast, including Isla del Coco, possesses a much higher diversity of birds (93 species) than the Caribbean coast (54 species). This difference is likely explained by higher fluctuation in tides, larger extension, and greater topographical complexity on the Pacific coast. There are only 15 coastal or marine birds that reproduce (but not exclusively) in the country. From a conservation perspective, coastal and marine birds have received little attention in Costa Rica. Consequently, contamination, caused by pesticides, sewage, and solid trash, and habitat destruction, due to the construction of tourist infrastructure, seriously threaten the coastal and marine avifauna in Costa Rica.

Introduction

A total of 857 species of birds have been recorded from Costa Rica (Barrantes *et al.* 2002). Of these, we consider 96 (11%) as coastal and marine species. This group includes the typical sea birds (Slud 1967; Harrison 1983; Stiles 1984), shorebirds, and their relatives (Stiles & Smith 1977; Barrantes & Pereira 1992; Table 41.1 and Species List 41.1 which are included on the CD-Rom). There is only a small number of coastal and marine species that reproduce in Costa Rica (15 species, Table 41.2); the rest of them are wintering and transient species, or regular and occasional visitors (Stiles & Skutch 1989). In marine environments, we include a variety of habitats such as open-ocean, seashores, mudflats, estuaries, mangroves, islands, and islets. These habitats serve as feeding and/or nesting grounds for coastal and marine avifauna.

The world marine avifauna is relatively well represented at the family level in Costa Rica. There are only three families of typical marine birds that are absent from the country: Spheniscidae (penguins), Gaviidae (loons), and Alcidae (auks and puffins). Other families, like Hydrobatidae, Procellariidae, and specially Scolopacidae and Laridae have large numbers of species in Costa Rica (Table 41.1 and Species List 41.1 which is included on the CD-Rom). The high diversity of coastal and marine birds is primarily explained by the topographical characteristics of the coasts, diversity of habitats, and a pelagic site, Isla del Coco.

The littoral in Costa Rica extends 212 km on the Caribbean (east) coast and 1,254 km on the Pacific (west) coast. The topography of the Caribbean coast is regular with a few river mouths that dissect the sandy beaches that characterize this seashore, and two small nearshore islands. The simple topography and almost absence of tides result in a landscape with few habitats available for coastal and marine birds (Stiles 1984). Contrarily, the complex topography and high tidal fluctuations of the Costa Rican Pacific coast have resulted in numerous habitats appropriate for

Table 41.1 Orders of birds that include species using marine habitats in Costa Rica. Number of families, genera, and species are given

Order	No. of families	No. of genera	No. of species
Procellariiformes	3	7	18
Pelecaniformes	5	5	11
Charadriiformes	8	28	67

Table 41.2 Coastal and marine birds that reproduce in Costa Rica. Coast and specific habitats for breeding activities are included

Species	English name	Coast	Habitat
Pelecanus occidentalis	Brown Pelican	Pacific	Islands
Sula dactylatra	Masked Booby	Pacific	Isla del Coco
Sula leucogaster	Brown Booby	Pacific, Caribbean	Islands
Sula sula	Red-footed Booby	Pacific	Isla del Coco
Phalacrocorax brasilianus	Neotropical Cormorant	Pacific	Freshwater swamps
Fregata magnificens	Magnificent Frigatebird	Pacific	Islands
Fregata minor	Great Frigatebird	Pacific	Isla del Coco
Himantopus mexicanus	Black-necked Stilt	Pacific	Sandy beaches
Charadrius wilsonia	Wilson's Plover	Pacific	Sandy beaches
Charadrius collaris	Collared Plover	Pacific, Caribbean	Beaches
Sterna anaethetus	Bridled Tern	Pacific	Small grassy islands
Sterna fuscata	Sooty Tern	Pacific	Isla del Coco
Anous stolidus	Brown Noddy	Pacific	Islets, Isla del Coco
Anous minutus	Black Noddy	Pacific	Isla del Coco
Gygis alba	White Tern	Pacific	Isla del Coco

different groups of coastal and marine birds (Stiles 1984). For example, great extensions of mudflats that serve as breeding and stopover grounds for a large number of shorebirds, terns, and seagulls are exposed during low tides, particularly on the shore of the Golfo de Nicoya (northwestern region of Costa Rica) and around the numerous river mouths (Stiles & Smith 1977). The varied topography of the Pacific coast also includes estuaries, mangroves, sandy beaches, and several small islands. These habitats are used for feeding and/or reproducing by pelicans, frigatebirds, boobies, shorebirds, gulls, and terns. Pelagic seabirds (e.g., petrels, shearwaters) are often observed mostly along the Costa Rican west coast (seldom in embayments) during their migration or when tropical storms occur (Stiles 1984).

Isla del Coco is a small (24 km^2) single volcanic island with extremely rainy climate and covered with a dense evergreen vegetation (Slud 1967; Grant 1986). It is located 500 km offshore Costa Rican mainland and 630 km from Galapagos in the Pacific Ocean. The island has a seabird fauna distinct from that of Costa Rican coast. *Oceanodroma markhami*, *Sula sula*, *Fregata minor*, and *Gygis alba* are some of the species sighted on this oceanic island (Figs. 41.1–41.3).

Fig. 41.1 Red-footed booby (*Sula sula*) from Islote Manuelita, Isla del Coco, Pacific Costa Rica (Photo: Felipe López Pozuelo)

Fig. 41.2 Brown booby (*Sula leucogaster*) from Islote Manuelita, Isla del Coco, Pacific Costa Rica (Photo: Felipe López Pozuelo)

The high diversity of habitats available for coastal and marine birds on the Pacific seaside, including Isla del Coco, is likely to be the main factor determining the higher species richness (93 species) compared to the avifauna present on the Caribbean side (54 species). Of these, 42 species of coastal and marine birds have been recorded exclusively on the Pacific coast, whereas, only three in the Caribbean side (Species List 41.1 is included on the CD-Rom). A representative of the Caribbean bird fauna of Costa Rica is shown in Fig. 41.4.

Fig. 41.3 White tern (*Gygil alba*) from Isla del Coco, Pacific Costa Rica (Photo: Felipe López Pozuelo)

Fig. 41.4 Whimbrel (*Numenius phaeopus*) from the Caribbean coast of Costa Rica (Photo: Jeffry Ortiz)

Most coastal birds are migrants that reproduce in North America and winter in Costa Rica or other tropical areas. Practically all shorebirds, terns, and gulls have this migratory pattern. Few of these species have migratory and resident populations in Costa Rica, e.g., *Charadrius wilsonia* and *Hymantopus mexicanus*. Similarly, pelagic seabirds visit Costa Rican coasts during their regular migration or erratic long-distance movements, which are likely determined by environmental conditions (e.g., storms) or following changes in food resource availability. Few typical seabirds have breeding colonies in Costa Rica, e.g., *Sula leucogaster* (Fig. 41.5), *G. alba* (Table 41.2). Therefore, relatively few species are year-around residents in the country, and there are no endemic coastal or marine birds in Costa Rica.

The large majority of species included in Species List 41.1 (on CD-Rom) have either wide distribution or extensive migratory ranges (Harrison 1983; Hayman *et al.* 1986; Morrison & Ross 1989). Consequently, there are few differences in species richness and composition with changes in latitude as shown for five families in Table 41.3. Thus, the number of species as well as the species composition (using Jaccard index) vary little between Costa Rica, Mexico (Peterson & Chalif 1973), and Ecuador (Ridgely & Greenfield 2001). Procellariidae is the family that varies most among the three countries. In Mexico, this family has almost twice as many species as Costa Rica, possibly because of the extensive shoreline, the large number of islands, and the latitudinal range that Mexico covers.

The wide distributions of most marine and coastal species are not related to their abundances. Probably some of these species have been naturally scarce, however

Fig. 41.5 Brown boobies (*Sula leucogaster*), a typical visitor of shrimp trawlers along the Pacific coast of Costa Rica (Photo: Ingo S. Wehrtmann)

Table 41.3 Species richness of five marine and coastal bird families in Costa Rica, Mexico, and Ecuador. In parentheses are included the comparison of species composition between Costa Rica and Mexico, and Costa Rica and Ecuador, using Jaccard index (0–100% of similarity). Species that do not use coastal or marine habitats were excluded from the analysis

	Costa Rica	Mexico	Ecuador
Procellariidae	9	15 (26)	9 (29)
Hydrobatidae	8	9 (70)	10 (64)
Charadriidae	6	8 (75)	10 (45)
Scolopacidae	26	27 (77)	23 (88)
Laridae	24	28 (62)	28 (48)

populations of these species have declined and continue declining as a result of habitat destruction and contamination in breeding and wintering areas (Roca *et al.* 1996). In Costa Rica, the construction of large tourist infrastructures on the Pacific coast has eliminated extensive areas of mangrove forests and mudflats. The contamination in habitats of coastal and marine birds in Costa Rica comes from three main sources: (1) wash down of pesticides mainly from banana and rice plantations, (2) solid material, and (3) sewage that pollute many feeding and breeding areas of coastal birds (Hidalgo-Calderón 1986; Burger *et al.* 1993; Burger & Gochfeld 1995). The elimination and contamination of such natural habitats threaten many species that use coastal habitats for feeding and reproducing. Shorebirds are most affected when mudflats are eliminated in northwestern Costa Rica, since the coasts along this region constitute one of the most important stopover areas in the Neotropics (Stiles 1983). Possibly, millions of individuals from different shorebird, gull, and tern species winter or stop during migration on the mudflats of northwestern Costa Rica (Smith & Stiles 1979; Barrantes & Pereira 1992).

Additionally, a large proportion of marine birds (e.g., Sulidae, Fregatidae) nest in colonies. This behavior makes these birds more susceptible to changes in their environment, since these nesting sites are scarce, and colonies are commonly on small islands, frequently located in regions with high risk of hurricanes and other environmental disturbances. Furthermore, recent invasions or introductions, e.g., rats and cats, in islands and other sites used by colonial birds have decreased the breeding success of these species (Harrison 1990). Therefore, it is imperative to monitor the destruction and contamination of the habitats used by marine and coastal birds to preserve this portion of our biological diversity (Croxall *et al.* 1984; for a more extensive discussion on these topics).

Specialists and Collections

The diversity of marine birds has been relatively well studied in Costa Rica. For more than 20 years F. Gary Stiles devoted a considerable amount of effort completing the list of marine birds of Costa Rica, as well as collecting these species for the Museo de Zoología of the Universidad de Costa Rica. In general, we consider that the list of marine species of the country is nearly complete. However, the presence of at least seven species that occur only offshore (e. g. Procellariidae, Hydrobatidae) needs to be confirmed (for examples see Stiles & Skutch 1989 and Barrantes *et al.* 2002). Even taking into account that the list of pelagic species is not complete, the occurrence of at least six species of seabirds in Central America is known only from their presence in Costa Rica (Species List 41.1 which is included on the CD-Rom). This indicates the lack of information on marine birds in other countries of Central America.

Most of the species recorded in Costa Rica have specimens deposited in two national museums (Museo Nacional and Museo de Zoología). Nevertheless, these species are generally poorly represented in terms of number of specimens. The

Museo de Zoología of the Universidad de Costa Rica and the Natural History Department of the Museo Nacional of Costa Rica have at least one individual of most species recorded in the country. Plovers and sandpipers are relatively well represented in the national collections, but other groups, in particular pelagic species, are represented by only one or two specimens. Furthermore, 21 species of marine birds that certainly occur in Costa Rica do not count with an official voucher, e.g., museum specimen deposited in national museums, song record, or photo (Barrantes *et al.* 2002). Some of the species without official voucher may be deposited in museums of other countries, but no information is available (J.E. Sánchez personal communication, 2002). Thus, it is necessary to improve the collection of marine and coastal birds in national museums, especially for future studies in systematics, evolution, and conservation of these groups.

Past publications on marine and coastal birds focused mainly on the addition of new species to the list of Costa Rica (Slud 1967, 1979; Dickerman 1971; Stiles & Smith 1977; Carr 1979; 1980; Stiles 1988). However, new studies on coastal and marine birds are focused on reproductive biology, ecology and conservation (Burger *et al.* 1993; Burger & Gochfeld 1995; Spear *et al.* 1995; Barrantes 1998; Chaves-Campos & Torres 2002), with few papers describing new records from Isla del Coco (Acevedo-Gutiérrez 1994; Dudzik 1996). Most studies in Costa Rica have been conducted with shorebirds (Fleischer 1983; Pereira 1990; Alvarado-Quesada & Moreno 1997), or seabirds that feed and breed close to the shore (e. g. pelicans; Orians 1969; McCoy & Vaughan 1981; Arnqvist 1992; and boobies; Chaves-Campos & Torres 2002). Few studies have dealt with offshore species (Duffy & Hoch 1995; Spear *et al.* 1995).

Currently, there are no specialists studying seabirds in the country. Nevertheless, C.J. Ralph and M. Widdowson have been conducting migration censuses of diurnal migrants along the coast of Tortuguero National Park during the last six years. This project will probably provide valuable information on the taxonomic composition and distribution of some migrant marine birds on the Caribbean coast of Costa Rica.

Acknowledgements We are grateful to D. Nettleship, J.E. Sánchez, and particularly F.G. Stiles for their helpful comments on an earlier draft of this paper.

References

1. Acevedo-Gutiérrez A (1994) First records of occurrence and nesting of three bird species at Isla del Coco, Costa Rica. Rev Biol Trop 42:762

Alvarado-Quesada GM, Moreno LI (1997) Use of salt evaporation ponds by aquatic birds and bird abundance. Tropical Diversity Origins, Maintenance, and Conservation. ATB & OTS Symp. Ann. Meet. Abstr. San José, Costa Rica

Arnqvist G (1992) Brown pelican foraging success related to age and height of dive. Condor 94:521–522

Barrantes G (1998) Reproductive activity of birds in a mangrove in northwest Costa Rica. Rev Biol Trop 46:1163–1166

Barrantes G, Pereira A (1992) Abundancia y fluctuaciones de aves limícolas (Charadriiformes) en una playa fangosa de Chomes, Costa Rica. Rev Biol Trop 40:303–307

2. Barrantes G, Chaves-Campos J, Sánchez J (2002) On the current status of Costa Rican birds. Official list of birds with conservation comments. Zeledonia (Bol Esp) 1:1–30

Burger J, Gochfeld M (1995) Biomonitoring of heavy metals in the Pacific Basin using avian feathers. Environm Toxic Chem 14:1233–1239

Burger J, Rodgers JA, Gochfeld M (1993) Heavy metal and selenium levels in endangered wood storks *Mycteria americana* from nesting colonies in Florida and Costa Rica. Arch Environm Cont Toxic 24:417–420

Carr GS (1979) Greater Shearwater in Costa Rica. Condor 8:323

Chaves-Campos J, Torres J (2002) Distribution of nests of the Brown Booby (*Sula leucogaster*) in relation to the inclination of terrain. Ornitol Neotrop 13:205–208

Croxall JP, Evans PGH, Schreiber RW (1984) Status and Conservation of the World's Seabirds. ICBP, Cambridge, England, 778 p

Dickerman RW (1971) Further Notes on Costa Rican Birds. Condor 73:252–253

Dickinson EC (2003) The Howard & Moore Complete Checklist of the Birds of the World. 3rd Edition. Princeton University Press, Princeton, NJ, 1056 p

3. Dudzik KJ (1996) First record of *Pelecanus occidentalis* (Aves: Pelecanidae) and *Phaeton lepturus* (Aves: Phaethontidae), at Isla del Coco, Costa Rica. Rev Biol Trop 44:303–304

Duffy DC, Hoch L (1995) Probable breeding of Red-billed Tropicbird in Costa Rica. Colonial Waterbirds 18:214–215

Fleischer RC (1983) Relationships between tidal oscillations and Ruddy Turnstone flocking, foraging, and vigilance behavior. Condor 85:22–29

Grant PR (1986) Ecology and Evolution of Darwin's Finches. Princeton University Press, Princeton, NJ, 492 p

4. Harrison P (1983) Seabirds. An Identification Guide. Houghton Mifflin Company, Boston, MA, 448 p

Harrison GS (1990) Seabirds of Hawaii. Natural History and Conservation. Cornell University Press, New York, 249 p

5. Hayman P, Marchant J, Prater T (1986) Shorebirds. An Identification Guide to Waders of the World. Croom Helm, London, 412 p

Hidalgo-Calderón C (1986) Determinación de residuos de plaguicidas organoclorados en huevos de ocho especies de aves acuáticas, colectados durante 1983–1984 en la Isla Pájaros, Guanacaste, Costa Rica. M.Sc. Tesis, University of Costa Rica, San José, Costa Rica

6. Howell SNG, Webb S (1995) A Guide to the Birds of Mexico and Northern Central America. Oxford University Press, New York, 298 p

McCoy MB, Vaughan C (1981) Datos preliminares sobre el éxito reproductivo del pelícano pardo en Costa Rica. *In*: McCoy MB (ed) Contribuciones de la Escuela de Ciencias Ambientales al Primer Congreso Nacional sobre Conservación de Fauna Silvestre. San José, Costa Rica, pp 39–51

Morrison RIG, Ross RK (1989) Atlas of Neartic Shorebirds on the Coast of South America. Vols. 1 & 2. Canadian Service Spec. Publ., Ottawa, 325 p

Orians GH (1969) Age and hunting success in the brown pelican (*Pelecanus occidentalis*). Animal Behav 17:316–319

Pereira A (1990) Ecología de la alimentación de los correlinos (Aves: Scolopacidae) de una playa fangosa del Golfo de Nicoya, Costa Rica. M.Sc. Thesis, University of Costa Rica, San José, Costa Rica

Peterson RT, Chalif EL (1973) A Field Guide to Mexican birds. Houghton Mifflin, Boston, MA, 298 p

Ridgely RS, Greenfield PJ (2001) The birds of Ecuador. Cornell University Press, New York, 740 p

7. Ridgely RS, Gwynne JA (1989) A guide to the birds of Panama with Costa Rica, Nicaragua, and Honduras. Princeton University Press, New Jersey, 534 p

Roca R, Adkins L, Wurschy MC, Skerl KL (1996) Wings From Afar. An Ecoregional Approach to Conservation of Neotropical Migratory Birds in South America. The Nature Conservancy, Arlington, Virginia, 180 p

8. Slud P (1967) The birds of Isla del Coco, Costa Rica. Bull Am Mus Nat Hist 134:53–148

Slud P (1979) First record of Leach's storm-petrel in Costa Rica: a correction. Condor 81:102–103

Smith S, Stiles FG (1979) Banding studies of migrant shorebirds in northwestern Costa Rica. Stud Avian Biol 2:41–48

Spear LB, Ainley DG, Nur N, Howell SNG (1995) Population size and factors affecting at-sea distributions of four endangered procellariids in the tropical Pacific. Condor 97:613–638

Stiles FG (1983) Birds. *In*: Janzen DH (ed) Costa Rican Natural History. University of Chicago Press, Illinois, pp 502–530

Stiles FG (1984) Status and conservation of seabirds in Costa Rican waters. *In*: Croxall JP, Evans PGH, Schreiber RW (eds). Status and Conservation of World's Seabirds. ICBP, Cambridge, England, pp 223–229

Stiles FG (1988) Notes on the distribution and status of certain birds in Costa Rica. Condor 90:931–933

9. Stiles FG, Skutch AF (1989) A Guide to the Birds of Costa Rica. Cornell University Press, New York, 511 p

Stiles FG, Smith SM (1977) New information on Costa Rican waterbirds. Condor 79:91–97

Stiles FG, Smith SM (1980) Notes on bird distribution in Costa Rica. Brenesia 17:137–156

Part 42
Marine Mammals

Laura May-Collado

Humpback whale (*Megaptera novaeangliae*) off Corcovado, Pacific Costa Rica (Photo: Frank Garita)

Abstract About 34 species of marine mammals have been documented in Costa Rican waters, representing approximately 26% of all marine mammals worldwide. The Costa Rican marine mammal fauna consist of 30 cetacean species, one manatee, and three pinnipeds, one of which went extinct since the 1950s. At least 31 of these species most likely also occur in other Central American countries. About 27 marine mammal species have been observed or are expected to occur in the Pacific

L. May-Collado
Department of Environmental Sciences & Policy, George Mason University: MSN 5F2, 4400 University Drive, Fairfax VA 22030, USA
e-mail: lmaycollado@gmail.com
www.delphinids.com

I.S. Wehrtmann, J. Cortés (eds.) *Marine Biodiversity of Costa Rica, Central America*,

(96.5% are confirmed) and 29 (only 28% are confirmed) in the Caribbean waters of Costa Rica. The majority of the marine mammals in Central America inhabit in neritic and pelagic waters. Some species have coastal resident populations, including *Megaptera novaeangliae*, *Stenella attenuata graffmani*, *Sotalia guianensis*, *Tursiops truncatus*, and *Trichechus manatus manatus*. Because of their coastal habits, some of these species are affected by gillnets, uncontrolled whale-watching industry, habitat loss, and direct hunting. In the past, a combination of some of these activities brought to extinction the Caribbean monk seal (*Monachus tropicalis*). The Antillean manatee is vulnerable to these same factors and as a result its population sizes are decreasing alarmingly. There is an urgent need to generate based-line information (e.g., estimates of abundance, distribution patterns, habitat use, and effect of human activities) to promote and support the establishment of appropriate conservation and management regulations in Costa Rica and other Central American countries.

Introduction

Living marine mammals are a compendium of several unrelated orders: Carnivora (polar bears, sea otters, seals, sea lions, and walruses), Cetacea (dolphins, porpoises, and whales) and Sirenia (dugons and manatees). Cetaceans are closely related to artiodactyls, sirenians to elephants, and to other subungulates, and marine carnivores presumably derived from ursid bear and mustelid ancestors (Reynolds *et al.* 1999, May-Collado & Agnarsson 2006, Agnarsson & May-Collado 2008). The first to appear were the cetaceans and sirenians approximately 50 million years ago, pinnipedians 29–23 million years ago, and modern sea otters 7 million years ago (Berta & Sumich 1999).

In spite of their independent histories, all marine mammals, to a certain degree share a suit of anatomical adaptations to an aquatic life style: streamlined-body (including reduction and modification of appendages, e.g., limbs), a thick layer of blubber or dense fur for thermal insulation, and a variety of physiological adaptations for diving, osmoregulation, communication, and navigation. Of all marine mammals, cetaceans are perhaps the most anatomically modified group. They have fusiform bodies, lack ears and external reproductive organs, the hair is only present during fetal development, the front limbs are modified into paddle-like flippers, and the hind limbs are powerful tail flukes; they have a dorsal fin or a ridge; the nares have migrated to the top of the head, the skull is telescoped, and a thick layer of blubber among other adaptations (Reynolds *et al.* 1999). They also vary greatly in size, ranging from the small harbor porpoise (*Phocoena phocoena*, 1.45–1.6 m) (Bjørge & Tolley 2002) to the gigantic blue whale (*Balenoptera musculus*, 26.8–33.6 m) (Sears 2002).

The geographical distribution of marine mammals ranges from cosmopolitan to small and localized areas. They have invaded a variety of habitats, ranging from ice packs, coastal and oceanic waters, to rivers and lakes (Forcada 2002). In some species environmental factors affect their prey abundance or availability and thus indirectly influence the distribution of cetaceans (Au & Perryman 1985).

In Central American waters, several marine mammal species maintain year-round populations. In Costa Rica, some of these species have resident populations (e.g., humpback whales, bottlenose, and spotted dolphins) that inhabit exclusively coastal waters where protection is lacking. Even though Costa Rica has been in the forefront of terrestrial conservation in the world, as most other countries, its marine conservation efforts lags far behind. Thus only a small fraction of the protected national territory includes marine environments, and these cover relatively small areas given the ranges of most marine mammals. The aim of this contribution is to provide a comprehensive list of the marine mammals of Central America with emphasis on those from Costa Rica. The list provides available basic information on the occurrence, habitat, and distribution that are essential for the establishment of adequate management and conservation strategies.

Species Richness

A total of 34 marine mammal species (1 sirenian, 3 pinnipeds, and 30 cetaceans) have been reported in Costa Rican waters (Rodríguez-Herrera *et al.* 2004), representing approximately 26% of the 128 described species of marine mammals worldwide (Reynolds *et al.* 2002b) and 45% of the 76 species known from Latin America (Vidal 1993). Most species occurring in Costa Rica breed locally, but some are seasonal migrants (e.g., *Megaptera novaeangliae*), occasional visitors (e.g., *Orcinus orca*), vagrants (e.g., *Zaluphus californianus*, *Z. wollebaeki*), and at least one has gone extinct (*Monachus tropicalis*). Six marine mammals are exclusively found in the Atlantic Ocean *Trichechus manatus manatus*, *Mesoplodon europaeus*, *Stenella clymene*, *S. frontalis*, and *Sotalia guianensis*. Another five only occur in the eastern tropical Pacific are *M. peruvianus*, *S. attenuata graffmani*, *S. longirostris centroamericana*, *Zalophus californianus*, and *Z. wollebaeki*.

Most Costa Rican marine mammals are cetaceans (5 families, 18 genera, and 30 species), representing about 36% of the 83 species known worldwide and 39% of all Latin American marine mammal species (Leatherwood & Reeves 1983; Jefferson *et al.* 1993; Vidal 1993). Delphinidae is the most diverse cetacean group in Costa Rica, with a total of 13 and 16 species of dolphins in the Pacific and Caribbean, respectively. This represents 43% and 53%, respectively, of the cetacean diversity of the country. Some of these species live in pelagic and neritic waters (e.g., *S. attenuata* [Fig. 42.1], *Steno bredanensis*) and others inhabit bays, gulfs (e.g., *Tursiops truncatus*) (May-Collado *et al.* 2005a) and in some cases estuaries and areas close to river mouths (e.g., *S. guianensis*) (Edwards & Schnell 2001a, b, Gamboa-Poveda & May-Collado 2006). Two species of dolphins are endemic to the tropical and warm-temperate Atlantic, including the Gulf of Mexico and Caribbean Sea, *S. clymene* and *S. frontalis* (Perrin 2002c; Jefferson & Curry 2003) and one is endemic to the Caribbean, *S. guianensis* (da Silva & Best 1996). Only 28% of the 29 marine mammal species listed for the Caribbean have been confirmed by more than one reliable source (Species List 42.1 is included on the CD-Rom). On

Fig. 42.1 Pantropical spotted dolphin (*Stenella attenuate*), off the Peninsula de Osa, southern Pacific of Costa Rica (Photo: Frank Garita)

the other hand 96.5% of the 28 species listed for the Pacific are confirmed from several reliable sources (Species List 42.2 is included on the CD-Rom). In that region, endemism occurs to the subspecies level, two are exclusively found in the waters of the eastern tropical Pacific: *S. longirostris centroamericana* and *S. attenuata graffmani* (Perrin 2001, 2002a, b).

Gerrodette & Palacios (1996) have provided preliminary estimates of the population sizes of cetacean groups in the Pacific Exclusive Economic Zone (PEEZ), for each country from Mexico to Ecuador (Guatemala, El Salvador, Honduras, and Nicaragua were clumped in one category due to their small maritime areas and lack of clear maritime boundaries between some of these countries). They found that the abundance and diversity of cetaceans decreased from Mexico to Ecuador. The most abundant species were *Stenella* spp. and *Delphinus delphis* followed by *T. truncatus*, *Grampus griseus*, *S. bredanensis*, *Globicephala macrorhynchus*, and ziphiids. Rorquals had the lowest abundance in the area. In Costa Rican waters the most common cetaceans are: *D. delphis*, *S. coeruleoalba*, *S. attenuata*, *T. truncatus*, and ziphiids (Gerrodette & Palacios 1996). Abundance estimates for all species in the region can be found in the Southwest Fisheries Science Center (US Department of Commerce, National Marine Fisheries Service) website http:// swfsc.nmfs.noaa.gov/publications (e.g., Wade & Gerrodette 1993; Gerrodette & Palacios 1996; Gerrodette & Forcada 2002). Updated distribution maps of most species in the PEEZ of Costa Rica can be found in May-Collado *et al.* (2005a) and for Central America in Kinzey *et al.* (2000, 2001).

The other four marine mammals included in the list are the Antillean manatee (*T. manatus manatus*), which sustains a resident population in Tortuguero National Park and possibly in other areas along the coast, but the status of this population is not well known; the Californian sea lion (*Zalophus californianus*), which has been observed occasionally in Isla del Coco, Golfo Dulce, Playa Tamarindo, Playas del

Coco, Caletas and Paquera (Acevedo-Gutierrez 1994, 1996; Cubero-Pardo & Rodríguez 1996); and vagrants of the Galápagos sea lion (*Z. wollebaeki*) have been also reported on Isla del Coco (Heath 2002). Finally, the Caribbean monk seal (*M. tropicalis*) was the only pinniped (*O. Carnivora*) native to Central America. The monk seal has been extinct since the 1950s (Kenyon 1977, 1986).

Present Status of the Marine Mammals of Costa Rica

Research on the resident populations of Costa Rican marine mammals started approximately 15 years ago (Acevedo & Würsig 1991; Steiger *et al.* 1991; Acevedo & Burkhart 1998; Cubero-Pardo 1998; May-Collado 2001; May-Collado & Forcada 2001; May-Collado & Morales-Ramírez 2005). These studies have produced baseline information on the seasonal distribution, relative abundance, use of habitat, and behavior of four cetacean species: (1) the bottlenose dolphins (*T. truncatus*) (Acevedo & Würsig 1991; Acevedo & Smultea 1995; Acevedo 1996; Acevedo & Burkhart 1998; Cubero-Pardo 1998; Acevedo-Gutierrez & Parker 2000; May-Collado & Oviedo 2007; Wartzok 2008); (2) the coastal spotted dolphins (*S. attenuata graffmani*) (Cubero-Pardo 1998; May-Collado 2001; May-Collado & Forcada 2001; May-Collado & Morales-Ramírez 2005); and (3) the humpback whales (*M. novaeangliae*) (Steiger *et al.* 1991; Calambokidis *et al.* 1999, 2000; Rasmussen *et al.* 2001a, b) (4) the Guyana dolphin (*S.guianensis*) (Gamboa-Poveda & May-Collado 2006, May-Collado & Gamboa-Poveda 2006, May-Collado & Wartzok 2008). In addition to this, new information has been generated for a handful of other species (e.g., Gerrodette & Palacios 1996; May-Collado *et al.* 2005 [small cetaceans]; Ferlt *et al.* 1996 [*O. orca*]; Acevedo-Gutierrez 1996 [*Ziphius cavirostris*]; Acevedo *et al.* 1997 [*Pseudorca crassidens*]).

Resident Populations

Cetaceans: Up until now, two resident populations of bottlenose dolphins, one in the Isla del Coco and another in Golfo Dulce are well documented (Acevedo & Würsig 1991; Acevedo & Burkhart 1998; Cubero-Pardo 1998; Acevedo & Parker 2000), and a third one in Manzanillo, Limón, which is being studied since 2003 by the author and the biologist Mónica Gamboa. In Golfo Dulce, the average group size of bottlenose dolphins is about 5.8 ± 4.14 (school size ranged from 1 to 25 individuals), they are frequently observed in shallow waters, close to shore, near rivers, and along steep marine slopes (Acevedo & Burkhart 1998). Feeding activities are more intense during the dry season (Cubero-Pardo 1998). During the beginning of the wet season, dolphins are likely to allocate more time to passive social activities, and to movement. Active social behaviors and milling are the predominant behaviors at the end of the rainy season (Cubero-Pardo 1998). In the Isla del Coco, bottlenose dolphins feed during the day in clear and near-shore waters

(Acevedo & Parker 2000). In Gandoca-Manzanillo Wildlife Refuge, bottlenose dolphins are usually found in associations with the Guyanese dolphin (*S. guianensis*) (Acevedo-Gutiérrez *et al.* 2005). Up to now, 46 individuals have been photo-identified based on natural marks of their dorsal fins (Gamboa-Poveda & May-Collado 2006). West of Gandoca-Manzanillo, in Bocas del Toro (Panama) there is a bottlenose resident population. About 90 animals have been photo-identified. However, there are not indications yet, that these two populations are interacting May-Collado (unpublished data).

According to photo-identification studies (May-Collado 2001), coastal spotted dolphins are possibly resident in Golfo de Papagayo (Islas Murciélagos and Bahía Culebra), in addition to one in Golfo Dulce (Acevedo 1996; Acevedo & Burkhart 1998; Cubero-Pardo 1998). In Papagayo, the average group size of spotted dolphins is about 9.95 ± 10.28 in the islands, and 8.44 ± 5.40 in the bay (school size ranged from 1 to 50 individuals). Their annual relative abundance seems to be correlated with local environmental factors, particularly water depth, water transparency, and dissolved oxygen concentration (May-Collado 2001; May-Collado & Forcada 2001). Spotted dolphins are more abundant in Golfo de Papagayo during the dry season (May-Collado & Forcada 2001). Seasonal changes in relative abundance are likely to be associated with food availability, as observed in the high number of groups involved in foraging activities during the dry season (May-Collado & Forcada 2001; May-Collado & Morales-Ramírez 2005). Social activities and movements are more frequent during the rainy season (May-Collado & Morales-Ramírez 2005). In contrast, the spotted dolphins of the Golfo Dulce, have an average group size of 37.60 ± 49.54 (school range 1–300 individuals), they are associated with deeper waters, increasing in numbers with increased distance from shore (Acevedo & Burkhart 1998). They forage, socialize, and move more intensely during the dry season, whereas in rainy season they spend more time milling and involve in active social behaviors (Cubero-Pardo 1998). Like in Golfo de Papagayo, the Panamanian spotted dolphins distribution patterns appear to be explained by depth, and also by the absence of bottlenose dolphins (García & Dawson 2003).

Overall, little is known about the worldwide distribution of the family Ziphiidae. Previous observations of the Cuvier's beaked whale (*Z. cavirostris*) around Isla del Coco suggest that it may resident (Acevedo 1996) and new information on its distribution in the Exclusive Economic Zone of the Pacific of Costa Rica indicate this species may be more common than previously thought (May-Collado *et al.* 2005).

Although little work has been done in the Caribbean, it is likely that bottlenose dolphins and marine tucuxi dolphins (*S. guianensis*) also maintain resident populations along the Costa Rican Caribbean coast (DiBerardinis *et al.* 1997). Previously, the marine Tucuxi was believed to be a single species, *S. fluviatilis* (Fig. 42.2), with two ecotypes (the riverine *S. f. fluviatilis* and the marine *S. f. guianensis*) (da Silva & Best 1996), that was limited from Santa Catalina, Brazil, to the Caribbean Sea off central Panama (Bössenecker 1978; Borobia *et al.* 1991). However, Carr & Bonde (2000) reported new records expanding the known range 800 km north, to the Leimus Lagoon in Nicaragua. It is not known if the range of the animals has changed recently or if the range extension is simply the result of added knowledge.

Fig. 42.2 The marine tucuxi dolphin (*Sotalia guianensis*), reported from the Caribbean coast of Costa Rica (Photo: Laura May-Collado)

Recently, two species of *Sotalia* are recognized: *S. fluviatilis* (common name tucuxi) endemic to the Orinoco and Amazon rivers and *S. guianensis* (common name Guyanese dolphin) endemic to the Caribbean coastal waters of Central and South America (Monteiro-Filho *et al.* 2002; Cunha *et al.* 2005). In Costa Rica, the Guyanese dolphin is commonly found in Gandoca-Manzanillo in groups of 1–15 individuals. Its distribution overlaps with that of the bottlenose dolphins were apparently, the two species forage and socialize together (DiBerardinis *et al.* 1997). In addition, sexual encounters between species occur and observations of bottlenose dolphins with features characteristic of the Guyanese dolphin suggest that the two may hybridize to some extent (Forestell *et al.* 1999). The Guyanese dolphin is considered one of the ten endemic species of Latin America (Vidal 1993), and it is listed as insufficiently known by the IUCN Red List of Threatened species (1996, 2000). Most of the information on the species comes from studies in the coast of Brazil (Geise 1991; Flores 2002; Rosas-Weber & Monteiro-Filho. 2002). Little information exists for Central America; few of the existing data have been published (Carr & Bonde 2000; Edwards & Schnell 2001a, b, May-Collado *et al.* 2005b).

Sirenians: The Antillean manatee (*T. manatus manatus*) occupies the waters of 19 countries (from Texas to South America), and its range may overlap with the Amazon River manatee (*T. inunguis*) (Reynolds & Powell 2002a). In addition to warm water, other factors such as freshwater and adequate food supply have affected its current and historical distribution (Reeves *et al.* 1992). The exact number of Antillean manatees is unknown, but recent aerial surveys in Belize and southern Quintana Roo (Mexico) counted more than 400 individuals. This corridor between Belize and Mexico is considered to be a last stronghold for the subspecies (O'Shea & Salisburg 1991; Reynolds & Powell 2002a). Even there, manatees were seriously exploited even before the arrival of the Spanish colonist. The middens data from AD 400 to 700 DC on Moko Cay near Belize City contains the largest number of manatee bones of any site examined in the Caribbean (Reeves *et al.* 1992).

In Central America, Nicaragua has the most extensive and favorable habitat for manatees (Reeves *et al.* 1992; Jiménez 1999, 2000, 2002). In 1994, 77 individuals were observed in a period of 18 h along Nicaragua's Miskito coast (Carr 1994). Unfortunately, this is not the case for the Honduran and Costa Rican manatees (Husar 1978). The most important localities for manatees in Honduras are Mosquitia, the rivers east of Trujillo, and lagoons and rivers west of the La Ceiba (Rathbun *et al.* 1983). Despite of the protection provided by the national laws manatees are traditionally hunted in Mosquitia, with modified fish/turtle harpoons and fishing nets (Rathbun *et al.* 1983).

The manatees of Costa Rica have suffered a similar fate. The clearing of forest for commercial purposes (e.g., banana and cattle) has shrunken most of their habitat (Smethurst & Nietschmann 1999). Although, favorable habitat still exists along the northeastern and southeastern coasts, including Tortuguero National Park, even there manatees are rarely observed (Reynolds & Odell 1991; Reynolds *et al.* 1995; Smethurst & Nietschmann 1999). Smethurst & Nietschmann (1999) reported a total of 29 sightings in surveys between Moín and the Nicaraguan border, during a period of 79 days. Despite the historical hunting by Spanish colonists during the sixteenth century (Reeves *et al.* 1992), manatees were still abundant in the northeastern coast of Costa Rica, particularly during the 1940s (O'Donnell 1982). Then a sharp decline occurred in their numbers as result of illegal hunting, net fishing, and habitat degradation and since the 1970s they became rare (O'Donnell 1982; Mou-Sue *et al.* 1990; Reynolds & Odell 1991). Jiménez (1999, 2000, 2002) in a more recent study described the manatees of Costa Rica and Nicaragua as surprisingly abundant. These manatees inhabit a relative protected system of rivers, lagoons, and channels. However, boat traffic and pouching are still important threats to this population. In Panama some manatees have survived in the slow-moving rivers of Bocas del Toro, protected lagoons and waters bordering the Panama Canal (Reeves *et al.* 1992). *T. manatus manatus* is the only Central American marine mammal listed to be in critical conditions by the IUCN, a view shared by most scientists (Mou-Sue *et al.* 1990; Reynolds *et al.* 2002a).

Carnivora: The Caribbean monk seal, also known as the West Indian monk seal, was the only marine carnivore with resident populations along Central America and the Caribbean Sea. Records of the species appear in Christopher Columbus reports during his second voyage to the New World in 1494 (Reeves *et al.* 1992). Even though, little was learned about its biology before it became extinct, it is clear that the Caribbean monk seal ranged from Gulf of Mexico, Bahamas, Yucatan Peninsula, Central America, and east to the northern Lesser Antilles (Allen 1887; Reeves *et al.* 1992). Although, a few reliable reports made in 1952 indicated the existence of a small colony on Serranilla Bank (halfway between Jamaica and Honduras), it is believed that the Caribbean monk has been extinct since the early 1950s (Kenyon 1977, 1986). The existing reports about its behavior suggest that the seals were remarkably tame, not even the human presence appeared to trigger any aggressive reaction (Ward 1887). No doubt this "docility" facilitated its massive exploitation by Europeans during the seventeenth and eighteenth centuries (Allen 1887).

Seasonal Visitors

Although rorquals are overall the least abundant cetaceans in Central American waters (Gerrodette & Palacios 1996), humpback whales (*M. novaeangliae*) are seasonally abundant in the Pacific coast of Costa Rica, where they arrive every year to breed and give birth (Steiger *et al.* 1991; Calambokidis *et al.* 1999, 2000; Rasmussen *et al.* 2001). Humpback whales have been known to fishermen for a long time, but their presence was first suggested in 1991, and confirmed 4 years later, since then, concern about their lack of protection has been rising (Steiger *et al.* 1991; Acevedo & Smultea 1995; Calambokidis *et al.* 1999, 2000; Rasmussen *et al.* 2001a, b). Every year, humpback whales arrive to the Península de Osa (Isla del Caño, Drake, Golfo Dulce), Cuajiniquil Bay, and Golfo de Papagayo (Islas Murciélago, Bahía Culebra, Isla Catalina) to reproduce and to give birth. Humpback whales from both hemispheres use the same locations at different times of the year (Rasmussen *et al.* 2001a) The northern hemisphere humpback whales (mainly from off South California) are commonly observed from January to April (Rasmussen *et al.* 2001a), whereas the southern hemisphere humpback whales from July to October (Rasmussen *et al.* 2001a, b, 2002). The northern hemisphere humpback whales have their feeding grounds off the coast of California (Steiger *et al.* 1991; Calambokidis *et al.* 2000) and southern hemisphere humpback whales off the Antarctic Peninsula (Rasmussen *et al.* 2001a, b, 2002). Because sightings are still reported between wintering seasons, it is possible that a temporal overlap of the different populations exist (Rasmussen *et al.* 2001a, b). Furthermore, the one-way migration between Antarctica Peninsula and Costa Rica (approximately distance between the locations is 8,400 km) may be the longest recorded migration by an individual mammal (Rasmussen *et al.* 2001b, 2002).

In addition to humpback whales, blue whales (*Balaenoptera musculus*) also migrate to Costa Rican waters. Blue whales from the eastern North Pacific subpopulation have been found as far north as Alaska, but are regularly observed from California in summer, south to Mexico, and Costa Rican waters in winter (Sears 2002).

Endemism

The only "endemic" cetacean in the area is the Central American spinner dolphin, *S. longirostris centroamericana*, previously called the Costa Rican spinner dolphin (Rice 1998). It ranges no farther than 92 km off the Pacific coast (Perrin *et al.* 1985). This subspecies can be distinguished from the other two subspecies, *S. longirostris longirostris* (white-belly spinner dolphin) and *S. longirostris orientalis* (eastern spinner dolphin) because the former lacks a bold pattern and has an extremely long snout (Perrin 2002b). In adult males of the eastern and Central American spinner dolphins, the dorsal fin may lean slightly forward. This is also related with the presence of a large postanal ventral hump. Both the dorsal fin and the ventral hump appear to be secondary sexual trait (Perrin 2002b). In general, little is known about the biology of this subspecies.

Status and Conservation Efforts

Costa Rican conservation and management efforts for the protection of marine mammals are few and isolated. The problem stems from the lack of basic knowledge on their status and biology (e.g., relative abundance, habitat use, and potential effect human activities). Marine mammal conservation efforts have been focused at species level. However, understanding local populations may have important implications for conservation and management strategies. In fact, Reynolds *et al.* (2002b) proposed that conservation efforts should focus more at the stock or population levels in order to maintain biological diversity. Genetic data from populations of coastal spotted dolphins along Central America appears to coincide with jurisdictional waters of each country (Escorza-Treviño *et al.* 2003) this indicates the importance of localized conservation efforts.

With the decrease of commercial whaling, it is generally believed that cetacean casualties in fisheries have become insignificant. Yet, this is still a major cause of death for several species in different parts of the world (Mitchell *et al.* 1996; Wade 1995; Clapham *et al.* 1999). The belief that human activities do not affect cetacean populations is predominant in Central America (personal observation). However, previous studies suggest that gillnets, the most common type of nets used by artisanal fishermen in Central America, might have an impact in small cetacean populations (Reyes & Oporto 1994; Vidal & Van-Waerebeek 1994; Palacios & Gerrodette 1996). Based on the limited information available on numbers of fishing boats, fishing effort, net sizes, and small cetacean capture rates, Palacios and Gerrodette (1996) predicted an annual mortality of small cetaceans by gillnets of about 16,600 in Costa Rica and 3,600 in Panama. This study provides evidence that incidental mortality by gillnets may have greater impact in the area than previously thought. Furthermore, a recent study by Gerrodette and Forcada (2005) on the recovery of the stocks of *S. attenuata attenuata* and *S. longirostris orientalis* showed no recovery. These two stocks are the most affected by the bycatch in pursue nets for yellowfin tuna.

Furthermore, whale-watching activities have increased in the recent years in Costa Rica (Fig. 42.3). There are at least 25 tour companies in several parts of the Pacific and Caribbean coasts (Cubero-Pardo 2001). According to Cubero-Pardo (2001), 85% of the time cetaceans and especially dolphins try to avoid encounters with whale-watching boats by swimming away, probably in response to improper approach by the boat drivers. Therefore, not only the direct, or accidental, kill affects local cetacean populations. Improper whale-watching activities could, in addition to fisheries, affect and change the behavior of resident populations.

For these reasons, the design and implementation of conservation strategies to protect marine mammals should be considered as a top priority in the marine conservation efforts in Costa Rica and Central American countries. This should be done by, establishing long-term monitoring studies about the species biology and their habitat. Furthermore, certain marine areas need rapid protection if we are not to see the Antillean manatee in Costa Rica be the next to suffer the fate of the Caribbean monk seal.

Fig. 42.3 Whale-watching (*Pseudorca crassidens*) at the Pacific coast of Costa Rica (Photo: Frank Garita)

Specialists

ALEJANDRO ACEVEDO-GUTIERREZ: Department of Biology, Western Washington University, Bellingham, WA 98225–9160 USA – Costa Rica. acevedo@biol.wwu.edu

JOHN CALAMBOKIDIS, GRETCHEN STEIGER, & KRISTIN RASMUSSEN: Cascadia Research Collective, Waterstreet Building 218, West Fourth Avenue, Olympia, Washington, USA–Washington, Oregon, California, and Central America. *Calambokidis@CascadiaResearch.org*, *GSteiger@CascadiaResearch.org* & *Krill@aol.com*

PRISCILLA CUBERO-PARDO & JAVIER RODRÍGUEZ: Foundation PROMAR. Apdo. 11709–1000 San Jose. Costa Rica. priscilla.cubero@promar.or.cr & Javier.rodríguez@promar.or.cr

JAUME FORCADA: NERC British Antarctic Survey. High Cross, Madingley Road, Cambridge CB30ET, UK – Eastern Tropical Pacific and Antarctica Peninsula. JFOR@bas.ac.uk

MÓNICA GAMBOA, JOSÉ D. MARTÍNEZ & JOSÉ PALACIOS: Fundación Keto, Programa regional en manejo de vida silvestre para Mesoamérica y el Caribe. Universidad Nacional, Costa Rica–info@fundacionketo.org; www.fundacionketo.org

CAROLINA GARCIA IMHOF. Carrera 4 No. 71–61 apt. 1001. Bogota-Colombia. Tel: 5713458377. garciaimhof@hotmail.com

FRANK GARITA ALPÍZAR: Asociación ambiental Vida. Apdo. 7–0350–1000, San José. Costa Rica. garita@vida.org; Fundación Keto. info@fundacionketo.org

TIM GERRODETTE: National Marine Fisheries Service, Southwest Fisheries Science (SWFS), National Oceanic and Atmospheric Administration (NOAA). Center 8604 La Jolla Shores Drive, California 92037, USA – Eastern Tropical Pacific. Tim.Gerrodette@noaa.gov

IGNACIO JIMÉNEZ PÉREZ, CARLOS ESPINOZA MARÍN & ALEX GÓMEZ LÉPIZ: Fundación Salvemos al manatí de Costa Rica. Apdo. 245–3015, San Rafael de Heredia, Costa Rica–contacto@fundacionmanati.org

LAURA MAY-COLLADO: Department of Biological Sciences, Florida International University, Ph.D. Student, 11200 SW 8th Street UP Miami-FL 33199. Centro de Investigación en Ciencias del Mar y Limnología (CIMAR), Universidad de Costa Rica – Costa Rica and Panama. lmayc001@fiu.edu; www.delphinids.com
MARCO TULIO SABORÍO: Apartado 292–2300, San José, Costa Rica – Costa Rica. msaborio@conexion.co.cr

Collections

Costa Rica marine mammal collections are poor. The Zoology Museum of the Universidad de Costa Rica holds only two skulls of dolphins (possibly *T. truncatus*), the lower jaw of a sperm whale (*Physeter macrocephalus*), and one complete skull of a manatee (*T. manatus manatus*). Other institutions outside the country, such as the Smithsonian Institution in Washington, DC, the Otago Museum in New Zealand, the California Academy of Sciences, the American Museum of Natural History in New York, and the Southwest Fisheries Science Center in La Jolla, California, hold skulls, skeletons, and skin biopsies of several cetacean species collected along Central America waters (refer to the Marine Mammal Database of Dr. James Mead, Smithsonian Institution, e-mail: mead.james@nmnh.si.edu). The National Museum of Natural History of the Smithsonian Institution contains in its marine mammal collection a few specimens collected throughout Central America. Among them is one skull of *D. delphis* (unknown gender), and several of *Peponocephala electra* (females, males, and unknown gender specimens) collected in Costa Rica (one was loaned to the National Museum of New Zealand); skulls and skeletons (females and males) of *S. attenuata*, one skull of *S. frontalis* (female), and two skulls and skeletons (male and unknown gender) of *T. truncatus* collected in Panama. Finally, two more skulls and skeletons (female and males) of *Stenella* spp., collected in Guatemala are located in the Academy of Sciences, San Francisco, and one *T. truncatus* skull (unknown gender) from Nicaragua is in the American Museum of Natural History, New York. All the information with respect to these specimens is available in the Marine Mammal Database (http://www.nmnh.si.edu/vert/mammals/mmp.html) of Dr. James Mead (mead.james@nmnh.si.edu) at the Natural History Museum, Smithsonian Institution.

Recommendations

In general, Central America is an excellent place to conduct baseline information studies (e.g. habitat use, distribution patterns, abundance estimates). There is a need for basic information to support the establishment of appropriate conservation and management regulations. It is urgent to establish long-term studies to generate demographic data (e.g. relative abundance, age distribution, reproduction rates), temporal and spatial distribution trends (e.g., how far do animals move? Where do they go?), habitat use (e.g., does habitat use vary between seasons? Are variations related to prey availability, predators, or to abiotic factors?), behavior (e.g., foraging behavior, social structure, allocation of time per activities according to daily and seasonal patterns). It is also important to investigate fisheries interactions with cetaceans (e.g., proportion of the artisanal fleet uses gillnets, and what effect this has on small cetacean populations), and to create workshops to train whale-watching boat pilots, as well as to keep developing studies to determine the impact of these activities. Anthropogenic noise is another treat associated to fisheries and whale-watching activities that can affect cetacean populations. Like many other marine organisms (e.g., shrimp, swim-bladded fishes) cetaceans rely on sound to survive. Thus increasing anthropogenic ocean noise may jeopardize an organism's survival by masking conspecific communication signals, interfering with active acoustics for prey detection and passive acoustics for predator avoidance, and causing temporary or permanent hearing losses (Wartzok *et al.* 2004). Therefore, noise may act much like other pollutants in rendering habitats unsuitable. Finally, swift action is needed to protect the most vulnerable species (*M. novaeangliae* and *T. manatus manatus*) by creating adequate regulations, conservation strategies, and management plans according to the status of the different species in each country and pressure from human activities. In Costa Rica, some actions to protect cetaceans have been taken by colleagues of PROMAR, but regional efforts still need to be addressed.

References

Acevedo A, Wursig B (1991) Preliminary observations on bottlenose dolphins, *Tursiops truncatus*, at Isla del Coco, Costa Rica. Aquat Mamm 17:148–151

1. Acevedo-Gutierrez A (1994) First record of a sea lion *Zalophus californianus*, at Isla del Coco, Costa Rica. Mar Mamm Sci 10:484–485
2. Acevedo A, Smultea MA (1995) First records of humpback whales including calves at Golfo Dulce and Isla del Coco, Costa Rica, suggesting geographical overlap of northern and southern Hemisphere populations. Mar Mamm Sci 11:554–560
3. Acevedo-Gutierrez A (1996) Lista de mamíferos marinos en Golfo Dulce e Isla del Coco, Costa Rica. Rev Biol Trop 44:933–934
4. Acevedo A, Brennan B, Rodríguez P, Thomas P (1997) Re-sightings and behavior of false killer whales (*Pseudorca crassidens*) in Costa Rica. Mar Mamm Sci 13:307–314

Acevedo A, Burkhart S (1998) Seasonal distribution of bottlenose (*Tursiops truncatus*) and pantropical spotted (*Stenella attenuata*) dolphins (Cetacea:Delphinidae) in Golfo Dulce, Costa Rica. Rev Biol Trop 46:91–101

Acevedo-Gutiérrez A, DiBerardinis A, Larkin S, Larkin K, Forestell P (2005) Social interactions between tucuxis and bottlenose dolphins in Gandoca-Manzanillo, Costa Rica. LAJAM 4:49–54

Acevedo-Gutierrez A, Parker EN (2000) Surface behavior of bottlenose dolphins is related to spatial arrangement of prey. Mar Mamm Sci 16:287–298

Agnarsson I, May-Collado, LJ (2008) The phylogeny of Cetartiodactyla: the importance of dense taxon sampling, missing data, and the remarkable promise of Cytochrome b to provide reliable species-level phylogenies. Molecular Evolution and Phylogenetics. doi:10.1016/j.ympev.2008.05.046

5. Allen JA (1887) The West Indian Seal (*Monachus tropicalis*). Bull Amer Mus Nat Hist II(I):1–34

Au DWK, Perryman WL (1985) Dolphin habitats in the Eastern Tropical Pacific. Fish Bull 83:623–644

Berta A, Sumich JL (1999) Marine mammals: evolutionary biology. Academic, San Diego, CA
Bjørge A, Tolley KA (2002) Harbor Porpoise. In: Perrin W, Würsig B, Thewissen JGM (eds) Encyclopedia of Marine Mammals, Academic, San Diego, CA, p 549
Borobia MS, Siciliano LL, Hoek W (1991) Distribution of the South American dolphin *Sotalia fluviatilis*. Can J Zool 69:1025–1039
Bössenecker PJ (1978) The capture and care of *Sotalia guianensis*. Aquat Mamm 6:13–17
6. Calambokidis J, Rasmussen K, Steiger GH (1999) Humpback whales and other marine mammals off Costa Rica, 1996–1999. Report of research during Oceanic Society Expeditions in 1999 in cooperation with Elderhostel volunteers. Cascadia Research Collective, p. 21
7. Calambokidis J, Steiger GH, Rasmussen K, Urbán JR, Balcomb KC, Ladrón de Guevara P, Salinas M, Jacobsern JK, Herman LM, Cerchio S, Darling JD (2000) Migratory destinations of humpback whales that feed off California, Oregon and Washington. Mar. Ecol. Prog. Ser. 192:295–304
8. Carr T (1994) The Manatees and Dolphins of the Miskito Coast Protected Area, Nicaragua. US Department of Commerce, National Technical Reports. Springfield, VA. PB94–170354, p 33
9. Carr T, Bonde RK (2000) Tucuxi (*Sotalia fluviatilis*) occurs in Nicaragua, 800 Km north of its previously known range. Mar Mamm Sci 16:452–459
10. Carwardine M (1995) Whales, Dolphins and Porpoises. Dorling Kindersley Handbooks, London/New York
Clapham PJ, Young SB, Brownell RL Jr (1999) Baleen whales: conservation issues and the status of the most endangered populations. Mamm Rev 29:35–60
Cubero-Pardo P (1998) Patrones de comportamiento diurnos y estacionales de *Tursiops truncatus* y *Stenella attenuata* (Mammalia:Delphinidae) en el Golfo Dulce, Costa Rica. Rev Biol Trop 46(Suppl 6):103–110
Cubero-Pardo P (2001) Assesing the impact of whale-watching activities on wild cetaceans in Costa Rica. 14th Biennial Conf. Biol. Mar. Mamm., Vancouver Nov. 28–Dec. 2, 2001. p. 50
11. Cubero-Pardo P, Rodríguez J (1996) Ampliación de ámbito de *Zalophus californianus* (Pinnipedia:Otariidae) en Costa Rica. Rev Biol Trop 44:273
da Silva VM, Best RC (1996) *Sotalia fluviatilis*. Mamm Spec 527:1–7
Cunha HA, da Silva VMF, Lailson-Brito JJr, Santos MCO, Flores PAC, Martin AR, Azevedo AF, Fragoso ABL, Zanelatto RC, Solé-Cava AM (2005) Riverine and marine ecotypes of *Sotalia* dolphins are different species. Mar Biol 148:449–457
12. DiBerardinis A, Larkin S, Schot V (1997) Identification of *Sotalia fluviatilis* (tucuxi) outside of previously reported range. Report to Ministerio del Ambiente y Energia, Costa Rica, p. 9
Edwards HH, Schnell GD (2001a) Body, length, swimming, speed, dive duration, and coloration of the dolphin *Sotalia fluviatilis* (tucuxi) in Nicaragua. Carib J Sci 37:371–272
Edwards HH, Schnell GD (2001b) Status and ecology of *Sotalia fluviatilis* in the Cayos Miskitos Reserve, Nicaragua. Mar Mamm Sci 17:445–472
Escorza-Trevino, S, A Lang, AE Dizon (2002) Genetic differentiation and intraspecific structure of eastern tropical Pacific spotted dolphins, *Stenella attenuata*, revealed by mitochondrial and microsatellite DNA analyses. SWFSC Admin. Rep., La Jolla, LJ-02–38, p 20
13. Fertl D, Acevedo A, Darly FL (1996) A report of killer whales (*Orcinus orca*) feeding on a carcharinid shark in Costa Rica. Mar Mamm Sci 12:606–611
Flores PAC (2002) Tucuxi, *Sotalia fluviatilis*. *In*: Perrin W, Würsig B, Thewissen JGM (eds) Encyclopedia of Marine Mammals, Academic, San Diego, CA, pp 1267–1269
Forcada J (2002) Distribution of Marine Mammals. *In*: Perrin W, Würsig B, Thewissen JGM (eds) Encyclopedia of Marine Mammals. Academic, San Diego, CA, pp 327–332
Forestell P, Wright A, DiBerardinis A, Larkin S, Schot V (1999) Sex and the single tucuxi: mating between bottlenose and tucuxi dolphins in Costa Rica. Abstracts 13th Biennial Conf. Biol. Mar. Mamm. Nov. 28 to Dec. 3. Wailea, Maui, Hawaii
Gamboa-Poveda M, May-Collado, LJ (2006) Insights on the occurrence, residency, and behavior of two coastal dolphins from Gandoca-Manzanillo, Costa Rica: *Sotalia fluviatilis* and

Tursiops truncatus (Family Delphinidae). Paper SC/58/SM4 presented to the IWC Scientific Committee, June 2006
García C, Dawson SM (2003) Distribution of pantropical spotted dolphins in Pacific coastal waters of Panama. LAJAM 2:29–38
Geise L (1991) *Sotalia guianensis* (Cetacea: Delphinidae) population in the Guabanara Say, Río de Janeiro, Brazil. Mammalia 55:371–374
14. Gerrodette T, Palacios DM (1996) Estimates of cetacean abundance in EEZ waters of the eastern tropical Pacific. NMFS, Southwest Fisheries Science Center, Administrative Report JL-96–10, p 25
Gerrodette T, Forcada J (2002) Estimates of abundance of the northeastern offshore spotted coastal and eastern spinner dolphins in the Eastern Tropical Pacific Ocean. SWFSC. Admin. Rep. La Jolla, CA. LJ-02–06, p 43
Gerrodette T, Forcada J (2005) Non-recovery of two spotted and spinner dolphin populations in the eastern tropical Pacific Ocean. Mar Ecol Prog Ser 291:1–21
Jiménez I (1999) Estado de conservación, ecología y conocimiento popular del manatí *(Trichechus manatus*) en Costa Rica. Vida Silvestre Neotropical 8:18–30
Jiménez I (2000) Los manatíes del río San Juan y los Canales de Tortuguero: ecología y conservación. Amigos de la Tierra, San José, Costa Rica, p. 120
Jiménez I (2002) Heavy poaching in prime habitat: the conservation status of the West Indian manatee in Nicaragua. Oryx 36:1–13
15. Heath CB (2002) California, Galápagos and Japanese Sea Lions (*Zalophus californianus, Z. wollebaeki,* and *Z. japonicus*). *In*: Perrin W, Würsig B, Thewissen JGM (eds) Encyclopedia of Marine Mammals. Academic, San Diego, CA, pp 180–186
Husar SL (1978) *Trichechus manatus*. Mamm Spec 93:1–5
Jefferson TA, Leatherwood S, Webber MA (1993) FAO Species Identification Guide. Marine Mammals of the World. FAO, Rome, p 320
Jefferson TA, Curry E (2003) *Stenella clymene*. Mamm Spec 726:1–5
16. Kenyon KW (1977) Caribbean Monk Seal extinct. J Mamm 58:97–98
Kenyon KW (1986) Caribbean Monk Seal extinct. Mar Mamm Sci 2:70–72
17. Kinzey D, Gerrodette T, Barlow J, Dizon A, Perryman W, Olson P (2000) Marine mammal data collected during a survey in the eastern tropical Pacific Ocean aboard the NOAA ships McArthur and David Starr Jordan. July 28–December 9, 1999. NOAA-TM-NMFS-SWFSC-293. US Department of Commerce, Southwest Fisheries Science Center, p 89
18. Kinzey D, Gerrodette T, Dizon A, Perryman W, Olson P, Rankin S (2001) Marine mammal data collected during a survey in the eastern tropical Pacific Ocean aboard the NOAA ships McArthur and David Starr Jordan. July 28–December 9, 2000. NOAA-TM-NMFS-SWFSC-303. US Department of Commerce, Southwest Fisheries Science Center, p 100
19. Leatherwood S, Reeves R (1983) The Sierra Club Handbook of Whales and Dolphins. Sierra Club, San Francisco, CA
May-Collado LJ (2001) Ecología y comportamiento del delfín manchado costero, *Stenella attenuata graffmani* (Cetacea: Delphinidae) del Pacifico norte de Costa Rica. M.Sc. Thesis, Universidad de Costa Rica, San Pedro
May-Collado LJ, Agnarsson I (2006) Cytochrome b and Bayesian interface of whale phylogeny. Mol Phylogen Evol 38:344–354
20. May-Collado LJ, Forcada J (2001) Abundance, occurrence and behavior of the coastal spotted dolphin (*Stenella attenuata graffmani*) in the northern Pacific of Costa Rica. 14th Biennial Conf. Biol. Mar. Mamm., Vancouver Nov. 28–Dec. 2, 2001, p 175
21. May-Collado LJ, Gerrodette T, Calambokidis J, Rasmussen K, Sereg I (2005) Patterns of cetacean sighting distribution in the Pacific Exclusive Economic Zone of Costa Rica: based on data collected from 1979–2001. Rev Biol Trop 53:249–263
May-Collado LJ, Morales-Ramírez A (2005) Presencia y Patrones de comportamiento del delfín manchado costero, *Stenella attenuata graffmani* (Cetacea: Delphinidae) en el Golfo de Papagayo, Costa Rica. Rev Biol Trop 53:265–276
May-Collado LJ, Wartzok D (2008) A comparison of bottlenose dolphin whistle in the Western Atlantic Ocean: insights on factors promoting whistle variation. J Mamm 89(5):in press

May-Collado LJ, Wartzok D, Gamboa M (2005) ¿Plasticidad en las vocalizaciones de delfines?: efecto del ruido antropogénico en Costa Rica. Primer Congreso de Vida Silvestre Latinoamericano, San Jose, Costa Rica, Feb 2005

Mitchell ED, Reeves RR, Evely A (eds) (1996) Introductory essay. *In*: Bibliography of Whale Killing Techniques. Rep. Int. Whal. Comm., Spec. Issue 7. Cambridge, pp 1–12

Monteiro-Filho ELA, Monteiro LR, dos Reis SF (2002) Skull shape and size divergence in dolphins of the genus *Sotalia*: a tridimensional morphometric analysis. J Mamm 83:125–134

22. Mou-Sue L, Chen DH, Bonde RK, O'Shea TJ (1990) Distribution and status of manatees (*Trichechus manatus*) in Panama. Mar Mamm Sci 6:234–241

O'Donnell DJ (1982) Manatees and Man in Central America. Dissertation Abstract International Biological Sciences and Engineering 42:3175

23. O'Shea TJ, Salisbury CAL (1991) Belize- a last stronghold for manatees in the Caribbean. Oryx 25:156–164

Oviedo L (2007) Dolphin sympatric ecology in a triopucal fjord: habitat partitoning by bathway and topography as a strategy to coexist. J. Mar. Biol. Ass. UK. 87:1327–1335

Palacios DM, Gerrodette T (1996) Potential impact of artisanal gillnet fisheries on small cetaceans populations in the eastern tropical Pacific. NMFS, Southwest Fisheries Science Center, Administrative Report JL-96–11, p 17

Perrin WP (2001) *Stenella attenuata*. Mamm Spec 683:1–8

Perrin WP (2002a) Pantropical spotted dolphins (*Stenella attenuata*). *In*: Perrin W, Würsig B, Thewissen JGM (eds) Encyclopedia of Marine Mammals. Academic, San Diego, CA, pp 865–867

Perrin WP (2002b) Spinner dolphins (*Stenella longirostris*). In: Perrin W, Würsig B, Thewissen JGM (eds) Encyclopedia of Marine Mammals. Academic, San Diego, CA, pp 1174–1178

Perrin WP (2002c) *Stenella frontalis*. Mamm Spec 702:1–6

Perrin WP, Scott PA, Walker GJ, Cass VL (1985) Review of geographical stocks of tropical dolphins (*Stenella* spp. and *Delphinus delphis*) in the eastern Pacific. US Department of Commerce. NOAA Technical Report. NMFS 28, p 28

24. Rasmussen K, Calambokidis J, Steiger G, Saborío M, May-Collado L, Gerrodette T (2001a) Extent of geographic overlap of North Pacific and South Pacific humpback whales on their Central America wintering grounds. 14th Biennial Conf. Biol. Mar. Mamm., Vancouver Nov. 28 to Dec. 2, 2001, p 138

Rasmussen K, Calambokidis J, Steiger G (2001b) Humpback whales and other marine mammals off Costa Rica and surrounding waters, 1996–2001. Rep. Oceanic Society 2001 field season in cooperation with Elderhostel volunteers. Dec. 2001, p. 22

Rasmussen K, Calambokidis J, Steiger G (2002) Humpback whales and other marine mammals off Costa Rica and surrounding waters, 1996–2002. Rep. Oceanic Society 2001 field season in cooperation with Elderhostel volunteers. Dec. 2001, p 21

Rathbun G, Powell JA, Cruz G (1983) Status of the west Indian manatee in Honduras. Biol Conserv 26:301–308

25. Rice DW (1998) Marine Mammals of the World: Systematics and Distribution. Special Publication Number 4. The Society for Marine Mammalogy. Allen, Lawrence, KS

26. Reeves RR, Steward BS, Leatherwood S (1992) The Sierra Club Handbook of Seals and Serenians. Sierra Club Books, San Francisco, CA

Reyes JC, Oporto JA (1994) Gillnet fisheries and cetaceans in the southeast Pacific. Rep Int Whal Comm (Spec Issue 15):467–474

Reynolds JE, Odell DK (1991) Manatee and Dugons. Press Facts on File, New York/Oxford

Reynolds JE, Szelistowski WA, León MA (1995) Status and conservation of manatees *Trichechus manatus manatus* in Costa Rica. Biol Cons 71:193–196

Reynolds JE III, Odell DK, Sommel SA (1999) Marine mammals of the world. *In*: Reynolds JE III, Rommel SA (eds) Biology of Marine Mammals. Smithsonian Institution Press, Washington/London, pp 1–14

27. Reynolds JE III, Powell JA (2002a) Manatees (*Trichechus manatus, T. senegalensis* and *T. inunguis*). *In*: Perrin W, Würsig B, Thewissen JGM (eds) Encyclopedia of Marine Mammals. Academic, San Diego, CA, pp 709–720

Reynolds JE III, DeMaster DP, Silber GK (2002b) Endangered species and populations. In: Perrin W, Würsig B, Thewissen JGM (eds) Encyclopedia of Marine Mammals. Academic, San Diego, CA, pp 373–382

Rodríguez-Herrera B, Chinchilla FA, May-Collado LJ (2004) Lista de especies, endemismo y conservación de los mamíferos de Costa Rica. Rev Mex Mastozool 6:19–41

Rosas-Weber CF, Monteiro-Filho ELA (2002) Reproduction of the estuarine dolphin (*Sotalia guianensis*) on the coast of Paraná southern Brazil. J Mamm 83:507–515

28. Sears R (2002) Blue Whale (*Balaenoptera musculus*). *In*: Perrin W, Würsig B, Thewissen JGM (eds) Encyclopedia of Marine Mammals. Academic, San Diego, CA, pp 112–116
29. Smethurst D, Nietschmann B (1999) The distribution of manatees (*Trichechus manatus*) in the coastal waterways of Tortuguero, Costa Rica. Biol Cons 89:267–274
30. Steiger GH, Calambokidis Sears J, Balcomb KC, Cubbage JC (1991) Movements of humpback whales between California and Costa Rica. Mar Mamm Sci 7:306–310

Vidal OK (1993) Aquatic mammal conservation in Latin America: problems and perspectives. Cons Biol 7:788–795

Vidal O, Van Waerebeek K (1994) Cetaceans and gillnets fisheries in Mexico, Central America and Wider Caribbean: a preliminary review. Rep Int Whal Comm, Spec Issue 15:221–233

Wade PR, Gerrodette T (1993) Estimates of cetacean abundance and distribution in the eastern tropical Pacific. Rep Int Whal Comm 43:477–493

31. Wade PR (1995) Revised estimates of incidental kills of dolphins (Delphinidae) by purse-seine tuna fishery in the eastern tropical Pacific 1959–1972. Fish Bull 93:345–354

Ward HL (1887) Notes on the life history of *Monachus tropicalis*, the West Indian Seal. Amer Nat 21:257–264

Wartzok D, Popper AN, Gordon J, Merrill J (2004) Factors affecting the response of marine mammals to acoustic disturbance. Mar Tech Soc J 37:6–15

Other Sources

32. Ann DiBerdinis. Unpublished written and video data, 1997–2002. Talamanca Dolphin Foundation, Manzanillo, Talamanca, Costa Rica

Part 43
Other Taxonomic Groups (Fungi, Kinorhynchs, Invertebrate Chordates)

Jorge Cortés

Rhopalaea birkelandi, a common tunicate from Pacific Costa Rica, surrounded by hydroids (Photo: Jorge Cortés)

Abstract This part summarizes published information concerning different taxonomic groups for which no experts were found to study especially the Costa Rican or Central American fauna. Five genera of fungi living on mangrove trees, 2 species of kinorhynchs, 15 species of ascideans, and 1 species of a cephalochordate from the Pacific coast of Costa Rica are listed in this part. Two species of salps from the

J. Cortés
Centro de Investigación en Ciencias del Mar y Limnología (CIMAR), and Escuela de Biología, Universidad de Costa Rica, 11501-2060, San José Costa Rica.
e-mail: jorge.cortes@ucr.ac.cr

Caribbean coast are also included. All the species herein are from single locations with few observations or collected specimens. There are many more species to be identified of these as well as of other groups (e.g., cerianthids, free-living flatworms and nematodes, nemerteans, oligochaetes, and hemichordates) that have been observed but most of them have never been identified even to the family level in Costa Rica.

Introduction

There are several taxonomic groups (fungi, kinorhynchs, invertebrate chordates) which have received little or no attention in the Central American area. Currently, no experts of these groups are working in the region, and visits of specialists have been extremely scarce. As a consequence, few publications exist regarding these organisms. Based on the few existing publications, in this part I summarize the information concerning the above-mentioned taxonomic groups. However, many more species of these groups do exist in Costa Rica than are reported here, but to date no specific studies of these groups have been undertaken.

Fungi: So far, no marine fungi have been reported from the Caribbean coast of Costa Rica. Regarding the Pacific, five genera of fungi (Species List 43.1 is included on the CD-Rom), living on mangrove trees in Punta Morales, were identified; one Chytridiomycota could not be identified (Ulken *et al.* 1990). The most common genera were *Schizochytrium* and *Thraustochytrium*, and the authors found more specimens in the wet season than in the dry season (Ulken *et al.* 1990).

Salps: Two species of salps are reported from the Caribbean (Species List 43.2 is included on the CD-Rom), *Thalia democratica* and *Weelia cylindrica*. They were collected during Cruise P-6811 of the Office of Naval Research of the US Navy at Station 10 (10°23'N 82°38'W), on 11 November 1968, from a depth of 1,708 m (Michel & Foyo 1976). There are no published reports of salps from the Pacific of Costa Rica.

Kinorhynchs: The animals included in this part from the Pacific are presented in Species List 43.3 (included on the CD-Rom). Two papers describing one new genus and two new species of kinorhynchs have been published, *Campyloderes* cf. *vanhoeffeni* (Neuhaus 2004) and *Fissuroderes thermoi* gen. et sp. nov. (Neuhaus & Blasche 2006). Neuhaus (2004) described one male and one female collected during the RV Sonne expedition to the eastern Pacific in 1999. The specimens were collected from the Cocos Ridge at Station 49 (7°00.28'N 83°54.14'W), leg SO144, using a TV-grab from 1,048 m. The new genus and species were described with specimens collected during the RV Sonne expedition SO 144-3 in 1999 (OBS 5: 8°45.49'N 84°00.51'W, 98 m; OBS 1: 8°58.59'N 83°53.57'W, 85 m) (Neuhaus & Blasche 2006).

Invertebrate chordates: Fifteen species of ascideans, collected only at Playas del Coco in 1970 (Species List, inclued on CD-Rom), have been reported in two papers (Tokioka 1971, 1972), including a new species, *Rhopalaea birkelandi*, described with specimens from Costa Rica (Tokioka 1971). One species of cephalochordate, *Branchiostoma californiense*, was collected from soft bottoms of the Golfo de Nicoya (Maurer & Vargas 1984; Vargas *et al.* 1985).

Conclusions

There are many more species to be identified of the above-mentioned taxa, consequently the species numbers reported do not reflect the actual species diversity of these groups in Costa Rica. Several publications mention these groups (kinorhynchs: De la Cruz & Vargas 1986, 1987; Vargas 1988; B. Neuhaus personal communication, 2007; ascideans: J. Cortés personal observation, 1980–2008; salps from the Pacific: J. Cortés personal observation, 1999–2008); however, specimens were identified at most to the Phylum-level. Other taxonomic groups have been collected in Costa Rica but not identified to genus or species; also, I have observed animals that have not been identified to lower taxonomic levels. The following are examples of these two cases: cerianthids (J. Cortés personal observation, 2004, 2006), free-living flatworms (Dexter 1974; Vargas 1987, J. Cortés personal observation 1980–2008), free-living nematodes (De la Cruz & Vargas 1986, 1987; Guzmán *et al.* 1987; Vargas 1988), nemerteans (Dexter 1974; Vargas *et al.* 1985, personal observation), oligochaetes (Dexter 1974), and hemichordates (Vargas *et al.* 1985). There are probably other taxa of which we are not aware. More research, qualified taxonomists, and additional samples from unstudied habitats are needed to complete our picture of the actual marine biodiversity of Costa Rica.

Acknowledgment Most of the information in this paper was compiled during two visits to the National Museum of Natural History (NMNH), Smithsonian Institution; one visit was funded by the Smithsonian Institution and the other by a Fulbright Scholar Fellowship. Linda Cole of the National Museum of Natural History, Smithsonian Institution, helped to gather the information and taxonomy of the ascidians. Birger Neuhaus of Museum für Naturkunde, Berlin, helped with papers on the kinorhynchs. I thank Richard Petersen and José A. Vargas for reviewing the paper.

References

Brusca R. C. & G. J. Brusca 2003. Invertebrates, 2nd ed. Sinauer Associates, Inc., Massachusetts, USA. 936 p.

De la Cruz EM, Vargas JA (1986) Estudio preliminar de la meiofauna de la playa fangosa de Punta Morales, Golfo de Nicoya, Costa Rica. Brenesia 25–26:89–97

De la Cruz E, Vargas JA (1987) Abundancia y distribución vertical de la meiofauna en la playa fangosa de Punta Morales, Golfo de Nicoya, Costa Rica. Rev Biol Trop 35:363–367

Dexter DM (1974) Sandy-beach fauna of the Pacific and Atlantic coasts of Costa Rica and Colombia. Rev Biol Trop 22:51–66

Guzmán HM, Obando VL, Cortés J (1987) Meiofauna associated with a Pacific coral reef in Costa Rica. Coral Reefs 6:107–112

Maurer D, Vargas JA (1984) Diversity of soft bottom benthos in a tropical estuary: Gulf of Nicoya, Costa Rica. Mar Biol 81:97–106

1. Michel HB, Foyo M (1976) Caribbean Zooplankton. Part I. Siphonophora, Heteropoda, Copepoda, Euphausiacea, Chaetognatha and Salpidae. Off. Naval Res., Dept. Navy, Washington, DC, 549 p

Myers P, Espinosa R, Parr CS, Jones T, Hammond GS, Dewey TA (2006) The Animal Diversity Web (online). Accessed several times during 2006, at http://animaldiversity.org

2. Neuhaus B (2004) Description of *Campyloderes* cf. *vanhoeffeni* (Kinorhyncha, Cyclorhagida) from the Central American East Pacific Deep with a review of the genus. Meiofauna Mar 13:3–20
3. Neuhaus B, Blasche T (2006) *Fissuroderes*, a new genus of Kinorhyncha (Cyclorhagida) from the deep sea and continental shelf of New Zealand and from the continental shelf of Costa Rica. Zool Anz 245:19–52
4. Tokioka T (1971) A new species of *Rhopalaea* from the Pacific coast of Costa Rica (Tunicata, Ascidiacea). Publ Seto Mar Biol Lab XIX (2/3):119–122
5. Tokioka T (1972) On a small collection of ascidians from the Pacific coast of Costa Rica. Publ Seto Mar Biol Lab XIX(6):383–408
6. Ulken A, Víquez R, Valiente C, Campos M (1990) Marine fungi (Chytridiomycetes and Thraustochytriales) from a mangrove area at Punta Morales, Golfo de Nicoya, Costa Rica. Rev Biol Trop 38:243–250

Vargas JA (1987) The benthic community of an intertidal mud flat in the Gulf of Nicoya, Costa Rica. Description of the community. Rev Biol Trop 35:299–316

Vargas JA (1988) A survey of the meiofauna of an eastern Pacific intertidal mudflat. Rev Biol Trop 36:541–544

7. Vargas JA, Dean HK, Maurer D, Orellana P (1985) Lista preliminar de los invertebrados asociados a los sedimentos del Golfo de Nicoya, Costa Rica. Brenesia 24:327–342

Part 44
Marine Fish Parasites

Jorge Cortés

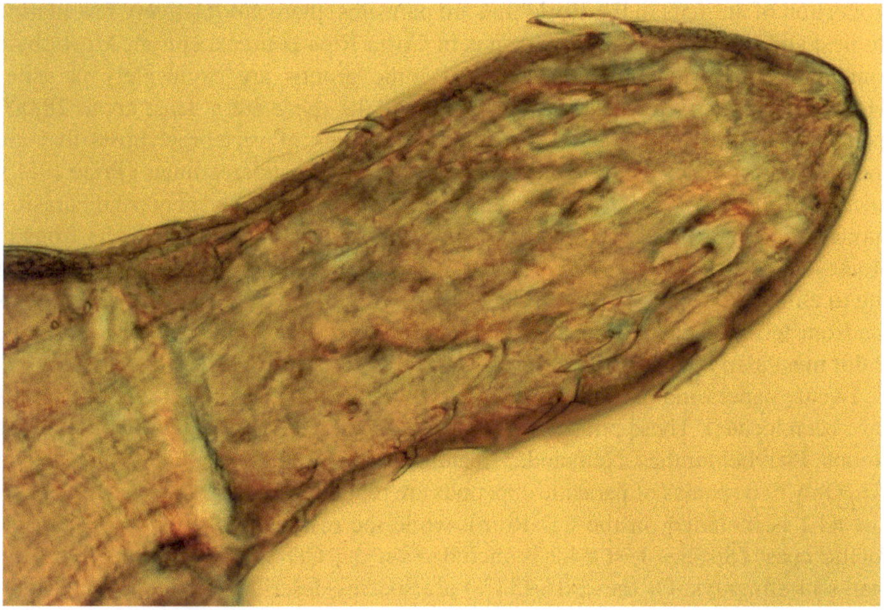

An unidentified species of *Acanthocephalus* obtained from a fish from Pacific Costa Rica. (Photo: Juan Alberto Morales)

Abstract Forty-seven species of parasites are listed, 46 from the Pacific coast, and 2 from the Caribbean of Costa Rica (one in common between both coasts but in different host fish). Sixteen (34%) were newly described species with specimens collected in Costa Rica. Forty species (85%) are worms (38 platyhelminths,

J. Cortés
Centro de Investigación en Ciencias del Mar y Limnología (CIMAR), and Escuela de Biología, Universidad de Costa Rica, 11501-2060, San José Costa Rica.
e-mail: jorge.cortes@ucr.ac.cr

1 acanthocephalan and 1 nematode), and the rest (15%) are parasitic copepods. Twenty papers dealing with this important group of the marine biodiversity of Costa Rica have been found. Only two species of parasites (copepods) have been reported from the Caribbean, and all reports are from fish, suggesting that there are probably many more species of marine parasites than the 47 species listed herein.

Introduction

Parasites constitute an important group of the marine biodiversity, but relative to the proportion of animals in the world that are parasites, there are relatively few marine studies, and the knowledge of parasites in Costa Rica is no exception. Most phyla contain members that are parasites and some groups are exclusively parasites (Hoberg 1997, Brusca & Bursca 2003). Potentially there are at least about 28,000 species of helminth parasites of the 45,000 species of vertebrate hosts that are known (Hoberg 1997), although this number may be an underestimate (Price 1980). As a result of global climate changes, the impact of pathogenic agents and parasites may also change, as was recently demonstrated with a fungus that infects frogs in Monteverde, Costa Rica (Pound *et al.* 2006). Life cycles of parasites are also changing due to climatic alteration (Blaustein & Dobson 2006). To date, all known examples are from terrestrial environments, but as yet undiscovered changes in the marine realm must also be occurring.

Twenty papers and one thesis on marine parasites collected in Costa Rican waters have been located. These publications include 47 species, belonging to the following groups: Platyhelminthes, Nematoda, Acanthocephala, and Copepoda; all parasites of fish. Only two species of parasitic copepods are reported from the Caribbean (Species List 44.1 is included on the CD-Rom), while the rest are mostly worms from the Pacific coast (Species List 44.2 is included on the CD-Rom), and most are from sharks or stingrays. Of these, 16 (34%) are species described from specimens collected in Costa Rica. The thesis by Contreras Angulo was finished in 2003, but has not been defended so the information therein is not considered in this part. Papers on parasitic isopods, Jiménez and Vargas (1990) for example, have been included in the Isopod part in this book.

Undoubtedly there are many more species of parasites yet to be discovered than the 47 species listed here. Many of the commercially important species still have not been examined for parasites, with the exception of the undefended thesis by Contreras Angulo (2003). Only a few collections have been made at a few sites on the Pacific coast and at only one site on the Caribbean. (Pérez-Ponce de León *et al.* (1998) incorrectly indicated that Playa Nacascolo was near Limón, on the Caribbean coast, and the coordinates given are also incorrect; Playa Nacascolo (10°37′45′N 85°40′29′W) is located in Bahía Culebra, Guanacaste Province, on the Pacific coast.) All parasites reported are from fish. The large portion of the studies of Costa Rican parasites contain descriptions of new material, clearly indicating that more research is needed on parasites of marine organisms.

Specialists

The following is a list of specialists that have worked on samples from Costa Rica. At present none is actively working collections from Costa Rica.

Dan R. Brooks: Dept. Zoology, Univ. Toronto, Canada – Tropical biodiversity, systematic biology. dbrooks@zoo.utoronto.ca

Luis García-Prieto: Laboratorio de Helmintología, Instituto de Biología, UNAM, México D.F., México – Helminths from Mexico. gprieto@ibiologia.unam.mx

Eric P. Hoberg: US National Parasite Collection, US Dept. Agriculture, Beltsville, Maryland, USA – Trematoda, Cestoda, Acanthocephala. ehoberg@anri.barc.usda.gov

Rafael Lamothe-Argumedo, Lab. helmintos, Inst. Biología, UNAM, México D.F., México – Helminths of vertebrates from Mexico. lamothe@servidor.unam.mx

Virginia León-Règagnon: Laboratorio de Helmintología, Instituto de Biología, UNAM, México D.F., México – Phylogenetic and molecular systematics; evolution of parasite–host systems. vleon@ibiologia.unam.mx

Scott Monks: Centro de Investigaciones Biológicas, Universidad Autónoma del Estado de Hidalgo, Pachuca, Hidalgo, México – Systematics and biogeography of helminth parasites, life cycles/ecology of Acanthocephala, Cestoda, Digenea. smonks@uaeh.edu.mx

Juan Alberto Morales Acuña: Escuela de Ciencias Veterinarias Tropicales, Universidad Nacional, Heredia, Costa Rica – Systematics of freshwater, marine and wild animals' parasites. jalb_morales@hotmail.com

Gerardo Pérez-Ponce de León: Laboratorio de Helmintología, Instituto de Biología, UNAM, México D.F., México – Systematics and biogeography of helminth parasites of wild vertebrates. ppdleon@servidor.unam.mx

Beatriz Rodríguez-Ortíz: Laboratorio de Helmintología, Facultad de Microbiología, Universidad de Costa Rica. Retired.

Collections

Canadian Museum of Nature, Invertebrate Zoology Collection, Ottawa, Canada. www.nature.ca

Colección de Helmintos de Costa Rica, Facultad de Microbiología, Universidad de Costa Rica. www.micro.ucr.ac.cr

Colección de Parásitos, Escuela de Ciencias Veterinarias Tropicales, Universidad Nacional, Heredia, Costa Rica.

Colección Nacional de Helmintos, Universidad Nacional Autónoma de México, México D.F., México. www.ibiologia.unam.mx/cnhe/

Harold W. Manter Laboratory (HWML), Division of Parasitology, University of Nebraska State Museum, Lincoln, Nebraska, USA. http://hwml.unl.edu/

US National Parasite Collection, Beltville, Maryland, USA. www.anri.barc.usda.gov/bnpcu/

Acknowledgments Most of the information in this part was compiled thanks to the help of Kay Hale, former Director of the RSMAS Library, University of Miami. Visits to the Smithsonian Tropical Research Institute, Panamá, and two stays at the National Museum of Natural History (NMNH), Smithsonian Institution (the first funded by the Smithsonian Institution and the second by a Fulbright Scholar Fellowship), increased the database enormously. Scott Monks of the CIB, Universidad Autónoma del Estado de Hidalgo, México and Dan Brooks of the Department of Zoology, University of Toronto, Canada helped with corrections and information on Costa Rican marine parasites. Juan Alberto Morales of the Universidad Nacional helped with photographs, comments and access to a thesis on marine parasites. Richard Petersen corrected the English.

References

1. Berman R, Brooks DR (1994) *Escherbothrium molinae* n. gen. et n. sp. (Eucestoda: Tetraphyllidea: Triloculariidae) in *Urotrygon chilensis* (Chondrichthyes: Myliobatiformes: Urolophidae) from the Gulf of Nicoya, Costa Rica. J Parasitol 80:775–780

Blaustein AR, Dobson A (2006) A message from the frogs. Nature 439:143–144

2. Bravo-Hollis M, Arroyo-Sánchez G (1962) Helmintos de peces costarricenses del Pacífico. An Inst Biol, UNAM Ser Zool 33:79–95
3. Brenes RR (1961) Catálogo de los helmintos parásitos de Costa Rica. Rev Biol Trop 9:67–95
4. Brooks DR, McCorquodale S (1995) *Acanthobothrium nicoyaense* n.sp. (Eucestoda: Tetraphyllidea: Onchobothriidae) in *Aetobatus narinari* (Euphrasen) (Chondrichthyes: Myliobatiformes: Myliobatidae) from the Gulf of Nicoya, Costa Rica. J Parasitol 81:244–246

Brusca RC, Brusca GJ (2003) Invertebrates, 2nd ed. Sinauer Assoc, Sunderland MA, 936 p

5. Caballero CE, Brenes-Madrigal R (1957) Helmintos de la República de Costa Rica VI: Algunos tremátodos de peces, reptiles y mamíferos. An Inst Biol, UNAM 28:217–240
6. Campbell RA, Beveridge I (1997) *Pterobothrioides*, a new genus of tapeworm (Cestoda: Trypanorhyncha: Pterobothriidae) from dasyatid stingrays in the eastern Atlantic and Pacific. Syst Parasitol 38:81–91

Contreras Angulo G (2003) Descripción, caracterización y cuantificación de la fauna simbionte de parásitos de cuatro familias de peces comerciales del Golfo de Nicoya, Costa Rica. Posgrado en Ciencias Veterinarias Tropicales, Universidad Nacional, Heredia, Costa Rica, 42 p

7. Cressey R, Cressey HL (1980) Parasitic copepods of mackerel- and tuna-like fishes (Scombridae) of the world. Smiths Contr Zool 311:1–186

Hoberg EP (1997) Phylogeny and historical reconstruction: host-parasite systems as keystones in biogeography and ecology. *In*: Reaka-Kudla ML, Wilson DE, Wilson EO (eds) Biodiversity II: Understanding and Protecting Our Biological Resources. Joseph Henry Press, Washington, DC, pp 243–261

8. Hoberg EP, Brooks DR, Molina-Ureña H, Erbe E (1998) *Echinocephalus janzeni* n.sp. (Nematoda: Gnathostomatidae) in *Himantura pacifica* (Chondrichthyes: Myliobatiformes) from the Pacific coast of Costa Rica and Mexico, with historial biogeographic analysis of the genus. J Parasitol 84:571–581

Jiménez MP, Vargas M, V (1990) *Probopyrus pandalicola* (Isopoda: Bopyridae) infesting *Palaemonetes hiltonii* (Crustacea: Caridae) along the Pacific coast of Costa Rica. Rev Biol Trop 38:457–462

9. Lamothe Argumedo R, Aranda C, Pérez-Ponce de León G (1997) Catálogo de la Colección Nacional de Helmintos. Publ Esp, Inst Biol UNAM-CONABIO, 271 p
10. Manter HW (1940) Digenetic trematodes of fishes from the Galapagos Islands and the neighboring Pacific. Allan Hancock Pac Exped 2(14):329–497
11. Marques F, Brooks DR, Monks S (1995) Five new species of *Acanthobothrium* van Beneden 1849 (Eucestoda: Tetraphyllidea: Onchobothriidae) in stingrays from the Gulf of Nicoya, Costa Rica. J Parasitol 81:942–951
12. Marques F, Brooks DR, Molina-Ureña H (1996) Two new species of tetraphyllidean cestodes in *Himantura pacifica* (Chondrichthyes: Myliobatiformes: Dasyatididae) from the northwest coast of Costa Rica. J Parasitol 82:302–306
13. Marques F, Centritto R, Stewart SA (1997) Two new species of *Acanthobothrium* in *Narcine entemedor* (Rajiformes: Narcinidae) from the northwest coast of Guanacaste Peninsula, Costa Rica. J Parasitol 83:927–931
14. Meserve FG (1938) Some monogenetic trematodes from the Galapagos Islands and the neighboring Pacific. Allan Hancock Pac Exped 2(5):31–89
15. Monks S, Marques F, León-Règagnon V, Pérez-Ponce de León G (1997) *Koronacantha pectinaria* n. comb. (Acanthocephala: Illiosentidae) from *Microlepidotus brevipinnis* (Haemulidae) and redescription of *Tegorhynchus brevis*. J Parasitol 83:485–494
16. Pérez-Ponce de León G, León-Règagnon V, Monks S (1998) *Theletrum lamothei* sp. nov. (Digenea), parasite of *Echidna nocturna* from Cuajiniquil. Guanacaste, and other digenes of marine fishes from Costa Rica. Rev Biol Trop 46:345–354
17. Ponciano-Rodríguez G (1986) Estudios taxonómicos de tremátodos de peces marinos y dulciacuícolas de México y América Central. Tesis Lic., UNAM, México, 128 p

Pound JA, Bustamante MR, Coloma LA, Consuegra JA, Fogden MPL, Foster PN, La Marca E, Masters KL, Merino-Viteri A, Puschendorf R, Ron SR, Sánchez-Azofeifa GA, Still CJ, Young BE (2006) Widespread amphibian extinction from epidemic disease driven by global warming. Nature 439:161–167

Price PW (1980) Evolutionary Biology of Parasites. Princeton University Press, Princeton, NJ, 237 p

18. Rodríguez-Ortíz B, García-Prieto L, Pérez-Ponce de León G (2004) Checklist of the helminth parasites of vertebrates in Costa Rica. Rev Biol Trop 52:313–354
19. Rojas JR, Rodríguez-Solano O, Morales-Ramírez A (2001) Size and distribution of *Pandarus satyrus* (Copepoda: Pandaridae) on the blue shark *Prionace glauca* (Carcharhiniformes: Carcharhinidae) in Costa Rica. Rev Biol Trop 49:199–201
20. Van Cleave HJ (1940) The Acanthocephala collected by the Allan Hancock Pacific Expedition, 1934. Allan Hancock Pac Exped 2(15):501–527
21. Wilson CB (1937) Parasitic copepods taken during the third Hancock expedition to the Galapagos Islands. Allan Hancock Pac Exped 2(4):23–31

Part 45
Sea Turtle Parasites

Mario Santoro and Simonetta Mattiucci

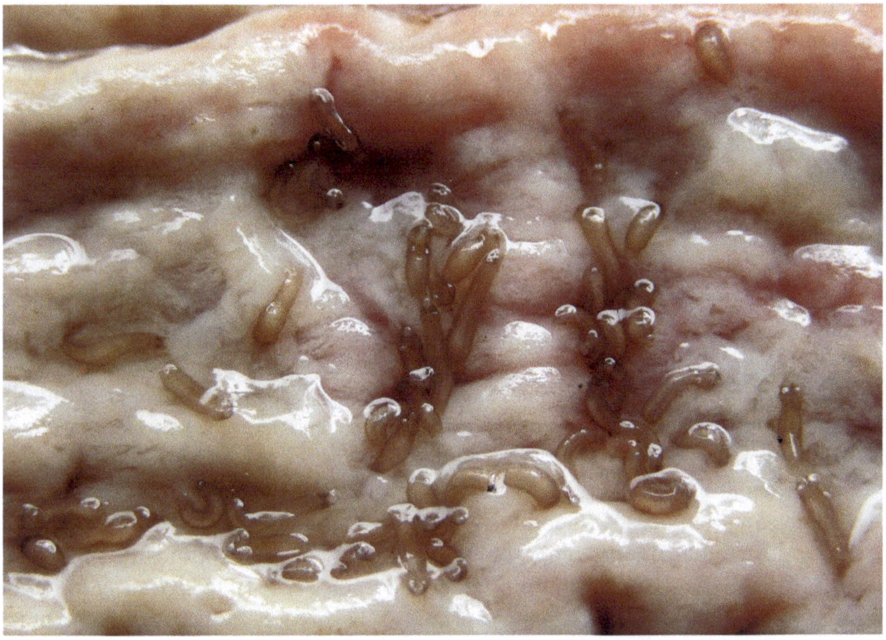

Charaxicephalus robustus, a parasite from the stomach of *Chelonia mydas*, Parque Nacional Tortuguero, Caribbean coast of Costa Rica (Photo: M. Santoro)

M. Santoro
Departamento de Patología, Escuela de Medicina Veterinaria,Universidad Nacional, PO Box 86, Heredia 3000, Costa Rica.
e-mail: marisant@libero.it

S. Mattiucci
Department of Public Health Sciences, Section of Parasitology, University of Rome "La Sapienza", P.le Aldo Moro, 5 00185 Rome, Italy.
e-mail: simonetta.mattiucci@uniroma1.it

I.S. Wehrtmann, J. Cortés (eds.) *Marine Biodiversity of Costa Rica, Central America*,

Abstract In the present review we report a total of 34 species of parasites from two sea turtle species (*Chelonia mydas* and *Lepidochelys olivacea*) from Costa Rican waters. Most of the parasites were digenetic trematodes except for one annelid, the leech *Ozobranchus branchiatus*. Thirty-one species have been identified from the Caribbean coast and only four from the Pacific coast. *O. branchiatus* occurs on both sea turtle species examined from the two coastal areas. Eleven digeneans represent new geographical records for the Atlantic Ocean. Three species are reported for the first time for the Caribbean Sea. The most represented family of parasites found was Pronocephalidae with 15 species (including a new species, *Pleurogonius tortugueroi*). On the Caribbean coast, Tortuguero National Park was the only locality from where sea turtles (*C. mydas*) were examined, while three specimens of *L. olivacea* stranded along the Pacific coast were studied. Due to the scarcity of sea turtle parasite studies (only two of the six species of sea turtles that occur along the Costa Rican coasts have been examined so far), it is reasonable to suppose that parasite species diversity in sea turtles of Costa Rica is considerably higher than reported to date.

Introduction

Of the eight species of sea turtles currently recognized in the marine environment, six have been so far reported from Costa Rican waters (Savage 2002): the green turtle (*Chelonia mydas*) and loggerhead (*Caretta caretta*) from the Caribbean coast; Pacific green turtle (*C. agassizii*) and olive ridley (*Lepidochelys olivacea*) from the Pacific coast; hawksbill (*Eretmochelys imbricata*) and leatherback (*Dermochelys coriacea*) from both coasts.

Sea turtles are parasitized by a rich variety of endoparasites, primarily digenetic trematodes and nematodes (George 1997). Different habitat use and feeding habits of sea turtles from different geographical areas influence the composition and richness of parasite communities (Pérez-Ponce De León *et al.* 1996; Aznar *et al.* 1998; Santoro *et al.* 2006). *Chelonia* spp. is primarily herbivore. In the Caribbean, green turtles graze on the sea grass, *Thalassia testudinum*, which constitutes almost 90% of their diet. Loggerheads and olive ridleys are omnivorous generalists, foraging principally on fish, crustaceans, and mollusks; hawksbill turtles forage most commonly on coral reefs and invertebrates, primarily sponges; leatherbacks feed on jellyfish, salps, and other gelatinous organisms (Bjorndal 1997).

Parasites are integral components of marine ecosystems. They could be indicators of the general ecosystem biodiversity (at population, species, and gene level), of the ecosystem functioning (such as food web stability) (Mattiucci & Nascetti 2007), of their host species migrations (such as fish stocks), but also of the phylogenetic history of their hosts (Mattiucci *et al.* 2008). At any host taxonomic level, parasite diversity can thus be understood as determined by two filters, i.e., the probability of encounter, and the physiological compatibility between host and parasite. Both filters contain evolutionary and ecological components (Holmes

1990; Hoberg & Klassen 2002). Aznar *et al.* (1998) suggested that the diversity of parasites in marine turtles is determined by historical reasons: although there might be regular contacts between parasites from other marine hosts and marine turtles (e.g., through trophic webs), the compatibility filter would ultimately impede the establishment of these parasites in the turtles. In other words, marine turtles are physiologically so distinct from other marine vertebrate hosts (fish, birds, mammals) that any marine turtle could exchange parasites only with other marine turtle species because all share a basic physiological similarity (derived from common ancestry). Depending on the diet of each turtle species, the probability of infestation by the same parasite could increase. Eight gastrointestinal helminth species, which have been reported previously only from sea turtles, were also found with other marine hosts; moreover, six supposedly host-specific species from other marine hosts were also encountered in sea turtles (Aznar *et al.* 1998). However, physiological specialization of parasites can become narrower, and many parasite taxa also found in some turtle species could not survive in other marine turtle species and/or they do not reach the adult stage.

These general similarities has led to the presence of the same parasite "morphospecies" even in different host species from the same geographical area, and/or in the same host. This parallelism and/or convergent phenomena in parasite morphology sometimes was found to hide the presence of different populations of sibling or cryptic species (biological species which are difficult or impossible to distinguish based on morphological characters). They are common in all major marine groups and habitats, including parasites of marine organisms. Indeed, this phenomenon has been discovered in the last 20 years with the use of genetic/molecular techniques. Sibling species are commonly found in several taxa of parasites of marine organisms, such as in the case of the anisakid nematodes of the genera *Anisakis*, *Contracaecum*, and *Pseudoterranova*, parasites of marine mammals (Nascetti *et al.* 1986, 1993; Paggi *et al.* 1991; Mattiucci *et al.* 1997, 2001; Mattiucci & Nascetti 2006, 2008). The presence of sibling species in these marine organisms has ecological as well as evolutionary implications, because they exhibit distinct habitat preference in the same host or preferences for different hosts. Moreover, phylogenetic relationships among related taxa of anisakid nematodes of the genus *Anisakis* (parasites at adult stage of marine mammals) are often parallel to those of their principal hosts (cetaceans) (Mattiucci & Nascetti 2006), suggesting that they have probably experienced similar coevolutionary forces, including both co-divergence and switching events (Mattiucci & Nascetti 2006, 2008).

Our knowledge of sibling species within parasite taxa of sea turtles is fairly limited. Recently, Rawson *et al.* (2003) studied the phylogeography of the epibiontic coronulid barnacle, *Chelonibia testudinaria*, from loggerheads. Their analysis based on results obtained from molecular studies indicated that two Pacific populations (Japan and Mexico) were not only highly divergent from Atlantic populations but were highly divergent from one another. The authors concluded that barnacles from these populations may represent cryptic species. In contrast, sequence divergence was greatly reduced among the barnacles collected from Florida *versus* those from the eastern Mediterranean Sea; this seems to suggest that the migration of loggerheads helps to maintain the gene flow among the western Atlantic populations of the barnacle, and it is responsible for the expansion of the barnacle range into the Mediterranean Sea (Rawson *et al.* 2003).

Sea Turtle Parasites

Within protozoans, *Entamoeba invadens* has been reported in captive greens and loggerheads (George 1997). *Caryospora cheloniae* has been found in captive and free-ranging greens (Leibovitz *et al.* 1978; Gordon *et al.* 1993). *Eimeria carettae* has been described from a stranded Atlantic loggerhead (Upton *et al.* 1990). *Cryptosporidium* sp. oocysts were identified in fecal and intestinal samples of free-ranging green turtles (Graczyk *et al.* 1997). All these protozoans inhabit the digestive tract. To date, there have been no reports in the literature of any hemoparasitic protozoan in marine chelonians.

Among the reptiles, turtles possess the richest and largest helminth communities (Aho 1990). Digeneans belonging to Pronocephalidae (gastrointestinal flukes) and Spirorchiidae (cardiovascular flukes) are the most important trematodes reported in sea turtles regarding species richness and pathological effects, respectively (Pérez-Ponce De León & Brooks 1995; George 1997). With the exception of a few marine sanguinicolids that use annelids, flukes require mollusks as obligate intermediate hosts (Esch *et al.* 2001). The specificity of digenetic trematodes for mollusks is strict, and in many cases only closely related species of snails are used as hosts. These parasites are most frequently found in sea turtles that have vegetarian diets. Sea turtles grazing on sea grass and algae may ingest more easily some mollusk intermediate hosts carrying the infection (see the experimental infection on *C. mydas* by Hunter [1967]). In addition, many digenean species that infect herbivorous marine hosts can be found on the vegetation after leaving the mollusks (Bartoli 1987; Blair & Hudson 1992). The aspidogastrean (Aspidogastridae) *Lophotaspis vallei*, infecting the oesophagus and stomach of the loggerhead (Yamaguti 1963), is the only non-digenean trematode reported in sea turtles.

In comparison with the high number of digeneans reported, few species of nematodes are known to occur in sea turtles. *Anisakis* sp. and *Sulcascaris sulcata* (Anisakidae) have been found in the gastrointestinal tract of green and loggerhead turtles (Lester *et al.* 1980; Burke & Rodgers 1982; Orós *et al.* 2004). Nematodes belonging to Kathlanidae (*Kathlania leptura* and *Tonaudia tonaudia*) have been reported in the gastrointestinal tract of green, loggerhead, and olive ridley turtles (Brooks & Frazier 1980; Lester *et al.* 1980). Anisakids and kathlanids develop their larval forms in bivalves, cephalopods, crustaceans, and fishes. Within sea turtles, the loggerhead as a carnivorous generalist is the main host in the life cycle of the anisakid nematode *S. sulcata* (Lichtenfels *et al.* 1978; Lester *et al.* 1980). Preliminary results of our molecular study on populations of *S. sulcata* from *C. caretta* collected from the western Atlantic Ocean (Florida coast) indicated the existence of a single panmictic population of this parasite (Mattiucci & Santoro, unpublished data, 2008). Further genetic studies (including genetic differentiation, gene flow estimates, and phylogeographic analysis of mtDNA *cox-2* sequences) on other populations of the same anisakid nematodes from the loggerhead might provide more information of historical migration patterns among the Mediterranean and Atlantic host populations.

Sea turtles act as paratenic hosts for larvae of *Anisakis* sp., because they are unable to complete their life cycle in the tissues of cold-blooded animals (Glazebrook & Campbell 1990). *Cucullanus carettae* (Cucullanidae) and larval forms of *Echinocephalus* sp. (Gnathostomatidae) have been reported also in the stomach and intestine of loggerheads (Lester *et al.* 1980).

Within cestodes, trypanorhynchan (Trypanorhynchidae) larvae (*Lacistorhynchus* or *Eutetrarhynchus*?) have been found only occasionally in loggerheads from the Egyptian coast (Sey 1977). With respect to ectoparasites, two species of leeches, *O. branchiatus* and *O. margoi*, have been reported in marine chelonians (George 1997).

Sea Turtle Parasite Diversity in Costa Rica

Studies on parasites of sea turtles from Costa Rican waters are scarce. The only location at the Caribbean coast where sea turtles have been examined for parasites is the Tortuguero National Park (10°32′27″N, 83°29′59″W – 10°21′17″N, 83°23′29″W). This site is the most important green turtle nesting ground of the Atlantic. One leech species and 30 species of digenetic trematodes have been reported (Species List 45.1 is included on the CD-Rom) from the nesting green turtles found dead along the park beach (Santoro *et al.* 2006, 2007a, b). Green turtle parasites from the Caribbean coast of Costa Rica are represented by two phyla: Annelida and Platyhelminthes. The phylum Annelida is represented by the family Ozobranchidae with only one species (*O. branchiatus*). The phylum Platyhelminthes is represented by one subclass (Digenea) with six families including Clinostomidae (1 sp.), Microscaphidiidae (5 sp.), Paramphistomidae (2 sp.), Pronocephalidae (15 sp.), Rhytidodidae (2 sp.), and Spirorchiidae (5 sp.) (Santoro *et al.* 2006). All digenean species collected were recorded from Costa Rica for the first time. *P. tortugueroi* (Pronocephalidae) was a new species described from the intestine of 2 green turtles (Santoro *et al.* 2007c). New geographical locality records for the Atlantic Ocean included: *Amphiorchis solus*, *Charaxicephalus robustus*, *Cricocephalus resectus*, *Deuterobaris intestinalis*, *Hapalotrema postorchis*, *Microscaphidium warui*, *Monticellius indicum*, *Octangium hiphalum*, *P. sindhii*, *Rhytidodoides intestinalis*, and *R. similis*. New geographical locality records for the Caribbean Sea included *Neospirorchis* sp., *P. longiusculus*, and *Pronocephalus obliquus*.

In 2004, the first author of this paper had the opportunity to examine for parasites three olive ridley turtles found stranded along the north Pacific coast of Costa Rica (Fig. 45.1). Three species of digeneans (Platyhelminthes) including one Gorgoderidae, one Plagiorchiidae, and one Pachypsolidae were obtained (Species

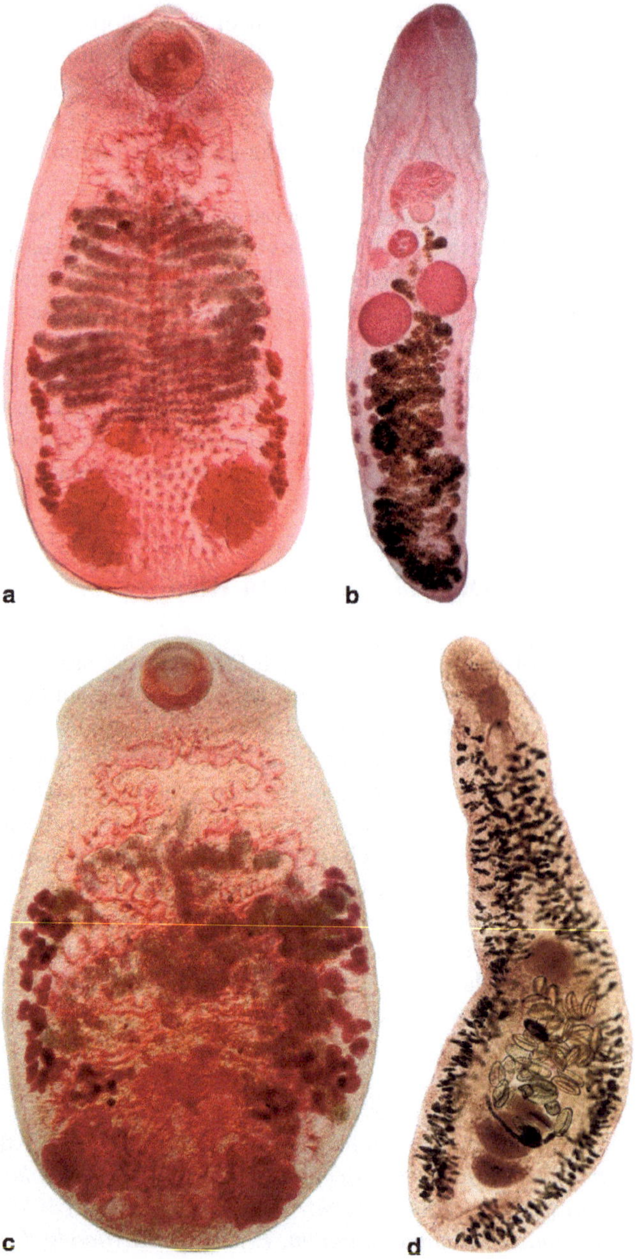

Fig. 45.1 Digenean trematodes from Costa Rican sea turtles. Specimens stained with Mayer's acid carmine. (a) *Cricocephalus resectus*, *C. mydas*. (b) *Enodiotrema megachondrus*, *L. olivacea*. (c) *Pyelosomum choclear*, *C. mydas*. (d) *Rhytidodoides intestinalis*, *C. mydas* (Photos: M. Santoro)

List 45.2 is included on the CD-Rom). Histopathological changes associated with eggs of cardiovascular flukes (Digenea: Spirorchiidae) were also observed from the lungs, spleen, and intestine of an olive ridley turtle (Santoro & Morales 2007). In addition, during clinical survey examinations carried out between 2002 and 2004 along the north Pacific coast of Costa Rica in Nancite nesting beach (10°48′17″N, 85°41′58″W) (Santa Rosa National Park) and Ostional National Wildlife Refuge (10° 00′00″N, 86°45′50″W), several specimens of *O. branchiatus* were found on the skin around the throat, dorsal surface of the neck, and axillary region of the flippers of live olive ridley turtles (Santoro, unpublished data). All digeneans from the olive ridleys represent the first record from Costa Rica (Santoro & Morales 2007). In the western hemisphere, these flukes have been reported in olive ridleys only from the Pacific coast of Mexico (Parra 1983; Pérez-Ponce De León *et al.* 1996).

According to Santoro *et al.* (2006), specialist parasites are defined as a parasite recorded only from one sea turtle species, while generalist species are parasites found in two or more turtle species or in other vertebrate species. Of the 33 species of digeneans obtained from sea turtles in Costa Rican waters, 15 are specialists in green turtles (Santoro *et al.* 2006). The remaining species are generalist in sea turtles, with the exception of *Clinostomum complanatum*, which is also a common generalist in birds. However, *Cricocephalus albus* has occasionally been recovered from the marine French angelfish *Pomacanthus arcuatus* (Pérez-Ponce De León & Brooks 1995), *C. resectus* from a star tortoise *Testudo elegans* (Yamaguti 1971), and *Polyangium linguatula* from the kidneys of the arctic loon *Colymbus arcticus* (= *Gavia arctica*) (Poche 1925). These latter three records, together with *C. complanatum* from Caribbean green turtles (Santoro *et al.* 2006), should be considered as accidental findings.

O. branchiatus (Fig. 45.2) is a generalist leech and the only turtle parasite reported from both coasts. The parasite community of the green turtles from the Caribbean Sea appears strongly dominated by pronocephalids (15 sp.). All parasites recorded from the Pacific olive ridley are generalist in sea turtles. Due to the scarce number of studies performed, which have included only two of the six sea turtles that occur along the Costa Rican coasts, it is reasonable to assume that the species diversity of the sea turtle parasites in Costa Rica is considerably higher than that reported by us.

Endoparasitic investigations are possible only on dead, moribund, or stranded individuals. The high temperature in the tropics as well as the large number of predators often occurring off-shore and/or on the beach, could limit the collection and preservation of turtle parasites. Sea turtles are endangered and protected species (IUCN/SSC 2002). Thus, the collection and storage of parasites from these hosts is particularly precious and valuable.

Future application of genetic molecular methods in addition to morphological studies of parasitic species from different sea turtles will shed light on the possible occurrence of cryptic species, elucidating parasite life cycles, mechanisms of host preference, marine food webs, host migration patterns and, finally, the general biodiversity of marine ecosystems inhabited by sea turtles.

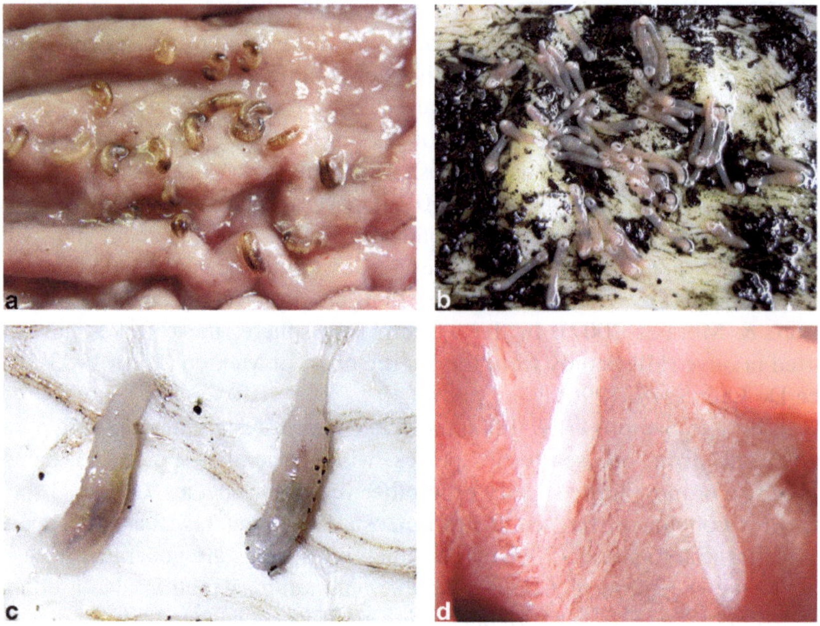

Fig. 45.2 Parasites from *Chelonia mydas* from Tortuguero National Park, Caribbean coast of Costa Rica. (a) *Charaxicephaloides* sp., stomach. (b) *Schizamphistomoides* sp., intestine. (c) *Ozobranchus branchiatus*. (d) *Learedius learedi*, heart (Photos: M. Santoro)

Recommendations

Due to the scarce number of olive ridley turtles examined, the authors recommend developing parasite investigations along the Pacific coast of Costa Rica. Field efforts should concentrate in the Refugio Nacional de Vida Silvestre Ostional where olive ridleys exhibit a characteristic synchronous massive nesting aggregation termed "arribada" with thousands of individuals arriving approximately one or two times per month all year round (G. Chaves, personal communication 2003). During these monthly events, it is common to find on the beach or near-shore several recently dead turtles which might be used for parasite investigations (personal observation 2003).

Specialists

The authors of this paper carried out the only sea turtle parasite project on both coasts of Costa Rica. No other sea turtle parasite specialist is working in Costa Rica at present. The expert on Costa Rican reptile parasites is Dr. Daniel R. Brooks of the Department of Zoology, University of Toronto, Canada (dbrooks@zoo.utoronto.ca).

Collections

The largest parasitological collections containing specimens from sea turtles are listed below:

- British Museum (Natural History), London
- Commonwealth Institute of Health, University of Sidney, Australia
- Colección de Helmintos de Costa Rica, Facultad de Microbiología, Universidad de Costa Rica
- Colección Nacional de Helmintos, Universidad Nacional Autónoma de México, México D.F., México
- Instituto Butantan, Sao Paolo, Brazil
- Instituto Oswaldo Cruz, Rio de Janeiro, Brazil
- Harold W. Manter Laboratory, Division of Parasitology, University of Nebraska State Museum, Lincoln, Nebraska, USA
- South Australian Museum, Adelaide, Australia
- U.S. National Parasite Collection, Beltville, Maryland, USA
- Zoologisches Museum, Berlin, Germany

Acknowledgments We thank our friend Garry Foster (College of Veterinary Medicine, University of Florida, USA) for his kind review of this paper and helpful suggestions and comments. We are also very grateful to Francisco Javier Aznar (Cavanilles Institute of Biodiversity and Evolutionary Biology, University of Valencia, Spain) for his fruitful review of the manuscript.

References

1. Aguirre AA, Spraker TR, Balazs GH, Zimmerman B (1998) Spirorchiidiasis and fibropapillomatosis in green turtles from the Hawaiian Islands. J Wildl Dis 34:91–98

Aho JM (1990) Helminth communities of amphibians and reptiles: comparative approaches to understanding patterns and processes. *In*: Esch GW, Bush AO, Aho JM (eds) Parasite Communities: Patterns and Processes. Chapman & Hall, New York, pp 157–195

2. Aznar FJ, Badillo FJ, Raga JA (1998) Gastrointestinal helminths of loggerhead turtles (*Caretta caretta*) from the western Mediterranean: constrains on community structure. J Parasitol 84:474–479
3. Badillo FJ, Raga JA (1996) Intestinal helminths of the loggerhead sea turtle, *Caretta caretta* (Linnaeus 1758) in the western Mediterranean. Parassitologia 38:107

Bartoli P (1987) Caracteres adaptifs originaux des digènes intestinaux de *Sarpa salpa* (Teleostei, Sparidae) et leur interprétation en termes d'évolution. Ann Parasitol Hum Comp 62:542–576

4. Blair D, Limpus CJ (1982) Some digeneans (Platyhelminthes) parasitic in the loggerhead turtle, *Caretta caretta* (L.), in Australia. Aust J Zool 30:653–680
5. Blair D (1983) Paramphistomes (Digenea: Paramphistomidae) parasitic in marine turtles (Reptilia: Chelonia). Aust J Zool 31:851–867
6. Blair D (1986) A revision of the subfamily Microscaphidiinae (Platyhelminthes: Digenea: Microscaphidiidae) parasitic in marine turtles (Reptilia: Chelonia). Aust J Zool 34:241–277
7. Blair D (1987) A revision of the subfamily Octangiinae (Platyhelminthes: Digenea: Microscaphidiidae) parasitic in marine turtles (Reptilia: Chelonia). Aust J Zool 35:75–92

Blair D, Hudson BET (1992) Population structure of *Lankatrematoides gardneri* (Digenea: Opisthotrematidae) in the pancreas of the dugong (*Dugong dugon*) (Mammalia: Sirenia). J Parasitol 78:1077–1079

Bjorndal KA (1997) Foraging ecology and nutrition of sea turtles. *In*: Lutz PL, Musick JA (eds) The Biology of Sea Turtles. CRC Press, Boca Raton, FL, pp 199–231

Brooks DR, Frazier J (1980) New host and locality for *Kathlania leptura* (Rudolphi) (Nematoda: Oxyurata: Kathlanidae). Proc Helm Soc Wash 47:267–268

Burke JB, Rodgers LJ (1982) Gastric ulceration associated with larval nematodes (*Anisakis* sp. type I) in pen reared green turtles (*Chelonia mydas*) from Torres Strait. J Wildl Dis 18:41–46

8. Caballero CE (1954) Helmintos de la República de Panamá. X. Algunos tremátodos de *Chelonia mydas* (L.) tortuga marina comestible del Océano Pacífico del norte. Anales Inst Biol Univ Nac Autón Méx 8:31–58
9. Chattopadhyaya DR (1972) Studies on the trematode parasite of reptiles found in India. Contribution to our knowledge of the family Pronocephalidae Looss 1902. Riv Parassitol 33:99–124
10. Coil WH, Reid WA (1965) *Desmogonius desmogonius* Stephens 1911 (Pronocephalidae: Digenea), a redescription with observation on egg filament formation. Zt Parasitenkd 25:506–509
11. Cordero-Tapia A, Gardner SC, Arellano-Blanco J, Inohuye-Rivera RB (2004) *Learedius learedi* infection in black turtles (*Chelonia mydas agassizii*), Baja California Sur, Mexico. J Parasitol 90:645–647
12. Cribb TH, Gordon AN (1998) *Hapalotrema* (Digenea: Spirorchiida) in the green turtle (*Chelonia mydas*) in Australia. J Parasitol 84:375–378
13. Dailey MD, Fast ML, Balazs GH (1992) A survey of the Trematoda (Platyhelminthes: Digenea) parasitic in green turtles (*Chelonia mydas*) from Hawaii. Bull South Calif Acad Sci 91:84–91
14. Dyer WG, Williams EH, Jr., Bunkley-Williams L (1991) Some digeneans (Trematoda) of the green turtle, *Chelonia mydas* (Testudines: Cheloniidae) from Puerto Rico. J Helm Soc Wash 58:176–180
15. Dyer WG, Williams EH, Jr., Bunkley-Williams L (1995) Digenea of the green turtle (*Chelonia mydas*) and the leatherback turtle (*Dermochelys coriacea*) from Puerto Rico. Carib J Sci 31:269–273

Esch GW, Curtis LA, Barger MA (2001) A perspective on the ecology of trematode communities in snails. Parasitology 123:57–75

16. Fischthal JH, Kuntz RE (1975) Some trematodes of amphibians and reptiles from Taiwan. Proc Helm Soc Wash 42:1–13
17. Fischthal JH, Acholonu AD (1976) Some digenetic trematodes from the Atlantic hawksbill turtle, *Eretmochelys imbricata imbricata* (L.) from Puerto Rico. Proc Helm Soc Wash 43:174–185
18. George RH (1997) Health problems and diseases of sea turtles. *In*: Lutz PL, Musick JA (eds) The Biology of Sea Turtles. CRC Press, Boca Raton, FL, pp 363–385

Glazebrook JS, Campbell RSF (1990) A survey of the diseases of marine turtles in northern Australia. I. Farmed turtles. Dis Aquat Org 9:83–95

Gordon AN, Kelly WR, Lester JG (1993) Epizootic mortality of free-living green turtles, *Chelonia mydas*, due to coccidiosis. J Wildl Dis 29:490–494

19. Gordon AN, Kelly WR, Cribb TH (1998) Lesions caused by cardiovascular flukes (Digenea: Spirorchiidae) in stranded green turtles (*Chelonia mydas*). Vet Pathol 35:21–30

Graczyk TK, Balazs GH, Work T, Aguirre AA, Ellis DM, Murakawa SKK, Morris R (1997) *Cryptosporidium* sp. infections in green turtles, *Chelonia mydas*, as a potential source of marine waterborne oocysts in the Hawaiian Islands. Appl Environ Microbiol 63: 2925–2927

20. Groschaft J, Otero AC, Tenora F (1977) Trematodes (Trematoda) from Cuban turtles *Chelonia mydas mydas* (L.) and *Eretmochelys imbricata imbricata* (L.) (Testudinata-Cheloniidae). Acta Univ Agric Fac Agron 25:155–167
21. Groschaft J, Tenora F (1978) *Charaxicephaloides polyorchis* gen. nov., sp. nov. (Trematoda: Charaxicephalinae) from *Chelonia mydas mydas* (Testudinata) in Cuba. Vestnik Cs Spol Zool 42:108–111

Hoberg EP, Klassen GJ (2002) Revealing the faunal tapestry: co-evolution and historical biogeography of hosts and parasites in marine systems. Parasitology 124:3–22

Holmes JC (1990) Helminth communities in marine fishes. *In*: Esch GW, Bush AO, Aho JM (eds) Parasite Communities: Patterns and Processes. Chapman & Hall, New York, pp 101–130

Hunter WS (1967) Notes on the life history of *Pleurogonius malaclemys* Hunter 1961 (Trematoda: Pronocephalidae) from Beaufort, North Carolina, with a description of the cercaria. Proc Helm Soc Wash 34:33–40

22. Inohuye-Rivera RB, Cordero-Tapia A, Arellano-Blanco J, Gardner S (2004) *Learedius learedi* Price 1934 (Trematoda: Spirorchiidae), in black turtle (*Chelonia mydas agassizii*) hearts from Magdalena Bay, Baja California Sur, Mexico. Comp Parasitol 71:37–41

IUCN/SSC 2002. IUCN Red List of Threatened Species. Downloaded: September 2005, www.redlist.org

23. Jacobson ER, Homer BL, Stacy BA, Greiner EC, Szabo NJ, Chrisman CL, Origgi F, Coberley S, Foley AM, Landsberg JH, Flewelling L, Ewing RY, Moretti R, Schaf S, Rose C, Mader DR, Harman GR, Manire CA, Mettee NS, Mizisin AP, Shelton GD (2006) Neurological disease in wild loggerhead sea turtles *Caretta caretta*. Dis Aquat Org 70:139–54

24. Kanev I, Radev V, Fried B (2002) Family Clinostomidae Lühe 1901. *In*: Gibson DI, Jones A, Brady RA (eds) Keys to the Trematodes, Vol. 1, CAB International, Oxon, pp 113–120

Leibovitz L, Rebell G, Boucher GC (1978) *Caryospora cheloniae* sp. n.: a coccidial pathogen of mariculture-reared green sea turtles (*Chelonia mydas*). J Wildl Dis 14:269–275

Lester RJG, Blair D, Heald D (1980) Nematodes from scallops and turtles from Shark Bay, Western Australia. Aust J Mar Freshwater Res 31:713–717

Lichtenfels JR, Bier JW, Madden PA (1978) Larval anisakid (*Sulcascaris*) nematodes from Atlantic molluscs with marine turtles as definitive hosts. Trans Am Micros Soc 97:199–207

25. Linton E (1910) Helminth fauna of Dry Tortugas. 2. Trematodes. Carnegie Inst Wash 4:11–99

26. Looss A (1901) Notizen zur Helminthologie Ägyptens. IV. Über Trematoden aus Seeschildkröten der ägyptischen Küsten. Centralbl f Bakt Paras Infekt 30:555–569, 618–625

27. Looss A (1902) Über neue und bekannte Trematoden aus Seeschildkröten. Nebst Erörterungen zur Systematik und Nomenklatur. Zool Jahrb Abt Syst 16:411–894

28. Manfredi MT, Piccolo G, Meotti C (1996) Parasites of Italian sea turtle. II. Loggerhead turtles (*Caretta caretta* Linnaeus 1758). Parassitologia 40:305–308

29. Manter HW (1954) Trematoda of the Gulf of Mexico. US Fish Wildl Serv, Fish Bull 55:335–350

Mattiucci S, Nascetti G (2006) Molecular systematics, phylogeny and ecology of anisakid nematodes of the genus *Anisakis* Dujardin 1845: an update. Parasité 13:99–112

Mattiucci S, Nascetti G (2007) Genetic diversity and infection leveles of anisakid nematodes parasitic in fish and marine mammals from Boreal and Austral hemispheres. Vet Parasitol 148:43–57

Mattiucci S, Nascetti G (2008) Advances and Trends in the Molecular Systematics of Anisakid Nematodes, with Implications for theie Ecology and Host-Parasite Co-evolutionary Processes. Adv Parasitol 66:47–148

Mattiucci S, Nascetti G, Cianchi R, Paggi L, Arduino P, Margolis L, Brattey J, Webb SC, D'Amelio S, Orecchia P, Bullini L (1997) Genetic and ecological data on the *Anisakis simplex* complex with evidence for a new species (Nematoda, Ascaridoidea, Anisakidae). J Parasitol 83:401–416

Mattiucci S, Paggi L, Nascetti G, Abollo E, Webb SC, Pascual S, Cianchi R, Bullini L (2001) Genetic divergence and reproductive isolation between *Anisakis brevispiculata* and *Anisakis physeteris* (Nematoda: Anisakidae). Int J Parasitol 31:9–14

Mattiucci S, Paoletti M, Webb SC, Sardella N, Timi J, Berland B, Nascetti G (2008) Genetic relationships among species of Contracaecum Railliet and Henry, 1912 and Phocascaris Host, 1932 (Nematoda: Anisakidae) from pinnipeds baesd on mitochondrial cox2 sequences and cogruence with allozyme data. Parasitè 15(3), in press.

30. Mehra HR (1939) New blood flukes of the family Spirorchiidae Stunkard (Trematoda) from the marine turtle *Chelonia mydas* of the Arabian Sea with observations on the synonymity of certain genera and classification of the family. Proc Nat Acad Sci India 9:155–167

31. Mehrotra V (1973) Digenea from some reptilian hosts in India. *In*: Proc 60th Indian Sci Cong, Part. IV, pp 46–47

Nascetti G, Paggi L, Orecchia P, Smith JW, Mattiucci S, Bullini L (1986) Electrophoretic studies on the *Anisakis simplex* complex (Ascaridida: Anisakidae) from the Mediterranean and North-East Atlantic. Int J Parasitol 16:633–640

Nascetti G, Cianchi R, Mattiucci S, D'Amelio S, Orecchia P, Paggi L, Brattey J, Berland B, Smith JW, Bullini L (1993) Three sibling species within *Contracaecum osculatum* (Nematoda, Ascaridida, Ascaridoidea) from the Atlantic Arctic-boreal region: reproductive isolation and host preferences. Int J Parasitol 23:105–120

Orós J, Calabuig P, Déniz S (2004) Digestive pathology of sea turtles stranded in the Canary Islands between 1993 and 2001. Vet Rec 155:169–174

32. Osorio-Sarabia D, Peñaflores-Salazar C, García-Hernández A, Ramírez-Lezama J (2006) Identificación de parásitos y epibiontes de la tortuga golfina (*Lepidochelys olivacea*) que arribó a playas de Michoacán y Oaxaca, México. Veterinaria México 37:431–440

Paggi L, Nascetti G, Cianchi R, Orecchia P, Mattiucci S, D'Amelio S, Berland B, Brattey J, Smith JW, Bullini L (1991) Genetic evidence for three species within *Pseudoterranova decipiens* (Nematoda, Ascaridida, Ascaridoidea) in the North Atlantic and Norwegian and Barents Seas. Int J Parasitol 21:195–212

33. Parra RL (1983) Estudio de algunos monogéneos y tremátodos parásitos de reptiles de México. Graduation Thesis, Fac Cienc, Univ Nac Autón México, 350 p

Pérez-Ponce De León G, Brooks DR (1995) Phylogenetic relationship of the genera of the Pronocephalidae Looss 1902 (Digenea: Paramphistomiformes). J Parasitol 81:267–277

34. Pérez-Ponce De León G, García-Prieto L, León-Régagnon V (1996) Gastrointestinal digenetic trematodes of olive ridley's turtle (*Lepidochelys olivacea*) from Oaxaca, Mexico. Taxonomy and infracommunity structure. Proc Helm Soc Wash 63:76–82

Poche F (1925) Das System der Platodaria. Arch Naturgesch A 91:1–458

35. Price EW (1939) A new genus and two new species of digenetic trematodes from a marine turtle. Proc Helm Soc Wash 6:24–25

36. Rao SL (1975) Studies on the trematode parasites of turtles from India. Further contribution to our knowledge of the family Pronocephalidae Looss 1902. Riv Parassitol 36:137–151

37. Rao SL (1976) On two new blood flukes (Trematoda: Spirorchiidae) from the heart of *Chelonia mydas* L. Acta Parasit Pol 24:119–124

38. Raidal SR, Ohara M, Hobbs RP, Prince RI (1998) Gram-negative bacterial infections and cardiovascular parasitism in green sea turtles (*Chelonia mydas*). Aust Vet J 76:415–417

Rawson PD, Macnamee M, Frick MG, Williams KL (2003) Phylogeography of the coronulid barnacle, *Chelonibia testudinaria*, from loggerhead sea turtles, *Caretta caretta*. Mol Ecol 12:2697–2706

Santoro M, Morales JA (2007) Some digenetic trematodes of the olive ridley sea turtle, *Lepidochelys olivacea* (Testudines, Cheloniidae) in Costa Rica. Helminthologia 44:25–28

Santoro M, Greiner EC, Morales JA, Rodríguez-Ortíz B (2006) Digenetic trematode community in nesting green sea turtles (*Chelonia mydas*) from Tortuguero National Park, Costa Rica. J Parasitol 92:1202–1206

Santoro M, Morales JA, Stacy B, Greiner EC (2007a) *Rameshwarotrema uterocrescens* trematode parasitism of the oesophageal glands in green sea turtles (*Chelonia mydas*). Vet Rec 159:59–60

39. Santoro M, Morales JA, Rodríguez-Ortíz B. (2007b) Spirorchiidiosis (Digenea: Spirorchiidae) and lesions associated with parasites in Caribbean green sea turtles (*Chelonia mydas*). Vet Rec. 161:482–486

40. Santoro M, Greiner EC, Morales JA, Rodríguez-Ortíz B (2007c) A new pronocephalid *Pleurogonius tortugueroi* n. sp. (Digenea), from the intestine of green sea turtles (*Chelonia mydas*) in Costa Rica. Parassitologia. 49:97–100

Savage JM (2002) The Amphibians and Reptiles of Costa Rica. The University of Chicago Press, Chicago, 934 p

41. Sey O (1977) Examination of helminth parasites of marine turtles caught along the Egyptian coast. Acta Zool Acad Sci Hung 23:387–394
42. Simha SS, Chattopadhyaya DR (1970) A new genus and species of a blood fluke, *Squaro-acetabulum solus* from the ventricle of the heart of a marine turtle. Zool Anz 184:290–294
43. Smith JW (1997) The blood flukes (Digenea: Sanguinicolidae and Spirorchiidae) of cold-blooded vertebrates: part. 2. Helminthol Abstr 66:329–344
44. Stephens JWW (1911) *Desmogonius desmogonius*, a new species and genus of monostome flukes. Ann Trop Med Parasit 5:497–500
45. Tandon V, Gupta NK (1981) On *Deuterobaris intestinalis* Mehrotra 1973 and *Octangium takanoi* Kobayashi 1921 (Trematoda: Microscaphidiidae) from a marine turtle in India. Proc Indian Acad Sci (Anim Sci) 90:553–560
Upton SJ, Odell DK, Walsh MT (1990) *Eimeria caretta* sp. nov. (Apicomplexa: Eimeriidae) from the loggerhead sea turtle *Caretta caretta* (Testudines). Can J Zool 68:1268–1269
46. Werneck MR, Becker JH, Gallo BG, Silva RJ (2006) *Learedius learedi*, Price 1934 (Digenea, Spirorchiidae) in *Chelonia mydas,* Linnaeus 1758 (Testudines, Chelonidae) in Brazil: case report. Arq Bras Med Vet Zootec 58:550–555
47. Work T, Balazs GH, Schumacher JL, Marie A (2005) Epizoology of spirorchiid infection in green turtles (*Chelonia mydas*) in Hawaii. J Parasitol 91:871–876
Yamaguti S (1963) Systema Helminthum, Vol. IV. Monogenea and Aspidocotylea. Interscience, New York/London, 699 p
48. Yamaguti S (1971) Synopsis of Digenetic Trematodes of Vertebrate. Keigaku, Tokyo, 1074 p

Chapter V
Marine Biodiversity of Costa Rica: Perspectives and Conclusions

Ingo S. Wehrtmann, Jorge Cortés, and Silvia Echeverría-Sáenz

Space view of Costa Rica (Credits: Kohlmann *et al.* 2002)

I.S. Wehrtmann
Unidad Investigación Pesquera y Acuicultura (UNIP) of the Centro de Investigación en Ciencias del Mar y Limnología (CIMAR), and Museo de Zoología, Escuela de Biología, Universidad de Costa Rica, 11501-2060 San José, Costa Rica.
e-mail: ingowehrtmann@gmx.de

J. Cortés
Centro de Investigación en Ciencias del Mar y Limnología (CIMAR), and Escuela de Biología, Universidad de Costa Rica, 11501-2060, San José Costa Rica.
e-mail: jorge.cortes@ucr.ac.cr

S. Echeverría-Sáenz
Unidad Investigación Pesquera y Acuicultura (UNIP) of the Centro de Investigación en Ciencias del Mar y Limnología (CIMAR), Universidad de Costa Rica, 11501-2060 San José, Costa Rica.
e-mail: sipama14@yahoo.com

I.S. Wehrtmann, J. Cortés (eds.) *Marine Biodiversity of Costa Rica, Central America*,

Marine Biodiversity of Costa Rica: A Summary

According to the numbers presented in the different taxonomic chapters of the present book, the marine biodiversity of Costa Rica comprises a total of 6,778 species, representing about 3.5% of all marine species reported worldwide (~190,000 spp.: Reaka-Kudla 1997, www.marinespecies.org). Of these species, 4,745 spp. are reported from the Pacific and 2,321 spp. from the Caribbean of Costa Rica. A total of 288 species have been found on both coasts of the country. Costa Rica harbors a total of 85 species so far encountered exclusively in its waters. Arthropoda is the phylum with the highest number of endemic species (37 spp.). Regarding endemism, once again Isla del Coco plays an outstanding role with 41.2% of the country's endemics. The most species rich groups are Mollusca (2,170 spp.), Chordata (1,605 spp.), and Arthropoda (1,066 spp.), followed by benthic algae (420 spp.), Annelida (318 spp.), Cnidaria (290 spp.), phytoplankton (268 spp.) and Echinodermata (229 spp.) (Table V.1). The total number of species from the Pacific is more than double the number reported for the Caribbean (Fig. V.1a), and this situation occurs in most taxonomic groups, but there are several exceptions in which there are higher species numbers for the Caribbean: benthic algae (297 vs 175 spp.), sponges (65 vs 62 spp.), Anthozoa (83 vs 77 spp.), Trematoda (30 vs 21 spp.), Isopoda (46 vs 34 spp.), and Thaliacea (2 spp. vs 0 sp.). On the other hand, there are numerous groups with substantially more species reported from the Pacific coast, such as Polychaeta (317 vs 4 spp.), Hydrozoa (118 vs 14 spp.), Decapoda (437 vs 119 spp.), Copepoda (171 vs 47 spp.), and Echinodermata (187 vs 44 spp.).

Isla del Coco is of special importance for the marine biodiversity of Costa Rica. This fairly well-studied island contains 1,142 species; 35 of which have been reported so far exclusively from Isla del Coco and thus can be considered as endemic. Similar to the situation on the mainland, the most specious groups on the island are Mollusca (441 spp.), Chordata (355 spp., including 285 spp. of fishes), Arthropoda (128 spp., including 104 decapod species), and Echinodermata (124 spp.). Interestingly, 66.3% of all echinoderms reported from Pacific Costa Rica are from Isla del Coco. It is noteworthy that several taxa are underrepresented according to the compiled data: for example, for phytoplankton, marine macroalgae, pelagic opisthobranchs, copepods, and flatworms not even a single species has been reported, although they certainly occur around this island. Regarding other taxa, the differences in species numbers between Isla del Coco and the Pacific mainland, respectively, are striking: examples are benthic algae (0 sp. vs 175 spp.), sea slugs (11 vs 155 spp.), bivalves (65 vs 380 spp.), hydroids (12 vs 118 spp.), sipunculids (1 vs 15 spp.), polychaets (8 vs 317 spp.), and decapods (104 vs 437 spp.). These differences might be due to the absence or scarcity of these groups in Isla del Coco, but more likely reflect the lack of detailed studies concerning these taxa.

In the following section we introduce two biodiversity indices which provide a more detailed and more accurate, i.e., realistic, picture of the biodiversity of the marine realm. These indices relate the species number to (1) the length of the coastline, the Linear Biodiversity Index (LBI: number of species per kilometer coastline)

Table V.1 A summary of Costa Rican marine biodiversity

Phylum or group	Caribbean species number	Pacific species number	Isla del Coco species number	Isla del Coco endemics	Species present at both coasts	Total species number	Costa Rican endemics
Foraminifera	4	80	8	0	0	84	0
Phytoplankton	53	258	n.a.	n.a.	43	268	n.a.
Benthic algae and Cyanobacteria	297	175	n.a.	n.a.	52	420	6
Fungi	n.a.	5	n.a.	n.a.	n.a.	5	n.a.
Plantae	8	10	0	0	4	14	0
Porifera	65	62	8	n.a.	0	127	n.a.
Cnidaria	97	203	49	3	10	290	3
Ctenophora	n.a.	n.a.	n.a.	n.a.	n.a.	n.a.	n.a.
Platyhelminthes	30	42	n.a.	n.a.	n.a.	72	n.a.
Nemertea	n.a.	n.a.	n.a.	n.a.	n.a.	n.a.	n.a.
Rotifera	n.a.	n.a.	n.a.	n.a.	n.a.	n.a.	n.a.
Gastrotrichia	n.a.	n.a.	n.a.	n.a.	n.a.	n.a.	n.a.
Kinorhyncha	n.a.	2	n.a.	n.a.	n.a.	2	n.a.
Nemata = Nematoda	n.a.	1	n.a.	n.a.	n.a.	1	n.a.
Acanthocephala	n.a.	1	n.a.	n.a.	n.a.	1	n.a.
Entoprocta	n.a.	n.a.	n.a.	n.a.	n.a.	n.a.	n.a.
Priapula	n.a.	n.a.	n.a.	n.a.	n.a.	n.a.	n.a.
Sipuncula	4	15	1	0	2	17	0
Echiura	0	1	n.a.	n.a.	0	1	0
Tardigrada	n.a.	n.a.	n.a.	n.a.	n.a.	n.a.	n.a.
Annelida	5	318	8	0	5	318	13
Arthropoda	287	837	128	8	58	1,066	37
Mollusca	712	1494	441	1	36	2,170	3
Phoronida	0	0	0	0	0	0	0
Ectoprocta	13	49	14	0	1	61	0
Brachiopoda	1	8	6	0	1	8	0
Echinodermata	44	187	124	3	2	229	3
Chaetognatha	6	16	n.a.	n.a.	3	19	n.a.
Hemichordata	n.a.	n.a.	n.a.	n.a.	n.a.	n.a.	n.a.
Chordata	695	981	355	20	71	1,605	20
TOTAL	2321	4745	1142	35	288	6778	85

and (2) to the area of the continental shelf, the Area Biodiversity Index (ABI: number of species per square kilometer of continental shelf, less than 200 m deep). The purpose of using these indices is to have unifying values which compensate for size differences of coasts, countries and/or marine areas. We believe that these indices represent a valuable tool for a better understanding of the marine biodiversity in different countries or regions. One significant limitation is the fact that these

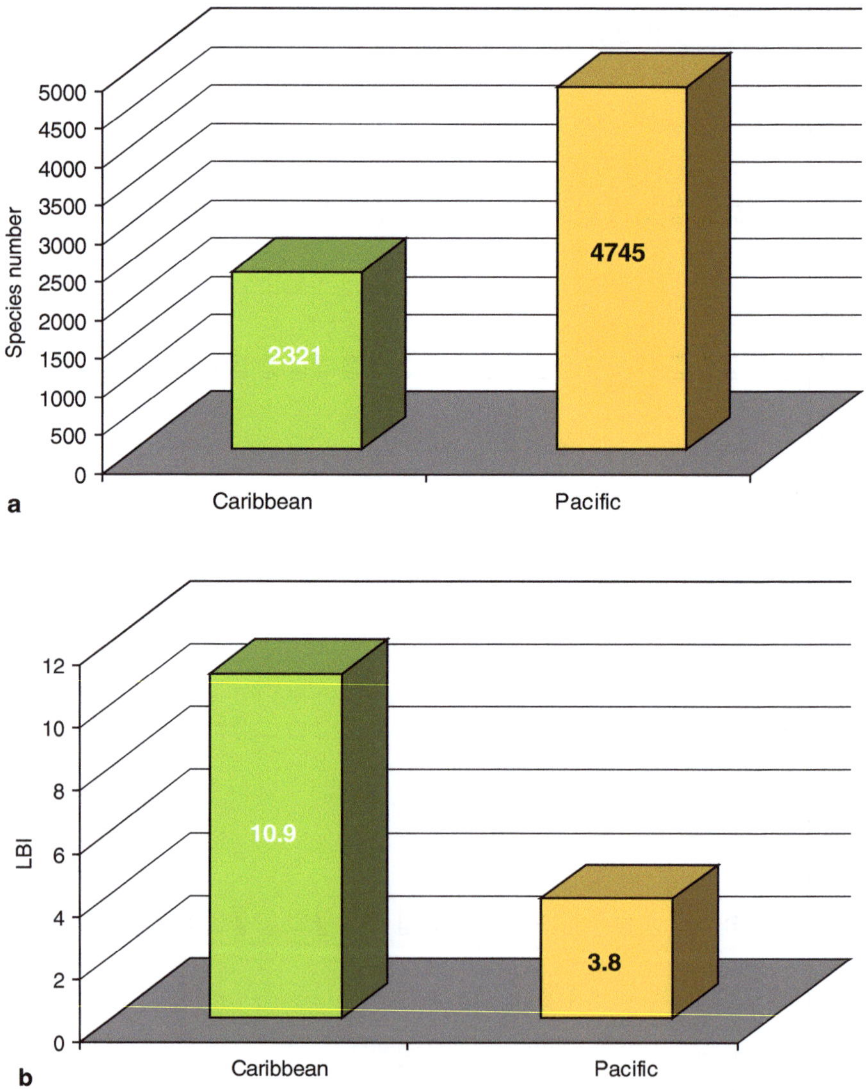

Fig. V.1 Comparison of the marine biodiversity of the Pacific and Caribbean coast of Costa Rica. a) Total number of species; b) Linear Biodiversity Index (LBI; species per kilometer); c) Area Biodiversity Index (ABI: number of species per square kilometer of continental shelf)

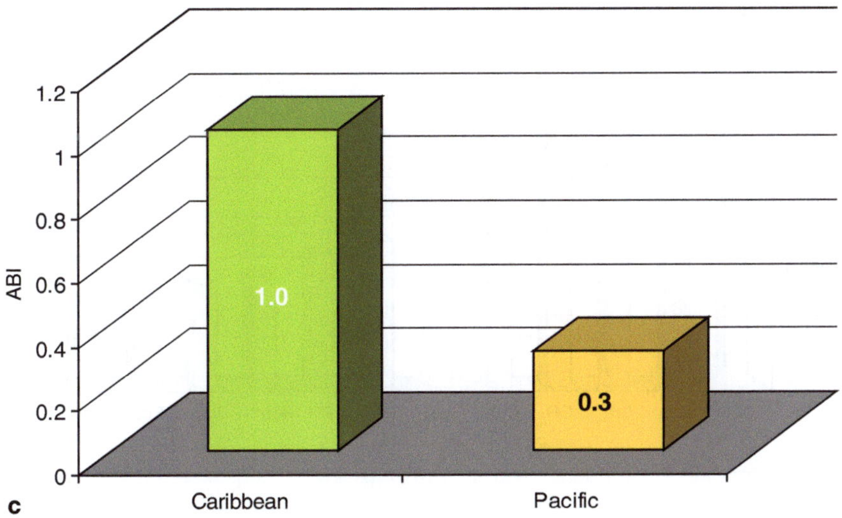

Fig. V.1 (continued)

data are not available for all countries. As a consequence, the application of these indices is limited to those countries where the data concerning species number, length of coastline, and area of the continental shelf were available.

In Costa Rica, the coastline of the Pacific is roughly six times longer than that of the Caribbean (1,254 and 212 km, respectively; IGN 2007). Although species number is substantially higher for the Pacific (Fig. V.1a), the results of the LBI provide a completely different picture: for the Caribbean the LBI is almost three times higher (10.93 spp./km) compared to that of the Pacific (3.79 spp./km; Fig. V.1b). According to the FAO (http://www.fao.org/fi/fcp/en/) Costa Rica has a continental shelf area (0–200 m) of approximately 15,800 km^2. Considering the 6,777 species reported for both coasts of Costa Rica, the ABI is 0.43 spp./km^2. Due to the fact that the FAO does not provide information separately for the shelf areas of Pacific and Atlantic coasts of Costa Rica, we used the data of TNC & CI (2007); Pacific = 15,295 km^2; Caribbean = 2,310 km^2. When applying these data for the ABI, we found a similar result as for the LBI: the Caribbean shelf area is much more diverse than that on the Pacific side (1.00 vs 0.31, respectively; Fig. V.1c).

Marine Biodiversity of Costa Rica Compared to Other Countries in the Region

Comparing the total number of marine species reported from countries in the region (Fig. V.2a), Costa Rica occupies the fourth place with 6,778 spp. after Chile (8,649 spp.), Colombia (7,217 spp.), and Brazil (7,083 spp.). However, when trying to apply the Linear Biodiversity Index, we encountered huge differences between

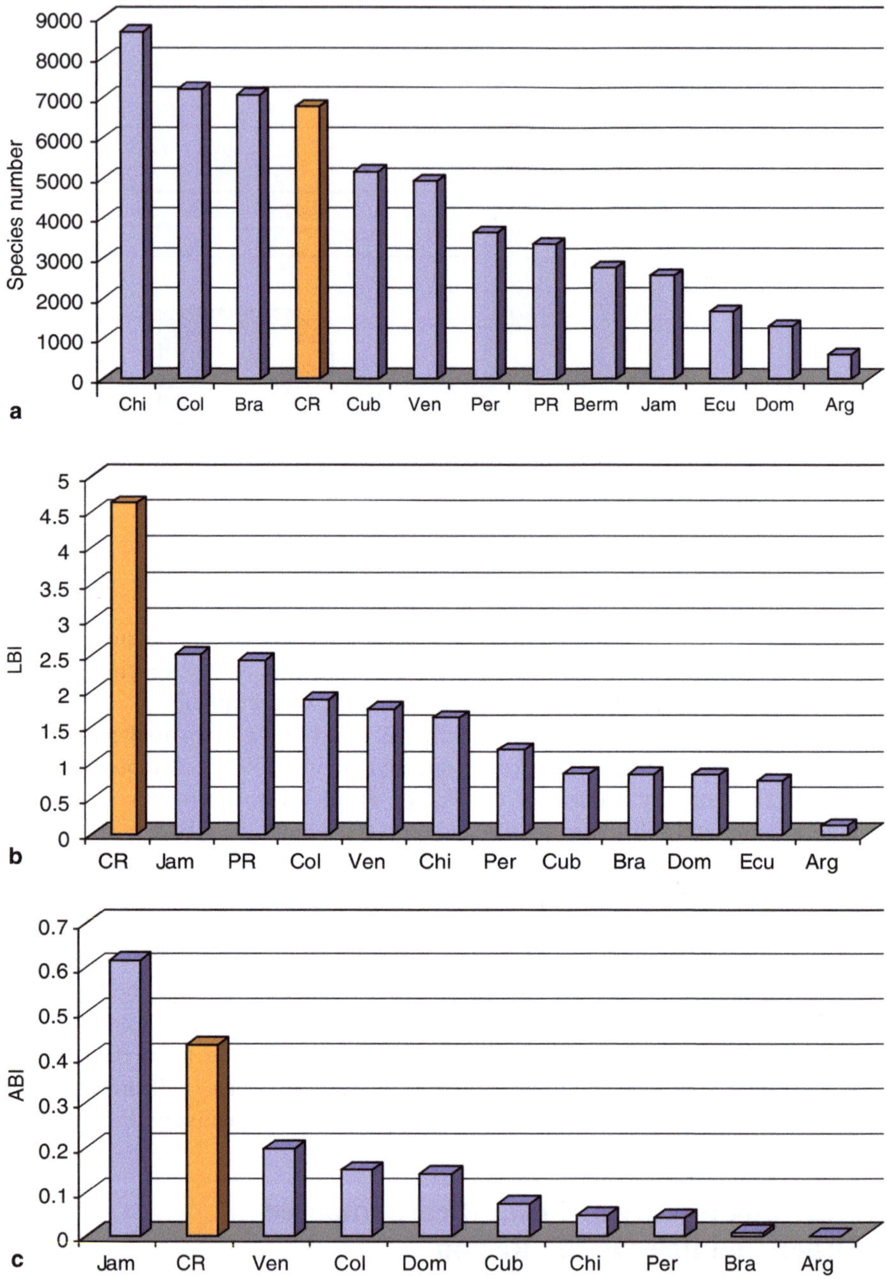

Fig. V.2 Comparison of the total marine biodiversity of Costa Rica and other countries in the region. a) Total number of species; b) Linear Biodiversity Index (LBI; species per kilometer); c) Area Biodiversity Index (ABI: number of species per square kilometer of continental shelf)

Arg: Argentina; Berm: Bermudas; Bra: Brasil; Chi: Chile; Col: Colombia; CR: Costa Rica; Cuba; Dom: Dominican Republic; Ecu: Ecuador; Jam: Jamaica; Mex: Mexico; Peru; PR: Puerto Rico; Ven: Venezuela

coast lengths provided by different institutions regarding the same country. An example is the length of the Chilean coastline: according to the FAO (http://www.fao.org/fi/fcp/en/CHL/profile.htm), the length is 5,300 km, while the World Resource Institute (http://earthtrends.wri.org/country_profiles/index.php?theme=1) indicates a length of 78,563 km. Such differences are most probably related to the scale of measurement (straight line vs detailed geomorphology of the coastline). Due to this situation, for our comparison of species numbers versus length of coastline, we used only the data provided by the FAO profile for each country.

The LBI for Costa Rica is 4.6, and is almost double the value of the following two, which are islands (Jamaica: 2.5; Puerto Rico: 2.4). All the other countries represent LBI values below 2.0 (Fig. V.2b), including the three countries with the highest species numbers reported (Chile, Colombia, and Brazil). Regarding the Area Biodiversity Index (ABI: number of species per square kilometer of continental shelf), Costa Rica occupies second place (0.4) after Jamaica (0.6). The ABI of the remaining countries is substantially lower than those of Jamaica and Costa Rica (Fig. V.2c). Both indices tend to be higher for well-studied small islands, such as Jamaica and Puerto Rico. This could probably be due to a better coverage of the entire coastline by taxonomic studies, resulting in a more detailed and complete picture of the biodiversity per area.

When comparing the marine biodiversity reported for the Atlantic (Fig. V.3), Brazil has the highest species numbers (7,083 spp.) followed by Cuba, Venezuela, and Colombia. Costa Rica is in eighth place with 2,321 spp. When considering the length of the coastline of the Atlantic (LBI), Costa Rica has a substantially higher index (10.9) than the subsequent localities: Colombia with 2.8; Jamaica (2.5), and Puerto Rico (2.4). The ABI renders a similar result with an index value of 1.0 for Costa Rica and 0.6 for Jamaica; all the other countries have indices below 0.2.

Along the Pacific coast of Latin America (Fig. V.4), Chile comprises the most diverse marine flora and fauna with a total of 8,649 species followed by Costa Rica (4,745 spp.) and Peru (3,640 spp.). However, when using the LBI, Costa Rica is the leading country (3.8) followed by Chile (1.6) and Peru (1.2). The ABI reveals the same pattern: Costa Rica has the highest index (0.3), followed by Colombia (0.2). The remaining countries have considerably lower values (all below 0.05).

The Marine Biodiversity of Costa Rica: The Known and the Unknown

The present book provides an overview about the available information of the marine biodiversity of Costa Rica. However, the lack of chapters regarding some faunal elements is equally important, because it clearly indicates the gaps in our knowledge of the marine biodiversity and does not necessarily mean that a particular taxonomic group is absent in Costa Rica. For example, there are no chapters about bacteria, flatworms, or nemerteans even though they have been collected. Unfortunately, no publications are available yet. Other widely distributed groups,

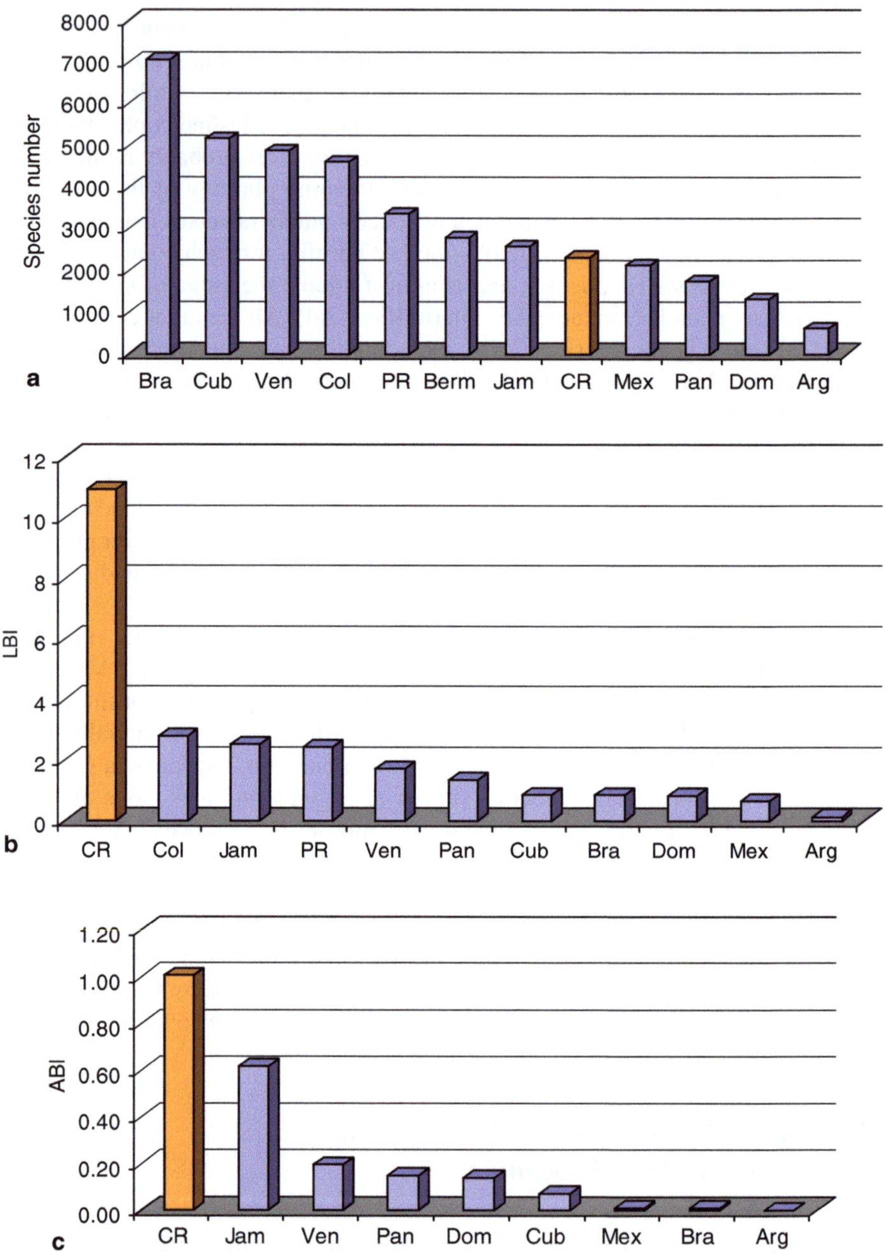

Fig. V.3 Comparison of the marine biodiversity of Caribbean coast of Costa Rica and other countries in the Caribbean/Atlantic region. a) Total number of species; b) Linear Biodiversity Index (LBI; species per kilometer); c) Area Biodiversity Index (ABI: number of species per square kilometer of continental shelf)

Arg: Argentina; Berm: Bermudas; Bra: Brasil; Col: Colombia; CR: Costa Rica; Cub: Cuba; Dom: Dominican Republic; Jam: Jamaica; Mex: Mexico; Pan: Panama; PR: Puerto Rico; Ven: Venezuela

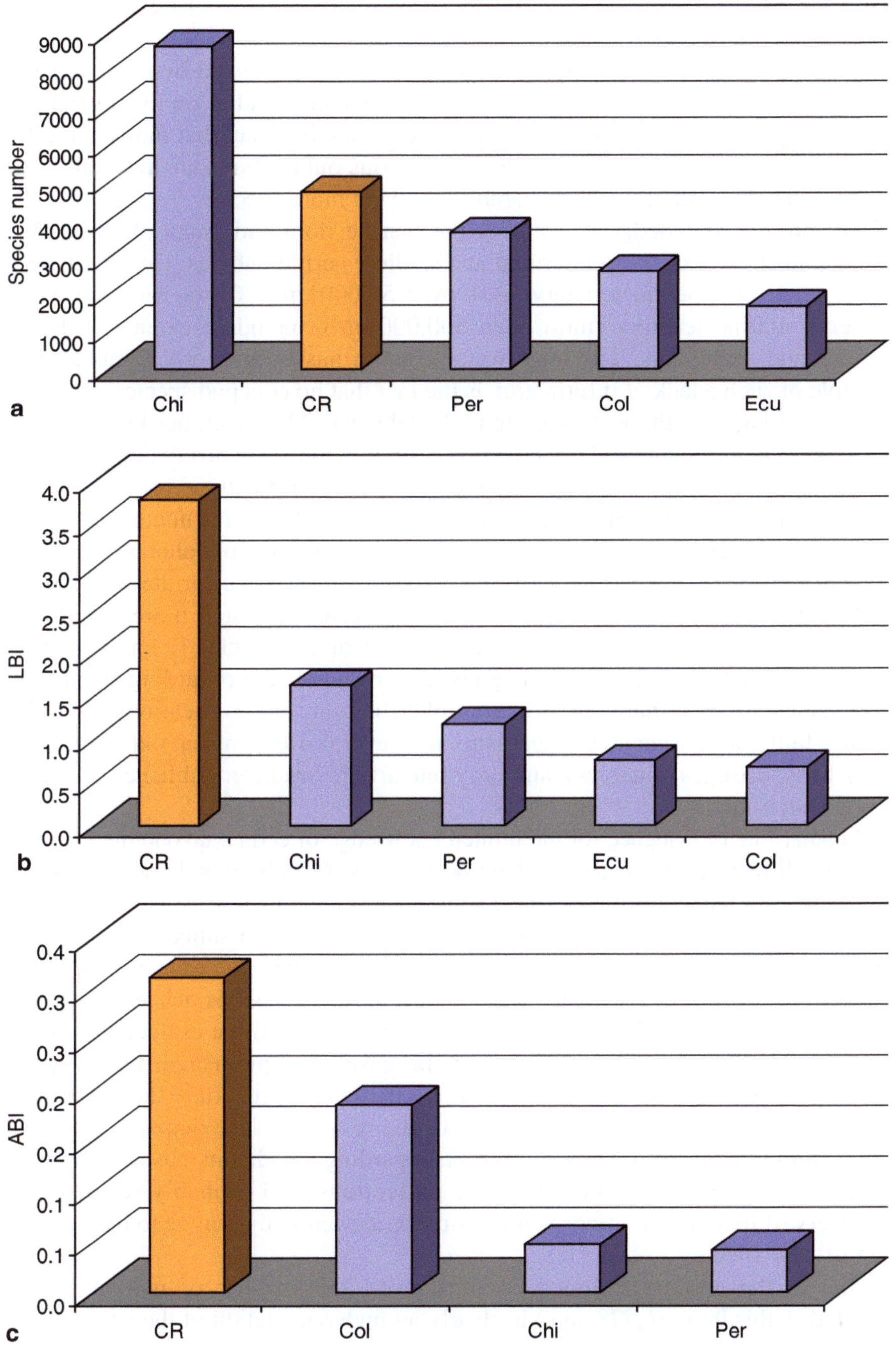

Fig. V.4 Comparison of the marine biodiversity of the Pacific coast in Costa Rica and other countries in the region. a) Total number of species; b) Linear Biodiversity Index (LBI; species per kilometer); c) Area Biodiversity Index (ABI: number of species per square kilometer of continental shelf)
Chi: Chile; Col: Colombia; CR: Costa Rica; Ecu: Ecuador; Mex: Mexico; Peru

such as hydroids, anemones, and tunicates, are only represented by few species in one or two publications. We are sure that more detailed studies on the diversity of these groups will lead to a much higher species number than that indicated in the present book. Even in those taxa where numerous publications are available (e.g., crustaceans, mollusks), future research will reveal more species.

Our present knowledge concerning the marine flora and fauna of Costa Rica refers almost exclusively to intertidal and shallow neritic habitats. Despite its relatively small continental territory (just over 50,000 km^2), Costa Rica covers an immense marine territory (more than 500,000 km^2), including extensive abyssal habitats and seamounts. The huge water column has been poorly sampled. One example of such a lack of information is the fact that no copepod species have been reported so far from the waters around Isla del Coco. Moreover, our knowledge of deepwater environments and organisms is extremely limited, mostly due to the lack of adequate equipment and sampling opportunities from these habitats. Future access to material collected in deep water will further increase the number of species reported for Costa Rica. One obvious example are the pogonophorans shown in Chapter I (Fig. I.23), where no published reports are available for these worms.

Although shallow water areas have been studied in much more detail, this does not mean that we have a complete picture of the species richness in these systems. In fact, there are obvious gaps in our knowledge regarding the marine biodiversity of these habitats. For example, there are no publications regarding cryptic fauna and associated organisms for several different habitats (e.g., seagrass beds, sponges, burrows), and only one article has been published about the sandy beach fauna.

Another line of evidence for the limited knowledge of certain taxonomic groups is the large difference of species numbers between the Caribbean and Pacific of Costa Rica within the same taxon. One outstanding case is the polychaetes with 317 species from the Pacific and just four species from the Caribbean; other similar examples have been mentioned above. These differences are most probably due to unbalanced sampling efforts along both coasts and do not reflect the actual species richness per coast.

Numerous historical and recent expeditions deposited the collected material in museums outside of Costa Rica. In the case of unpublished material, one needs to locate the collection and to initiate the contacts to study the specimens. In the case of published historical records, the revision of the deposited material will be important to clarify possible doubts regarding the identity based on updated literature. In any case, access to these collections will probably result in the discovery of new species records for Costa Rica, eventually maybe to the description of new species.

Due to the above-mentioned circumstances, the number of marine species reported in this book (6,778 spp.) is clearly an underestimation of the actual diversity. Even during the editing process of the book, it was necessary to update already submitted manuscripts, which indicates that the inventory of the marine flora and fauna of Costa Rica is an ongoing process and, thus, is not even close to be complete. Additional funding for basic research as well as capacity building together with a closer collaboration on a regional and international level will certainly result in a substantial increase of the known species numbers.

Limiting Factors and Perspectives

As mentioned in Chapter II, several international expeditions (usually without any participation of Costa Rican scientists) collected material in Costa Rican waters, but no samples were deposited in any Costa Rican institution. In several cases, the results of these expeditions have not been published yet (e.g., *Velero* IV, *Searcher*). Another case refers to international expeditions, where Costa Rican scientists were invited as observers to join the cruise. These colleagues could not develop their own sampling program, and often returned without any specimens. Finally, the most desirable collaboration between national and foreign scientists is having jointly-developed research projects. One example of such a fruitful collaboration was between CIMAR and the University of Delaware in which the RV *Skimmer* was stationed in Costa Rica for a couple of years, and many projects developed in collaboration (Vargas 1995; History, Chapter II, this book). Another example is the RV *Victor Hensen* 1993–1994 expedition, where Costa Rican and German colleagues jointly studied the Golfo Dulce and adjacent areas (Vargas & Wolff 1996).

The founding of the Centro de Investigación en Ciencias del Mar y Limnología (CIMAR) at the Universidad de Costa Rica about 25 years ago represented a milestone for the marine biodiversity research in Costa Rica, and demonstrated that the country recognized the need to study the marine environments along both coasts. A national research vessel has never been available, and the reduced funding available further complicated the efforts to study the marine flora and fauna of the country. Thus, during the last decades, most of the collecting has been carried out from the shore, by utilizing small boats or by scuba diving up to 30–40 m. Despite these limitations, the deposited material in the Museo de Zoología as well as the number of publications regarding the marine biodiversity of Costa Rica is increasing steadily.

Well-managed collections are an indispensable tool for marine biodiversity research. Fortunately, in Costa Rica we have the Museo Nacional and, even more important for the marine studies, the extensive collections of aquatic organisms deposited in the Museo de Zoología and the Herbario of the Universidad de Costa Rica. These collections have been proven to be extremely valuable not only for national students and scientists, but also for a large group of foreign taxonomists interested in tropical biodiversity, systematics, and phylogeny. The support provided by the Universidad de Costa Rica is essential for the maintenance of the collections and the continuity of taxonomic research. However, the lack of sufficient space for collections and research laboratories, restricted funding for the enlargement of the collections, training, and recruitment of qualified personnel are current limitations for the museum and herbarium. Together with traditional taxonomy, the application of molecular techniques is necessary to advance in the study of marine biodiversity.

Therefore, capacity building should include the training of the new generation of taxonomists combining both morphological and genetic approaches.

Considering the scarcity of funds for taxonomic research, a more regional and/or international approach is needed to facilitate the capacity building, including the exchange of students and scientists in Central America. The sharing of valuable experiences in identifying certain taxonomic groups should be promoted, for example by creating a regional network of taxonomists and by developing a series of workshops, where specialists, students, and the interested public can come together to tackle special taxonomic groups and the application of emerging techniques. Such workshops could serve to standardize (1) the design and implementation of regional sampling programs as well as (2) the adequate preservation of organisms, DNA and tissue samples, subsequent identification of the collected flora and fauna, plus (3) data storage and management. Moreover, these workshops will be an excellent opportunity to develop strategies and proposals for a more regional approach to study and protect the marine biodiversity of Central America in an integrated manner.

Conclusions

The present book represents a snapshot concerning the available information about the marine biodiversity of Costa Rica. Our knowledge regarding the marine biodiversity in Central America and especially in Costa Rica is growing constantly, and we know that more species will be encountered in the region by the time this book has been published. Even so, it is unquestionable that Costa Rica is a country rich in species, not only regarding its terrestrial systems, but as documented in this book, it is highly diverse concerning its marine environments. We assume that a similar situation may also occur in most of the neighboring countries. However, published information concerning the marine biodiversity of these countries is extremely limited, with the exception of Panama and Belize regarding some taxonomic groups. Therefore, we hope that the present book will encourage more research not only in Costa Rica, but also in the neighboring countries. In order to obtain a more complete picture and understanding of the marine biodiversity in the region, it will be indispensable to improve and in some cases provide adequate infrastructure necessary for such studies. An important step toward this goal will be the development of a joint and coordinated strategy for obtaining sufficient funding. This effort must be accompanied by a capacity building program, preferably at a regional level.

For most people, the marine realm is far away and outside of their personal experiences. Due to this situation, it is extremely difficult to convince, even the coastal populations, of the importance of protecting, conserving, and rationally using the marine environment and its resources. After many years of educational programs, scientific studies and the promotion of environmental laws, such awareness has been reached to some extent for the terrestrial biodiversity and ecosystems, at least in Costa Rica. There is an obvious and urgent need to reach a similar level of consciousness for the marine environment. Therefore, it is our hope that this book may promote awareness and interest for the marine biodiversity of Costa Rica and the Central American region.

References

IGN (Instituto Geográfico Nacional de Costa Rica) (2007) Downloaded from web page: http://www.mopt.go.cr/ign/geografia_litorales_fronteras.html

Kohlmann B, Wilkinson J, Lulla K (2002) Costa Rica from Space. EARTH, NASA & UNESCO. Neotrópica Foundation, San José, Costa Rica, 227 p

Reaka-Kudia M (1997) The global biodiversity of coral reefs: a comparison with rain forests, pp 83–108. *In*: Reaka-Kudla ML, Wilson DE, Wilson EO (eds) Biodiversity II: Understanding and Protecting our Biological Resources. Joseph Henry Prees, Washington, DC

TNC (The Nature Conservancy) & CI (Conservación Internacional) (2007) Revisando las prioridades de conservación de la biodiversidad en Mesoamérica: Pacífico tropical oriental (Costa Rica, Panamá y Colombia) y zonas económicas exclusivas del Carine de Costa Rica y Panamá. Programa Regional de Ciencias de TNC: Región de Mesoamérica y el Caribe. San José, Costa Rica

Vargas JA (1995) The Gulf of Nicoya estuary, Costa Rica: past, present, and future cooperative research. Helgol Meeresunters 49:821–828

Vargas JA, Wolff M (1996) Preface. Pacific coastal ecosystems of Costa Rica with emphasis on the Golfo Dulce and adjacent areas: a synoptic view based on the R.V. Victor Hensen – expedition 1993/94 and previous studies. Rev Biol Trop 44 (Suppl 3):iii–vi

Index

MIX
Papier aus verantwortungsvollen Quellen
Paper from responsible sources
FSC® C105338

If you have any concerns about our products,
you can contact us on
ProductSafety@springernature.com

In case Publisher is established outside the EU,
the EU authorized representative is:
Springer Nature Customer Service Center GmbH
Europaplatz 3, 69115 Heidelberg, Germany

Printed by Libri Plureos GmbH
in Hamburg, Germany